Continuous System Simulation

Continuous System Simulation

By

François E. Cellier
Ernesto Kofman

 Springer

Francois E. Cellier
ETH - Zentrum
Inst. Informatik
HRS G25
8092 ZÜRICH
SWITZERLAND

Ernesto Kofman
Universidad Nacional de Rosario
Laboratory for System Dynamics and Signal Processing
School of Electronic Engineering – FCEIA
Riobamba 245 bis
2000 ROSARIO
ARGENTINA

Continuous System Simulation

e-ISBN 0-387-30260-3
ISBN 978-1-4419-3863-3   e-ISBN 978-0-387-30260-7

Printed on acid-free paper.

Printed in the United States of America.

9 8 7 6 5 4 3 2 1

springer.com

Preface

The book *Continuous System Simulation* is the long overdue sequel to the book *Continuous System Modeling* that had been published with Springer–Verlag in 1991.

Whereas the book *Continuous System Modeling* dealt with the abstraction from a physical system to its mathematical description, the book *Continuous System Simulation* concerns itself with the transition from such a mathematical description, usually formulated as either a set of ordinary differential equations (ODEs) or a set of differential and algebraic equations (DAEs), to the trajectory behavior.

Consequently, the companion book was essentially a book of *theoretical physics*, whereas this book is a book of *applied mathematics*. It introduces the concepts behind numerical ODE and DAE solvers, as well as symbolic preprocessing algorithms that may be used to precondition a model in such a way as to improve the run–time efficiency of the resulting simulation code.

Why do we need yet another book on numerical ODE solutions? Haven't there been written enough books on that topic already? This surely must be a rather mature field of research by now.

In order to provide an answer to this question, the reader may allow us to return in time to the mid seventies. In those days, one of the authors of this book used to be a graduate student of control engineering at ETH Zurich. As he thought of writing a Ph.D. dissertation concerning the simulation of continuous systems with heavy discontinuities in the models, a good research topic at that time, he asked for an appointment with Peter Henrici, who headed the applied mathematics group at ETH in those years. He told him about his plans, and asked him, whether he would consent to serve as a co–advisor on this dissertation. Peter Henrici's answer was that "he didn't know anything about simulation, but would be glad to learn about the topic from the dissertation." Yet, the truth was that Peter Henrici already knew *a lot* about simulation . . . he only didn't know that he knew a lot about this topic.

The problem is that whereas engineers talk about "simulation," applied mathematicians write about "numerical ODEs." While mathematicians speak of "ODE solvers," engineers refer to these same tools as "integration algorithms." When engineers concern themselves with "discontinuity handling," mathematicians ponder about "root solvers." The two communities don't read each other's literature, and have developed not only different terminologies, but also different mathematical notations.

An engineer is likely to represent a dynamical system in the form of a state–space model:

$$\dot{\mathbf{x}} = \mathbf{f}(\mathbf{x}, \mathbf{u}, t); \quad \mathbf{x}(t_0) = \mathbf{x_0}$$

where $\mathbf{x}(t)$ is the state vector, $\mathbf{u}(t)$ is the input vector, and t is the independent variable, always considered to be *time*.

A mathematician is more likely to write:

$$y' = f(x, y)$$

calling the independent variable x, using a prime symbol instead of the dot symbol for the derivative, and placing the variables in reverse order. Mathematicians rarely worry about distinguishing between scalars and vectors in the way engineers do, they hardly ever think about input variables, and they mention initial conditions explicitly only, when they need them for a particular purpose.

The difference between the two representations is influenced by different goals. An engineer is eager to specify a complete problem that can be simulated to generate a specific trajectory behavior, whereas a mathematician is more inclined to look at the problem from the angle of finding general solution techniques that will work for all initial conditions.

Other mathematicians even prefer a notation, such as:

$$x \in X \subset \mathbb{R}; \quad y \in Y \subset \mathbb{R}^n; \quad y' \in Y' \subset \mathbb{R}^n$$
$$f : X \times Y \to Y'$$

which looks utterly foreign to most engineers.

This book is written by engineers for engineers. It uses a notation that is common to engineers, which should make this book much more accessible to engineers than the average numerical ODE book written by a mathematician for an audience of mathematicians.

Also, most mathematicians consider it "elegant" to write their books as much as possible in mathematical language: definition, lemma, proof. They use English prose as sparingly as they can get away with. Also this book is full of formulae and equations. However, much text is placed in between these equations, providing plenty of rationale for what these equations are supposed to mean. In this book, we use mathematical apparatus only for the purpose of making a statement either more precise or more concise or both. Mathematical apparatus is never being used as a goal in its own right.

Yet, although this book caters to engineering and computer science majors primarily, it should still contain plenty of material of interest to applied mathematicians as well. It introduces many exciting new algorithms that were developed by the authors, and that cannot be found elsewhere. It relies much more on graphical techniques than traditional ODE books do for illustrating the characteristics of particular algorithms, such as their numerical stability, accuracy, and damping properties. Some of the known algorithms were derived in novel, and maybe more elegant, ways than had been explored before. Finally, the last two chapters of the book revolutionize in fundamental ways the manner that numerical ODEs are being looked at. They offer a paradigm shift opening the door to an entirely new theory of numerical ODEs that is based on state discretization in place of time discretization.

The material covered in this book is clearly graduate–level material, even for a mathematics department. An undergraduate curriculum would hardly find the space necessary to dealing with numerical ODEs in such depth and breadth. Yet, the material is presented in a fairly self–contained manner. Thus, the book can as easily be used for self–study as in a class–room setting.

In some ways, as happens frequently in science, the book wrote itself in the end. Material wanted to be written. The authors were driven more by their own curiosity and inner drive than by conscious design. The story wanted to get out, and here it is.

When the companion book was written, we thought that this book would contain three sections: one section on numerical ODEs, one section on parameter estimation, and one section on simulation in the presence of noise.

Yet, as we were researching the issues surrounding numerical ODEs and DAEs, each new answer that we found led to at least two new questions that wanted to be researched, and so, we ended up with a book on numerical ODEs and DAEs only.

Maybe, one of these days, we may sit down and think once more deep and hard about the remaining problems: about parameter estimation and state identification, off–line methods and adaptive algorithms, supervised and unsupervised learning; and a third volume may emerge, probably just as broad and deep as this one, probably spread over just as many pages as the first two volumes were; but for now, we are at peace. We are content that the story is out. A story, that has kept us in its grip for a dozen years, has finally been told; and we, the executors of that story, have been released.

François E. Cellier and Ernesto Kofman
Zurich, Switzerland and Rosario, Argentina
September 2005

About This Book

This text introduces concepts of simulating physical systems that are mathematically described by sets of differential and algebraic equations (DAEs). The book is written for modeling and simulation (M&S) practitioners, who wish to learn more about the "intestines" of their M&S environments. Modern physical systems M&S environments are designed to relieve the occasional user from having to understand in detail, what the environment does to their models. Simulation results appear *magically* upon execution of the model.

Magic has its good and its bad sides. On the one hand, it enables us to separate the discussion of the tasks of *modeling* from those of *simulation*. The occasional user of M&S environments may be perfectly happy to only learn about modeling, leaving the gruesome details of numerical DAE solvers to the specialist.

Yet, for those among our readers, who are not in the habit of leaving the railway station through platform $9\,{}^3/_4$, this book may be helpful, as it explains, in lots of detail, how M&S environments operate. Thanks to this knowledge, our readers will understand what they need to do, when the magic fails, i.e., the simulation run is interrupted prematurely with an error message. They will also be able to understand, why their simulation program is consuming an unreasonable amount of execution time. Finally, they will feel more comfortable with the simulation results obtained, as they understand, how these results have been produced. "Magic" is awfully difficult to explain to your boss.

The text contains 12 chapters that are unfortunately rather heavily dependent on each other. Thus, reading one chapter of the book, because it discusses a topic that you are currently interested in, may not get you very far. Each chapter assumes the knowledge presented in previous chapters.

Chapters 1–4 introduce the concepts of numerical ODEs in a fairly classical way. After a general introduction to the topics that this book concerns itself with, presented in Chapter 1, Chapter 2 offers an introduction to the basic properties of numerical ODE solvers: numerical stability and accuracy. These are introduced by means of the two most basic explicit and implicit ODE solvers to be found: the forward and backward Euler algorithms.

Chapter 3 offers a discussion of single–step integration algorithms. New concepts introduced include a new stability definition, called F–stability or *faithful* stability, denoting algorithms, whose border of numerical stability coincides with the imaginary axis of the complex eigenvalue plane. Another new concept introduced is the *frequency order star*, leading to an attractive new definition of an *accuracy domain*. New ODE solvers include the backinterpolation algorithms, which can be designed to be either F–stable or L–stable.

Chapter 4 offers a discussion of linear multi–step integration algorithms. All of these algorithms are derived by means of Newton–Gregory polynomials, which offer a much more elegant way of introducing these algorithms, than those found in most other numerical ODE textbooks. New ODE solvers introduced in this chapter include a set of higher–order stiffly stable BDF methods that are based on least squares extrapolation.

Chapters 5 and 6 complete the discussion of numerical ODEs. These chapters can be skipped without making the subsequent chapters more difficult to understand.

Chapter 5 discusses special–purpose ODE solvers for second–derivative models, as they occur naturally in the mathematical description of mechanical systems. This topic has been discussed in the past in a few mechanics books, but it is hardly ever covered in the numerical ODE literature.

Chapter 6 offers a fairly classical discussion of the method–of–lines approximation to partial differential equations (PDEs). Thereby PDEs are converted to sets of ODEs. This topic is not usually covered in the numerical ODE literature, but has been dealt with in the past in more specialized textbooks on numerical PDEs. New in this chapter is the derivation of the formulae for computing spatial derivatives by means of Newton–Gregory polynomials. Also innovative is the use of Richardson extrapolation methods, previously introduced in Chapter 3 in their normal context, for the computation of spatial derivatives.

Chapters 7 and 8 deal with the issues surrounding numerical DAEs. Chapter 7 concerns itself with the symbolic conversion of sets of DAEs to sets of ODEs that can subsequently be dealt with numerically using the techniques introduced in earlier chapters of the book. Chapter 8, on the other hand, deals with the numerical solution of DAEs without previous conversion to explicit ODE form.

The symbolic tools presented in Chapter 7 are the result of a collaboration between one of the authors of this book with *Hilding Elmqvist* of *Dynasim*, a Swedish company specialized in the development of modern physical systems M&S environments, and *Martin Otter* of the *German Aerospace Center (DLR)* in Oberpfaffenhofen, Germany.

The numerical tools presented in Chapter 8 are a bit more classical. Some of these concepts can be found in the numerical DAE literature. However, the concepts presented previously in Chapter 7 help in presenting these algorithms in a clear and easily understandable fashion, which is not true for much of the existing numerical DAE literature.

We are convinced that the material presented in Chapters 7 and 8 makes a significant contribution to advancing the maturity of understanding of the relatively recent research field of numerical DAEs. New in Chapter 8 are the discussion of *inline integration*, and the way, in which we deal with the problem of overdetermined DAEs. The problem of overdetermined DAEs has only recently been recognized in the numerical DAE literature, and furthermore, the techniques for tackling them proposed by other authors, such as *Ernst Hairer* and *Gerhard Wanner*, are quite different from ours.

Chapter 9 discusses the problems surrounding the numerical simulation across discontinuities. This is a topic that both authors of this book were centrally concerned with in their respective Ph.D. dissertations. Chapter 9 presents the tools and technique developed by the first author, whereas those used by the second author are postponed to Chapter 12.

Chapter 10 introduces the reader to the problems of performing simulation runs in real time, i.e., synchronizing the numerical ODE solvers with the real-time clock. Interesting in this context may be the discussion of the linearly-implicit integration algorithms. More results, and more fundamental results concerning real-time simulation are provided in Chapter 12 of the book.

Up to this point, the book follows fairly classical approaches to numerical ODE, PDE, and DAE solutions. All of the techniques presented discretize the time axis, and perform the numerical simulation by means of extrapolations that are approximations of Taylor-series expansions. All of the techniques presented are *synchronous* algorithms, as all differential equations are simulated synchronously, in step with the temporal discretization.

Chapters 11 and 12 represent a radical departure from these concepts. In these chapters, the state variables themselves are being discretized. We call this a *spatial discretization* instead of a *temporal discretization*. In contrast, the time axis is no longer discretized. Furthermore, these algorithms proceed *asynchronously*, i.e., each state variable carries its own simulation clock that it updates as needed. The simulation engine ensures that the state variable that is currently the one most behind always gets updated next. The different state variables communicate with each other by means of state interpolation.

As these are the last two chapters in the book, they can be skipped without any problem. Yet, these are easily the most interesting chapters in the entire book, as they revolutionize the way of looking at numerical ODEs, offer an exciting new theory of numerical stability, and lend themselves to plenty of fascinating open research questions.

Acknowledgments

A work of this size and ambition cannot be completed without the help of numerous individuals. We wish to take this opportunity to thank the following colleagues of ours for their critical review of and helpful comments and suggestions concerning aspects of this work:

> Pawel Bujakiewicz
> Mike Carver
> Hilding Elmqvist
> Walter Gander
> Jürgen Halin
> Sergio Junco
> Werner Liniger
> Hans Olsson
> Martin Otter
> Hans Schlunegger
> Gustaf Söderlind
> Michael Vetsch

Their help is highly appreciated.

Of course, 12 years (which is, how long it took us to put it all together) is a very long time. Meanwhile, both of us experienced several system crashes that deleted our respective mail files. Thus, we can only hope that we have not excluded anyone from this list who did contribute to the endeavor, but if this should have happened nevertheless, we beg these persons' forgiveness for not remembering.

We also wish to thank the following of our students, who have contributed, in one form or another, to the work presented in this book:

> Chris Beamis
> Jürgen Greifeneder
> Klaus Hermann
> Luoan Hu
> Matthias Krebs
> Marcelo Lapadula
> Robert McBride
> Gustavo Migoni
> Wes Morgan
> Esteban Pagliero
> Michael Schweisguth

Miguel Soto
Vicha Treeaporn
Wei Xie

Without the help of these students, we wouldn't ever have made it.

Finally, we wish to thank our scientific editors at Springer–Verlag, U.S.A. for not losing their trust in us, in spite of the incredibly long time, it took us to complete the work. We wore out two of them: Zvi Ruder and Thomas von Foerster. The work was finally completed under the guidance of Alex Greene. Thank you guys for your patience and your support of this project.

Contents

XX

1

Introduction, Scope, Definitions

Preview

The purpose of this chapter is to provide a framework for what this book is to cover. Which are the types of questions that it aspires to answer, and what are the kinds of knowledge that you, the reader, can expect to gain by working through the material presented in this book? What are the relations between real physical systems and their mathematical models? What are the characteristics of mathematical descriptions of physical systems? We shall then talk about simulation as a problem solving tool, and finally, we shall offer a classification of the basic characteristics of simulation software systems.

1.1 Modeling and Simulation: A Circuit Example

Let us begin by modeling a simple electrical circuit. The *circuit diagram* of this circuit is provided in Fig.1.1.

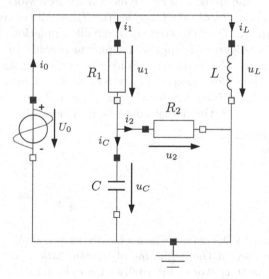

FIGURE 1.1. Circuit diagram of electrical RLC circuit.

Figure 1.1 was produced using *Dymola* [1.11, 1.13], which is currently the most advanced among all of the commercially available physical system modeling and simulation environments. The circuit diagram of Fig.1.1 *is* a mathematical model that can be used to simulate the circuit. It was composed by dragging icons from the graphical electrical component library into the graphical modeling window, dropping them there, and interconnecting them graphically. Associated with each of the icons is the mathematical description of the properties of that particular component model.

The diagram was then edited using a graphical editor to remove the numerical values of the components, and to add names and directions for all currents and voltages. Dymola creates its own names and direction conventions, but does not show them on the circuit diagram using the standard graphical electrical circuit library (this could be changed easily by modifying the component definitions in the library accordingly).

What does Dymola do with the *graphical model* of the circuit? The model is first captured in an alphanumerical form using a modeling language called *Modelica* [1.21]. In the process of compiling the model, the Dymola model compiler performs a lot of symbolic preprocessing on the original mathematical representation. We shall learn more about the symbolic formulae manipulation algorithms that Dymola employs in later chapters of this book. Once a suitable *simulation model* has been derived, it is translated into a C program that then gets compiled further. The compiled model is then simulated by making calls to the numerical run–time library that forms part of the overall Dymola modeling and simulation environment.

What if we were to use a professional circuit simulator, such as *PSpice* [1.19], instead of the more general Dymola software? Modern versions of Spice also offer a *Graphical user Interface (GUI)*, usually called a *schematic capture* program [1.17] in the context of circuit simulation. In the case of Spice, the circuit diagram is captured alphanumerically in the form of a so–called *netlist*. In older versions of Spice, the netlist constituted the user interface, just like older versions of Dymola used a language similar to Modelica as the input language for the description of models.

For the given circuit, the netlist could take the following form:

```
Vin 1 0 DC 10Volts
R1 1 2 00Ohms
R2 2 0 20Ohms
C 2 0 1uF
L 1 0 1.5mH
.END
```

Spice, contrary to Dymola, performs hardly any symbolic preprocessing. The netlist is parsed at the beginning of the simulation, and the information contained in it is stored internally in a data structure that is then interpreted at run time.

How about using MATLAB [1.15] to simulate this circuit? MATLAB is of particular interest to us, since it is a wonderful language to describe algorithms in, and since this book is all about algorithms, we shall use MATLAB exclusively in this book for the documentation of these algorithms, as well as for the homework problems that accompany each of the chapters.

MATLAB is not geared toward simulation at all. It is a general purpose programming language supporting high–level data structures that are particularly powerful for the description of algorithms. Since MATLAB wasn't designed to support modeling and simulation, the user will have to perform considerably more work manually, before the circuit description can be fed into MATLAB for the purpose of simulation.

As the circuit contains five separate components in five distinct branches of the circuit, the dynamics of this circuits can be described by 10 variables, namely the five voltages across each of the branches, and the five currents flowing through them. Hence we shall need 10 separate and mutually independent equations to describe the model dynamics in terms of these variables.

The 10 equations can be read out of the circuit diagram easily. Five of them are the *constitutive equations* of the circuit components, relating the voltage across and the current through each of the branches to each other:

$$u_0 = 10 \tag{1.1a}$$

$$u_1 - R_1 \cdot i_1 = 0 \tag{1.1b}$$

$$u_2 - R_2 \cdot i_2 = 0 \tag{1.1c}$$

$$i_C - C \cdot \frac{du_C}{dt} = 0 \tag{1.1d}$$

$$u_L - L \cdot \frac{di_L}{dt} = 0 \tag{1.1e}$$

Three additional equations can be obtained by applying *Kirchhoff's Voltage Law (KVL)* to the circuit, which states that the voltages around a mesh must add up to zero. These are therefore often called the *mesh equations*.

$$u_0 - u_1 - u_C = 0 \tag{1.2a}$$

$$u_L - u_1 - u_2 = 0 \tag{1.2b}$$

$$u_C - u_2 = 0 \tag{1.2c}$$

The final two equations can be obtained by applying *Kirchhoff's Current Law (KCL)* to the circuit, which states that the currents flowing into a node must add up to zero. These are therefore often called the *node equations*. One of the node equations is always redundant, i.e., not linearly independent, and must therefore be omitted. It has become customary to omit the node equation of the ground node. The two remaining node equations can be written as:

$$i_0 - i_1 - i_L = 0 \tag{1.3a}$$
$$i_1 - i_2 - i_C = 0 \tag{1.3b}$$

These 10 equations together form another equivalent mathematical description of the circuit. They consist of a set of implicitly described partly algebraic and partly differential equations. We call this mathematical description an *implicit differential and algebraic equation (DAE) model*.

We can make the model explicit by deciding, which variable to solve for in each of the equations, and by arranging the equations in such a manner that no variable is being used before it has been defined. We call this the process of *horizontally and vertically sorting* the set of equations. In Chapter 7 of this book, you shall learn how equations can be sorted algorithmically. For now, let us simply present one possible solution to the sorting process.

$$u_0 = 10 \tag{1.4a}$$
$$u_2 = u_C \tag{1.4b}$$
$$i_2 = \frac{1}{R_2} \cdot u_2 \tag{1.4c}$$
$$u_1 = u_0 - u_C \tag{1.4d}$$
$$i_1 = \frac{1}{R_1} \cdot u_1 \tag{1.4e}$$
$$u_L = u_1 + u_2 \tag{1.4f}$$
$$i_C = i_1 - i_2 \tag{1.4g}$$
$$\frac{di_L}{dt} = \frac{1}{L} \cdot u_L \tag{1.4h}$$
$$\frac{du_C}{dt} = \frac{1}{C} \cdot i_C \tag{1.4i}$$
$$i_0 = i_1 + i_L \tag{1.4j}$$

In this model, the equal signs have assumed the role of assignments rather than equalities, which was the case with the previous model. Each unknown appears exactly once to the left of the equal sign, and all variables used in the expressions of the right hand sides have been assigned values, before they are being used.

Notice that the variables u_C and i_L are not treated as unknowns. Since they are the outputs of integrators, they are computed by the integration algorithm used in the simulation, and don't need to be computed by the model. Such variables are referred to as *state variables* in the literature.

We are still confronted with a mixture of algebraic and differential equations, but the model has now become explicit. We call this an *explicit DAE model*.

Sometimes, the explicit DAE model is also called *simulation model*, since the traditional simulation languages, such as *ACSL* [1.18], were able to deal with this type of mathematical description directly.

Although MATLAB can deal with simulation models, this is still not the preferred form to be used when simulating linear systems with MATLAB.

We can now plug the explicit equations into each other, substituting the unknowns on the right hand side by the expressions defining these unknowns, until we end up with equations for the variables du_C/dt and di_L/dt, the so–called *state derivatives*, that depend only on the state variables, u_C and i_L, as well as the input variable, u_0. These equations are:

$$\frac{du_C}{dt} = -\frac{R_1 + R_2}{R_1 \cdot R_2 \cdot C} \cdot u_C + \frac{1}{R_1 \cdot C} \cdot u_0 \tag{1.5a}$$

$$\frac{di_L}{dt} = \frac{1}{L} \cdot u_0 \tag{1.5b}$$

We can add one or several *output equations* for those variables that we wish to plot as simulation results. Let i_2 be our output variable. We can obtain an equation for i_2 that depends only on state variables and input variables in the same fashion:

$$i_2 = \frac{1}{R_2} \cdot u_C \tag{1.6}$$

This mathematical representation is called an *explicit ordinary differential equation (ODE) model*. In the control literature, it is usually referred to as the *state–space model*.

If the state–space model is linear, as in the given case, it can be written in a matrix–vector form:

$$\begin{pmatrix} \frac{du_C}{dt} \\ \frac{di_L}{dt} \end{pmatrix} = \begin{pmatrix} -\frac{R_1+R_2}{R_1 \cdot R_2 \cdot C} & 0 \\ 0 & 0 \end{pmatrix} \cdot \begin{pmatrix} u_C \\ i_L \end{pmatrix} + \begin{pmatrix} \frac{1}{R_1 \cdot C} \\ \frac{1}{L} \end{pmatrix} \cdot u_0 \tag{1.7a}$$

$$i_2 = \begin{pmatrix} \frac{1}{R_2} & 0 \end{pmatrix} \cdot \begin{pmatrix} u_C \\ i_L \end{pmatrix} \tag{1.7b}$$

This model finally is in an appropriate form for feeding it into MATLAB. The following MATLAB code may be used to simulate the circuit:

```
% Enter parameter values
%
R₁ = 100;
R₂ = 20;
L = 0.0015;
C = 1e-6;
%
% Generate system matrices
%
R₁C = 1/(R₁ * C);
R₂C = 1/(R₂ * C);
a₁₁ = -(R₁C + R₂C);
A = [ a₁₁ , 0 ; 0 , 0 ];
b = [ R₁C ; 1/L ];
c = [ 1/R₂ , 0 ];
d = 0;
%
% Make a system and simulate
%
S = ss(A, b, c, d);
t = [ 0 : 1e-6 : 1e-4 ];
u = 10 * ones(size(t));
x₀ = zeros(2, 1);
y = lsim(S, u, t, x₀);
%
% Plot the results
%
subplot(2, 1, 1)
plot(t, y, 'k − ')
grid on
title('\tex{Electrical RLC Circuit}')
xlabel('\tex{time}')
ylabel('\tex{$i_2$}')
print −deps fig1_2.eps
return
```

The simulation results are presented in Fig.1.2.

Clearly, MATLAB employs a considerably lower–level user interface than either Dymola or PSpice, but maybe that is good, since the purpose of this book is to teach simulation methods.

Do we now understand more about how the simulation was performed using MATLAB? Unfortunately, this question must be answered in the negative. The entire simulation takes place inside the *lsim* box, which we haven't opened yet. The main purpose of this book is to open up the *lsim* box, and understand, how it has been built, but more about that later.

To be fair to MATLAB, it must be mentioned that also MATLAB, just like its competitors, offers a graphical user interface, called *SIMULINK* [1.7]. However, SIMULINK is not a schematic capture program. It is only a *block diagram editor*.

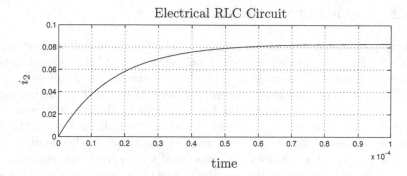

FIGURE 1.2. Simulation results of electrical RLC circuit.

SIMULINK is thus located at the level of the explicit DAE model. Given that model, we can start by drawing the two integrator boxes, and then work ourselves backward toward the input variable, and forward toward the output variable. The resulting block diagram is shown in Fig.1.3.

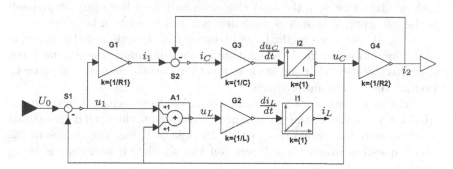

FIGURE 1.3. Block diagram of electrical RLC circuit.

Figure 1.3 was not drawn using SIMULINK, but instead, we chose to draw the figure in Dymola, using Dymola's graphical block diagram library. We then edited the graph manually by adding the names of the variables to each of the signals.

What was gained by representing the explicit DAE model graphically as a block diagram? The only advantage of doing so is that it becomes evident from the block diagram that the integrator computing the variable i_L could have been pruned away, as it does not contribute at all to computing the output variable.

Block diagrams are useful tools for representing control systems. They are not useful, however, for representing electrical circuits.

1.2   Modeling vs. Simulation

In the previous section, we have shown a full modeling and simulation cycle, starting out with a physical system, an electrical RLC circuit, and ending with the display of the trajectory behavior of the output variable, i_2.

The process of *modeling* concerns itself with the extraction of knowledge from the physical plant to be simulated, organizing that knowledge appropriately, and representing it in some unambiguous fashion. We call the end product of the modeling cycle the *model* of the system to be simulated.

The process of *simulation* concerns itself with performing experiments on the model to make predictions about how the real system would behave if these very same experiments were performed on it.

At the University of Arizona, we offer currently two senior/graduate level classes dealing with the issues of modeling and simulating physical systems. One of them, *Continuous System Modeling*, deals with the issues of creating suitable models of physical systems. For it, the companion book of this textbook, also entitled *Continuous System Modeling* [1.6], was developed. The other, *Continuous System Simulation*, concerns itself with the issues of simulating these models accurately and efficiently. For that class, this textbook has been written.

A question that you, the reader, may already have begun to ask yourself is the following: Where does modeling end and simulation begin?

In the old days, we might have answered that question in the following way: Simulation is what is being done by the computer, whereas modeling concerns the steps that the modeler has to undertake manually in order to prepare the simulation program.

Yet, this answer is not very satisfactory. We have seen that, when using MATLAB to simulate the circuit, the modeler had to do much more manual preprocessing than when using either Dymola or Spice. The answer to the above question would thus depend on the simulation tool that is being used. This is not very useful.

A more gratifying answer may be obtained by looking at Fig.1.4.

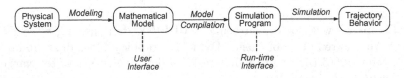

FIGURE 1.4. Modeling *vs.* simulation.

The whole purpose of the *mathematical model* is to provide the human user of the modeling and simulation environment with a means to represent knowledge about the physical system to be simulated in a way that is as *convenient* to him or her as possible. Modeling is thus indeed always done manually. The mathematical model represents the user interface. It has absolutely nothing to do with considerations of how that model is going to be used by the simulation engine.

Which is the most appropriate mathematical model of a system to be simulated depends on the nature of the physical system itself, and maybe also on the types of experiments that are to be performed on the model.

We have already mentioned that a *block diagram* may be a suitable tool to represent the knowledge needed to simulate a control system. It is certainly not a convenient tool to represent the knowledge needed to simulate an electrical circuit. A *circuit diagram*, on the other hand, may be the most natural way to represent an electrical circuit, as long as the experiment to be performed on the model does not concern itself with non–electrical phenomena, such as the heating of the device that results from current flowing through resistors, and the temperature dissipation of the package, in which the circuit has been integrated. In that case, a *bond graph* may be a much better choice for representing the physical knowledge needed to simulate the circuit.

The bond graph of the above circuit is shown in Fig.1.5.

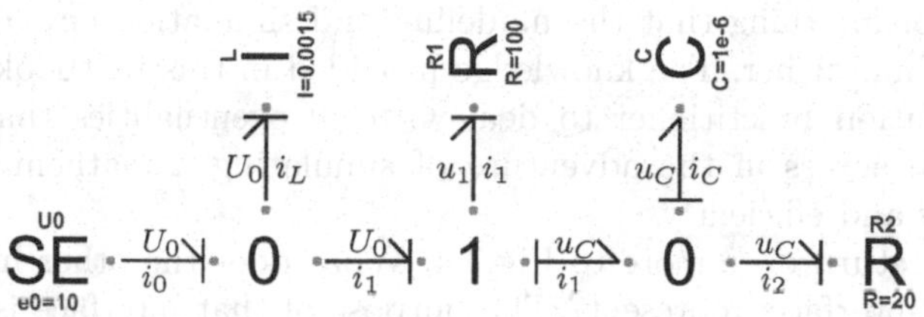

FIGURE 1.5. Bond graph of electrical RLC circuit.

Figure 1.5 was produced using Dymola's graphical bond graph library [1.3]. Bond graphs play an important role in the companion book, *Continuous System Modeling* [1.6], to this text. They are of no concern to this class, since they are only used to the left of the mathematical model in Fig.1.4. In this textbook, we do not concern ourselves with issues to the left of the mathematical model.

Once the mathematical model has been formulated, the modeling and simulation environment can make use of that model to perform simulations, and produce simulation results. For models as simple as our electrical circuit, either of the three representations: the circuit diagram, the block diagram, or the bond graph, can be simulated equally easily, accurately, and efficiently. The user simply instructs Dymola to simulate the model, Dymola then performs the necessary model compilations, executes the simulation run, and prepares the variables in a data base, such that the user can then pick the output variable(s) he or she is interested in, and plot them.

Since modeling of a physical system is always done manually, it is evident that we need to offer a class, teaching the students, how to generate a model of a physical system that is suitable for performing a given set of experiments on it. Yet, if everything to the right of the mathematical model in Fig.1.4 can be fully automated, why should an engineering student concern him- or herself at all with simulation issues? Why not leave these issues to the experts, i.e., the applied mathematicians?

Unfortunately, things are not going always as smoothly as in this simple electrical RLC circuit. It happens more often than not that a simulation does not produce the desired results the first time around. A user who only understands modeling and uses the simulation environment as a black box will most likely be at a total loss as to what went wrong and why, and he or she will have no inkling as to how the problems can be overcome. In fact, the more complex the symbolic formulae manipulation algorithms are that are being employed by the modeling and simulation environment as part of the model compilation, the less likely it is that an uninformed user of that environment will be able to make sense out of error messages that result from mishaps happening at run time, the so–called *run–time exceptions.*

The main purpose of this class and this textbook are to prepare the student for anything that the modeling and simulation environment may throw at him or her. The knowledge provided in this textbook will enable the simulation practitioner to deal with all eventualities that he or she may come across in the adventure of simulating a mathematical model effectively and efficiently.

Let us return once more to Fig.1.4. What does the other interface, the run–time interface, represent? The purpose of that interface is to define a simulation model that can be simulated *efficiently* and *accurately.*

It was already mentioned that Spice essentially simulates the netlist directly, whereas Dymola performs a lot of symbolic preprocessing on the model, i.e., the distance between the mathematical model and the simulation program is very small in Spice, whereas it is impressively wide in Dymola.

You shall learn in this class that it actually matters, which way we proceed. The algorithms underlying Spice simulations only work because the possible structures of an electronic circuit are very well defined and don't change much from one circuit to the next. On the other hand, if we were to simulate how a circuit heats up during simulation, and simultaneously wanted to simulate how the electrical parameter values (the resistances and capacitances) change in function of the current device temperature, the algorithms underlying the Spice simulation, the so–called *sparse tableau equations* that are used in a *modified nodal analysis*, would break down, because the so modified model would contain additional algebraic loops that these algorithms could not possibly handle.

Thus, the most appropriate run–time interface is also a function of the system to be simulated, and possibly of the experiment or set of experiments to be performed on the model. Yet, this interface only concerns itself with the way, the simulation algorithms work. It has no bearing whatsoever on how the user represents his or her mathematical model.

In which book are the model compilation issues to be discussed? Since both interfaces move around, i.e., they are sometimes a little further to the right, and sometimes a little further to the left, it is important to look at these issues both from the perspective of a modeler and from that of a simulation practitioner. Hence there is a certain degree of overlap and redundance between the two textbooks as far as model compilation algorithms are concerned. This decision was taken on purpose to allow the students to take the two classes in any sequence. Neither of them depends on the knowledge provided in the other.

1.3   Time and Again

In the real world, time simply happens. We can measure it, but we cannot influence it. Every morning, when we wake up, we have aged by precisely one day since the previous morning. There is nothing to be done about. If we are slow, in getting something done, we have to hurry up, as we cannot slow time down.

In simulation, time does not simply happen. We need to make it happen. When we simulate a system, it is our duty to manage the simulation clock, and how effectively we are able to manage the simulation clock will ultimately decide upon the efficiency of our simulation run.

In the previous two sections of this chapter, we have looked at different ways for representing a model. At the bottom of the hierarchy, we encountered the *explicit ODE model*, which we also called the *state–space model*. We simulated a simple electrical RLC circuit, represented as a linear state–space model, by use of MATLAB, and obtained a trajectory for the output variable, i_2, as a function of time. That output trajectory was depicted graphically in Fig.1.2.

The trajectory $i_2(t)$ seems to be a real–valued function of one real–valued argument. For any value of t, we can obtain the appropriate value of i_2. Yet, this is only an illusion, created to make us believe that the simulation is a faithful image of how we perceive the real system to work.

A digital computer has no means of computing numerically any real–valued function of a real–valued argument. To do so would require an infinite amount of real time. Instead, the time axis in the simulation must be *discretized*, such that the total number of discrete time points within the range of simulated time remains finite, and the simulation must proceed by jumping from one discrete time point to the next. The coarser we can choose the discretization in time, the smaller the total number of discrete time points will be, and consequently, the less work needs to be done in the simulation to evaluate the model at the output points. The discretization in time directly influences the efficiency of the simulation run.

Consequently, neither of the previously introduced model types can be simulated directly. Inside the simulation box, the model gets converted once more by reducing differential equation models to *difference equation (ΔE) models*. Thereby, an explicit ODE model is converted to an explicit ΔE model, whereas an implicit DAE model is converted to an implicit ΔAE model, etc.

The illusion of a continuous $i_2(t)$ curve was created by making the plot routine connect neighboring data points using a straight–line approximation. How often do we need to actually compute values of i_2? We need to do so sufficiently often that the straight–line approximation looks smooth to the naked eye. We call the distance between two neighboring computed output data points the *communication interval*. When we simulate a system, the simulation software asks us to provide that information to it. In the MATLAB code, we created a vector:

```
t = [ 0 : 1e-6 : 1e-4 ];
```

of communication points. It states that we wish to compute the output variable once every 10^{-6} seconds up until the final time of 10^{-4} seconds, giving us a result vector of 101 data points.

Does this mean that the simulation proceeds at the pace dictated by the communication grid? Absolutely not. The communication grid was only created to please the user, such that he or she can enjoy the illusion of a smoothly looking output variable. The simulation pace, however, is dictated by the numerical needs of the algorithm. The more accurately we wish to simulate, the smaller the time steps of the simulation must be chosen.

Thus, the simulation clock can advance either more slowly than the communication clock by allowing multiple simulation steps to occur within a single communication interval, or it could proceed more rapidly. In the latter case, the intermediate output points are obtained not by *simulation*, but by *interpolation*. If the interpolation routine can produce an interpolation of the same order of approximation accuracy as the integration, this is a perfectly valid way of computing output points.

Figure 1.6 depicts the relationship between the different types of *time* that we have to deal with in a simulation.

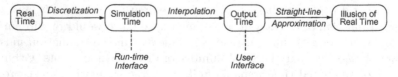

FIGURE 1.6. The different faces of *time*.

Whereas the communication grid is usually equidistantly spaced, the simulation grid is not. The step size, h, of the simulation is usually allowed to adjust itself, such that the accuracy requirements are met. A simulation user knows how to set the *communication interval* or *sampling rate*, t_s, but he or she wouldn't know how to set the step size, h, of the simulation. Consequently, most simulation software systems will ask the user to specify an *accuracy requirement* instead. The integration algorithm uses some formula to estimate the numerical *integration error*, and then uses a control scheme to adjust the step size such that the integration error is kept as large as possible, while not exceeding the specified maximum error.

Does the simulation clock at least advance monotonously with real time, i.e., will the time difference, Δt, of the simulation clock between two subsequent evaluations of the model be always positive? Unfortunately, also this question must usually be answered in the negative for three separate reasons.

1. The step size, h, is not necessarily identical with the time advance, Δt, of model evaluations. Many integration algorithms, such as the famous Runge–Kutta algorithms, which we shall meet in Chapter 3 of this book, perform multiple model evaluations within a single time step. Thus, each time step, h, contains several micro–steps, Δt, whereby Δt is not a fixed divider of h. Instead, the simulation clock may jump back and forth within each individual time step.

2. Even if the integration algorithm used is such that Δt remains positive at all times, the simulation clock does not necessarily advance monotonously with real time. There are two types of error–controlled integration algorithms that differ in the way they handle steps that exhibit an error estimate that is too large. *Optimistic algorithms* simply continue, in spite of the exceeded error tolerance, while reducing the step size for the subsequent step. In contrast, *conservative algorithms* reject the step, and repeat it with a smaller step size. Thus, whenever a step is rejected, the simulation clock in a conservative algorithm turns back to repeat the step, while not committing the same error. Wouldn't it be nice if we could do the same in the real world?

3. Even if an optimistic algorithm with positive Δt values is being employed, the simulation clock may still not advance monotonously with real time. The reason is that integration algorithms cannot integrate across discontinuities in the model. Thus, if a discontinuity is encountered somewhere inside an integration step, the step size must be reduced and the step must be repeated, in order to place the discontinuity in between subsequent steps. These issues shall be discussed in Chapter 9 of this book.

Hence the flow chart shown in Fig.1.6 is still somewhat oversimplified, as it does not account for the micro–management of time within a single integration step.

The issues surrounding *time management* as part of the simulation algorithms shall haunt us throughout the various chapters of this book.

1.4   Simulation as a Problem Solving Tool

Simulation has become the major analysis tool in essentially all of engineering, and much of science. Industry nowadays demands that companies providing parts for their products ship their parts with simulation models that can be assembled in just about the same fashion as the real system is. For example, when you buy these days an all–American car, you may not want to check too closely what is under the hood, because you may quickly discover that your car comes equipped with a German engine and a Japanese transmission.

Car manufacturers these days allow two years from the conception of a new model, until the first cars roll off the production line. During the first year, the car itself is designed and its performance is optimized by means of continuous system simulation; during the second year, the production process of the car is designed, again involving a lot of simulation, though mostly of a discrete event nature.

This can only work if the parts come equipped with ready–to–use simulation models that can be plugged together quickly and painlessly. This is only possible if the modeling methodology in use is *object oriented*, which invariably leads to large sets of implicitly defined DAE systems.

To this end, the *Modeling and Simulation (M&S) environment* must be able to deal with implicit DAE descriptions, either by simulating such descriptions directly, or by automatically converting them to explicit ODE descriptions beforehand. The days of 10,000 lines of spaghetti FORTRAN code to e.g. simulate the flight of a missile, taking into account such gory details as the seeker and its gyroscopically stabilized platform, as well as the flopping around of the liquid fuel in the fuel tank, are thus finally over.

Whereas the issues surrounding object–oriented modeling are not the aim of this book[1], issues surrounding the symbolic model transformations to precondition the simulation code for efficient run–time performance are being dealt with in later chapters of this textbook.

Modern M&S environments, such as Dymola [1.11, 1.13], are capable of automatically generating simulation code from an object–oriented mathematical model that runs as efficiently as, if not more efficiently than, the best among the hand–coded spaghetti simulation programs of the past. The translation of the model is usually accomplished within seconds of real time.

---

[1] These issues are discussed both extensively and intensively in the companion book of this text, *Continuous System Modeling* [1.6].

In the past, the life cycle of a simulation program often extended beyond that of its designer. The engineer who originally designed and wrote the spaghetti simulation code retired before the program itself had reached the end of its usefulness. Maintaining these programs, after the original designer could no longer be consulted, was an absolute nightmare. Also these days are luckily over.

1.5   Simulation Software: Today and Tomorrow

We published an article with the same title, *Simulation Software: Today and Tomorrow*, a little over 20 years ago [1.5], because at that time, we felt that the earlier article discussing similar topics [1.4] had meanwhile outlived its usefulness.

Reading through the 1983 paper once more, we recognize and happily acknowledge how hopelessly outdated that article has meanwhile become. This discovery is a cause of excitement, not depression, because it shows us how incredibly active this research area has been over the past 20 years, and how wonderfully dynamic this research area continues to be to this day.

Although the principles of *object–oriented modeling* had been developed already in the sixties [1.8], *Simula 67* had only been designed for discrete event simulation, not for continuous system simulation, and these concepts could not easily be carried over to modeling physical systems. The reason is that, in discrete event simulation, we always know what are the causes, and which are their effects. In physical system modeling, this is not the case. The *computational causality* of physical laws can therefore not be predetermined, but depends on the particular use of that law. We cannot conclude whether it is the current flowing through a resistor that causes a voltage drop, or whether it is the difference between the potentials at the two ends of the resistor that causes current to flow. Physically, these are simply two concurrent aspects of one and the same physical phenomenon. Computationally, we may have to assume at times one position, and at other times the other.

First attempts at dealing with the problems of physical system modeling in an object–oriented fashion were developed simultaneously in two seminal Ph.D. dissertations By Elmqvist [1.11] and Runge [1.20]. Whereas Elmqvist focused his attention on symbolic formulae manipulation as a tool for preconditioning the model equations to obtain efficiently executing simulation code, Runge attempted to solve implicit DAE models directly.

Whereas a first prototypical implementation of *Dymola* had been implemented by Hilding Elmqvist already as part of his Ph.D. dissertation [1.11], Dymola was not yet capable of dealing with large–scale engineering models in those days. The code got stuck, as soon as it encountered either an *algebraic loop* or a *structural singularity*, which happened invariably in most large–scale engineering models.

First attempts at tackling the algebraic loop and structural singularity problems in a completely generic fashion were undertaken by Hilding Elmqvist in 1993 [1.2]. This research was followed up in 1994 by an important paper on *symbolic tearing* methods [1.10]. By 1997, heuristic procedures had been developed to automatically identify a suitable set of tearing variables. By that time, we finally had available a tool that could reduce, in a fully automated fashion, any implicit DAE model to explicit ODE form. We shall talk much more about these algorithms in Chapter 7 of this book.

A first prototype of a *Graphical User Interface (GUI)* for Dymola was created by Hilding Elmqvist as early as 1982 [1.12]. The graphical software *HIBLIZ* [1.12] had a number of interesting features, yet it was far ahead of its time, as the computer hardware of those days wasn't ready yet for these types of applications. Elmqvist resumed his work on a GUI for Dymola in 1993, which resulted in a very powerful modern graphical software environment, of which you have already seen some samples earlier in this chapter.

Elmqvist proved to be one of the most innovative and visionary researchers in M&S methodology and technology of the last quarter of a century, and his *Dynamic Modeling Laboratory, Dymola*, has become the de facto industry standard by now. No other tool on the market comes even close to Dymola in terms of flexibility and generality of its use.

On the numerical front, progress has been a bit less spectacular. The 4th–order Runge–Kutta algorithms in use today are still the same algorithms that were known and used in 1983. However, the development of production–grade direct DAE solvers [1.1], a direct outflow of Runge's earlier work, fell in this time frame, and stirred quite a bit of excitement among applied mathematicians.

Furthermore, a lot has happened in terms of the development of better software aiding the design of new numerical algorithms. *MATLAB* [1.15] has become the de facto industry standard for the description of numerical algorithms. All of the algorithms described in this book are explained in terms of snippets of MATLAB code, and most of the homework problems are designed to be solved using MATLAB.

In the same context, the quite impressive advances in the development of tools for *computational algebra* deserve to be mentioned as well. Applied mathematicians like to present the coefficients of their algorithms symbolically as rational expressions, rather than numerically as numbers with many digits after the dot, because in this way, the numerical accuracy of the algorithm can fully exploit the available mantissa length of the computer on which the algorithm is being implemented. Tools, such as *MAPLE* [1.16] and *Mathematica* [1.22] have made the design of new algorithms considerably less painful than in the past, and indeed, several errors were recently discovered using computational algebra tools in a number of numerical algorithms that had been around for decades [1.14]. When developing this book, we made frequent use of MATLAB's *symbolic toolbox*, which is based on MAPLE, to derive correct rational expressions for the coefficients of new algorithms.

The advent of ever more powerful computer hardware made it possible to search for new algorithms much more efficiently than in the past. For example, we could not have developed the higher–order stiffly–stable linear multi–step methods that are described in Chapter 4 of this book as little as 10 years ago, since several of the search algorithms used in the process milled for more than 30 minutes of real time on a 2.5 GHz personal computer, whereas 10 years ago, we had to rely on a 1 MHz VAX computer for all of our computations.

Finally, the automatic preconditioning of models by means of symbolic formulae manipulation made it possible to employ highly promising numerical algorithms that could not have been used previously, because they would have forced the users to manually convert the models in a manner, which would have been far too cumbersome for them. A good example of this are the *inline integration* algorithms [1.9] that are discussed in Chapter 8 of this book.

For these reasons, we expect that a good number of exciting new numerical algorithms will appear in the open literature at a much more rapid pace over the next few years.

What are tools that are still missing or unsatisfactory in Dymola? A first issue to be improved is the mechanism, by which run–time exceptions are reported back to the user. Advanced *reverse engineering mechanisms* ought to be put in place to translate run–time exceptions back to terms that are related to the original model, i.e., terms that the user of the Dymola M&S environment can understand. Right now, the debugging of Dymola models can be quite challenging.

A second issue to be looked into concerns Dymola's way of handling table–lookup functions. The treatment of tabular functions is unsatisfactory on several counts.

1. If an input variable is provided to the simulation engine in the form of a table, sampled once per communication interval, Dymola uses linear interpolation to estimate intermediate values of the input variable. Yet, the simulation engine may simulate the model using a higher–order algorithm, possibly subdividing the communication interval into several steps. This situation can be remedied easily by use of the Nordsieck vector approach that is discussed in Chapter 4 of this book.

2. If the independent variable of a table–lookup function is not *time*, but a dependent variable of the model, the situation gets more complicated. Yet, the necessary history information could be traced back also in this case. Furthermore, the effects of reduced–order numerical

approximations of table–lookup functions on the overall simulation accuracy ought to be properly studied. This has not happened to date. This could be a nice research topic for a young aspiring applied mathematician.

3. The treatment of large tables, as currently implemented in Dymola, is highly inefficient. This is a compiler issue that will need to be addressed.

4. Large multi–dimensional tables need to be interpolated directly on the storage medium, rather than loading them into the model, and manipulating them at compile time. This is not currently the case. However, the use of *Modelica* as the underlying alphanumerical model representation helps in this respect. Modelica is a full–fledged language, in which adequate table–lookup mechanisms could easily be implemented [1.21].

A third and very interesting research issue concerns the automated assembly of models. For example, if we wish to model a chemical reaction system, we ought to be able to automatically extract the necessary parameters and table–lookup functions from the open literature.

How do we go about such modeling issues today? We probably would use *Google* to find the missing information on the web. Google has become the de facto standard for finding the answer to pretty much any question that we may have. Google has become our most important interface to the accumulated world knowledge.

Yet in order to use Google effectively, we must first come up with the right keywords to find the most suitable articles on the web, and it will be furthermore our task to manually sift through the articles returned to find what we need.

We foresee the need to automate these two current user interfaces as part of a future *distributed M&S environment*. The M&S environment ought to be able to automatically query a distributed data base for the availability of entire models, model parameter values, and table–lookup functions. This demand could provide challenging and exciting research topics for several Ph.D. students of computer science.

1.6   Summary

In this chapter, we started out with a set of different ways how a mathematical model of a physical system can be formulated. We demonstrated that it is important to distinguish the mathematical model (the user interface) from the simulation program (the run–time interface), such that the mathematical model can be defined to maximize the convenience for the human user of the tool, whereas the simulation program can be defined to optimize run–time efficiency of the simulation code.

We looked at the important issue of time management during execution of a continuous system simulation program with a bird's eye's view. Whereas all of these issues will be revisited throughout the chapters of this book, we considered it useful to bring these issues to the reader's attention early on.

The chapter ended with a discussion of where we stand today in terms of modeling and simulation environments, and what additional features we expect will be required in the near future.

1.7 References

[1.1] Kathryn E. Brenan, Stephen L. Campbell, and Linda R. Petzold. *Numerical Solution of Initial–Value Problems in Differential–Algebraic Equations*. North–Holland, New York, 1989. 256p.

[1.2] François E. Cellier and Hilding Elmqvist. Automated Formula Manipulation Supports Object–oriented Continuous System Modeling. *IEEE Control Systems*, 13(2):28–38, 1993.

[1.3] François E. Cellier and Robert T. McBride. Object–oriented Modeling of Complex Physical Systems Using the Dymola Bond–graph Library. In François E. Cellier and José J. Granda, editors, *Proceedings of the 2003 SCS Intl. Conf. on Bond Graph Modeling and Simulation*, pages 157–162, Orlando, Fl., 2003. The Society for Modeling and Simulation International.

[1.4] François E. Cellier. Continuous System Simulation by Use of Digital Computers: A State–of–the–Art Survey and Perspectives for Development. In Mohamed H. Hamza, editor, *Proceedings Simulation'75*, pages 18–25, Zurich, Switzerland, 1975. ACTA Press.

[1.5] François E. Cellier. Simulation Software: Today and Tomorrow. In Jacques Burger and Yvon Varny, editors, *Proceedings of the IMACS Symposium on Simulation in Engineering Sciences*, pages 3–19, Nantes, France, 1983. North–Holland Publishing.

[1.6] François E. Cellier. *Continuous System Modeling*. Springer Verlag, New York, 1991. 755p.

[1.7] James B. Dabney and Thomas L. Harman. *Mastering SIMULINK 4*. Prentice–Hall, Upper Saddle River, N.J., 2001. 432p.

[1.8] Ole-Johan Dahl, Bjørn Myhrhaug, and Kristen Nygaard. Simula 67 Common Base Language. Technical report, Norwegian Computing Center, Oslo, Norway, 1968.

[1.9] Hilding Elmqvist, Martin Otter, and François E. Cellier. Inline Integration: A New Mixed Symbolic/Numeric Approach for Solving Differential–Algebraic Equation Systems. In *Proceedings European Simulation Multiconference*, pages xxiii–xxxiv, Prague, Czech Republic, 1995.

[1.10] Hilding Elmqvist and Martin Otter. Methods for Tearing Systems of Equations in Object–oriented Modeling. In *Proceedings European Simulation Multiconference*, pages 326–332, Barcelona, Spain, 1994.

[1.11] Hilding Elmqvist. *A Structured Model Language for Large Continuous Systems*. PhD thesis, Dept. of Automatic Control, Lund Institute of Technology, Lund, Sweden, 1978.

[1.12] Hilding Elmqvist. A Graphical Approach to Documentation and Implementation of Control Systems. In *Proceedings 3rd IFAC/IFIP Symposium on Software for Computer Control (SOCOCO'82)*, Madrid, Spain, 1982.

[1.13] Hilding Elmqvist. *Dymola — Dynamic Modeling Language, User's Manual, Version 5.3*. DynaSim AB, Research Park Ideon, Lund, Sweden, 2004.

[1.14] Walter Gander and Dominik Gruntz. Derivation of Numerical Methods Using Computer Algebra. *SIAM Review*, 41(3):577–593, 1999.

[1.15] Duane Hanselman and Bruce Littlefield. *Mastering MATLAB 6*. Prentice–Hall, Upper Saddle River, N.J., 2001. 832p.

[1.16] André Heck. *Introduction to Maple*. Springer Verlag, New York, 2nd edition, 1996. 525p.

[1.17] Marc E. Herniter. *Schematic Capture with Cadence PSpice*. Prentice–Hall, Upper Saddle River, N.J., 2nd edition, 2002. 656p.

[1.18] Edward E. L. Mitchell and Joseph S. Gauthier. *ACSL: Advanced Continuous Simulation Language — User Guide and Reference Manual*. Mitchell & Gauthier Assoc., Concord, Mass., 1991.

[1.19] Franz Monssen. *OrCAD PSpice with Circuit Analysis*. Prentice–Hall, Upper Saddle River, N.J., 3rd edition, 2001. 400p.

[1.20] Thomas F. Runge. *A Universal Language for Continuous Network Simulation*. PhD thesis, Dept. of Computer Science, University of Illinois, Urbana–Champaign, Ill., 1977.

[1.21] Michael M. Tiller. *Introduction to Physical Modeling with Modelica*. Kluwer Academic Publishers, Boston, Mass., 2001. 368p.

[1.22] Stephen Wolfram. *The Mathematica Book.* Wolfram Media, Inc, Champaign, Ill., 5th edition, 2003. 1488p.

1.8 Homework Problems

[H1.1] Different Mathematical Models

Given the electrical circuit shown in Fig.H1.1a.

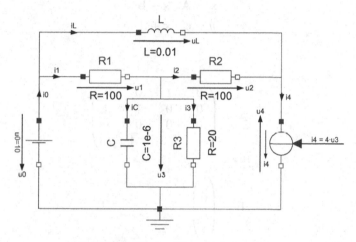

FIGURE H1.1a. Electrical circuit.

The circuit contains a constant voltage source, u_0, and a dependent current source, i_4, that depends on the voltage across the capacitor, C, and the resistor, R_3.

Write down the element equations for the seven circuit elements. Since the voltage u_3 is common to two circuit elements, these equations contain 13 rather than 14 unknowns. Add the voltage equations for the three meshes and the current equations for three of the four nodes. One current equation is redundant. Usually, the current equation for the ground node is therefore omitted.

Formulate an implicit DAE model of this circuit by placing all unknowns to the left of the equal sign, and all known expressions to the right of the equal sign.

Sort the equations both horizontally and vertically. Since you haven't learnt yet a systematic algorithm for doing this (such an algorithm shall be presented in Chapter 7 of this book), use intuition to come up with the sorted set of equations.

Formulate an explicit DAE model of this circuit using the sorted equations.

Use variable substitution to derive a state–space model of this circuit in matrix–vector form. We shall assume that u_3 is our output variable.

Simulate the circuit across 50 μsec using MATLAB's *lsim* function. Store 101 equidistantly spaced output values, and plot the output variable as a function of time.

[H1.2] Discretization of State Equations

Given the following explicit ODE model:

$$\dot{\mathbf{x}} = \mathbf{A} \cdot \mathbf{x} + \mathbf{b} \cdot u \tag{H1.2a}$$
$$y = \mathbf{c}' \cdot \mathbf{x} + d \cdot u \tag{H1.2b}$$

where:

$$\mathbf{A} = \begin{pmatrix} 0 & 1 & 0 & 0 \\ 0 & 0 & 1 & 0 \\ 0 & 0 & 0 & 1 \\ -2 & -3 & -4 & -5 \end{pmatrix} \tag{H1.2c}$$

$$\mathbf{b} = \begin{pmatrix} 0 \\ 0 \\ 0 \\ 1 \end{pmatrix} \tag{H1.2d}$$

$$\mathbf{c}' = \begin{pmatrix} 1 & 0 & 0 & 0 \end{pmatrix} \tag{H1.2e}$$

$$d = 10 \tag{H1.2f}$$

Engineers would usually call such a model a *linear single–input, single–output (SISO) continuous–time state–space model.*

We wish to simulate this model using the following integration algorithm:

$$\mathbf{x_{k+1}} = \mathbf{x_k} + h \cdot \dot{\mathbf{x}}_k \tag{H1.2g}$$

which is known as the *Forward Euler (FE) integration algorithm.* If $\mathbf{x_k}$ denotes the state vector at time t^*:

$$\mathbf{x_k} = \mathbf{x}(t) \Big|_{t=t^*} \tag{H1.2h}$$

then $\mathbf{x_{k+1}}$ represents the state vector one time step later:

$$\mathbf{x_{k+1}} = \mathbf{x}(t) \Big|_{t=t^*+h} \tag{H1.2i}$$

Obtain an explicit ΔE model by substituting the state equations into the integrator equations. You obtain a model of the type:

$$\mathbf{x_{k+1}} = \mathbf{F} \cdot \mathbf{x_k} + \mathbf{g} \cdot u_k \qquad \text{(H1.2j)}$$

$$y_k = \mathbf{h}' \cdot \mathbf{x_k} + i \cdot u_k \qquad \text{(H1.2k)}$$

which engineers would normally call a *linear single–input, single–output (SISO) discrete–time state–space model*.

Let $h = 0.01 \; sec$, $t_f = 5 \; sec$, $u(t) = 5 \cdot sin(2t)$, $x_0 = ones(4,1)$, where t_f denotes the final time of the simulation.

Simulate the ΔE model using MATLAB by iterating over the difference equations. Plot the output variable as a function of time.

[H1.3] Time Reversal

Given a state–space model of the form:

$$\dot{\mathbf{x}}(t) = \mathbf{f}(\mathbf{x}(t), \mathbf{u}(t), t) \quad ; \quad \mathbf{x}(t = t_0) = \mathbf{x_0} \quad ; \quad t \in [t_0, t_f] \qquad \text{(H1.3a)}$$

which generates the trajectory behavior $\mathbf{x}(t)$.

The state–space model:

$$\dot{\mathbf{y}}(\tau) = -\mathbf{f}(\mathbf{y}(\tau), \mathbf{u}(\tau), \tau) \quad ; \quad \mathbf{y}(\tau = t_f) = \mathbf{x_f} \quad ; \quad \tau \in [t_f, t_0] \qquad \text{(H1.3b)}$$

generates the trajectory behavior $\mathbf{y}(\tau)$.

Show that:

$$\mathbf{y}(\tau) = \mathbf{x}(t_0 + t_f - t) \qquad \text{(H1.3c)}$$

In other words, any state–space model can be simulated backward through time by simply placing a minus sign in front of every state equation.

[H1.4] Van–der–Pol Oscillator and Time Reversal

Given the following nonlinear system:

$$\ddot{x} - \mu(1 - x^2)\dot{x} + x = 0 \qquad \text{(H1.4a)}$$

This system exhibits an oscillatory behavior. It is commonly referred to as the *Van–der–Pol oscillator*. We wish to simulate this system with $\mu = 2.0$ and $x_0 = \dot{x}_0 = 0.1$.

Draw a block diagram of this system. The output variable is x. The system is autonomous, i.e., it doesn't have an input variable.

Derive a state–space description of this system. To this end, choose the outputs of the two integrators as your two state variables.

Simulate the system across $2 \; sec$ of simulated time. Since the system is nonlinear, you cannot use MATLAB's *lsim* function. Use function *ode45* instead.

At time $t = 2.0$ sec, apply the time reversal algorithm, and simulate the system further across another 2 sec of simulated time. This is best accomplished by adjusting the model such that it contains a factor c in front of each state equation. $c = +1$ during the first 2 sec of simulated time, and $c = -1$ thereafter. You can interpret c as an input variable to the model. Make sure that $t = 2.0$ sec defines an output point.

As you simulate the system backward through time for the same time period that you previously used to simulate the system forward through time, the final values of your two state variables ought to be identical to the initial values except for numerical inaccuracies of the simulation. Verify that this is indeed the case. How large is the accumulated error of the final values? The accumulated simulation error is defined as the norm of the difference between final and initial values.

Plot $x(t)$ and $\dot{x}(t)$ on the same graph.

Repeat the previous experiment, this time simulating the system forward during 20 sec of simulated time, then backward through another 20 sec of simulated time. What do you conclude?

1.9   Projects

[P1.1] Definitions

Get a number of simulation and/or system theory textbooks from your library and compile a list of definitions of "What is a System"? Write a term paper in which these definitions are critically reviewed and classified. (Such a compilation has actually been published once.)

2

Basic Principles of Numerical Integration

Preview

In this chapter, we shall discuss some basic ideas behind the algorithms that are used to numerically solve sets of ordinary differential equations specified by means of a state–space model. Following a brief introduction into the concept of numerical extrapolation that is at the heart of all numerical integration techniques, and after analyzing the types of numerical errors that all these algorithms are destined to exhibit, the two most basic algorithms, Forward Euler (FE) and Backward Euler (BE), are introduced, and the fundamental differences between explicit and implicit integration schemes are demonstrated by means of these two algorithms.

The reader is then introduced to the concept of numerical stability as opposed to analytical stability. The numerical stability domain is introduced as a tool to characterize an integration algorithm, and a general procedure to find the numerical stability domain of any integration scheme is presented. The numerical stability domain of an integration method is a convenient tool to assess some of its most important numerical characteristics.

2.1 Introduction

Given a state–space model of the form:

$$\dot{\mathbf{x}}(t) = \mathbf{f}(\mathbf{x}(t), \mathbf{u}(t), t) \tag{2.1}$$

where $\mathbf{x}$ is the *state vector*, $\mathbf{u}$ is the *input vector*, and t represents time, with a set of initial conditions:

$$\mathbf{x}(t = t_0) = \mathbf{x_0} \tag{2.2}$$

Let $x_i(t)$ represent the ith state trajectory as a function of simulated time, t. As long as the state–space model does not contain any discontinuities in either $f_i(\mathbf{x}, \mathbf{u}, t)$ or any of its derivatives, $x_i(t)$ is itself a continuous function of time. Such function can be approximated with any desired precision by a Taylor–Series expansion about any given point along its trajectory, as long as the function does not exhibit a finite escape time, i.e., approaches infinity for any finite value of time. Let t^* denote the point in time, about which we wish to approximate the trajectory using a Taylor Series, and let t^*+h be the point in time, at which we wish to evaluate the approximation. The value of the trajectory at that point can then be given as follows:

$$x_i(t^* + h) = x_i(t^*) + \frac{dx_i(t^*)}{dt} \cdot h + \frac{d^2 x_i(t^*)}{dt^2} \cdot \frac{h^2}{2!} + \dots \qquad (2.3)$$

Plugging the state–space model into (2.3), we find:

$$x_i(t^* + h) = x_i(t^*) + f_i(t^*) \cdot h + \frac{df_i(t^*)}{dt} \cdot \frac{h^2}{2!} + \dots \qquad (2.4)$$

Different integration algorithms vary in how they approximate the higher state derivatives, and in the number of terms of the Taylor–Series expansion that they consider in the approximation.

2.2   The Approximation Accuracy

Evidently, the accuracy with which the higher order derivatives are approximated should match the number of terms of the Series that are considered. If $n+1$ terms of the Taylor Series are considered, the approximation accuracy of the second state derivative $d^2 x_i(t^*)/dt^2 = df_i(t^*)/dt$ should be of order $n-2$, since this factor is multiplied with h^2. The accuracy of the third state derivative should be of order $n-3$, since this factor is multiplied with h^3, etc. In this way, the approximation is correct up to h^n. n is therefore called the *approximation order* of the integration method, or, more simply, the integration method is said to be of nth order.

The approximation error that is made because of the truncation of the Taylor Series after a finite number of terms is called *truncation error*. The truncation error contains terms in h^{n+1}, h^{n+2}, etc. It does not contain any terms in powers of h smaller than $n + 1$. However, since the magnitude of the remaining terms usually decreases rapidly with increasing powers of h, the truncation error itself is often approximated by a single term, namely the term in h^{n+1}. In order to be able to assess the accuracy of the numerical integration, it is essential to be aware of this term. Therefore, many numerical integration codes actually estimate this term, and use this information for such purposes as step–size control.

The higher the approximation order of a method, the more accurate will be the estimation of $x_i(t^* + h)$. Consequently, when using a high–order method, we can afford to integrate with a large step size. On the other hand, the smaller the step size that we employ, the faster decreases the importance of the higher–order terms in the Taylor Series, and therefore, when using a small step size, we can afford to truncate the Taylor Series early.

The cost of integrating a state–space model across a single integration step depends heavily on the order of the method in use. High–order algorithms are much more expensive than low–order methods in this respect. However, this cost may be offset by the fact that we can use a much larger step size, and therefore require a considerably smaller overall number of integration steps to complete the simulation run. We have therefore always a choice between employing a low–order algorithm with a small step size integrating the system over many such steps, or using a high–order algorithm with a large step size integrating the system over much fewer steps.

Which of these choices is more economical in a given situation, depends on various factors. However, for now, the following simple rule of thumb may be used as an often quite decent indicator [2.4]:

> If the local relative accuracy required by an application, i.e., the largest error tolerated within a single integration step, is 10^{-n}, then it is best to choose at least a nth order algorithm for the numerical integration.

For this reason, the simulation of problems from celestial dynamics requires the highest–order algorithms. We usually apply eighth–order algorithms to such problems. On the other hand, most simulations of economic systems call for first– or second–order methods, since the parameters of the models themselves are not more accurate than that. It makes no sense whatsoever to waste a superb integration algorithm on a garbage model. Garbage integrated with high precision still remains garbage.

Many engineering simulation applications require a global relative accuracy of approximately 0.001. We usually make the following assumption:

> If the local integration error, i.e., the error made during a single integration step, is proportional to h^{n+1}, then the global integration error, i.e., the error of the results at the end of the simulation run, is proportional to h^n.

This assumption is correct for a sufficiently small step size, h, i.e., in the so–called *asymptotic region* of the algorithm.

The above heuristic can be justified by the following observation. If the *local integration error* is of size e_ℓ, then the *per–unit–step integration error* assumes a value of $e_{p.u.s} = e_\ell/h$. The *global integration error* is proportional to the per–unit–step integration error, as long as the integration error does not accumulate excessively across multiple steps.

A global relative error of 0.001, as required by most engineering applications, calls for an algorithm with an approximation order of h^3 for the global error. In accordance with the previously made observation, this corresponds to an algorithm with an approximation order of h^4 for the local integration error. Therefore, we should require a local accuracy of 0.0001. This means that a fourth–order algorithm is about optimal, and, since engineers are the most highly valued customers of continuous–system simulation software designers, this is what most such simulation software systems offer as their default integration algorithm, i.e., as the algorithm that is used by the system if the user doesn't specify explicitly, which technique he or she wishes to be employed.

A second type of approximation error to be looked at is caused by the finite word length of the computer, on which the simulation is performed. On a digital computer, real numbers can only be represented with a finite precision. This type of error is called the *roundoff error*. The roundoff error is important since, in numerical integration, invariably very small numbers are added to very large numbers.

For example, let us assume we employ a third-order algorithm:

$$x(t^* + h) \approx x(t^*) + f(t^*) \cdot h + \frac{df(t^*)}{dt} \cdot \frac{h^2}{2!} + \frac{d^2 f(t^*)}{dt^2} \cdot \frac{h^3}{3!} \qquad (2.5)$$

to integrate a scalar state–space model:

$$\dot{x} = f(x, u, t) \qquad (2.6)$$

across one second of simulated time. Let us assume for simplicity that the magnitude of x and its first three time derivatives is in the order of 1.0, thus:

$$\|x\| \approx \|f\| \approx \|\dot{f}\| \approx \|\ddot{f}\| \approx 1.0 \qquad (2.7)$$

Let us further assume that a constant step size of $h = 0.001$ is employed throughout the simulation. The simulation is performed on a computer with a word length of 32 bits in single precision. Such machines usually offer a mantissa length of 24 bits, and an exponent of eight bits. On such a machine, the roundoff error is approximately:

$$\varepsilon_{\text{roundoff}} = 2^{-24} \approx 10^{-6} \qquad (2.8)$$

Thus, a real number in single precision carries approximately six significant decimals. Applying this information to the process of numerical integration, we find:

$$\|x(t^* + h)\| \approx \|x(t^*)\| + \|f(t^*) \cdot h\| + \|\frac{df(t^*)}{dt} \cdot \frac{h^2}{2!}\| + \|\frac{d^2 f(t^*)}{dt^2} \cdot \frac{h^3}{3!}\|$$
$$\approx 1.0 + 0.001 + 10^{-6} + 10^{-9} \qquad (2.9)$$

Thus, while the constant term contributes six significant digits to the result of the addition, already the linear term contributes only three digits to the result, and the second–order term does not contribute anything of significance at all. We might just as well never have computed it in the first place. This fact is illustrated in Fig.2.1.

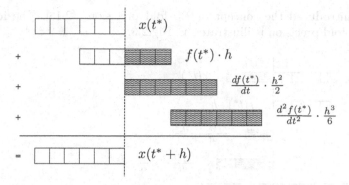

FIGURE 2.1. Effects of roundoff on numerical integration.

Consequently, using single precision on a 32 bit machine for numerical integration algorithms of order higher than two may be quite problematic. In reality, the effects of shiftout will not necessarily be as dramatic as shown in the above example, since higher–order algorithms allow use of a larger step size. Yet, double precision algorithms will be definitely more robust due to their reduced risk of shiftout, and they can meanwhile be implemented quite efficiently also. Therefore, there is no good reason anymore to use single precision on any integration algorithm but Euler. In this context, it is interesting to notice that many commercially available simulation software systems, such as ACSL [2.20], use a single–precision fourth–order variable–step Runge–Kutta algorithm as their default integration method, integrating happily –and dangerously– along.

A double precision representation will take care of roundoff errors as shown in Fig.2.2. A double precision representation on a 32 bit machine provides about 14 significant digits, since double precision words usually offer a 52 bit mantissa and a 12 bit exponent on such a machine.

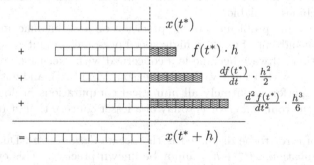

FIGURE 2.2. Roundoff in double precision.

Unfortunately, double precision operations are more time–consuming and therefore more expensive than single precision operations. This may still cause a problem especially in real–time applications. For this reason, Korn and Wait introduced the concept of 1.5–fold precision [2.18]. The idea behind 1.5–fold precision is illustrated in Fig.2.3.

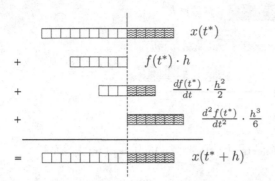

FIGURE 2.3. Roundoff in 1.5–fold precision.

It may not be necessary to store *all* real numbers in double precision. It may suffice to store only the state vector itself in double precision. In particular, this makes it possible to evaluate the nonlinear state–space model in single precision. Thereby, some of the accuracy of the state vector is compromised. The lost digits are shaded in Fig.2.3. However, these errors will not migrate to the left, i.e., a sufficiently large number of digits remains significant. The accuracy of the state vector is indeed roughly half way between that of single precision and that of double precision, but the overall price of the computation is closer to that of single precision.

Simulations of celestial dynamics problems should be performed in full double precision on a 64 bit machine, or, if only 32 bit machines are available, full four–fold precision should be used. It hardly ever makes sense to employ an algorithm of order higher than eight, since otherwise, the roundoff errors will dominate over the truncation errors even on the highest precision machines available.

For most engineering problems, double precision on a 32 bit machine is sufficient. With the advent of modern high–speed personal computers, producers of simulation software became less concerned with execution speed and more concerned with accuracy. For this reason, MATLAB, and with it also SIMULINK, perform routinely all numerical computations in double precision. Hence the roundoff error is today of a lesser concern than it used to be in the past.

A third type of error to be discussed is the *accumulation error*. Due to roundoff and truncation, $x(t^* + h)$ cannot be known precisely. This error will be inherited by the next integration step as an error in its initial conditions. Thus, errors accumulate when numerical integration proceeds over many steps. Fortunately, it can be observed that the effects of the initial conditions will eventually die out in the analytical solution of an analytically stable system. Consequently, it can be expected that a numerically stable numerical integration (we shall present a proper definition of this term in due course) will dampen out the effects of initial conditions as well, and will thereby, as a side effect, also get rid of inaccuracies in the initial conditions.

This is very fortunate, since it indicates that errors in initial conditions of an integration step don't usually affect the overall simulation too much. However, this assumption holds only for analytically stable systems. This is the reason why numerical integration algorithms have a tendency to stall when confronted with analytically unstable systems even before any trajectory of the analytical solution has grown alarmingly large. While simulating an analytically unstable system, it can no longer be assumed that the global integration error is proportional to the per–unit–step integration error, since the integration error can accumulate excessively across multiple steps. In such a case, it may be better to start from the end, and integrate the system backward through simulated time.

On top of all these errors, the simulation practitioner is confronted with inaccuracies of the model itself. These can be decomposed into *parametric model errors*, i.e., errors that reflect inaccurately estimated model parameters, and *structural model errors*, i.e., unmodeled dynamics.

To summarize the above, the modeler and the simulation practitioner must deal with five different types of errors. Modeling errors can be subdivided into structural and parametric errors. The modeler must verify that his model reflects reality sufficiently well for the purpose of the study at hand. This process is commonly referred to as *model validation*. Techniques for model validation are discussed in detail in the companion book to this text: *Continuous System Modeling* [2.5]. Once it has been asserted that the model reflects reality sufficiently well, the simulation practitioner must now verify that the numerical trajectories obtained by means of a numerical simulation of the model decently replicate the analytical trajectories that would result if the model were computed with infinite precision. This process is referred to as *simulation verification*. Simulation verification plays a central role in this textbook. Simulation errors can be classified into truncation errors, roundoff errors, and accumulation errors. It is the conglomerate of all of these errors that makes the life of an applied mathematician interesting indeed.

2.3   Euler Integration

Let us now look at some actual numerical integration algorithms.

The simplest integration algorithm is obtained by truncating the Taylor Series after the linear term.

$$\mathbf{x}(t^* + h) \approx \mathbf{x}(t^*) + \dot{\mathbf{x}}(t^*) \cdot h \qquad (2.10a)$$

or:

$$\mathbf{x}(t^* + h) \approx \mathbf{x}(t^*) + \mathbf{f}(\mathbf{x}(t^*), t^*) \cdot h \qquad (2.10b)$$

It is obviously possible to write the integration algorithm in vector form, i.e., the entire state vector can be integrated in parallel. The above scheme is particularly simple, since it doesn't require the approximation of any higher–order derivatives. The linear term is readily available from the state–space model. This integration scheme is called *Forward Euler* algorithm, and will, from now on, be abbreviated as FE algorithm. Figure 2.4 depicts graphically how the FE integration method approximates a state trajectory.

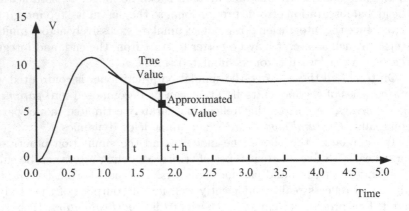

FIGURE 2.4. Numerical integration using Forward Euler.

Simulation using the FE algorithm is straightforward. Since the initial conditions, $\mathbf{x}(t = t_0) = \mathbf{x_0}$ are given, we can proceed as follows:

step 1a:	$\dot{\mathbf{x}}(t_0)$	$=$	$\mathbf{f}(\mathbf{x}(t_0), t_0)$
step 1b:	$\mathbf{x}(t_0 + h)$	$=$	$\mathbf{x}(t_0) + h \cdot \dot{\mathbf{x}}(t_0)$
step 2a:	$\dot{\mathbf{x}}(t_0 + h)$	$=$	$\mathbf{f}(\mathbf{x}(t_0 + h), t_0 + h)$
step 2b:	$\mathbf{x}(t_0 + 2h)$	$=$	$\mathbf{x}(t_0 + h) + h \cdot \dot{\mathbf{x}}(t_0 + h)$
step 3a:	$\dot{\mathbf{x}}(t_0 + 2h)$	$=$	$\mathbf{f}(\mathbf{x}(t_0 + 2h), t_0 + 2h)$
step 3b:	$\mathbf{x}(t_0 + 3h)$	$=$	$\mathbf{x}(t_0 + 2h) + h \cdot \dot{\mathbf{x}}(t_0 + 2h)$

etc.

Simulation becomes a straightforward and quite procedural matter, since the numerical integration algorithm depends only on past values of state variables and state derivatives. An integration scheme that exhibits this property is called *explicit integration algorithm*. Most integration algorithms employed in today's general–purpose continuous–system simulation

languages, such as ACSL [2.20], are of this nature. However, this statement does not hold for special–purpose simulation software, such as electric
circuit simulators.

Let us now introduce a different integration algorithm. Figure 2.5 depicts
a slightly modified scheme.

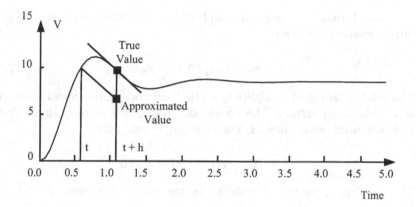

FIGURE 2.5. Numerical integration using Backward Euler.

In this scheme, the solution $\mathbf{x}(t^* + h)$ is approximated using the values
of $\mathbf{x}(t^*)$ and $\mathbf{f}(\mathbf{x}(t^* + h), t^* + h)$ using the formula:

$$\mathbf{x}(t^* + h) \approx \mathbf{x}(t^*) + \mathbf{f}(\mathbf{x}(t^* + h), t^* + h) \cdot h \qquad (2.11)$$

This scheme is commonly referred to as the *Backward Euler* integration
rule. It will, from now on, be abbreviated as BE algorithm.

As can be seen, this integration formula depends on current as well as
past values of variables. This fact causes problems. In order to compute
$\mathbf{x}(t^* + h)$ from Eq.(2.11), we need to know $\mathbf{f}(\mathbf{x}(t^* + h), t^* + h))$, however,
in order to compute $\mathbf{f}(\mathbf{x}(t^* + h), t^* + h))$ from Eq.(2.1), we need to know
$\mathbf{x}(t^* + h)$. Thus, we are confronted with a nonlinear *algebraic loop*. Algorithms that are of this type are referred to as *implicit integration techniques*.
The integration algorithms that are employed in electronic circuit simulators, such as PSpice [2.21], are of this type. Although implicit integration
techniques are advantageous from a numerical point of view (as we shall
learn later), the additional computational load created by the necessity to
solve simultaneously a set of nonlinear algebraic equations at least once every integration step may make them undesirable for use in general–purpose
simulation software except for specific applications, such as stiff systems.

2.4   The Domain of Numerical Stability

Let us now turn to the solution of autonomous, time–invariant linear systems of the type:

$$\dot{\mathbf{x}} = \mathbf{A} \cdot \mathbf{x} \tag{2.12}$$

with initial conditions as specified in Eq.(2.2). The solution of such a system can be analytically given:

$$\mathbf{x}(t) = \exp(\mathbf{A} \cdot t) \cdot \mathbf{x_0} \tag{2.13}$$

The solution is called *analytically stable* if all trajectories remain bounded as time goes to infinity. The system of Eq.(2.12) is analytically stable if and only if all eigenvalues of $\mathbf{A}$ have negative real parts:

$$\mathbb{Re}\{\mathrm{Eig}(\mathbf{A})\} = \mathbb{Re}\{\lambda\} < 0.0 \tag{2.14}$$

The domain of analytical stability in the complex λ–plane is shown in Fig.2.6.

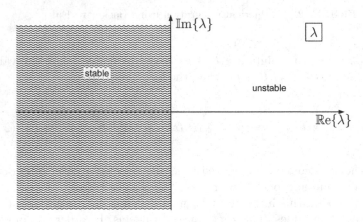

FIGURE 2.6. Domain of analytical stability.

Let us now apply the FE algorithm to the numerical solution of this problem. Plugging the system of Eq.(2.12) into the algorithm of Eq.(2.10), we obtain:

$$\mathbf{x}(t^* + h) = \mathbf{x}(t^*) + \mathbf{A} \cdot h \cdot \mathbf{x}(t^*) \tag{2.15}$$

which can be written in a more compact form as:

$$\mathbf{x}(k + 1) = [\mathbf{I^{(n)}} + \mathbf{A} \cdot h] \cdot \mathbf{x}(k) \tag{2.16}$$

$\mathbf{I}^{(n)}$ is an identity matrix of the same dimensions as $\mathbf{A}$, i.e., $n \times n$. Instead of referring to the simulation time explicitly, we simply index the time, i.e., k refers to the kth integration step.

By plugging the state equations into the integration algorithm, we have converted the former continuous–time system into an "equivalent" discrete–time system:

$$\mathbf{x_{k+1}} = \mathbf{F} \cdot \mathbf{x_k} \qquad (2.17)$$

where the discrete state matrix, $\mathbf{F}$, can be computed from the continuous state matrix, $\mathbf{A}$, and the step size, h, as:

$$\mathbf{F} = \mathbf{I}^{(n)} + \mathbf{A} \cdot h \qquad (2.18)$$

The term "equivalence" is defined in the sense of the employed numerical integration algorithm. It does not mean that the converted discrete–time system behaves identically to the original continuous–time system. The two systems are "equivalent" in the same sense as the numerical trajectory is "equivalent" to its analytical counterpart.

For simplicity, we shall consistently employ the following notation in this book:

$$\dot{\mathbf{x}} = \mathbf{A} \cdot \mathbf{x} + \mathbf{B} \cdot \mathbf{u} \qquad (2.19a)$$
$$\mathbf{y} = \mathbf{C} \cdot \mathbf{x} + \mathbf{D} \cdot \mathbf{u} \qquad (2.19b)$$

denotes the continuous–time linear system, where $\mathbf{A}$ is the state matrix, $\mathbf{B}$ is the input matrix, $\mathbf{C}$ is the output matrix, and $\mathbf{D}$ is the input/output matrix. $\mathbf{x}$ denotes the state vector. It is of length n ($\mathbf{x} \in \mathbb{R}^n$). $\mathbf{u}$ is the input vector, $\mathbf{u} \in \mathbb{R}^m$, and $\mathbf{y}$ is the output vector, $\mathbf{y} \in \mathbb{R}^p$.

The equivalent discrete–time linear system is written as:

$$\mathbf{x_{k+1}} = \mathbf{F} \cdot \mathbf{x_k} + \mathbf{G} \cdot \mathbf{u_k} \qquad (2.20a)$$
$$\mathbf{y_k} = \mathbf{H} \cdot \mathbf{x_k} + \mathbf{I} \cdot \mathbf{u_k} \qquad (2.20b)$$

where $\mathbf{F}$ now denotes the state matrix, $\mathbf{G}$ is the input matrix, $\mathbf{H}$ is the output matrix, and $\mathbf{I}$ is the input/output matrix.

The discrete–time system of Eq.(2.17) is analytically stable if and only if all its eigenvalues are located inside a circle of radius 1.0 about the origin, the so–called *unit circle*. From Eq.(2.18), we can conclude that all eigenvalues of $\mathbf{A}$ multiplied by the step size, h, must lie inside a circle of radius 1.0 about the point -1.0.

We define that the linear time–invariant continuous–time system integrated using a given fixed–step integration algorithm is *numerically stable* if and only if the "equivalent" linear time–invariant discrete–time system (the term equivalence meant in the sense of the same integration algorithm) is analytically stable.

Figure 2.7 shows the domain of numerical stability of the FE algorithm.

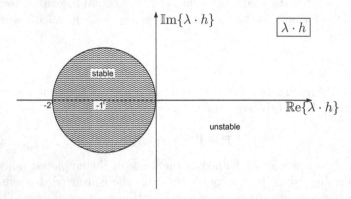

FIGURE 2.7. Domain of numerical stability of Forward Euler.

Notice that the numerical stability domain is, in a rigorous sense, only defined for linear time–invariant continuous–time systems, and applies only to fixed–step algorithms. Nevertheless, it is appealing that the numerical stability domain of an integration algorithm can be computed and drawn once and for all, and does not depend on any system properties other than the location of its eigenvalues.

The numerical stability domain of the FE algorithm tries to approximate the analytical stability domain, but evidently does a quite poor job at that.

Let us now try the following experiment. We simulate the scalar system $\dot{x} = a \cdot x$ with initial condition $x_0 = 1.0$ and a fixed step size of $h = 1.0$ over ten steps, i.e., from time $t = 0.0$ to time $t = 10.0$ using the FE algorithm. We repeat the experiment four times with different values of the parameter a. The results of this experiment are shown in Fig.2.8 The solid lines represent the analytical solutions, whereas the dashed lines represent the numerically found solutions. In the first case with $a = -0.1$, there exists a good correspondence between the two solutions. In the second case with $a = -1.0$, the numerical solution is still stable but bears little resemblance with the analytical solution, i.e., is very inaccurate. In the third case with $a = -2.0$, the numerical solution is marginally stable, and in the fourth case with $a = -3.0$, the numerical solution is unstable.

This result is in agreement with the numerical stability domain shown in Fig.2.7. We would have had to multiply the eigenvalue $\lambda = a = -3.0$ with a step size of $h = 2/3$, in order to obtain an even marginally stable solution, i.e., in order to get the eigenvalue into the stable region of the $\lambda \cdot h$–plane. In order to obtain an accurate result, a considerably smaller step size would have been needed. A 10% integration accuracy requires a step size of approximately $h = 0.1$ when applied to the system with $a = -3.0$, a 1%

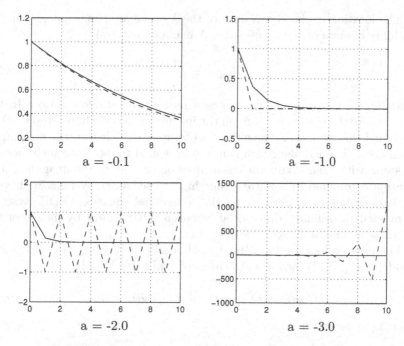

FIGURE 2.8. Numerical experiment using Forward Euler.

accuracy forces us to reduce the step size to $h = 0.01$, and a 0.1% accuracy calls for a step size of $h = 0.001$. In this case, we need already 10,000 steps to integrate this trivial system across 10 seconds. Quite obviously, the FE algorithm is not suitable if such high an accuracy is desired.

Moreover, the above experiment tested the FE algorithm on a very benign example. Systems with pairs of conjugate complex stable eigenvalues close to the imaginary axis are much worse. This fact is demonstrated in Fig.2.9.

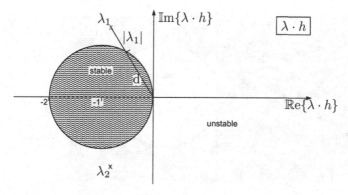

FIGURE 2.9. Determination of maximum step size with Forward Euler.

The location of the eigenvalues of the λ–plane are superimposed on the stability domain of the $\lambda \cdot h$–plane. A maximum step size of:

$$h_{max} = \frac{d}{|\lambda_1|} \qquad (2.21)$$

must be used in order to guarantee a numerically stable solution. In the case, where the eigenvalues are on the imaginary axis itself, no step size can be found that will make the numerical solution exhibit the true undamped oscillation. The FE algorithm is not at all suited to integrate such models. Systems with their dominant eigenvalues on or close to the imaginary axis are quite common. They are either highly oscillatory systems with very little damping, or hyperbolic partial differential equation (PDE) systems converted to ordinary differential equation (ODE) form by means of the method–of–lines approximation.

Let us now look at the BE algorithm. We shall plug the state–space model of Eq.(2.12) into the algorithm of Eq.(2.10). We obtain:

$$\mathbf{x}(t^* + h) = \mathbf{x}(t^*) + \mathbf{A} \cdot h \cdot \mathbf{x}(t^* + h) \qquad (2.22)$$

which can be rewritten as:

$$[\mathbf{I^{(n)}} - \mathbf{A} \cdot h] \cdot \mathbf{x}(t^* + h) = \mathbf{x}(t^*) \qquad (2.23)$$

or:

$$\mathbf{x}(k + 1) = [\mathbf{I^{(n)}} - \mathbf{A} \cdot h]^{-1} \cdot \mathbf{x}(k) \qquad (2.24)$$

Thus:

$$\mathbf{F} = [\mathbf{I^{(n)}} - \mathbf{A} \cdot h]^{-1} \qquad (2.25)$$

Figure 2.10 shows the stability domain of this technique.

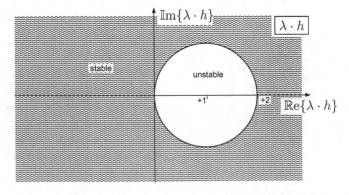

FIGURE 2.10. Stability domain of Backward Euler.

As in the case of the FE algorithm, BE tries to approximate the analytical stability domain, and does an equally poor job.

Let us repeat our previous experiment, this time with values of $a = -3.0$, and $a = +3.0$. It is of interest to us to simulate the system also in an analytically unstable configuration. Figure 2.11 shows the results of our efforts.

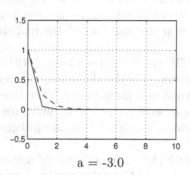
$a = -3.0$

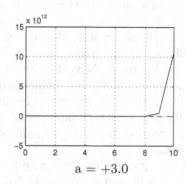
$a = +3.0$

FIGURE 2.11. Numerical experiment using Backward Euler.

The results could have been predicted easily from the stability domain shown in Fig.2.10.

The BE algorithm does a fairly decent job on the problem with $a = -3.0$. The results are not very accurate with $h = 1.0$, but, at least, they bear some resemblance with reality. This type of algorithm is therefore better suited than the FE type to solve problems with eigenvalues far out on the negative real axis of the λ–plane. Systems with eigenvalues whose real parts are widespread along the negative real axis are called *stiff systems*. Stiff systems are quite common. In particular, they often result from converting parabolic PDEs to sets of ODEs using the method–of–lines approximation. Contrary to the situation when the FE algorithm is used, the step size will, in the case of the BE algorithm, be dictated solely by *accuracy requirements* of the system, and not by the *numerical stability domain* of the method.

The problem with $a = +3.0$ reveals yet a different type of problem. The analytical solution is unstable, but the numerical simulation suggests that the system is perfectly stable. This can be quite dangerous. Imagine that a nuclear reactor has been designed and simulated using the BE algorithm. The simulation makes the engineers believe that everything is fine, but in reality, the reactor will blow up on the first occasion. Traditionally, researchers have focused more on the simulation of analytically stable systems, and therefore, many simulation practitioners aren't fully aware of the dangers that might result from using implicit algorithms, such as BE, to simulate systems that are potentially unstable in the analytical sense.

The lesson to be learnt is the following: When it really matters, it may be a good idea to simulate the system twice, once with an algorithm that exhibits a stability domain comparable to that of the FE algorithm, and once with an algorithm that behaves like the BE algorithm. If both simulations produce similar trajectories, the engineer may assume that the results are true to the model, though not necessarily to the physical plant. This is the most valuable simulation verification technique that exists, and the importance of this recommendation cannot be overestimated.

As in the case of the FE algorithm, BE has not much luck with marginally stable systems, i.e., with systems whose dominant eigenvalues are located on the imaginary axis. As before, no step size will predict the undamped oscillation of the true system.

How has the stability domain for the BE algorithm been found? Although there exist analytical techniques to determine the domain of numerical stability, they are somewhat cumbersome and error prone. Therefore, we prefer to go another route and devise a general–purpose computer program that can determine the domain of numerical stability of any integration algorithm.

We start out with a scalar problem with $|\lambda| = 1.0$, i.e., with its eigenvalue anywhere along the unit circle. In order to avoid complex numbers, we may alternatively use a second–order system with a complex conjugate pair of eigenvalues on the unit circle.

$$\mathbf{A} = \begin{pmatrix} 0 & 1 \\ -1 & 2\cos(\alpha) \end{pmatrix} \tag{2.26}$$

is a matrix with a pair of conjugate complex eigenvalues on the unit circle, where α denotes the angle of one of the two eigenvalues counted in the mathematically positive (i.e., counterclockwise) sense away from the positive real axis.

The following MATLAB routine computes $\mathbf{A}$ for any given value of α.

```
function [A] = aa (alpha)
   radalpha = alpha * pi/180;
   x = cos(radalpha);
   A = [ 0 , 1 ; −1 , 2 * x ];
return
```

We then compute the $\mathbf{F}$–matrix for this system. The ff–function accepts the $\mathbf{A}$–matrix, the step size, h, and a number representing the integration algorithm, $algor$, as input arguments, and returns the respective $\mathbf{F}$–matrix as output argument. The routine is here only shown with the code for the first two algorithms, the FE and BE algorithms.

```
function [F] = ff(A, h, algor)
    Ah = A * h;
    [n, n] = size(Ah);
    I = eye(n);
    %
    % algor = 1 : Forward Euler
    %
    if algor == 1,
        F = I + Ah;
    end
    %
    % algor = 2 : Backward Euler
    %
    if algor == 2,
        F = inv(I - Ah);
    end
    return
```

Now, we compute the largest possible value of h, for which all eigenvalues of $\mathbf{F}$ are inside the unit circle. The hh–function calls upon the aa– and ff–functions internally. It accepts α and $algor$ as input arguments. It also requires lower and upper bounds for the step size, h_{lower} and h_{upper}, such that the solution of the discretized problem is stable for one of them, and unstable for the other. The function returns the value of the step size, h_{max}, for which the discretized problem is marginally stable.

```
function [hmax] = hh(alpha, algor, hlower, hupper)
    A = aa(alpha);
    maxerr = 1.0e-6;
    err = 100;
    while err > maxerr,
        h = (hlower + hupper)/2;
        F = ff(A, h, algor);
        lmax = max(abs(eig(F)));
        err = lmax - 1;
        if err > 0,
            hupper = h;
        else
            hlower = h;
        end,
        err = abs(err);
    end
    hmax = h;
    return
```

The hh–function, as shown above, works only for algorithms with stability domains similar to that of the FE algorithm. The logic of the if-statement must be reversed for algorithms of the BE type, but we didn't want to make the code poorly readable by including too many implementational details.

Finally, we need to sweep over a selected range of α values, and plot h_{max} as a function of α in polar coordinates. There certainly exist more efficient curve tracking algorithms than the one outlined above, but for the time being, this algorithm will suffice.

2.5   The Newton Iteration

One additional problem needs to be discussed. In the above example, it was easy to perform the simulation using the BE algorithm. Since the system to be simulated is linear, we were able to compute the $\mathbf{F}$–matrix explicitly by means of matrix inversion.

This cannot be done in a nonlinear case. We need to somehow solve the implicit set of nonlinear algebraic equations that are formed by the state–space model and the implicit integration algorithm. To this end, we need an iteration procedure.

The first idea that comes to mind is to employ a *predictor–corrector technique*. The idea is quite simple. We start out with an explicit FE step, and use the result of that step (the predictor) for the unknown state derivative of the implicit BE step. We repeat by iterating on the BE step.

$$\text{predictor:} \qquad \dot{\mathbf{x}}_{\mathbf{k}} = \mathbf{f}(\mathbf{x}_{\mathbf{k}}, t_k)$$
$$\mathbf{x}_{\mathbf{k+1}}^{P} = \mathbf{x}_{\mathbf{k}} + h \cdot \dot{\mathbf{x}}_{\mathbf{k}}$$

$$\text{1}^{\text{st}} \text{ corrector:} \qquad \dot{\mathbf{x}}_{\mathbf{k+1}}^{P} = \mathbf{f}(\mathbf{x}_{\mathbf{k+1}}^{P}, t_{k+1})$$
$$\mathbf{x}_{\mathbf{k+1}}^{C1} = \mathbf{x}_{\mathbf{k}} + h \cdot \dot{\mathbf{x}}_{\mathbf{k+1}}^{P}$$

$$\text{2}^{\text{nd}} \text{ corrector:} \qquad \dot{\mathbf{x}}_{\mathbf{k+1}}^{C1} = \mathbf{f}(\mathbf{x}_{\mathbf{k+1}}^{C1}, t_{k+1})$$
$$\mathbf{x}_{\mathbf{k+1}}^{C2} = \mathbf{x}_{\mathbf{k}} + h \cdot \dot{\mathbf{x}}_{\mathbf{k+1}}^{C1}$$

$$\text{3}^{\text{rd}} \text{ corrector:} \qquad \dot{\mathbf{x}}_{\mathbf{k+1}}^{C2} = \mathbf{f}(\mathbf{x}_{\mathbf{k+1}}^{C2}, t_{k+1})$$
$$\mathbf{x}_{\mathbf{k+1}}^{C3} = \mathbf{x}_{\mathbf{k}} + h \cdot \dot{\mathbf{x}}_{\mathbf{k+1}}^{C2}$$

etc.

The iteration is terminated when two consecutive approximations of $\mathbf{x}_{\mathbf{k+1}}$ differ less than a prescribed tolerance. Since the predictor step is explicit, the overall algorithm is explicit as well. This iteration scheme is called *fixed–point iteration*.

If we apply the linear system of Eq.(2.12) to this algorithm, and insert all the equations into each other, we find:

$$\mathbf{F}^{P} = \mathbf{I}^{(n)} + \mathbf{A} \cdot h$$
$$\mathbf{F}^{C1} = \mathbf{I}^{(n)} + \mathbf{A} \cdot h + (\mathbf{A} \cdot h)^2$$
$$\mathbf{F}^{C2} = \mathbf{I}^{(n)} + \mathbf{A} \cdot h + (\mathbf{A} \cdot h)^2 + (\mathbf{A} \cdot h)^3$$
$$\mathbf{F}^{C3} = \mathbf{I}^{(n)} + \mathbf{A} \cdot h + (\mathbf{A} \cdot h)^2 + (\mathbf{A} \cdot h)^3 + (\mathbf{A} \cdot h)^4$$

For infinitely many iterations, we obtain:

$$\mathbf{F} = \mathbf{I}^{(n)} + \mathbf{A} \cdot h + (\mathbf{A} \cdot h)^2 + (\mathbf{A} \cdot h)^3 + \ldots \tag{2.27}$$

Thus:

$$(\mathbf{A} \cdot h) \cdot \mathbf{F} = \mathbf{A} \cdot h + (\mathbf{A} \cdot h)^2 + (\mathbf{A} \cdot h)^3 + (\mathbf{A} \cdot h)^4 + \ldots \tag{2.28}$$

and subtracting Eq.(2.28) from Eq.(2.27), we find:

$$[\mathbf{I}^{(n)} - \mathbf{A} \cdot h] \cdot \mathbf{F} = \mathbf{I}^{(n)} \tag{2.29}$$

or:

$$\mathbf{F} = [\mathbf{I}^{(n)} - \mathbf{A} \cdot h]^{-1} \tag{2.30}$$

Thus, we are hopeful that we just found a (very expensive) explicit integration algorithm that behaves like the BE method. Unfortunately, nothing could be farther from the truth. Figure 2.12 depicts the resulting stability domain when plugging the $\mathbf{F}$–matrix of Eq.(2.27) into the algorithm that generates stability domains.

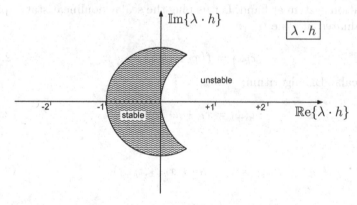

FIGURE 2.12. Stability domain of predictor–corrector FE–BE technique.

The reason for the half–moon domain obtained in this way is that the infinite series of Eq.(2.27) converges only if all eigenvalues of $\mathbf{A} \cdot h$ are inside the unit circle. The subtraction of the two infinite series of Eq.(2.27) and Eq.(2.28) is only legal if this is the case. Thus, the stability domain approaches that of BE only for sufficiently small values of $|\mathrm{Eig}(\mathbf{A} \cdot h)|$.

Let us try something else. Figure 2.13 shows how a zero–crossing of a function can be found using Newton iteration.

Given an arbitrary function $\mathcal{F}(x)$. We want to assume that we know the value of the function and its derivative $\partial \mathcal{F} / \partial x$ at some point x^ℓ. We notice

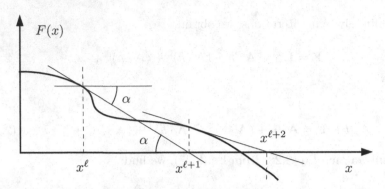

FIGURE 2.13. Newton iteration.

that:

$$\tan \alpha = \frac{\partial \mathcal{F}^\ell}{\partial x} = \frac{\mathcal{F}^\ell}{x^\ell - x^{\ell+1}} \qquad (2.31)$$

Thus:

$$x^{\ell+1} = x^\ell - \frac{\mathcal{F}^\ell}{\partial \mathcal{F}^\ell / \partial x} \qquad (2.32)$$

Let us apply this technique to the problem of iterating the nonlinear algebraic equation system at hand. Let us plug the scalar nonlinear state–space model evaluated at time t_{k+1}:

$$\dot{x}_{k+1} = f(x_{k+1}, t_{k+1}) \qquad (2.33)$$

into the scalar BE algorithm:

$$x_{k+1} = x_k + h \cdot \dot{x}_{k+1} \qquad (2.34)$$

We find:

$$x_{k+1} = x_k + h \cdot f(x_{k+1}, t_{k+1}) \qquad (2.35)$$

or:

$$x_k + h \cdot f(x_{k+1}, t_{k+1}) - x_{k+1} = 0.0 \qquad (2.36)$$

Equation (2.36) is in the desired form to apply Newton iteration. It describes a nonlinear algebraic equation in the unknown variable x_{k+1}, the zero–crossing of which we wish to determine. Thus:

$$x_{k+1}^{\ell+1} = x_{k+1}^\ell - \frac{x_k + h \cdot f(x_{k+1}^\ell, t_{k+1}) - x_{k+1}^\ell}{h \cdot \partial f(x_{k+1}^\ell, t_{k+1}) / \partial x - 1.0} \qquad (2.37)$$

where k is the integration step count, and ℓ is the Newton iteration count.

The matrix extension of the Newton iteration algorithm looks as follows:

$$\mathbf{x}^{\ell+1} = \mathbf{x}^\ell - \left(\mathcal{H}^\ell\right)^{-1} \cdot \mathcal{F}^\ell \tag{2.38}$$

where:

$$\mathcal{H} = \frac{\partial \mathcal{F}}{\partial \mathbf{x}} = \begin{pmatrix} \partial \mathcal{F}_1/\partial x_1 & \partial \mathcal{F}_1/\partial x_2 & \ldots & \partial \mathcal{F}_1/\partial x_n \\ \partial \mathcal{F}_2/\partial x_1 & \partial \mathcal{F}_2/\partial x_2 & \ldots & \partial \mathcal{F}_2/\partial x_n \\ \vdots & \vdots & \ddots & \vdots \\ \partial \mathcal{F}_n/\partial x_1 & \partial \mathcal{F}_n/\partial x_2 & \ldots & \partial \mathcal{F}_n/\partial x_n \end{pmatrix} \tag{2.39}$$

is the *Hessian matrix* of the iteration problem.

Applying this iteration scheme to the vector state–space model and the vector BE algorithm, we obtain:

$$\mathbf{x}_{k+1}^{\ell+1} = \mathbf{x}_{k+1}^\ell - [h \cdot \mathcal{J}_{k+1}^\ell - \mathbf{I}^{(n)}]^{-1} \cdot [\mathbf{x}_k + h \cdot \mathbf{f}(\mathbf{x}_{k+1}^\ell, t_{k+1}) - \mathbf{x}_{k+1}^\ell] \tag{2.40}$$

where:

$$\mathcal{J} = \frac{\partial \mathbf{f}}{\partial \mathbf{x}} = \begin{pmatrix} \partial f_1/\partial x_1 & \partial f_1/\partial x_2 & \ldots & \partial f_1/\partial x_n \\ \partial f_2/\partial x_1 & \partial f_2/\partial x_2 & \ldots & \partial f_2/\partial x_n \\ \vdots & \vdots & \ddots & \vdots \\ \partial f_n/\partial x_1 & \partial f_n/\partial x_2 & \ldots & \partial f_n/\partial x_n \end{pmatrix} \tag{2.41}$$

is the *Jacobian matrix* of the dynamic system.

Any implementation of this iteration scheme requires, in general, the computation of at least an approximation of the Jacobian matrix, as well as an inversion (refactorization) of the Hessian matrix. Since both operations are quite expensive, different implementations vary in how often they recompute the Jacobian (so–called modified Newton iteration). The more nonlinear the problem, the more frequently the Jacobian must be recomputed. Notice further that a modification of the step size does not require the computation of a new Jacobian, but it forces us to refactorize the Hessian.

Let us now analyze how this iteration scheme will affect the solution of linear problems, and, in particular, how it will influence the stability domain of the method.

The Jacobian of the linear state–space model is simply its state matrix:

$$\mathcal{J} = \mathbf{A} \tag{2.42}$$

Consequently, the Jacobian of a linear time–invariant model never needs to be updated, although a new inverse Hessian will still be required whenever the step size of the algorithm is modified.

Plugging the linear system into Eq.(2.40), we find:

$$\mathbf{x}_{k+1}^{\ell+1} = \mathbf{x}_{k+1}^\ell - [\mathbf{A} \cdot h - \mathbf{I}^{(n)}]^{-1} \cdot [(\mathbf{A} \cdot h - \mathbf{I}^{(n)}) \cdot \mathbf{x}_{k+1}^\ell + \mathbf{x}_k] \tag{2.43}$$

or:

$$x_{k+1}^{\ell+1} = [I^{(n)} - A \cdot h]^{-1} \cdot x_k \qquad (2.44)$$

Evidently, Newton iteration does not influence the stability properties of the linear system. This is generally true for all integration algorithms, not only when the Newton iteration is applied to the BE algorithm.

2.6   Semi–analytic Algorithms

As we have seen, numerical integration algorithms call at various places for the computation of derivatives. Time derivatives of f are needed for the higher–order terms of the Taylor–Series expansion. Spatial derivatives of f are required by the Newton iteration algorithm. More uses of derivatives will be met in due course.

However, the numerical computation of derivatives by explicit algorithms is notoriously ill–conditioned. An inaccurate evaluation of the Jacobian is relatively harmless. This will simply slow down the convergence of the Newton iteration. However, numerical errors in the higher–order terms of the Taylor Series are devastating. Therefore, numerical analysts have learnt to reformulate the problem so that a direct computation of the higher-order Taylor–Series terms can be avoided. We shall talk about this more in the following chapters of this book.

However, for now, we shall explore another avenue. The Taylor Series could easily be evaluated directly if only we had available analytical expressions for the higher derivatives. While analytical expressions for the higher derivatives can be derived fairly easily, it is painful for the user to have to manually derive those expressions. If the model is even only modestly complex, the user will probably make mistakes on the way.

However, techniques for algorithmic formulae manipulation have meanwhile been developed. In fact, this branch of computer science has been met with quite remarkable success over the past few years. Algorithmic differentiation of formulae has become a standard feature offered by many symbolic processing programs. However, many of these systems generate derivative formulae that expand, i.e., are much longer than the original formulae. This pitfall can be avoided. Joss developed a technique that avoids formulae expansion in symbolic differentiation [2.16]. The idea behind his technique is surprisingly simple. The original formulae are decomposed into primitives, each of which can be differentiated separately. The example shown below illustrates how the algorithm works in practice. Given the following function:

$$\dot{x} = \sin^2(\sqrt{x} + \frac{x^2 \cdot t}{2}) \qquad (2.45)$$

Its algebraic differentiation can be computed in the following way:

$$
\begin{aligned}
\dot{x} &= c_1^2 & \Rightarrow & \quad \ddot{x} &= 2 \cdot c_1 \cdot \dot{c}_1 \\
c_1 &= \sin(c_2) & \Rightarrow & \quad \dot{c}_1 &= \cos(c_2) \cdot \dot{c}_2 \\
c_2 &= c_3 + c_4 & \Rightarrow & \quad \dot{c}_2 &= \dot{c}_3 + \dot{c}_4 \\
c_3 &= \sqrt{x} & \Rightarrow & \quad \dot{c}_3 &= \dot{x}/(2 \cdot \sqrt{x}) \\
c_4 &= 0.5 \cdot c_5 \cdot c_6 & \Rightarrow & \quad \dot{c}_4 &= 0.5 \cdot (c_5 \cdot \dot{c}_6 + \dot{c}_5 \cdot c_6) \\
c_5 &= x^2 & \Rightarrow & \quad \dot{c}_5 &= 2 \cdot x \cdot \dot{x} \\
c_6 &= t & \Rightarrow & \quad \dot{c}_6 &= 1.0
\end{aligned}
$$

It can easily be verified that equations are available to compute all the unknown variables. The equations only need to be sorted into an executable sequence. The second time derivative of x is indeed being evaluated correctly. Since all possible primitive expressions can be tabulated together with their derivatives, the process of algorithmically generating derivatives is a fairly simple task. No formulae expansion takes place when differentiation is implemented in this fashion.

Joss also discovered that it is possible to compute derivatives not only of formulae, but even of entire programs. He developed an ALGOL program that can differentiate any ALGOL procedure or set of ALGOL procedures with respect to any variable or set of variables, generating new ALGOL procedures for the derivatives. Unfortunately, his dissertation was never translated into English. However, there exist newer references in English that can be consulted [2.17, 2.19, 2.22]. Kurz [2.19] used the algorithm of Joss for developing a PASCAL program that computes the derivative of any FORTRAN subroutine or set of FORTRAN subroutines with respect to any variable or set of variables, generating new FORTRAN subroutines for the derivatives. A treatise of these issues can be found in [2.11, 2.12].

In the context of simulation, symbolic differentiation was first employed by Halin [2.14, 2.15]. Halin was mostly concerned with real–time simulation, and therefore automatically generated code for a parallel multiprocessor. The run–time performance of his system was amazingly fast taking into account the primitive nature of the individual processors that he employed in his multiprocessor system. Moreover, his architecture is still valid. All that needs to be done is to replace the individual processors of his system by more modern architectures. One disadvantage of his approach to real–time simulation is that real–time simulators should be able to process external inputs, i.e., signals produced from a real plant by real–time sensors. Quite obviously, symbolic differentiation cannot find analytical expressions for time derivatives of such signals, since no formulae for the original signals are provided.

Modern modeling software, such as Dymola [2.5, 2.3], is able to choose from a rich palette of formulae manipulation algorithms when preprocessing the model in preparation of a simulation run. Algebraic differentiation is one of the tools that is being offered, and it is being used for a variety of different purposes.

This is clearly the current trend. Symbolic and numeric processing have both their strengths and weaknesses. A well–engineered combination of the two types of processing can preserve the best of both worlds, and can provide us with faster, more robust, and more user–friendly modeling and simulation environments.

2.7   Spectral Algorithms

Obviously, a Taylor–Series expansion is not the only way to approximate an analytic trajectory. Alternatively, the trajectory could be decomposed into a Fourier Series, and, at least in the case of marginally stable models, as they result from highly oscillatory systems and method–of–lines approximations to hyperbolic PDEs, this might even make a lot of sense.

Such techniques were investigated quite early by Brock and Murray [2.2]. However, at that time, no efficient techniques were known that would have allowed to generate algorithms that could compete with Taylor–Series methods in terms of run–time efficiency. However, the advent of the Fast Fourier Transform (FFT) and newly available FFT chips gave rise to a renewed interest in such techniques [2.10, 2.24]. New theoretical results were also reported by Bales et al. [2.1] and Tal–Ezer [2.23].

However, it is a fact that all numerical integration algorithms that are employed in today's commercially available simulation software make use of Taylor Series as a basis for their approximations, and therefore, we shall ignore other techniques in this book.

2.8   Summary

In this chapter, we have introduced the basic concepts of accuracy and stability as they relate to differential equation solvers. It turns out that, whenever we dealt with questions of *accuracy*, we were looking at nonlinear state–space models, whereas, whenever we were discussing *stability*, we were looking at linear state–space models. This is somewhat unsatisfactory. After all, accuracy is a local property of the algorithm, whereas stability is a more global facet of it. The reason for this inconsistency is simple. We dealt with accuracy in nonlinear terms, because it was easy to do, whereas we restricted our discussion of stability to the linear problem, since a general nonlinear treatise of stability issues is a very difficult subject indeed.

Linear stability considerations cannot always be extended to the nonlinear case, or, if they are, they may yield misleading answers. In fact, even linear time–variant systems may behave in surprising ways. To demonstrate this fact, let us look at the linear time–variant autonomous continuous–time system:

$$\dot{\mathbf{x}} = \mathbf{A}(t) \cdot \mathbf{x} = \begin{pmatrix} -2.5 & 1.5 \cdot \exp(8t) \\ -0.5 \cdot \exp(-8t) & -0.5 \end{pmatrix} \cdot \mathbf{x} \qquad (2.46a)$$

with initial conditions:

$$\mathbf{x_0} = \begin{pmatrix} 23 \\ 11 \end{pmatrix} \qquad (2.46b)$$

The eigenvalues of the $\mathbf{A}$–matrix are -1.0 and -2.0, i.e., they are constant and negative real.

$$\begin{aligned} \mathrm{Eig}(\mathbf{A}(t)) &= \mathrm{Root}(\det(\lambda \cdot \mathbf{I}^{(n)} - \mathbf{A})) \\ &= \mathrm{Root}((\lambda + 2.5) \cdot (\lambda + 0.5) + 0.75) = \mathrm{Root}(\lambda^2 + 3 \cdot \lambda + 2) \end{aligned}$$

Yet, the analytical solution is:

$$\begin{aligned} x_1(t) &= 5 \cdot \exp(7t) + 18 \cdot \exp(6t) & (2.47a) \\ x_2(t) &= 5 \cdot \exp(-t) + 6 \cdot \exp(-2t) & (2.47b) \end{aligned}$$

Evidently, $x_1(t)$ is unstable, although both eigenvalues of the system are in the left half λ–plane.

A similar discrete–time example can easily be constructed also. Let us look at the linear time–variant autonomous discrete–time system:

$$\mathbf{x_{k+1}} = \mathbf{F}(t) \cdot \mathbf{x_k} = \begin{pmatrix} -1 & 1.5 \cdot 8^k \\ -0.5 \cdot 8^{-k} & 1 \end{pmatrix} \cdot \mathbf{x_k} \qquad (2.48a)$$

with initial conditions:

$$\mathbf{x_0} = \begin{pmatrix} 11 \\ 5 \end{pmatrix} \qquad (2.48b)$$

The eigenvalues of the $\mathbf{F}$–matrix are $+0.5$ and -0.5, i.e., they are constant and within the unit circle.

$$\begin{aligned} \mathrm{Eig}(\mathbf{F}(t)) &= \mathrm{Root}(\det(\lambda \cdot \mathbf{I}^{(n)} - \mathbf{F})) \\ &= \mathrm{Root}((\lambda + 1) \cdot (\lambda - 1) + 0.75) = \mathrm{Root}(\lambda^2 - 0.25) \end{aligned}$$

Yet, the analytical solution is:

$$\begin{aligned} x_1(k) &= 2 \cdot 4^k + 9 \cdot (-4)^k & (2.49a) \\ x_2(k) &= 2 \cdot 0.5^k + 3 \cdot (-0.5)^k & (2.49b) \end{aligned}$$

Evidently, $x_1(k)$ is unstable, although both eigenvalues of the system are within the unit circle of the λ–plane.

This is bad news. In fact, let us assume that an analytically stable linear time–invariant continuous–time system is being integrated with an obscure variable–step integration algorithm, whose stability region contains the entire left half $(\lambda \cdot h)$–plane (such a method is called *A–stable*.) Since the **F**–matrix is a function of the step size, h, it is entirely feasible that a sequence of h values can be chosen such that the numerical solution will blow up anyway. Such anomalies were reported in [2.6].

A general discussion of numerical stability in the nonlinear sense does exist. A major breakthrough in this research area was achieved by Dahlquist in two seminal papers published in the mid seventies [2.8, 2.7]. A mature discussion of the topic can be found in [2.9, 2.13]. The main idea behind Dahlquist's approach to nonlinear stability was to focus on a side effect of stability. In a stable system, trajectories that start out from neighboring initial conditions contract with time. Dahlquist focused on formulating conditions for when trajectories contract. Therefore, it has become customary to refer to nonlinear stability as *contractivity*. However, the theory is too involved to be dealt with in this book. The (linear) stability domain, that was introduced in this chapter, serves our purposes perfectly well, since our major goals are to help the simulation practitioner with this book to attain a feel for when which technique might have a decent chance of success, and if a technique fails to succeed, why this is, and what can be done about it.

2.9 References

[2.1] Laurence A. Bales. Cosine Methods for Second–Order Hyperbolic Equations with Time–Dependent Coefficients. *Mathematics of Computation*, 45(171):65–89, 1985.

[2.2] Paul Brock and Francis J. Murray. The Use of Exponential Sums in Step–by–Step Integration. *Math. Tables Aids Comput.*, 6:63–78, 1952.

[2.3] François E. Cellier and Hilding Elmqvist. Automated Formula Manipulation Supports Object–Oriented Continuous–System Modeling. *IEEE Control Systems*, 13(2):28–38, 1993.

[2.4] François E. Cellier and Peter J. Möbius. Toward Robust General Purpose Simulation Software. In Robert D. Skeel, editor, *Proceedings of the 1979 SIGNUM Meeting on Numerical Ordinary Differential Equations*, pages 18:1–5, Urbana, Ill., 1979. Dept. of Computer Science, University of Illinois at Urbana–Champaign.

[2.5] François E. Cellier. *Continuous System Modeling*. Springer Verlag, New York, 1991. 755p.

[2.6] Germund G. Dahlquist, Werner Liniger, and Olavi Nevanlinna. Stability of Two–Step Methods for Variable Integration Steps. *SIAM J. Numerical Analysis*, 20(5):1071–1085, 1983.

[2.7] Germund G. Dahlquist. Error Analysis for a Class of Methods for Stiff Nonlinear Initial Value Problems. In G. Alistair Watson, editor, *Proceedings 6th Biennial Dundee Conference on Numerical Analysis*, volume 506 of *Lecture Notes in Mathematics*, pages 60–72. Springer–Verlag, Berlin, 1975.

[2.8] Germund G. Dahlquist. On Stability and Error Analysis for Stiff Nonlinear Problems. Technical Report TRITA–NA–7508, Dept. of Information Processing, Royal Institute of Technology, Stockholm, Sweden, 1975.

[2.9] Kees Dekker and Jan G. Verwer. *Stability of Runge–Kutta Methods for Stiff Nonlinear Differential Equations.* North–Holland, Amsterdam, The Netherlands, 1984. 307p.

[2.10] David Gottlieb and Steven A. Orszag. *Numerical Analysis of Spectral Methods: Theory and Applications*, volume 26. SIAM Publishing, Philadelphia, Penn., 1977. 172p.

[2.11] Andreas Griewank. On Automatic Differentiation. In Masao Iri and Kunio Tanabe, editors, *Mathematical Programming: Recent Developments and Applications*, pages 83–108. Kluwer Academic Press, 1989.

[2.12] Andreas Griewank. *User's Guide for ADOL–C, Version 1.0.* Mathematics and Computer Science Division, Argonne National Laboratory, Argonne, Ill., 1990.

[2.13] Ernst Hairer and Gerhard Wanner. *Solving Ordinary Differential Equations II: Stiff and Differential–Algebraic Problems*, volume 14 of *Series in Computational Mathematics.* Springer–Verlag, Berlin, Germany, 2nd edition, 1996. 632p.

[2.14] Hans Jürgen Halin, Richard Bürer, Walter Hälg, Hans Benz, Bernard Bron, Hans-Jörg Brundiers, Anders Isacson, and Milan Tadian. The ETH Multiprocessor Project: Parallel Simulation of Continuous Systems. *Simulation*, 35(4):109–123, 1980.

[2.15] Hans Jürgen Halin. The Applicability of Taylor Series Methods in Simulation. In *Proceedings 1983 Summer Computer Simulation Conference*, volume 2, pages 1032–1076, Vancouver, Canada, July 11–13, 1983. SCS Publishing, San Diego, Calif.

[2.16] Johann Joss. *Algorithmisches Differenzieren.* PhD thesis, Diss ETH 5757, Swiss Federal Institute of Technology, Zürich, Switzerland, 1976. 69p.

[2.17] Gershon Kedem. Automatic Differentiation of Computer Programs. *ACM Trans. Mathematical Software*, 6(2):150–165, 1980.

[2.18] Granino A. Korn and John V. Wait. *Digital Continuous–System Simulation*. Prentice–Hall, Englewood Cliffs, N.J., 1978. 212p.

[2.19] Eberhard Kurz. Algebraic Differential Processor. Technical report, Department of Electrical and Computer Engineering, University of Arizona, Tucson, Ariz., 1986.

[2.20] Edward E. L. Mitchell and Joseph S. Gauthier. *ACSL: Advanced Continuous Simulation Language — User Guide and Reference Manual*. Mitchell & Gauthier Assoc., Concord, Mass., 1991.

[2.21] James W. Nilsson and Susan A. Riedel. *Introduction to PSpice for Electric Circuits*. Prentice–Hall, Upper Saddle River, N.J., 6th edition, 2002. 132p.

[2.22] Louis B. Rall. *Automatic Differentiation: Techniques and Applications*, volume 120 of *Lecture Notes in Computer Science*. Springer–Verlag, Berlin, 1981. 165p.

[2.23] Hillel Tal-Ezer. Spectral Methods in Time for Hyperbolic Equations. *SIAM J. Numerical Analysis*, 23(1):11–26, 1986.

[2.24] Robert Vichnevetsky and John B. Bowles. *Fourier Analysis of Numerical Approximations of Hyperbolic Equations*, volume 5 of *SIAM Studies in Applied Mathematics*. SIAM Publishing, Philadelphia, Penn., 1982. 140p.

2.10   Bibliography

[B2.1] George F. Corliss, Christèle Faure, Andreas Griewank, Laurent Hascoët, and Uwe Naumann, editors. *Automatic Differentiation of Algorithms: From Simulation to Optimization*. Springer–Verlag, Berlin, Germany, 2002. 459p.

[B2.2] C. William Gear. *Numerical Initial Value Problems in Ordinary Differential Equations*. Series in Automatic Computation. Prentice–Hall, Englewood Cliffs, N.J., 1971. 253p.

[B2.3] Curtis F. Gerald and Patrick O. Wheatley. *Applied Numerical Analysis*. Addison–Wesley, Reading, Mass., 6th edition, 1999. 768p.

[B2.4] John D. Lambert. *Numerical Methods for Ordinary Differential Systems: The Initial Value Problem*. John Wiley, New York, 1991. 304p.

2.11   Homework Problems

[H2.1] Marginal Stability

Given the following linear time–invariant continuous–time system:

$$
\dot{\mathbf{x}} = \begin{pmatrix}
1250 & -25113 & -60050 & -42647 & -23999 \\
500 & -10068 & -24057 & -17092 & -9613 \\
250 & -5060 & -12079 & -8586 & -4826 \\
-750 & 15101 & 36086 & 25637 & 14420 \\
250 & -4963 & -11896 & -8438 & -4756
\end{pmatrix} \cdot \mathbf{x} + \begin{pmatrix}
5 \\ 2 \\ 1 \\ -3 \\ 1
\end{pmatrix} \cdot u
$$

$$
\mathbf{y} = \begin{pmatrix} -1 & 26 & 59 & 43 & 23 \end{pmatrix} \cdot \mathbf{x} \tag{H2.1a}
$$

with initial conditions:

$$
\mathbf{x_0} = \begin{pmatrix} 1 & -2 & 3 & -4 & 5 \end{pmatrix}^T \tag{H2.1b}
$$

Determine the step size, h_{marg}, for which FE will give marginally stable results.

Simulate the system across 10 seconds of simulated time with step input using the FE algorithm with the following step sizes: (i) $h = 0.1 \cdot h_{\mathrm{marg}}$, (ii) $h = 0.95 \cdot h_{\mathrm{marg}}$, (iii) $h = h_{\mathrm{marg}}$, (iv) $h = 1.05 \cdot h_{\mathrm{marg}}$, and (v) $h = 2 \cdot h_{\mathrm{marg}}$. Discuss the results.

[H2.2] Integration Accuracy

For the system of Hw.[H2.1], determine the largest step size that will give you a global accuracy of 1%.

For this purpose, it is necessary to find the analytical solution of the given system. The easiest way to achieve this is to use the *spectral decomposition method*. The MATLAB statement:

$$
[\mathbf{V}, \mathbf{\Lambda}] = \mathrm{eig}(\mathbf{A}) \tag{H2.2a}
$$

generates two matrices. $\mathbf{\Lambda}$ is the *eigenvalue matrix*, i.e., a diagonal matrix with the eigenvalues of $\mathbf{A}$ placed along its diagonal, and $\mathbf{V}$ is the *right modal matrix*, i.e., a matrix that consists of the right eigenvectors of $\mathbf{A}$ horizontally concatenated to each other. The ith column of $\mathbf{V}$ contains the eigenvector associated with the eigenvalue located at the ith diagonal element of the $\mathbf{\Lambda}$–matrix.

Apply a *similarity transformation*:

$$
\xi(t) = \mathbf{T} \cdot \mathbf{x}(t) \tag{H2.2b}
$$

with:

$$
\mathbf{T} = \mathbf{V}^{-1} \tag{H2.2c}
$$

This will put the system into diagonal form, from which the analytical solution can be read out easily.

If you don't trust the accuracy of the numerical algorithm, you can compute the transfer function of the system using:

$$\textbf{Sys} \;=\; \text{ss}(\textbf{A}, \textbf{B}, \textbf{C}, \textbf{D}) \tag{H2.2d}$$

$$\textbf{G} \;=\; \text{tf}(\textbf{Sys}) \tag{H2.2e}$$

The numerator and denominator polynomials of the transfer function can then be extracted by means of:

$$[\textbf{p}, \textbf{q}] = \text{tfdata}(\textbf{G}, 'v') \tag{H2.2f}$$

Finally, the roots of the denominator polynomial can be found through:

$$\lambda = \text{roots}(\textbf{q}) \tag{H2.2g}$$

You can then perform a *partial fraction expansion* on the transfer function, and read the analytical solution out by taking the *inverse Laplacian* thereof.

Simulate the original system using the FE algorithm across 10 seconds of simulated time. Repeat the simulation with different step sizes, until you obtain agreement between the analytical and the numerical solution with an accuracy of 1%:

$$\varepsilon_{\text{global}} = \frac{\|\textbf{x}_{\text{anal}} - \textbf{x}_{\text{num}}\|}{\|\textbf{x}_{\text{anal}}\|} \le 0.01 \tag{H2.2h}$$

Repeat the same experiment with the BE algorithm. Since the system is linear, you are allowed to compute the $\textbf{F}$–matrix using matrix inversion.

[H2.3] Method Blending

Given the following linear time–invariant continuous–time system:

$$\dot{\textbf{x}} \;=\; \begin{pmatrix} 0 & 1 \\ -9.01 & 0.2 \end{pmatrix} \cdot \textbf{x} + \begin{pmatrix} 0 \\ 1 \end{pmatrix} \cdot u$$

$$\textbf{y} \;=\; \begin{pmatrix} 1 & 1 \end{pmatrix} \cdot \textbf{x} + 2 \cdot u \tag{H2.3a}$$

with initial conditions:

$$\textbf{x}_0 = \begin{pmatrix} 1 & -2 \end{pmatrix}^T \tag{H2.3b}$$

Find the analytical solution using one of the techniques described in Hw.[H2.2]. Simulate the system across 25 seconds of simulated time using the FE algorithm. Determine the largest step size that will lead to a global accuracy of 1%. Repeat the experiment with the BE algorithm. You may compute the $\textbf{F}$–matrix using matrix inversion. What do you conclude?

Let us now design another algorithm. This time, we shall repeat each single integration step once with FE and once with BE, and we shall use the arithmetic mean of the two answers as the initial condition for the next step. Such an algorithm is called a *blended algorithm*. Determine again the maximum step size that will provide a 1% accuracy. Compare your results with those obtained by FE or BE alone.

[H2.4] Cyclic Method

Repeat Hw.[H2.3]. However, this time, we shall design another algorithm. Instead of using the mean value of FE and BE to continue, we shall simply toggle between one step of FE followed by one step of BE, followed by another step of FE, etc. Such an algorithm is called a *cyclic algorithm*.

Determine again the maximum step size that will provide a 1% accuracy. Compare your results with those obtained by FE or BE alone.

[H2.5] Stability Domain

For the *predictor–corrector method* of Eq.(2.27), find the stability domains if: (i) no corrector is used, (ii) one corrector is used, (iii) two correctors are used, (iv) three correctors are used, and (v) four correctors are used. Plot the five stability domains on top of each other, and discuss the results.

[H2.6] Stability Domain: Blended and Cyclic Methods

Find the stability domain for the blended method of Hw.[H2.3]. What do you conclude when comparing the stability domain of that method with those of FE and BE? How does the stability domain of the blended method explain the result of Hw.[H2.3]?

Find the stability domain for the cyclic method of Hw.[H2.4]. Instead of interpreting this method as switching to another algorithm after each step, we can think of this technique as one that is described by a single *macro–step* consisting of two *semi–steps*. Thus:

$$\mathbf{x}(k+0.5) = \mathbf{x}(k) + 0.5 \cdot h \cdot \dot{\mathbf{x}}(k) \qquad (H2.6a)$$
$$\mathbf{x}(k+1) = \mathbf{x}(k+0.5) + 0.5 \cdot h \cdot \dot{\mathbf{x}}(k+1) \qquad (H2.6b)$$

Don't despair, this one is tricky. What do you conclude when comparing the stability domain of that method with those of FE and BE? How does the stability domain of the cyclic method explain the result of Hw.[H2.4]?

[H2.7] Stability Domain Shaping: Blended Method

We wish to construct yet another method. It is derived from the previously discussed blended algorithm. Instead of using the mean value of the FE and BE steps, we use a weighted average of the two:

$$\mathbf{x}(k+1) = \vartheta \cdot \mathbf{x_{FE}}(k+1) + (1-\vartheta) \cdot \mathbf{x_{BE}}(k+1) \qquad (H2.7a)$$

Such a method is called a ϑ–*method*. Plot the stability domains of these methods for:

$$\vartheta = \{0, 0.1, 0.2, 0.24, 0.249, 0.25, 0.251, 0.26, 0.3, 0.5, 0.8, 1\} \qquad \text{(H2.7b)}$$

Interpret the results. For this problem, it may be easier to use MATLAB's *contour* plot, than your own stability domain tracking routine.

[H2.8] Stability Domain Shaping: Cyclic Method

We shall now design another ϑ–method. This time, we start out with the cyclic method. The parameter that we shall vary is the step length of the two semi–steps. This is done in the following way:

$$\begin{align}
\mathbf{x}(k + \vartheta) &= \mathbf{x}(k) + \vartheta \cdot h \cdot \dot{\mathbf{x}}(k) & \text{(H2.8a)} \\
\mathbf{x}(k + 1) &= \mathbf{x}(k + \vartheta) + (1 - \vartheta) \cdot h \cdot \dot{\mathbf{x}}(k + 1) & \text{(H2.8b)}
\end{align}$$

Determine the ϑ parameter of the method such that the overall method exhibits a stability domain similar to BE, but where the border of stability on the positive real axis of the $(\lambda \cdot h)$–plane is located at $+10$ instead of $+2$. Plot the stability domain of that method.

2.12 Projects

[P2.1] ϑ–Methods

For the two ϑ–methods described in Hw.[H2.7] and Hw.[H2.8], determine optimal values of the ϑ parameter as a function of the location of the eigenvalues of the $\mathbf{A}$–matrix (for linear time–invariant systems). To this end, vary the ϑ parameter until you get a maximum value of h that guarantees 1% accuracy. Repeat for different locations of the eigenvalues of $\mathbf{A}$, and come up with a recipe of how to choose ϑ for any given linear system.

[P2.2] Cyclic Methods

Do a library search on cyclic methods, and come up with a decision tree that characterizes the various cyclic methods that have been proposed.

2.13 Research

[R2.1] Simulation Verification

Study the problem of simulation verification. What techniques could a robust simulation run–time library offer to support the user in asserting the correctness of his or her simulation results?

3

Single–step Integration Methods

Preview

This chapter extends the ideas of numerical integration by means of a Taylor–Series expansion from the first–order (FE and BE) techniques to higher orders of approximation accuracy. The well–known class of explicit *Runge–Kutta techniques* is introduced by generalizing the predictor–corrector idea.

The chapter then explores special classes of single–step techniques that are well suited for the simulation of stiff systems and for that of marginally stable systems, namely the *extrapolation methods* and the *backinterpolation algorithms*. The stability domain serves as a good vehicle for analyzing the stability properties of these classes of algorithms.

We are then delving more deeply into the question of approximation accuracy. The *accuracy domain* is introduced as a simple tool to explore this issue, and the *order star* approach is subsequently introduced as a more refined and satisfying alternative.

The chapter ends with a discussion of the ideas behind *step–size control* and *order control*, and the techniques used to accomplish these in the realm of single–step algorithms.

3.1 Introduction

In Chapter 2, we have seen that predictor–corrector techniques can be used to merge explicit and implicit algorithms into more complex entities that are overall of the explicit type, while inheriting some of the desirable numerical properties of implicit algorithms.

In particular, we introduced the following predictor–corrector method:

$$
\text{predictor:} \quad
\begin{aligned}
\dot{\mathbf{x}}_{\mathbf{k}} &= \mathbf{f}(\mathbf{x}_{\mathbf{k}}, t_k) \\
\mathbf{x}_{\mathbf{k+1}}^{\mathrm{P}} &= \mathbf{x}_{\mathbf{k}} + h \cdot \dot{\mathbf{x}}_{\mathbf{k}}
\end{aligned}
$$

$$
\text{corrector:} \quad
\begin{aligned}
\dot{\mathbf{x}}_{\mathbf{k+1}}^{\mathrm{P}} &= \mathbf{f}(\mathbf{x}_{\mathbf{k+1}}^{\mathrm{P}}, t_{k+1}) \\
\mathbf{x}_{\mathbf{k+1}}^{\mathrm{C}} &= \mathbf{x}_{\mathbf{k}} + h \cdot \dot{\mathbf{x}}_{\mathbf{k+1}}^{\mathrm{P}}
\end{aligned}
$$

Let us now perform a nonlinear error analysis of this simple predictor–corrector technique. To this end, we plug all the equations into each other.

We obtain:

$$\mathbf{x_{k+1}} = \mathbf{x_k} + h \cdot \mathbf{f}(\mathbf{x_k} + h \cdot \mathbf{f_k}, t_k + h) \qquad (3.1)$$

We wish to pursue the error analysis up to the quadratic term. Let us thus develop the expression $\mathbf{f}(\mathbf{x_k} + h \cdot \mathbf{f_k}, t_k + h)$ into a multidimensional Taylor Series around the point $< \mathbf{x_k}, t_k >$. Since this term in Eq.3.1 is multiplied by h, we may truncate the Taylor Series after the linear term.

Remember that:

$$f(x + \Delta x, y + \Delta y) \approx f(x, y) + \frac{\partial f(x, y)}{\partial x} \cdot \Delta x + \frac{\partial f(x, y)}{\partial y} \cdot \Delta y \qquad (3.2)$$

Thus:

$$\mathbf{f}(\mathbf{x_k} + h \cdot \mathbf{f_k}, t_k + h) \approx \mathbf{f}(\mathbf{x_k}, t_k) + \frac{\partial \mathbf{f}(\mathbf{x_k}, t_k)}{\partial \mathbf{x}} \cdot (h \cdot \mathbf{f_k}) + \frac{\partial \mathbf{f}(\mathbf{x_k}, t_k)}{\partial t} \cdot h \quad (3.3)$$

where $\partial \mathbf{f} / \partial \mathbf{x}$ is the meanwhile well–known *Jacobian* of the system. Plugging Eq.(3.3) into Eq.(3.1), we find:

$$\mathbf{x_{k+1}} \approx \mathbf{x_k} + h \cdot \mathbf{f}(\mathbf{x_k}, t_k) + h^2 \cdot \left(\frac{\partial \mathbf{f}(\mathbf{x_k}, t_k)}{\partial \mathbf{x}} \cdot \mathbf{f_k} + \frac{\partial \mathbf{f}(\mathbf{x_k}, t_k)}{\partial t} \right) \qquad (3.4)$$

Let us compare this with the true Taylor Series of $\mathbf{x_{k+1}}$ truncated after the quadratic term:

$$\mathbf{x_{k+1}} \approx \mathbf{x_k} + h \cdot \mathbf{f}(\mathbf{x_k}, t_k) + \frac{h^2}{2} \cdot \dot{\mathbf{f}}(\mathbf{x_k}, t_k) \qquad (3.5)$$

where:

$$\dot{\mathbf{f}}(\mathbf{x_k}, t_k) = \frac{d\mathbf{f}(\mathbf{x_k}, t_k)}{dt} = \frac{\partial \mathbf{f}(\mathbf{x_k}, t_k)}{\partial \mathbf{x}} \cdot \frac{d\mathbf{x_k}}{dt} + \frac{\partial \mathbf{f}(\mathbf{x_k}, t_k)}{\partial t} \qquad (3.6)$$

and:

$$\frac{d\mathbf{x_k}}{dt} = \dot{\mathbf{x}}_\mathbf{k} = \mathbf{f_k} \qquad (3.7)$$

Comparing the true Taylor–Series expansion of $\mathbf{x_{k+1}}$ with the results obtained from the predictor–corrector method, we find that we almost got a match. Only the factor 2 in the denominator of the quadratic term is missing. Thus, the predictor–corrector technique can be written as:

$$\mathbf{x_{PC}}(k + 1) \approx \mathbf{x_k} + h \cdot \mathbf{f}(\mathbf{x_k}, t_k) + h^2 \cdot \dot{\mathbf{f}}(\mathbf{x_k}, t_k) \qquad (3.8)$$

We notice at once that a simple blending of FE and PC will give us a method that is second order accurate:

$$\mathbf{x}(k+1) = 0.5 \cdot (\mathbf{x_{PC}}(k+1) + \mathbf{x_{FE}}(k+1)) \qquad (3.9)$$

or, in other words:

$$\begin{aligned}
\text{predictor:} \quad \dot{\mathbf{x}}_\mathbf{k} &= \mathbf{f}(\mathbf{x_k}, t_k) \\
\mathbf{x}^\mathbf{P}_{\mathbf{k}+1} &= \mathbf{x_k} + h \cdot \dot{\mathbf{x}}_\mathbf{k}
\end{aligned}$$

$$\begin{aligned}
\text{corrector:} \quad \dot{\mathbf{x}}^\mathbf{P}_{\mathbf{k}+1} &= \mathbf{f}(\mathbf{x}^\mathbf{P}_{\mathbf{k}+1}, t_{k+1}) \\
\mathbf{x}^\mathbf{C}_{\mathbf{k}+1} &= \mathbf{x_k} + 0.5 \cdot h \cdot (\dot{\mathbf{x}}_\mathbf{k} + \dot{\mathbf{x}}^\mathbf{P}_{\mathbf{k}+1})
\end{aligned}$$

which is *Heun's method*. This method is sometimes also referred to under the name *modified Euler method*.

In the following section, we want to generalize the idea behind Heun's method by parameterizing the search strategy for higher–order algorithms of this kind.

3.2   Runge–Kutta Algorithms

Heun's method uses an FE step as a predictor, and then a blend of an FE and a BE step as a corrector. Let us generalize this idea somewhat:

$$\begin{aligned}
\text{predictor:} \quad \dot{\mathbf{x}}_\mathbf{k} &= \mathbf{f}(\mathbf{x_k}, t_k) \\
\mathbf{x}^\mathbf{P} &= \mathbf{x_k} + h \cdot \beta_{11} \cdot \dot{\mathbf{x}}_\mathbf{k}
\end{aligned}$$

$$\begin{aligned}
\text{corrector:} \quad \dot{\mathbf{x}}^\mathbf{P} &= \mathbf{f}(\mathbf{x}^\mathbf{P}, t_k + \alpha_1 \cdot h) \\
\mathbf{x}^\mathbf{C}_{\mathbf{k}+1} &= \mathbf{x_k} + h \cdot (\beta_{21} \cdot \dot{\mathbf{x}}_\mathbf{k} + \beta_{22} \cdot \dot{\mathbf{x}}^\mathbf{P})
\end{aligned}$$

This set of methods contains four different parameters. The β_{ij} parameters are weighting factors of the various state derivatives that are computed during the step, and the α_1 parameter specifies the time instant at which the first stage of the technique is evaluated.

Plugging the parameterized equations into each other and developing functions that are not evaluated at time t_k into Taylor Series, we obtain:

$$\mathbf{x}^\mathbf{C}_{\mathbf{k}+1} = \mathbf{x_k} + h \cdot (\beta_{21} + \beta_{22}) \cdot \mathbf{f_k} + \frac{h^2}{2} \cdot [2 \cdot \beta_{11} \cdot \beta_{22} \cdot \frac{\partial \mathbf{f_k}}{\partial \mathbf{x}} \cdot \mathbf{f_k} + 2 \cdot \alpha_1 \cdot \beta_{22} \cdot \frac{\partial \mathbf{f_k}}{\partial t}]$$
$$(3.10)$$

The Taylor Series of $\mathbf{x}_{\mathbf{k}+1}$ truncated after the quadratic term can be written as:

$$\mathbf{x}_{\mathbf{k}+1} \approx \mathbf{x_k} + h \cdot \mathbf{f_k} + \frac{h^2}{2} \cdot [\frac{\partial \mathbf{f_k}}{\partial \mathbf{x}} \cdot \mathbf{f_k} + \frac{\partial \mathbf{f_k}}{\partial t}] \qquad (3.11)$$

A comparison of Eq.(3.10) and Eq.(3.11) yields three nonlinear equations in the four unknown parameters:

$$\beta_{21} + \beta_{22} \ = 1 \tag{3.12a}$$
$$2 \cdot \alpha_1 \cdot \beta_{22} \ = 1 \tag{3.12b}$$
$$2 \cdot \beta_{11} \cdot \beta_{22} \ = 1 \tag{3.12c}$$

Thus, there exist infinitely many such algorithms. Clearly, Heun's method belongs to this set of algorithms. Heun's method can be characterized by:

$$\alpha = \begin{pmatrix} 1 \\ 1 \end{pmatrix}; \quad \beta = \begin{pmatrix} 1 & 0 \\ 0.5 & 0.5 \end{pmatrix} \tag{3.13}$$

α_2 characterizes the time when the corrector is evaluated, which obviously always happens at t_{k+1}, thus, $\alpha_2 = 1.0$. β is a lower triangular matrix.

Many references represent the method in a slightly different form:

$$\begin{array}{c|cc} 0 & 0 & 0 \\ 1 & 1 & 0 \\ \hline x & 1/2 & 1/2 \end{array}$$

which is called the *Butcher tableau* of the method. The first row of the Butcher tableau here indicates the function evaluation at time t_k. The second row represents the predictor, and the third row denotes the corrector.

Another commonly used algorithm of this family of methods is characterized by the following α–vector and β–matrix:

$$\alpha = \begin{pmatrix} 0.5 \\ 1 \end{pmatrix}; \quad \beta = \begin{pmatrix} 0.5 & 0 \\ 0 & 1 \end{pmatrix} \tag{3.14}$$

with the Butcher tableau:

$$\begin{array}{c|cc} 0 & 0 & 0 \\ 1/2 & 1/2 & 0 \\ \hline x & 0 & 1 \end{array}$$

This method is sometimes referred to as *explicit midpoint rule*. It can be implemented as:

$$\text{predictor:} \quad \begin{aligned} \dot{\mathbf{x}}_k &= \mathbf{f}(\mathbf{x}_k, t_k) \\ \mathbf{x}_{k+\frac{1}{2}}^P &= \mathbf{x}_k + \tfrac{h}{2} \cdot \dot{\mathbf{x}}_k \end{aligned}$$

$$\text{corrector:} \quad \begin{aligned} \dot{\mathbf{x}}_{k+\frac{1}{2}}^P &= \mathbf{f}(\mathbf{x}_{k+\frac{1}{2}}^P, t_{k+\frac{1}{2}}) \\ \mathbf{x}_{k+1}^C &= \mathbf{x}_k + h \cdot \dot{\mathbf{x}}_{k+\frac{1}{2}}^P \end{aligned}$$

This technique evaluates the predictor at time $t_k + h/2$. It is a little cheaper than Heun's algorithm due to the additional zero in the β–matrix.

The entire family of such methods is referred to as second–order Runge–Kutta methods, abbreviated as RK2.

The idea can be further generalized by adding more stages. The general explicit Runge–Kutta algorithm can be described as follows:

0th stage: $\dot{\mathbf{x}}^{\mathbf{P_0}} = \mathbf{f}(\mathbf{x_k}, t_k)$

j^{th} stage: $\mathbf{x}^{\mathbf{P_j}} = \mathbf{x_k} + h \cdot \sum_{i=1}^{j} \beta_{ji} \cdot \dot{\mathbf{x}}^{\mathbf{P_{i-1}}}$
$\qquad\qquad\;\; \dot{\mathbf{x}}^{\mathbf{P_j}} = \mathbf{f}(\mathbf{x}^{\mathbf{P_j}}, t_k + \alpha_j \cdot h)$

last stage: $\mathbf{x_{k+1}} = \mathbf{x_k} + h \cdot \sum_{i=1}^{\ell} \beta_{\ell i} \cdot \dot{\mathbf{x}}^{\mathbf{P_{i-1}}}$

where ℓ denotes the number of stages of the method. The most popular of these methods is the following fourth–order accurate Runge–Kutta (RK4) technique:

$$\alpha = \begin{pmatrix} 1/2 \\ 1/2 \\ 1 \\ 1 \end{pmatrix} ; \quad \beta = \begin{pmatrix} 1/2 & 0 & 0 & 0 \\ 0 & 1/2 & 0 & 0 \\ 0 & 0 & 1 & 0 \\ 1/6 & 1/3 & 1/3 & 1/6 \end{pmatrix} \qquad (3.15)$$

or:

0th stage: $\dot{\mathbf{x}}_\mathbf{k} = \mathbf{f}(\mathbf{x_k}, t_k)$

1st stage: $\mathbf{x}^{\mathbf{P_1}} = \mathbf{x_k} + \frac{h}{2} \cdot \dot{\mathbf{x}}_\mathbf{k}$
$\qquad\qquad\;\; \dot{\mathbf{x}}^{\mathbf{P_1}} = \mathbf{f}(\mathbf{x}^{\mathbf{P_1}}, t_{k+\frac{1}{2}})$

2nd stage: $\mathbf{x}^{\mathbf{P_2}} = \mathbf{x_k} + \frac{h}{2} \cdot \dot{\mathbf{x}}^{\mathbf{P_1}}$
$\qquad\qquad\;\; \dot{\mathbf{x}}^{\mathbf{P_2}} = \mathbf{f}(\mathbf{x}^{\mathbf{P_2}}, t_{k+\frac{1}{2}})$

3rd stage: $\mathbf{x}^{\mathbf{P_3}} = \mathbf{x_k} + h \cdot \dot{\mathbf{x}}^{\mathbf{P_2}}$
$\qquad\qquad\;\; \dot{\mathbf{x}}^{\mathbf{P_3}} = \mathbf{f}(\mathbf{x}^{\mathbf{P_3}}, t_{k+1})$

4th stage: $\mathbf{x_{k+1}} = \mathbf{x_k} + \frac{h}{6} \cdot [\dot{\mathbf{x}}_\mathbf{k} + 2 \cdot \dot{\mathbf{x}}^{\mathbf{P_1}} + 2 \cdot \dot{\mathbf{x}}^{\mathbf{P_2}} + \dot{\mathbf{x}}^{\mathbf{P_3}}]$

or yet more simply:

0th stage: $\mathbf{k_1} = \mathbf{f}(\mathbf{x_k}, t_k)$

1st stage: $\mathbf{k_2} = \mathbf{f}(\mathbf{x_k} + \frac{h}{2} \cdot \mathbf{k_1}, t_k + \frac{h}{2})$

2nd stage: $\mathbf{k_3} = \mathbf{f}(\mathbf{x_k} + \frac{h}{2} \cdot \mathbf{k_2}, t_k + \frac{h}{2})$

3rd stage: $\mathbf{k_4} = \mathbf{f}(\mathbf{x_k} + h \cdot \mathbf{k_3}, t_k + h)$

4th stage: $\mathbf{x_{k+1}} = \mathbf{x_k} + \frac{h}{6} \cdot [\mathbf{k_1} + 2 \cdot \mathbf{k_2} + 2 \cdot \mathbf{k_3} + \mathbf{k_4}]$

This RK4 algorithm is particularly attractive due to the many zero elements in its β–matrix. As it is a four–stage algorithm, it involves four function evaluations. These are taken at t_k, $t_{k+1/2}$, $t_{k+1/2}$, and t_{k+1}. Thus, it is possible to think of this RK4 algorithm as a macro–step consisting of four micro–steps, two of length $h/2$, and two of length 0.

The Butcher tableau of this method can be written as:

$$
\begin{array}{c|cccc}
0 & 0 & 0 & 0 & 0 \\
1/2 & 1/2 & 0 & 0 & 0 \\
1/2 & 0 & 1/2 & 0 & 0 \\
1 & 0 & 0 & 1 & 0 \\
\hline
x & 1/6 & 1/3 & 1/3 & 1/6
\end{array}
$$

In general:

$$
\begin{array}{c|c}
\mathbf{c} & \mathbf{A} \\
\hline
x & \mathbf{b}'
\end{array}
$$

The $\mathbf{c}$–vector contains the time instants when the various function evaluations are performed, the $\mathbf{A}$–matrix contains the weights of the various predictor stages, and the $\mathbf{b}'$–vector contains the weights of the corrector stage.

Notice that the number of stages and the approximation order are not necessarily identical. Higher–order RK algorithms require a larger number of stages to achieve a given order of accuracy. Table 3.1 provides a historic overview of the development of RK algorithms.

Developer	Year	Order	# of Stages
Euler [3.8]	1768	1	1
Runge [3.21]	1895	4	4
Heun [3.14]	1900	2	2
Kutta [3.17]	1901	5	6
Huťa [3.15]	1956	6	8
Shanks [3.22]	1966	7	9
Curtis [3.4]	1970	8	11

TABLE 3.1. History of Runge–Kutta Algorithms.

It is interesting to notice that, although the general mechanism for designing such algorithms had been known for quite some time, higher–order RK algorithms were slow in coming. This is due to the fact that the setting up of the nonlinear equations and their subsequent solution is an utterly tedious process. The original algorithm by Kutta contained an error that went unnoticed until it was corrected by Nyström [3.20] in 1925. Curtis finally had to deal with a large number of very awkward nonlinear equations in more than 200 unknowns. Symbolic formulae manipulation programs, such as Mathematica or Maple, would make it much easier today to set up and solve these sets of equations without making errors on the way, but such programs were unavailable at the time, and so, at least for these researchers, mathematics wasn't always fun ... but required lots of patience, perseverance, and suffering.

It is possible to design RK algorithms in the same number of stages as the approximation order only up to fourth order. It can be shown that no five–stage RK method can be found that is fifth–order accurate. Of course, it is important to keep the number of stages as small as possible, since each additional stage requires an extra function evaluation.

Additional requirements are usually formulated that are not inherent in the technique itself, but make a lot of practical sense. Obviously, we want to request that, in an ℓ–stage algorithm:

$$\alpha_\ell = 1.0 \tag{3.16}$$

since we wish to end the step at t_{k+1}. Also, we usually want to make sure that:

$$\alpha_i \in [0.0, 1.0] \quad ; \quad i = \{1, 2, \ldots, \ell\} \tag{3.17}$$

that is, all function evaluations are performed at times that lie between t_k and t_{k+1}.

If we want to prevent the algorithm from ever "integrating backward through time," we shall add the constraint that:

$$\alpha_j \geq \alpha_i \quad ; \quad j \geq i \tag{3.18a}$$

If we want to disallow micro–steps of length 0, we make this condition even more stringent:

$$\alpha_j > \alpha_i \quad ; \quad j > i \tag{3.18b}$$

The previously introduced classical RK4 algorithm violates Eq.(3.18b).

Why is this last condition important? Modelers sometimes wish to explicitly use derivative operations in their models. This is generally a bad idea, but it may not always be avoidable. For example, if u is a real–time input that stems from a measurement sensor, and the model requires $\dot{u}$, there is nothing in the world that can save us from actually having to differentiate the input. The typical simulationist would then approximate the derivative by:

$$\dot{u} \approx \frac{u - u_{\text{last}}}{t - t_{\text{last}}} \tag{3.19}$$

where t_{last} is the time of the previous function evaluation, and u_{last} is the value of the input u at that time. Therefore, if it should ever happen that $t = t_{\text{last}}$, the numerical differentiation algorithm would get itself into trouble.

Two caveats are called for. While we were able to develop Heun's method using a matrix–vector notation, this technique won't work anymore as we proceed to third–order algorithms. Let us explain.

We found that:

$$\frac{d\mathbf{f}}{dt} = \frac{\partial \mathbf{f}}{\partial \mathbf{x}} \cdot \frac{d\mathbf{x}}{dt} + \frac{\partial \mathbf{f}}{\partial t} \tag{3.20}$$

or, in shorthand notation:

$$\dot{\mathbf{f}} = \mathbf{f_x} \cdot \mathbf{f} + \mathbf{f}_t \tag{3.21}$$

When we proceed to third–order algorithms, we need an expression for the second absolute derivative of $\mathbf{f}$ with respect to time. Thus, we are inclined to write formally:

$$
\begin{aligned}
\ddot{\mathbf{f}} &= (\mathbf{f_x} \cdot \mathbf{f} + \mathbf{f}_t)^{\cdot} \\
&= \dot{\mathbf{f}}_\mathbf{x} \cdot \mathbf{f} + \mathbf{f_x} \cdot \dot{\mathbf{f}} + \dot{\mathbf{f}}_t \\
&= (\dot{\mathbf{f}})_\mathbf{x} \cdot \mathbf{f} + \mathbf{f_x} \cdot (\dot{\mathbf{f}}) + (\dot{\mathbf{f}})_t \\
&= (\mathbf{f_x} \cdot \mathbf{f} + \mathbf{f}_t)_\mathbf{x} \cdot \mathbf{f} + \mathbf{f_x} \cdot (\mathbf{f_x} \cdot \mathbf{f} + \mathbf{f}_t) + (\mathbf{f_x} \cdot \mathbf{f} + \mathbf{f}_t)_t \\
&= \mathbf{f_{xx}} \cdot (\mathbf{f})^2 + 2 \cdot (\mathbf{f_x})^2 \cdot \mathbf{f} + 2 \cdot \mathbf{f}_{\mathbf{x}t} \cdot \mathbf{f} + 2 \cdot \mathbf{f_x} \cdot \mathbf{f}_t + \mathbf{f}_{tt} \quad (3.22)
\end{aligned}
$$

but it is not clear, what this is supposed to mean. Obviously, $\ddot{\mathbf{f}}$ is a vector, and so is $\mathbf{f}_{tt}$, but what is $\mathbf{f_{xx}} \cdot (\mathbf{f})^2$ supposed to mean? Is it a tensor multiplied by the square of a vector? Quite obviously, the formal differentiation mechanism doesn't extend to higher derivatives in the sense of familiar matrix–vector multiplications. Evidently, we must treat the expression $\mathbf{f_{xx}} \cdot (\mathbf{f})^2$ differently.

Butcher [3.3] developed a new syntax and a set of rules for how these higher derivatives must be interpreted. In essence, it turns out that, in this new syntax:

1. sums remain commutative and associative,

2. derivatives can still be computed in any order, i.e., $(\dot{\mathbf{f}})_\mathbf{x} = (\mathbf{f_x})^{\cdot}$, and

3. the multiplication rule can be generalized, thus: $(\mathbf{f_x} \cdot \mathbf{f})_\mathbf{x} = \mathbf{f_{xx}} \cdot \mathbf{f} + (\mathbf{f_x})^2$.

It is not necessary for us to learn Butcher's new syntax. It is sufficient to know that we can basically proceed as before, but must abstain from interpreting terms involving higher derivatives as consisting of factors that are combined by means of the familiar matrix–vector multiplication.

Prior to Butcher's work, all higher–order RK algorithms had simply been derived for the scalar case, and were then blindly applied to integrate entire state vectors. And here comes the second caveat. Butcher discovered that several of the previously developed and popular higher–order RK algorithms drop one or several orders of accuracy when applied to a state vector instead of a scalar state variable.

The reason for this somewhat surprising discovery is very simple. Already when computing the third absolute derivative of $\mathbf{f}$ with respect to time, the two terms $f_x \cdot f_{xx} \cdot (f)^2$ and $f_{xx} \cdot f \cdot f_x \cdot f$ appear in the derivation. In the scalar case, these two terms are identical, since:

$$a \cdot b = b \cdot a \qquad (3.23)$$

Unfortunately —and not *that* surprisingly after all— Eq.(3.23) does not extend to the vector case. Our new animals in the mathematical zoo of data structures and operations exhibit a property that we are already quite familiar with from matrix calculus, namely that multiplications are no longer commutative:

$$\mathbf{A} \cdot \mathbf{B} = (\mathbf{B}' \cdot \mathbf{A}')' \neq \mathbf{B} \cdot \mathbf{A} \qquad (3.24)$$

where $\mathbf{A}'$ denotes the transpose of $\mathbf{A}$. So, algorithms that had been developed without grouping such terms together continued to work properly also in the vector case, whereas algorithms that had made use of the commutative nature of scalar multiplications did work well for scalar problems, but dropped one or several approximation orders when exposed to vector problems.

The details of the Butcher syntax are of no immediate concern to us, since we never plan to actually perform these new operations. All we need in order to develop new RK algorithms is to be able to extract their coefficients. To this end, we can pretend that the normal rules of matrix and vector calculus still apply.

3.3   Stability Domains of RK Algorithms

Since all the previously presented RK algorithms are explicit algorithms, we expect their stability domains to look qualitatively like that of the FE algorithm, or more precisely, we expect the contours of marginal stability to bend into the left half $(\lambda \cdot h)$–plane.

Let us plug the linear system of Eq.(2.12) into Heun's algorithm. We find:

$$
\begin{aligned}
\text{predictor:} \quad & \dot{\mathbf{x}}_k = \mathbf{A} \cdot \mathbf{x}_k \\
& \mathbf{x}_{k+1}^P = \mathbf{x}_k + h \cdot \dot{\mathbf{x}}_k
\end{aligned}
$$

$$
\begin{aligned}
\text{corrector:} \quad & \dot{\mathbf{x}}_{k+1}^P = \mathbf{A} \cdot \mathbf{x}_{k+1}^P \\
& \mathbf{x}_{k+1}^C = \mathbf{x}_k + 0.5 \cdot h \cdot (\dot{\mathbf{x}}_k + \dot{\mathbf{x}}_{k+1}^P)
\end{aligned}
$$

or:

$$\mathbf{x}_{k+1}^C = [\mathbf{I}^{(n)} + \mathbf{A} \cdot h + \frac{(\mathbf{A} \cdot h)^2}{2}] \cdot \mathbf{x}_k \qquad (3.25)$$

i.e.,

$$\mathbf{F} = \mathbf{I}^{(n)} + \mathbf{A} \cdot h + \frac{(\mathbf{A} \cdot h)^2}{2} \qquad (3.26)$$

Since a two–stage algorithm contains only two function evaluations, no powers of h larger than two can appear in the $\mathbf{F}$–matrix. Since the technique is second–order accurate, it must approximate the analytical solution:

$$\mathbf{F} = \exp(\mathbf{A} \cdot h) = \mathbf{I}^{(n)} + \mathbf{A} \cdot h + \frac{(\mathbf{A} \cdot h)^2}{2!} + \frac{(\mathbf{A} \cdot h)^3}{3!} + \cdots \qquad (3.27)$$

up to the quadratic term. Consequently, all two–stage RK2 algorithms share the same stability domain, and the same holds true for all three–stage RK3s, and for all four–stage RK4s. The situation becomes more complicated in the case of the fifth–order algorithms, since there doesn't exist a five–stage RK5. Consequently, the $\mathbf{F}$–matrices of RK5s necessarily contain a term in h^6 (with incorrect coefficient), and since there is no reason why these sixth–order terms should carry the same coefficient in different RK5s, their stability domains will look slightly different one from another.

Let us apply our general–purpose stability domain plotting algorithm that was presented in Chapter 2. We find:

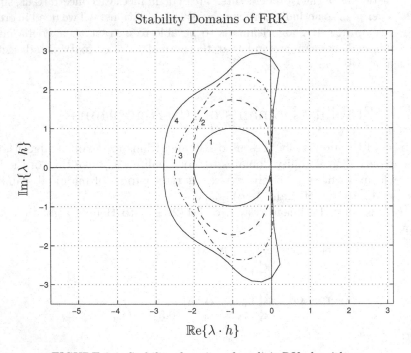

FIGURE 3.1. Stability domains of explicit RK algorithms.

Some of the RK5s are among those algorithms with small stable islands somewhere out in the unstable right half $(\lambda \cdot h)$–plane.

The reader may notice that these algorithms try indeed (and not surprisingly) to approximate the analytical stability domain, i.e., higher–order

RKs follow the imaginary axis better and better. Also, the stability domains grow with increasing approximation order. This is very satisfying, since higher–order algorithms call for larger step sizes.

3.4   Stiff Systems

Although the term "stiff system" has been popular at least since Gear's 1971 book [3.10] appeared, the numerical ODE literature still doesn't provide a crisp definition of what a stiff system really is. Even the 1991 book by Lambert [3.18] treats *"the nature of stiffness"* on as many as nine pages. Lambert observes that:

> *Statement # 1:* "A linear constant coefficient system is stiff if all of its eigenvalues have negative real part and the stiffness ratio is large."

> *Statement # 2:* "Stiffness occurs when stability requirements, rather than those of accuracy, constrain the step-length."

> *Statement # 3:* "Stiffness occurs when some components of the solution decay much more rapidly than others."

> *Statement # 4:* "A system is said to be stiff in a given interval of time if in that interval the neighboring solution curves approach the solution curve at a rate which is very large in comparison with the rate at which the solution varies in that interval."

The first statement is not overly useful since it relates to linear systems only. The second statement is not very precise since the accuracy requirements are not specified. Thus, one and the same system may be stiff, according to this statement, if the accuracy requirements are loose, and non–stiff if the accuracy requirements are tight. The third statement indirectly refers to the superposition principle, and is therefore, in a strict sense, again limited to linear systems. The fourth statement is basically a reformulation of the third.

Lambert concludes his exposé of the matter with the following definition:

> *Definition # 1:* "If a numerical method with a finite region of absolute stability, applied to a system with any initial conditions, is forced to use in a certain interval of integration a steplength which is excessively small in relation to the smoothness of the exact solution in that interval, then the system is said to be **stiff** in that interval."

Again, what exactly means "excessively small"?

Our remarks may sound critical of Lambert's work. They are not meant to be. Lambert's 1991 book represents a significant contribution to the numerical ODE literature. All we want to convey is that here is a term that has been around for more than a quarter of a century, and yet, the term is still fuzzy.

Let us attempt a more crisp definition of the term "stiff system":

> *Definition #2:* "An ODE system is called stiff if, when solved with any n^{th}–order accurate integration algorithm and a local error tolerance of 10^{-n}, the step size of the algorithm is forced down to below a value indicated by the local error estimate due to constraints imposed on it by the limited size of the numerically stable region."

Our definition comes closest to Lambert's statement #2, except that we added a definition of what we mean by accuracy requirements. Our definition is still somewhat fuzzy since it is possible that a system may fall under the category *stiff* when solved with one n^{th}–order accurate integration algorithm, and doesn't when solved with another. Yet, as we shall see, the grey zone of "marginally stiff" systems is fairly narrow, and moreover, this is exactly what these systems are: *marginally stiff*.

It is treacherous to rely on the eigenvalues of the Jacobian of a nonlinear or even linear but time–variant system to conclude anything about stiffness. Let us explain.

Given the system:

$$\dot{\mathbf{x}} = \mathbf{A}(t) \cdot \mathbf{x} = \begin{pmatrix} -2.5 & 1.5 \cdot \exp(-100t) \\ -0.5 \cdot \exp(100t) & -0.5 \end{pmatrix} \cdot \mathbf{x} \qquad (3.28a)$$

with initial conditions:

$$\mathbf{x_0} = \begin{pmatrix} 23 \\ 11 \end{pmatrix} \qquad (3.28b)$$

Its analytical solution is:

$$x_1(t) = 5 \cdot \exp(-101t) + 18 \cdot \exp(-102t) \qquad (3.29a)$$
$$x_2(t) = 5 \cdot \exp(-t) + 6 \cdot \exp(-2t) \qquad (3.29b)$$

Therefore, the system is awfully stiff. Yet, the eigenvalues of its Jacobian are -1.0 and -2.0, i.e., they are perfectly tame at all times.

If we plug this system into the FE algorithm (or RK1, which is the same algorithm), we find:

$$\mathbf{F}(t) = \mathbf{I^{(n)}} + \mathbf{A}(t) \cdot h = \begin{pmatrix} -2.5 \cdot h + 1.0 & 1.5 \cdot h \cdot \exp(-100t) \\ -0.5 \cdot h \cdot \exp(100t) & -0.5 \cdot h + 1.0 \end{pmatrix} \qquad (3.30)$$

Thus, the eigenvalues of the discrete–time system are at:

$$\lambda_1 = 1 - h \quad ; \quad \lambda_2 = 1 - 2h \tag{3.31}$$

which is what we would have expected from the locations of the eigenvalues of the continuous–time system and the stability domain of Fig.2.7. Thus, the stability domain doesn't indicate any foul play in this case. Codes that rely on the Jacobian for computing local error estimates will be fooled by this problem.

However, this is a particularly malignant problem, and fortunately one that physics doesn't usually prescribe. We may not truly want to simulate this system anyway since, already at simulated time $t = 10.0$, the element a_{21} has acquired a value of $-0.5 \cdot \exp(1000)$, something our simulator will most certainly complain bitterly about.

It is therefore still useful to search for methods that include in their numerically stable region the entire left half $(\lambda \cdot h)$–plane, or at least a large portion thereof.

> *Definition:* A numerical integration scheme that contains the entire left half $(\lambda \cdot h)$–plane as part of its numerical stability domain is called *absolute stable*, or, more simply, *A–stable*.

One way to obtain A–stable algorithms is to modify the recipe for developing RK algorithms by allowing non–zero elements also above the diagonal of the β–matrix [3.2] [3.13]. Such algorithms are invariably implicit. They are therefore called *implicit Runge–Kutta schemes*, abbreviated as IRK. A special role among those algorithms employ methods that limit the non-zero elements in their respective β–matrices to the first super–diagonal. Using the Butcher tableau representation, its **A**–matrix is still lower triangular, but contains nonlinear elements along its diagonal. Such algorithms are called *diagonally implicit Runge–Kutta schemes*, abbreviated as DIRK. They are implicit in each stage, but each stage can be iterated separately, and it is therefore fairly easy to implement a Newton iteration on them.

However, rather than looking at the problem of defining general IRK and DIRK algorithms through their α–vectors and β–matrices, we want to turn to two special classes of such algorithms that have interesting properties: the *extrapolation techniques*, and the *backinterpolation techniques*. We shall discuss some other classes of implicit Runge–Kutta algorithms in Chapter 8 in the context of solving sets of mixed differential and algebraic equations (DAEs).

3.5   Extrapolation Techniques

The idea behind the (Richardson) extrapolation techniques is quite straight-forward. We repeat the same integration step with several low–order techniques and blend the results to get a higher–order technique. Let us explain the concept by means of the linear system:

$$\dot{\mathbf{x}} = \mathbf{A} \cdot \mathbf{x} \tag{3.32}$$

We shall integrate the system four times across one macro–step of length h each time using the FE algorithm with different micro–step sizes: $\eta_1 = h$, $\eta_2 = h/2$, $\eta_3 = h/3$, and $\eta_4 = h/4$. Accordingly, we need only one micro–step of length η_1, but we need four micro–steps of length η_4. The corresponding discrete–time systems are:

$$
\begin{aligned}
\mathbf{x}^{\mathbf{P_1}}(k+1) &= [\mathbf{I}^{(\mathbf{n})} + \mathbf{A} \cdot h] \cdot \mathbf{x}(k) \\
\mathbf{x}^{\mathbf{P_2}}(k+1) &= [\mathbf{I}^{(\mathbf{n})} + \frac{\mathbf{A} \cdot h}{2}]^2 \cdot \mathbf{x}(k) \\
\mathbf{x}^{\mathbf{P_3}}(k+1) &= [\mathbf{I}^{(\mathbf{n})} + \frac{\mathbf{A} \cdot h}{3}]^3 \cdot \mathbf{x}(k) \\
\mathbf{x}^{\mathbf{P_4}}(k+1) &= [\mathbf{I}^{(\mathbf{n})} + \frac{\mathbf{A} \cdot h}{4}]^4 \cdot \mathbf{x}(k)
\end{aligned}
\tag{3.33a}
$$

with the corrector:

$$\mathbf{x}^{\mathbf{C}}(k+1) = \alpha_1 \cdot \mathbf{x}^{\mathbf{P_1}}(k+1) + \alpha_2 \cdot \mathbf{x}^{\mathbf{P_2}}(k+1) + \alpha_3 \cdot \mathbf{x}^{\mathbf{P_3}}(k+1) + \alpha_4 \cdot \mathbf{x}^{\mathbf{P_4}}(k+1) \tag{3.33b}$$

Multiplying the predictor formulae out, we find:

$$
\begin{aligned}
\mathbf{x}^{\mathbf{P_1}} &= [\mathbf{I}^{(\mathbf{n})} + \mathbf{A} \cdot h] \cdot \mathbf{x}(k) \\
\mathbf{x}^{\mathbf{P_2}} &= [\mathbf{I}^{(\mathbf{n})} + \mathbf{A} \cdot h + \frac{(\mathbf{A} \cdot h)^2}{4}] \cdot \mathbf{x}(k) \\
\mathbf{x}^{\mathbf{P_3}} &= [\mathbf{I}^{(\mathbf{n})} + \mathbf{A} \cdot h + \frac{(\mathbf{A} \cdot h)^2}{3} + \frac{(\mathbf{A} \cdot h)^3}{27}] \cdot \mathbf{x}(k) \\
\mathbf{x}^{\mathbf{P_4}} &= [\mathbf{I}^{(\mathbf{n})} + \mathbf{A} \cdot h + \frac{(\mathbf{A} \cdot h)^2}{8} + \frac{(\mathbf{A} \cdot h)^3}{16} + \frac{(\mathbf{A} \cdot h)^4}{256}] \cdot \mathbf{x}(k)
\end{aligned}
\tag{3.34a}
$$

and for the corrector, we obtain:

$$
\begin{aligned}
\mathbf{x}^{\mathbf{C}}(k+1) = &[(\alpha_1 + \alpha_2 + \alpha_3 + \alpha_4) \cdot \mathbf{I}^{(\mathbf{n})} \\
&+ (\alpha_1 + \alpha_2 + \alpha_3 + \alpha_4) \cdot \mathbf{A} \cdot h \\
&+ (\frac{\alpha_2}{4} + \frac{\alpha_3}{3} + \frac{3\alpha_4}{8}) \cdot (\mathbf{A} \cdot h)^2 \\
&+ (\frac{\alpha_3}{27} + \frac{\alpha_4}{16}) \cdot (\mathbf{A} \cdot h)^3 + \frac{\alpha_4}{256} \cdot (\mathbf{A} \cdot h)^4] \cdot \mathbf{x_k}
\end{aligned}
\tag{3.34b}
$$

Comparing Eq.(3.34b) with the correct Taylor Series truncated after the fourth–order term, we obtain four linear equations in the four unknown α_i parameters:

$$
\begin{pmatrix} 1 & 1 & 1 & 1 \\ 0 & 1/4 & 1/3 & 3/8 \\ 0 & 0 & 1/27 & 1/16 \\ 0 & 0 & 0 & 1/256 \end{pmatrix} \cdot \begin{pmatrix} \alpha_1 \\ \alpha_2 \\ \alpha_3 \\ \alpha_4 \end{pmatrix} = \begin{pmatrix} 1 \\ 1/2 \\ 1/6 \\ 1/24 \end{pmatrix} \tag{3.35}
$$

which can be solved directly. We find:

$$
\alpha_1 = -\frac{1}{6} \;;\; \alpha_2 = 4 \;;\; \alpha_3 = -\frac{27}{2} \;;\; \alpha_4 = \frac{32}{3} \tag{3.36}
$$

Thus, we just discovered another way to construct an RK4 algorithm since the extrapolation technique is fourth–order accurate ... at least for linear systems. We didn't bother to check whether the algorithm is also accurate up to fourth order for nonlinear systems, since unfortunately, the technique is quite inefficient. It took 10 function evaluations to complete a single macro–step. Compare this with the four function evaluations needed when performing an ordinary RK4 step.

Let us try another idea. The order of the algorithm wouldn't be all that important if we only could make the step size sufficiently small. Unfortunately, this would mean that we would have to perform many such steps in order to complete the simulation run ... or doesn't it?

We can write:

$$
\mathbf{x_{k+1}}(\eta) = \mathbf{x_{k+1}} + \mathbf{e_1} \cdot \eta + \mathbf{e_2} \cdot \frac{\eta^2}{2!} + \mathbf{e_3} \cdot \frac{\eta^3}{3!} + \ldots \tag{3.37}
$$

where $\mathbf{x_{k+1}}$ is the true (yet unknown) value of $\mathbf{x}$ at time $t_k + h$, whereas $\mathbf{x_{k+1}}(\eta)$ is the numerical value that we find when we integrate the system from time t_k to time $t_k + h$ using the micro–step size η. Obviously, this value contains an error. We now develop the numerical value into a Taylor Series in η around the (unknown) correct value. The $\mathbf{e_i}$ vectors are error vectors [3.6].

We truncate the Taylor Series after the cubic term, and write Eq.(3.37) down for the same values of η_i as before. We find:

$$
\begin{aligned}
\mathbf{x^{P_1}}(\eta_1) &\approx \mathbf{x_{k+1}} + \mathbf{e_1} \cdot h + \frac{\mathbf{e_2}}{2!} \cdot h^2 + \frac{\mathbf{e_3}}{3!} \cdot h^3 \\
\mathbf{x^{P_2}}(\eta_2) &\approx \mathbf{x_{k+1}} + \mathbf{e_1} \cdot \frac{h}{2} + \frac{\mathbf{e_2}}{2!} \cdot (\frac{h}{2})^2 + \frac{\mathbf{e_3}}{3!} \cdot (\frac{h}{2})^3 \\
\mathbf{x^{P_3}}(\eta_3) &\approx \mathbf{x_{k+1}} + \mathbf{e_1} \cdot \frac{h}{3} + \frac{\mathbf{e_2}}{2!} \cdot (\frac{h}{3})^2 + \frac{\mathbf{e_3}}{3!} \cdot (\frac{h}{3})^3 \\
\mathbf{x^{P_4}}(\eta_4) &\approx \mathbf{x_{k+1}} + \mathbf{e_1} \cdot \frac{h}{4} + \frac{\mathbf{e_2}}{2!} \cdot (\frac{h}{4})^2 + \frac{\mathbf{e_3}}{3!} \cdot (\frac{h}{4})^3
\end{aligned} \tag{3.38}
$$

or in matrix notation:

$$
\begin{pmatrix} x^{P_1} \\ x^{P_2} \\ x^{P_3} \\ x^{P_4} \end{pmatrix} \approx \begin{pmatrix} h^0 & h^1 & h^2 & h^3 \\ (h/2)^0 & (h/2)^1 & (h/2)^2 & (h/2)^3 \\ (h/3)^0 & (h/3)^1 & (h/3)^2 & (h/3)^3 \\ (h/4)^0 & (h/4)^1 & (h/4)^2 & (h/4)^3 \end{pmatrix} \cdot \begin{pmatrix} x_{k+1} \\ e_1 \\ e_2/2 \\ e_3/6 \end{pmatrix} \tag{3.39}
$$

By inverting the Van–der–Monde matrix, we can solve for the unknown x_{k+1} and the three error vectors. Since we aren't interested in the errors, we only look at the first row of the inverted Van–der–Monde matrix. It turns out that the values in this row don't depend at all on the step size h. We find:

$$
x_{k+1} \approx \begin{pmatrix} -\frac{1}{6} & 4 & -\frac{27}{2} & \frac{32}{3} \end{pmatrix} \cdot \begin{pmatrix} x^{P_1} \\ x^{P_2} \\ x^{P_3} \\ x^{P_4} \end{pmatrix} \tag{3.40}
$$

Obviously, x_{k+1} is no longer the truly correct solution since we had truncated the Taylor Series in η after the cubic term. However, the algorithm did the best it could to estimate the true value given the available data ... by raising the approximation order of the method to four.

Thus, we got precisely the same answers as before. We just found another way to derive the extrapolation method. Both approaches have their pros and cons. The first technique unveiled that the resulting method is indeed fourth–order accurate (at least for linear systems). The second method didn't show this fact explicitly ... it was more like swinging a magic wand. On the other hand, the first approach made explicitly use of the fact that each micro–step was performed by means of the FE algorithm. The second approach was not based on any such assumption.

Thus, we could now replace each of the micro–steps by a BE step of the same length, e.g. using Newton iteration if the system to be simulated is nonlinear, and still use the same corrector. The overall implicit extrapolation (IEX) technique then presents itself as:

$$1^{st} \text{ predictor:} \quad \mathbf{k_1} \quad = \mathbf{x_k} + h \cdot \mathbf{f}(\mathbf{k_1}, t_{k+1})$$

$$2^{nd} \text{ predictor:} \quad \mathbf{k_{2a}} = \mathbf{x_k} + \frac{h}{2} \cdot \mathbf{f}(\mathbf{k_{2a}}, t_{k+\frac{1}{2}})$$
$$\mathbf{k_2} \quad = \mathbf{k_{2a}} + \frac{h}{2} \cdot \mathbf{f}(\mathbf{k_2}, t_{k+1})$$

$$3^{rd} \text{ predictor:} \quad \mathbf{k_{3a}} = \mathbf{x_k} + \frac{h}{3} \cdot \mathbf{f}(\mathbf{k_{3a}}, t_{k+\frac{1}{3}})$$
$$\mathbf{k_{3b}} = \mathbf{k_{3a}} + \frac{h}{3} \cdot \mathbf{f}(\mathbf{k_{3b}}, t_{k+\frac{2}{3}})$$
$$\mathbf{k_3} \quad = \mathbf{k_{3b}} + \frac{h}{3} \cdot \mathbf{f}(\mathbf{k_3}, t_{k+1})$$

$$4^{th} \text{ predictor:} \quad \mathbf{k_{4a}} = \mathbf{x_k} + \frac{h}{4} \cdot \mathbf{f}(\mathbf{k_{4a}}, t_{k+\frac{1}{4}})$$
$$\mathbf{k_{4b}} = \mathbf{k_{4a}} + \frac{h}{4} \cdot \mathbf{f}(\mathbf{k_{4b}}, t_{k+\frac{1}{2}})$$
$$\mathbf{k_{4c}} = \mathbf{k_{4b}} + \frac{h}{4} \cdot \mathbf{f}(\mathbf{k_{4c}}, t_{k+\frac{3}{4}})$$
$$\mathbf{k_4} \quad = \mathbf{k_{4c}} + \frac{h}{4} \cdot \mathbf{f}(\mathbf{k_4}, t_{k+1})$$

$$\text{corrector:} \quad \mathbf{x_{k+1}} = -\frac{1}{6} \cdot \mathbf{k_1} + 4 \cdot \mathbf{k_2} - \frac{27}{2} \cdot \mathbf{k_3} + \frac{32}{3} \cdot \mathbf{k_4}$$

A complete analysis of the nonlinear accuracy order of this technique is quite involved, and we have not attempted it. However, by following our initial approach at deriving the extrapolation method now for BE steps in place of FE steps, it is a simple exercise to verify that the method indeed carries fourth-order accuracy for solving linear systems. As was to be expected from the latter way of reasoning, the α_i–parameters turn out to be exactly the same for BE steps as for FE steps.

Let us look at the stability domain of this method. It is presented in Fig.3.2. The method is A–stable, and has a nicely large unstable region in the right half $(\lambda \cdot h)$–plane.

Implicit extrapolation techniques, such as the IEX4 technique explained above, have, in comparison with IRK or DIRK algorithms, the distinct advantage that they are easy to construct. They have the disadvantages that no formal nonlinear accuracy analysis is currently available, and that they are still fairly inefficient. IEX4 is a 10–stage algorithm. In contrast, a fourth–order fully-implicit IRK algorithm can be constructed with only two stages, as shall be demonstrated in Chapter 8.

3.6 Marginally Stable Systems

We have seen in Chapter 2 that neither the FE nor the BE algorithm will do a decent job when confronted with eigenvalues on or in the vicinity of the imaginary axis. Unfortunately, this situation occurs quite frequently, and there exists an entire class of applications, namely the hyperbolic PDEs that, when converted to sets of ODEs, exhibit this property as we shall see later. It seems thus justified to analyze what can be done to tackle such problems. Let us start with a proper definition:

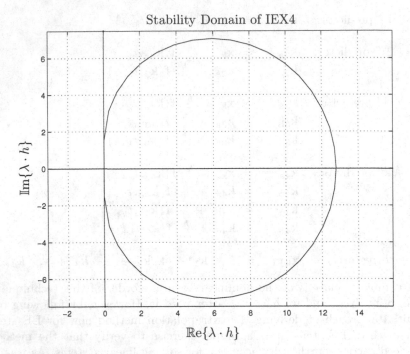

FIGURE 3.2. Stability domain of implicit extrapolation method.

Definition: A dynamical system whose Jacobian has its dominant eigenvalues on or in the vicinity of the imaginary axis is called *marginally stable.*

The dominant eigenvalues of a matrix are those eigenvalues that have the most positive real parts, i.e., that are located most to the right in the λ–plane.

In order to tackle such problems decently, we require integration algorithms that approximate the imaginary axis particularly well. Such algorithms do exist, and, in fact, algorithms of arbitrary order can be constructed whose borders of numerical stability coincide with the imaginary axis. We shall study these algorithms in due course.

Definition: A numerical integration scheme that contains the entire left half $(\lambda \cdot h)$–plane and nothing but the left half $(\lambda \cdot h)$–plane as its numerical stability domain is called *faithfully stable,* or, more simply, *F–stable.*

The reader may now be inclined to think that F–stable algorithms must be the answer to all our prayers. Unfortunately, this is not so. F–stable algorithms will perform poorly when asked to integrate stiff systems. The reason for this surprising disclosure is the following: If we think of the complex $(\lambda \cdot h)$–plane as an infinitely large plane, we are inclined to assume that each point at infinity is infinitely far away from each other point at infinity. However, it turns out to be more accurate to think of the complex $(\lambda \cdot h)$–plane as an infinitely large globe. From wherever we stand on that globe, infinity is the single one spot that is farthest away from us. Thus, infinity is a "single spot" in the sense that the numerical properties of any integration algorithm based on Taylor–Series expansion will be exactly the same irrespective of the direction from which we approach infinity.

Consequently, since (by definition) the entire imaginary axis belongs to the margin of stability, so does the infinity "spot" itself. This means that, although the entire left half $(\lambda \cdot h)$–plane is indeed stable, as we approach infinity along the negative real axis, points along the negative real axis will become less and less stable until, at point infinity, stability is lost. Similarly, although the entire right half $(\lambda \cdot h)$–plane is indeed unstable, as we approach infinity along the positive real axis, points along the positive real axis will become less and less unstable until, at point infinity, stability is reconquered.

The λ–plane has different properties. As we move along a line parallel to the real axis to the left, the damping of an eigenvalue located at that position increases constantly until it reaches a value of infinity at point infinity. In fact, the damping of an eigenvalue is identical with its distance from the imaginary axis.

An F–stable algorithm can obviously not mimic this facet of the λ–plane, and consequently, it will perform poorly when exposed to eigenvalues located far out to the left on the λ–plane. The time response due to these eigenvalues will not properly be dampened out. The F–stably simulated system with eigenvalues at such locations will therefore behave more sluggishly than the real system.

> *Definition:* A numerical integration scheme that is A–stable, and, in addition, whose damping properties increase to infinity as $\mathbb{Re}\{\lambda\} \rightarrow -\infty$, is called *L-stable*.

The various numerical stability definitions are, in a strict sense, only meaningful for linear time–invariant systems, but they are often good indicators when applied to nonlinear systems as well.

When dealing with stiff systems, it is not sufficient to demand A–stability from the integration algorithm. We need to look more closely at the damping behavior. L–stability may be a desirable property. Evidently, all F–stable algorithms are also A–stable, but never L–stable.

A system that is both marginally stable and stiff is difficult to cope with. Such systems exist, and we shall provide an example of one such system in due course. As of now, we are somewhat at a loss when asked which algorithm we recommend in this situation. What might possibly work best is an L–stable algorithm with an extra large unstable region in the right half $(\lambda \cdot h)$–plane, but best may still not be very good. We shall demonstrate how such algorithms can be constructed.

3.7   Backinterpolation Methods

We shall now look at yet another class of special IRK methods, called *backinterpolation techniques*, abbreviated as BI algorithms. Similar to the previously discussed *extrapolation techniques*, BI methods are easy to construct. However, they offer much better control over the accuracy order in comparison with the IEX algorithms even when applied to nonlinear systems. Also, they are considerably more efficient than IEX algorithms. BI algorithms can be made F–stable, L–stable, or anything in between, depending on the current needs of the user, and they lend themselves conveniently to stability domain shaping.

Let us look once more at the BE algorithm:

$$\mathbf{x_{k+1}} = \mathbf{x_k} + h \cdot \dot{\mathbf{x}}_{k+1} \tag{3.41}$$

We can rearrange Eq.(3.41) as follows:

$$\mathbf{x_k} = \mathbf{x_{k+1}} - h \cdot \dot{\mathbf{x}}_{k+1} \tag{3.42}$$

Thus, a step *forward* through time from time t_k to time t_{k+1} using the BE algorithm with a step size of h can also be interpreted as a step *backward* through time from time t_{k+1} to time t_k using the FE algorithm with a step size of $-h$.

Thus, one way to implement the BE algorithm is to start out from an estimate of the yet unknown value $\mathbf{x_{k+1}}$, and integrate backward through time to t_k. We then iterate on the unknown "initial" condition $\mathbf{x_{k+1}}$ until we hit the known "final" value $\mathbf{x_k}$ accurately. We accept the last guess of $\mathbf{x_{k+1}}$ as the correct value, and estimate $\mathbf{x_{k+2}}$. Now we integrate again backward through time to t_{k+1} until we hit $\mathbf{x_{k+1}}$.

This idea can, of course, be extended to any RK algorithm. For example, we can take any off–the–shelf RK4 algorithm to replace the former FE algorithm in integrating backward through time. This is the basic idea behind backinterpolation. These simplest of all BI algorithms are therefore sometimes called *backward Runge–Kutta methods*, or, abbreviated, BRK methods.

This gives us a series of algorithms of increasing order with the **F**–matrices:

$$\mathbf{F_1} = [\mathbf{I^{(n)}} - \mathbf{A} \cdot h]^{-1} \tag{3.43a}$$

$$\mathbf{F_2} = [\mathbf{I^{(n)}} - \mathbf{A} \cdot h + \frac{(\mathbf{A} \cdot h)^2}{2!}]^{-1} \tag{3.43b}$$

$$\mathbf{F_3} = [\mathbf{I^{(n)}} - \mathbf{A} \cdot h + \frac{(\mathbf{A} \cdot h)^2}{2!} - \frac{(\mathbf{A} \cdot h)^3}{3!}]^{-1} \tag{3.43c}$$

$$\mathbf{F_4} = [\mathbf{I^{(n)}} - \mathbf{A} \cdot h + \frac{(\mathbf{A} \cdot h)^2}{2!} - \frac{(\mathbf{A} \cdot h)^3}{3!} + \frac{(\mathbf{A} \cdot h)^4}{4!}]^{-1} \tag{3.43d}$$

Their numerical stability domains are shown in Fig.(3.3). Evidently, these stability domains are the mirror images of the stability domains of the explicit RK algorithms. This is not further surprising, since they are the same algorithms with h replaced by $-h$.

Stability Domains of BRK

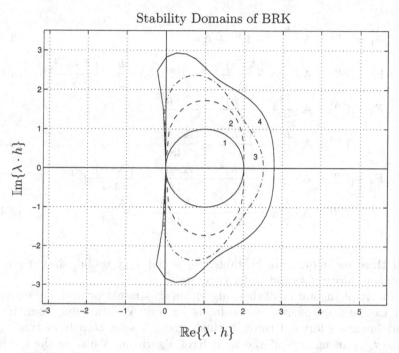

FIGURE 3.3. Stability domains of basic backinterpolation methods.

Let us now discuss whether we can exploit the backinterpolation idea to generate a set of F–stable algorithms of increasing order.

Several F–stable algorithms have been known for a long time. One of those algorithms is the *trapezoidal rule*:

$$1^{\text{st}} \text{ stage:} \quad \mathbf{x}_{k+\frac{1}{2}} = \mathbf{x}_k + \frac{h}{2} \cdot \dot{\mathbf{x}}_k$$

$$2^{\text{nd}} \text{ stage:} \quad \mathbf{x}_{k+1} = \mathbf{x}_{k+\frac{1}{2}} + \frac{h}{2} \cdot \dot{\mathbf{x}}_{k+1}$$

The trapezoidal rule is an implicit algorithm that can be envisaged as a cyclic method consisting of a semi–step of length $h/2$ using FE followed by another semi–step of length $h/2$ using BE.

Its **F**–matrix is thus:

$$\mathbf{F_{TR}} = [\mathbf{I}^{(n)} - \mathbf{A} \cdot \frac{h}{2}]^{-1} \cdot [\mathbf{I}^{(n)} + \mathbf{A} \cdot \frac{h}{2}] \tag{3.44}$$

The trapezoidal rule exploits the symmetry of the stability domains of its two semi–steps.

Since we can implement the BE semi–step as a BRK1 step using the backinterpolation method, we can extend this idea also to higher–order algorithms. Their $\mathbf{F}$–matrices will be:

$$\mathbf{F_1} = [\mathbf{I}^{(n)} - \mathbf{A} \cdot \frac{h}{2}]^{-1} \cdot [\mathbf{I}^{(n)} + \mathbf{A} \cdot \frac{h}{2}] \qquad (3.45a)$$

$$\mathbf{F_2} = [\mathbf{I}^{(n)} - \mathbf{A} \cdot \frac{h}{2} + \frac{(\mathbf{A} \cdot h)^2}{8}]^{-1} \cdot [\mathbf{I}^{(n)} + \mathbf{A} \cdot \frac{h}{2} + \frac{(\mathbf{A} \cdot h)^2}{8}] \qquad (3.45b)$$

$$\mathbf{F_3} = [\mathbf{I}^{(n)} - \mathbf{A} \cdot \frac{h}{2} + \frac{(\mathbf{A} \cdot h)^2}{8} - \frac{(\mathbf{A} \cdot h)^3}{48}]^{-1} \cdot$$

$$[\mathbf{I}^{(n)} + \mathbf{A} \cdot \frac{h}{2} + \frac{(\mathbf{A} \cdot h)^2}{8} + \frac{(\mathbf{A} \cdot h)^3}{48}] \qquad (3.45c)$$

$$\mathbf{F_4} = [\mathbf{I}^{(n)} - \mathbf{A} \cdot \frac{h}{2} + \frac{(\mathbf{A} \cdot h)^2}{8} - \frac{(\mathbf{A} \cdot h)^3}{48} + \frac{(\mathbf{A} \cdot h)^4}{384}]^{-1} \cdot$$

$$[\mathbf{I}^{(n)} + \mathbf{A} \cdot \frac{h}{2} + \frac{(\mathbf{A} \cdot h)^2}{8} + \frac{(\mathbf{A} \cdot h)^3}{48} + \frac{(\mathbf{A} \cdot h)^4}{384}] \qquad (3.45d)$$

All these techniques are F–stable. $\mathbf{F_2}$ is not very useful, since $\mathbf{F_1}$ is, by accident, already second–order accurate.

The implementation of these algorithms is straightforward. For example, $\mathbf{F_4}$ can be implemented in the following way. We start out from time t_k and integrate forward through time across a semi–step from time t_k to time $t_{k+\frac{1}{2}}$ using any off–the–shelf RK4 algorithm. We store the resulting state $\mathbf{x}_{k+\frac{1}{2}}^{\text{left}}$ for later reuse. We then estimate the value $\mathbf{x}_{k+1}$, e.g. by letting $\mathbf{x}_{k+1} = \mathbf{x}_{k+\frac{1}{2}}^{\text{left}}$, and integrate backward through time across the second semi–step from t_{k+1} to $t_{k+\frac{1}{2}}$ using the same off–the–shelf RK4 algorithm. The resulting state is $\mathbf{x}_{k+\frac{1}{2}}^{\text{right}}$. We now iterate on the unknown state $\mathbf{x}_{k+1}$, until $\mathbf{x}_{k+\frac{1}{2}}^{\text{right}} = \mathbf{x}_{k+\frac{1}{2}}^{\text{left}}$. We then use the final value of $\mathbf{x}_{k+1}$ as the initial condition for the next integration macro–step.

We need to analyze the iteration process somewhat more. Chapter 2 taught us that a poor choice of the iteration algorithm can foul up our stability domain.

When applying Newton iteration to the BI1 algorithm, we can set:

$$\mathcal{F}(\mathbf{x}_{k+1}) = \mathbf{x}_{k+\frac{1}{2}}^{\text{right}} - \mathbf{x}_{k+\frac{1}{2}}^{\text{left}} = 0.0 \qquad (3.46)$$

and find:

$$\mathbf{x}^{\text{left}}_{k+\frac{1}{2}} = \text{FE}(\mathbf{x}_k, t_k, \tfrac{h}{2})$$
$$\mathbf{x}^0_{k+1} = \mathbf{x}^{\text{left}}_{k+\frac{1}{2}}$$

$$\mathbf{J}^0_{k+1} = \mathcal{J}(\mathbf{x}^0_{k+1}, t_{k+1})$$
$$\mathbf{x}^{\text{right\_1}}_{k+\frac{1}{2}} = \text{FE}(\mathbf{x}^0_{k+1}, t_{k+1}, -\tfrac{h}{2})$$
$$\mathbf{H}^1 = \mathbf{I}^{(n)} - \tfrac{h}{2} \cdot \mathbf{J}^0_{k+1}$$
$$\mathbf{x}^1_{k+1} = \mathbf{x}^0_{k+1} - \mathbf{H}^{1^{-1}} \cdot (\mathbf{x}^{\text{right\_1}}_{k+\frac{1}{2}} - \mathbf{x}^{\text{left}}_{k+\frac{1}{2}})$$
$$\varepsilon^1_{k+1} = \|\mathbf{x}^1_{k+1} - \mathbf{x}^0_{k+1}\|_\infty$$

$$\mathbf{J}^1_{k+1} = \mathcal{J}(\mathbf{x}^1_{k+1}, t_{k+1})$$
$$\mathbf{x}^{\text{right\_2}}_{k+\frac{1}{2}} = \text{FE}(\mathbf{x}^1_{k+1}, t_{k+1}, -\tfrac{h}{2})$$
$$\mathbf{H}^2 = \mathbf{I}^{(n)} - \tfrac{h}{2} \cdot \mathbf{J}^1_{k+1}$$
$$\mathbf{x}^2_{k+1} = \mathbf{x}^1_{k+1} - \mathbf{H}^{2^{-1}} \cdot (\mathbf{x}^{\text{right\_2}}_{k+\frac{1}{2}} - \mathbf{x}^{\text{left}}_{k+\frac{1}{2}})$$
$$\varepsilon^2_{k+1} = \|\mathbf{x}^2_{k+1} - \mathbf{x}^1_{k+1}\|_\infty$$

etc.

where $\mathcal{J}$ denotes the Jacobian. For Heun's method (BI2), we find:

$$\mathbf{x}^{\text{left}}_{k+\frac{1}{2}} = \text{Heun}(\mathbf{x}_k, t_k, \tfrac{h}{2})$$
$$\mathbf{x}^0_{k+1} = \mathbf{x}^{\text{left}}_{k+\frac{1}{2}}$$

$$\mathbf{J}^0_{k+1} = \mathcal{J}(\mathbf{x}^0_{k+1}, t_{k+1})$$
$$\mathbf{x}^{\text{right\_1}}_{k+\frac{1}{2}} = \text{Heun}(\mathbf{x}^0_{k+1}, t_{k+1}, -\tfrac{h}{2})$$
$$\mathbf{J}^0_{k+\frac{1}{2}} = \mathcal{J}(\mathbf{x}^{\text{right\_1}}_{k+\frac{1}{2}}, t_{k+\frac{1}{2}})$$
$$\mathbf{H}^1 = \mathbf{I}^{(n)} - \tfrac{h}{4} \cdot (\mathbf{J}^0_{k+1} + \mathbf{J}^0_{k+\frac{1}{2}} \cdot (\mathbf{I}^{(n)} - \tfrac{h}{2} \cdot \mathbf{J}^0_{k+1}))$$
$$\mathbf{x}^1_{k+1} = \mathbf{x}^0_{k+1} - \mathbf{H}^{1^{-1}} \cdot (\mathbf{x}^{\text{right\_1}}_{k+\frac{1}{2}} - \mathbf{x}^{\text{left}}_{k+\frac{1}{2}})$$
$$\varepsilon^1_{k+1} = \|\mathbf{x}^1_{k+1} - \mathbf{x}^0_{k+1}\|_\infty$$

$$\mathbf{J}^1_{k+1} = \mathcal{J}(\mathbf{x}^1_{k+1}, t_{k+1})$$
$$\mathbf{x}^{\text{right\_2}}_{k+\frac{1}{2}} = \text{Heun}(\mathbf{x}^1_{k+1}, t_{k+1}, -\tfrac{h}{2})$$
$$\mathbf{J}^1_{k+\frac{1}{2}} = \mathcal{J}(\mathbf{x}^{\text{right\_2}}_{k+\frac{1}{2}}, t_{k+\frac{1}{2}})$$
$$\mathbf{H}^2 = \mathbf{I}^{(n)} - \tfrac{h}{4} \cdot (\mathbf{J}^1_{k+1} + \mathbf{J}^1_{k+\frac{1}{2}} \cdot (\mathbf{I}^{(n)} - \tfrac{h}{2} \cdot \mathbf{J}^1_{k+1}))$$
$$\mathbf{x}^2_{k+1} = \mathbf{x}^1_{k+1} - \mathbf{H}^{2^{-1}} \cdot (\mathbf{x}^{\text{right\_2}}_{k+\frac{1}{2}} - \mathbf{x}^{\text{left}}_{k+\frac{1}{2}})$$
$$\varepsilon^2_{k+1} = \|\mathbf{x}^2_{k+1} - \mathbf{x}^1_{k+1}\|_\infty$$

etc.

The algorithm stays basically the same, except that we now need two Jacobians evaluated at different points in time, and the formula for the Hessian becomes a little more involved.

If we assume that the Jacobian remains basically unchanged during one integration step (modified Newton iteration), we can compute both the Jacobian and the Hessian at the beginning of the step, and we find for BI1:

$$\mathbf{J} = \mathcal{J}(\mathbf{x_k}, t_k) \tag{3.47a}$$

$$\mathbf{H} = \mathbf{I}^{(n)} - \frac{h}{2} \cdot \mathbf{J} \tag{3.47b}$$

and for BI2:

$$\mathbf{J} = \mathcal{J}(\mathbf{x_k}, t_k) \tag{3.48a}$$

$$\mathbf{H} = \mathbf{I}^{(n)} - \frac{h}{2} \cdot \mathbf{J} + \frac{h^2}{8} \cdot \mathbf{J}^2 \tag{3.48b}$$

We recognize the pattern. Clearly, the sequence of $\mathbf{H}$–matrices is:

$$\mathbf{H_1} = \mathbf{I}^{(n)} - \mathbf{J} \cdot \frac{h}{2} \tag{3.49a}$$

$$\mathbf{H_2} = \mathbf{I}^{(n)} - \mathbf{J} \cdot \frac{h}{2} + \frac{(\mathbf{J} \cdot h)^2}{8} \tag{3.49b}$$

$$\mathbf{H_3} = \mathbf{I}^{(n)} - \mathbf{J} \cdot \frac{h}{2} + \frac{(\mathbf{J} \cdot h)^2}{8} - \frac{(\mathbf{J} \cdot h)^3}{48} \tag{3.49c}$$

$$\mathbf{H_4} = \mathbf{I}^{(n)} - \mathbf{J} \cdot \frac{h}{2} + \frac{(\mathbf{J} \cdot h)^2}{8} - \frac{(\mathbf{J} \cdot h)^3}{48} + \frac{(\mathbf{J} \cdot h)^4}{384} \tag{3.49d}$$

We may even decide to keep the same Jacobian for several steps in a row, and, in that case, we won't need to compute a new Hessian either, unless we decide to change the step size in between.

Evidently, the sequence in which we execute the forward and the backward semi–steps can be interchanged. The $\mathbf{F}$–matrix of the interchanged BI1 algorithm is:

$$\mathbf{F_{MP}} = [\mathbf{I}^{(n)} + \mathbf{A} \cdot \frac{h}{2}] \cdot [\mathbf{I}^{(n)} - \mathbf{A} \cdot \frac{h}{2}]^{-1} \tag{3.50}$$

which corresponds to the algorithm:

$$\mathbf{x_{k+1}} = \mathbf{x_k} + h \cdot \dot{\mathbf{x}}_{k+\frac{1}{2}} \tag{3.51}$$

which is the well–known *implicit midpoint rule*, the one–legged twin of the *trapezoidal rule*. In the same manner, it is possible to generate algorithms of higher orders as well. The two twins are identical in their linear properties, but they behave differently with respect to their nonlinear characteristics. The original BI algorithms are a little more *accurate* than their one–legged twins, since we read out the value of the state at the end of the iteration rather than after the forward semi–step. On the other hand, the one–legged variety has somewhat better *nonlinear stability* (contractivity) properties, as shown in [3.5].

BI techniques have a certain resemblance with *Padé approximation methods*, which we shall abbreviate as PA methods. The idea behind PA methods is the following: Every numerical ODE solver tries to somehow approximate the analytical **F**–matrix, which would be:

$$\mathbf{F} = \exp(\mathbf{A} \cdot h) \tag{3.52}$$

Equation (3.52) can be rewritten as:

$$\mathbf{F} = \exp(\mathbf{A}\frac{h}{2}) \cdot \exp(\mathbf{A}\frac{h}{2}) = [\exp(\mathbf{A}(-\frac{h}{2}))]^{-1} \cdot \exp(\mathbf{A}\frac{h}{2}) \tag{3.53}$$

According to [3.19], this can be approximated by:

$$\mathbf{F} \approx \mathbf{D}(p,q)^{-1} \cdot \mathbf{N}(p,q) \tag{3.54}$$

with:

$$\mathbf{D}(p,q) = \sum_{j=0}^{q} \frac{(p+q-j)! \, q!}{(p+q)! \, j! \, (q-j)!} \cdot (-\mathbf{A}h)^j \tag{3.55a}$$

$$\mathbf{N}(p,q) = \sum_{j=0}^{p} \frac{(p+q-j)! \, p!}{(p+q)! \, j! \, (p-j)!} \cdot (\mathbf{A}h)^j \tag{3.55b}$$

which, for $p = q$, leads to the following set of **F**–matrices:

$$\mathbf{F_2} = [\mathbf{I}^{(n)} - \mathbf{A} \cdot \frac{h}{2}]^{-1} \cdot [\mathbf{I}^{(n)} + \mathbf{A} \cdot \frac{h}{2}] \tag{3.56a}$$

$$\mathbf{F_4} = [\mathbf{I}^{(n)} - \mathbf{A} \cdot \frac{h}{2} + \frac{(\mathbf{A} \cdot h)^2}{12}]^{-1} \cdot [\mathbf{I}^{(n)} + \mathbf{A} \cdot \frac{h}{2} + \frac{(\mathbf{A} \cdot h)^2}{12}] \tag{3.56b}$$

$$\mathbf{F_6} = [\mathbf{I}^{(n)} - \mathbf{A} \cdot \frac{h}{2} + \frac{(\mathbf{A} \cdot h)^2}{10} - \frac{(\mathbf{A} \cdot h)^3}{120}]^{-1} \cdot$$
$$[\mathbf{I}^{(n)} + \mathbf{A} \cdot \frac{h}{2} + \frac{(\mathbf{A} \cdot h)^2}{10} + \frac{(\mathbf{A} \cdot h)^3}{120}] \tag{3.56c}$$

$$\mathbf{F_8} = [\mathbf{I}^{(n)} - \mathbf{A} \cdot \frac{h}{2} + \frac{3(\mathbf{A} \cdot h)^2}{28} - \frac{(\mathbf{A} \cdot h)^3}{84} + \frac{(\mathbf{A} \cdot h)^4}{1680}]^{-1} \cdot$$
$$[\mathbf{I}^{(n)} + \mathbf{A} \cdot \frac{h}{2} + \frac{3(\mathbf{A} \cdot h)^2}{28} + \frac{(\mathbf{A} \cdot h)^3}{84} + \frac{(\mathbf{A} \cdot h)^4}{1680}] \tag{3.56d}$$

As the indices indicate, these formulae are all accurate to the double order, i.e., while the individual semi–steps are no longer proper Runge–Kutta steps (they are themselves only first–order accurate), the overall method attains a considerably higher order of linear accuracy. Due to the symmetry between $\mathbf{D}(p,q)$ and $\mathbf{N}(p,q)$, i.e., due to selecting $p = q$, all these methods are still F–stable.

PA techniques have been intensively studied in [3.13, 3.16]. The problem with them is that an accuracy analysis is only available for the linear case. When exposed to nonlinear systems, the methods may drop several orders of accuracy. Thereby, $\mathbf{F_8}$ may degenerate to an algorithm of merely second order.

The BI algorithms don't share this problem. While they are less accurate than their corresponding PA counterparts for the same computational effort when solving linear problems, their order of accuracy never drops. When using BI4, we shall retain fourth–order accuracy even when solving nonlinear problems, since each of its semi–steps itself is fourth–order accurate for nonlinear as well as linear problems.

Let us check whether we can transform our F–stable BI techniques into a set of more strongly stable BI techniques. The previous set of F–stable backinterpolation techniques did exploit the symmetry of the stability domains of its two semi–steps. However, there is no compelling reason why the two semi–steps have to meet exactly in the middle. The explicit semi–step could span a distance of $\vartheta \cdot h$, and the implicit semi–step could span the remaining distance $(1 - \vartheta) \cdot h$. Such a technique is called ϑ–method. The resulting algorithm would still be accurate to the same order as its two semi–steps. Using this technique, the stability domain can be shaped.

The case with $\vartheta > 0.5$ is of not much interest, but the case with $\vartheta < 0.5$ is very useful. It produces a series of techniques with ever increasing stability until, at $\vartheta = 0.0$, we obtain a set of L–stable algorithms. The F–matrices of the ϑ–methods are:

$$\mathbf{F_1} = [\mathbf{I}^{(n)} - \mathbf{A}(1 - \vartheta)h]^{-1} \cdot [\mathbf{I}^{(n)} + \mathbf{A}\vartheta h] \tag{3.57a}$$

$$\mathbf{F_2} = [\mathbf{I}^{(n)} - \mathbf{A}(1 - \vartheta)h + \frac{(\mathbf{A}(1 - \vartheta)h)^2}{2!}]^{-1} \cdot$$
$$[\mathbf{I}^{(n)} + \mathbf{A}\vartheta h + \frac{(\mathbf{A}\vartheta h)^2}{2!}] \tag{3.57b}$$

$$\mathbf{F_3} = [\mathbf{I}^{(n)} - \mathbf{A}(1 - \vartheta)h + \frac{(\mathbf{A}(1 - \vartheta)h)^2}{2!} - \frac{(\mathbf{A}(1 - \vartheta)h)^3}{3!}]^{-1} \cdot$$
$$[\mathbf{I}^{(n)} + \mathbf{A}\vartheta h + \frac{(\mathbf{A}\vartheta h)^2}{2!} + \frac{(\mathbf{A}\vartheta h)^3}{3!}] \tag{3.57c}$$

$$\mathbf{F_4} = [\mathbf{I}^{(n)} - \mathbf{A}(1 - \vartheta)h + \frac{(\mathbf{A}(1 - \vartheta)h)^2}{2!} - \frac{(\mathbf{A}(1 - \vartheta)h)^3}{3!} +$$
$$\frac{(\mathbf{A}(1 - \vartheta)h)^4}{4!}]^{-1} \cdot [\mathbf{I}^{(n)} + \mathbf{A}\vartheta h + \frac{(\mathbf{A}\vartheta h)^2}{2!} + \frac{(\mathbf{A}\vartheta h)^3}{3!} +$$

$$\frac{(\mathbf{A}\vartheta h)^4}{4!}] \tag{3.57d}$$

Figure 3.4 shows the stability domains of the BI algorithms that result for:

$$\vartheta = 0.4 \tag{3.58}$$

These methods result in very nice stability domains with large unstable

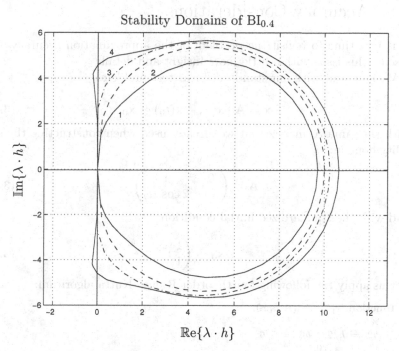

FIGURE 3.4. Stability domains of backinterpolation ϑ–methods.

regions in the right half $(\lambda \cdot h)$–plane. The selection of a good value for ϑ is a compromise. ϑ should be chosen large enough to generate meaningfully large unstable regions in the right half $(\lambda \cdot h)$–plane, yet small enough to dampen out the high frequency components appropriately in the left half $(\lambda \cdot h)$–plane. The fourth–order algorithm of Fig.3.4 is no longer A–stable. Its unstable region reaches slightly into the left half $(\lambda \cdot h)$–plane. Such a method is called (A,α)–stable, where α denotes the largest angle away from the negative real axis that contains only stable territory. BI$4_{0.4}$ is $(A,86°)$–stable.

The previously mentioned BRK algorithms are special cases of this new class of ϑ–methods with $\vartheta = 0$, and the explicit RK algorithms are special cases of this class of ϑ–methods with $\vartheta = 1$. The F–stable BI algorithms are special cases with $\vartheta = 0.5$.

A set of L–stable BI algorithms with $\vartheta > 0$ can be constructed by choosing the approximation order of the implicit semi–step one order higher than that of the explicit semi–step. For stiff engineering problems, a BI4 algorithm using an RK4 method for its explicit semi–step and a BRK5 method for its implicit semi–step together with $\vartheta = 0.45$ turns out to be generally an excellent choice [3.24]. This method will be abbreviated as $BI4/5_{0.45}$.

3.8   Accuracy Considerations

It is now time to revisit the problem of the approximation accuracy, and discuss this issue with a little more insight and detail.

We start out with our standard linear test problem:

$$\dot{\mathbf{x}} = \mathbf{A} \cdot \mathbf{x} \quad ; \quad \mathbf{x}(t_0) = \mathbf{x_0} \tag{3.59}$$

with the same $\mathbf{A}$–matrix that we already used when constructing the stability domain:

$$\mathbf{A} = \begin{pmatrix} 0 & 1 \\ -1 & 2\cos(\alpha) \end{pmatrix} \tag{3.60}$$

and with the standardized initial condition:

$$\mathbf{x_0} = \begin{pmatrix} 1 \\ 1 \end{pmatrix} \tag{3.61}$$

Let us apply the following fourth–order Runge–Kutta algorithm:

```
function [x] = rk4(A, h, x0)
   %
   h2 = h/2;   h6 = h/6;
   x(:, 1) = x0;
   %
   for i = 1 : 10/h,
      xx = x(:, i);
      k1 = A * xx;
      k2 = A * (xx + h2 * k1);
      k3 = A * (xx + h2 * k2);
      k4 = A * (xx + h * k3);
      x(:, i + 1) = xx + h6 * (k1 + 2 * k2 + 2 * k3 + k4);
   end
return
```

to simulate this system across 10 seconds of simulated time.

We want to compare the simulated solution $\mathbf{x_{simul}}$ with the analytical solution:

$$\mathbf{x_{anal}} = \exp(\mathbf{A} \cdot (t - t_0)) \cdot \mathbf{x_0} \tag{3.62}$$

We define the global error as follows:

$$\varepsilon_{\text{global}} = \|\mathbf{x_{anal}} - \mathbf{x_{simul}}\|_\infty \tag{3.63}$$

and vary the step size, h, until the global error matches a prescribed tolerance:

```
function [hmax] = hh2(alpha, hlower, hupper, tol)
%
A = aa(alpha);
x0 = ones(2, 1);
maxerr = 1E-6;   err = 100;
while err > maxerr,
    h = (hlower + hupper)/2;
    xsimul = rk4(A, h, x0);
    for i = 0 : 10/h,
        xanal(:, i + 1) = expm(A * h * i) * x0;
    end,
    eglobal = norm(xanal - xsimul,'inf');
    err = eglobal - tol;
    if err > 0,
        hupper = h;
    else
        hlower = h;
    end,
    err = abs(err);
end
hmax = h;
return
```

This routine looks very similar to the one that was presented in Chapter 2 for the computation of the stability domain.

We again sweep over a range of α values, and plot h_{max} as a function of α in polar coordinates. Figure 3.5 shows the results of our efforts.

The chosen error tolerance was $tol = 10^{-4}$.

Just like the stability domain, the *accuracy domain* can be plotted in the $(\lambda \cdot h)$–plane. If we were to select a pair of eigenvalues in the λ–plane twice as far away from the origin and use a step size, h, that is half as large, we would get exactly the same accuracy. This happens because both the analytical $\mathbf{F}$–matrix:

$$\mathbf{F_{anal}} = \exp(\mathbf{A} \cdot h) \tag{3.64}$$

and its numerical counterpart:

$$\mathbf{F_{RK4}} = \mathbf{I^{(n)}} + \mathbf{A} \cdot h + \frac{(\mathbf{A} \cdot h)^2}{2!} + \frac{(\mathbf{A} \cdot h)^3}{3!} + \frac{(\mathbf{A} \cdot h)^4}{4!} \tag{3.65}$$

are functions of $\mathbf{A} \cdot h$.

On first sight, this seems to be an important discovery — accuracy can be treated in the same way as stability. Unfortunately, this is not quite true.

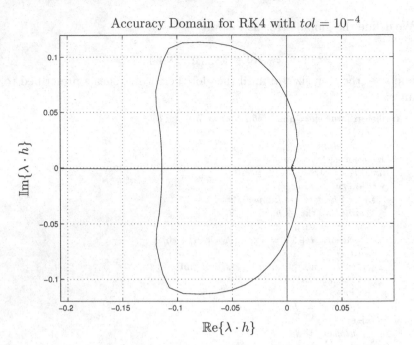

FIGURE 3.5. Accuracy domain of explicit fourth–order Runge–Kutta method.

The problem is that the accuracy domain depends heavily on the selected initial condition. The largest step size that can be chosen for a prescribed tolerance is approximately inverse proportional to the largest gradient in the simulation, and since:

$$\|\dot{\mathbf{x}}\|_\infty \approx \|\mathbf{A}\|_\infty \cdot \|\mathbf{x}\|_\infty \tag{3.66}$$

$h_{\max}$ is also approximately inverse proportional to any norm of the initial condition for stable systems. Notice the asymmetry of the accuracy domain with respect to the imaginary axis. If the poles are located in the analytically stable left half λ–plane, the transients die out with time, and the largest errors are committed early on in the game. On the other hand, if the poles are located in the analytically unstable right half λ–plane, the transients grow larger and larger, and the committed errors grow accordingly for any fixed step size. This is the major reason why an accurate simulation of analytically unstable systems is a quite expensive enterprize as we have to fight accumulation errors in that case.

Figure 3.6 shows the accuracy domains of the RK4 algorithm for three different error tolerances: $tol_1 = 10^{-4}$ (as before), $tol_2 = 10^{-3}$, $tol_3 = 10^{-2}$, and finally, $tol_4 = 10^{-1}$. The stability domain has been plotted on top of the three accuracy domains.

It can be noticed that all three accuracy domains are safely within the numerically stable region, at least as far as the left–half $\lambda \cdot h$–plane is

Accuracy Domains for RK4 in Function of *tol*

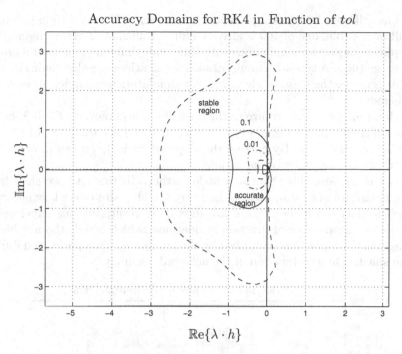

FIGURE 3.6. Accuracy domains of RK4 for different tolerance values.

concerned. However, this is deceiving. As the norm of the state vector decays for analytically stable eigenvalue locations, the step size could be chosen larger and larger to achieve the same tolerance.

The solution decays as $\exp(-\sigma \cdot t)$. Let us assume that the two eigenvalues are located at -1.0. In this case, the damping, σ, is 1.0. The norm of the chosen initial condition is $\|\mathbf{x_0}\| = 1.0$. With this initial condition, the accuracy domain intersects the negative real axis at about -0.1. Thus, when the norm of the state vector has decayed to roughly $\|\mathbf{x}\| = 0.04$, the accuracy domain has grown to the size of the stability domain, and from then onward, the step size will actually be controlled by the numerical stability requirements, and no longer by accuracy requirements. This happens after about 3.2 seconds of simulated time ... and this is approximately, how long we would usually simulate such a system before the trajectories become utterly uninteresting, as long as no input function adds to the "excitement."

If we now simulate a system, in which fast phenomena are superposed to slow phenomena, then the simulation run length will be determined by the slowest time constant whereas the step size will be dictated by the numerical stability requirements of the fastest component. This is the problem that was addressed under the heading *stiff system*. As the above calculation shows, it doesn't take a very large difference in time scales before stiffness becomes a problem. A time scale factor of 10 is probably still acceptable, one of 100 is already quite problematic. Many engineering applications call for simulations with time scale differences of several orders of magnitude. A typical example of such applications are electronic circuits. This is one of the reasons why *all* circuit simulators use implicit integration schemes.

However, let us now return to the question of accuracy. As Fig.3.6 shows, the difference in step size needed to improve the accuracy by a factor of 10 is not very large. By cutting the step size in half, we can improve the accuracy by one order of magnitude.

Let us explore this thought a little further. To this end, we shall keep the eigenvalues at −1.0, and we shall vary the step size to see how accurate the simulation will be. Figure 3.7 shows the results of this experiment. I plotted the number of function evaluations, which equals the number of stages of the algorithm multiplied by the number of steps performed during the simulation as a function of the achieved accuracy.

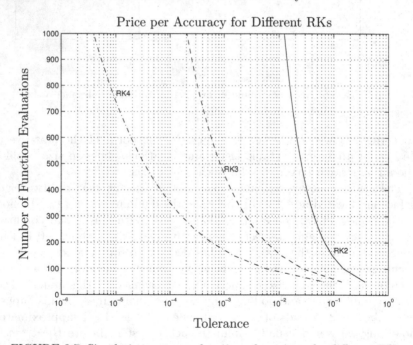

FIGURE 3.7. Simulation cost as a function of accuracy for different RKs.

It turns out that RK4 was cheaper than RK3, which in turn was cheaper than RK2 for *all* error tolerance values. This is not very surprising. By cutting the step size in half, we double the number of function evaluations needed to complete the simulation run. For the same "money," we could have kept the step size the same and instead doubled the accuracy order (at least for low–order algorithms). Thereby we would have gained two orders of magnitude in improved accuracy as opposed to only one order by reducing the step size.

Euler's performance is not shown in Fig.3.7. We also tried RK1 (FE), but we couldn't get a global accuracy of 10% (corresponding to a local accuracy of roughly 1%) for below 1000 function evaluations.

For a required global accuracy of 10%, the three algorithms shown in Fig.3.7 have a quite comparable price. For a 1% accuracy, RK2 is already out of the question, whereas both RK3 and RK4 still perform decently. For a 0.1% global accuracy, RK3 has become expensive, while RK4 still performs acceptably well.

Remembering that we usually control the *local* integration error rather than the *global* integration error, which is roughly one order of magnitude better, we see that indeed RK4 will work well for local errors of up to about 10^{-4}. If we want to compute more accurately than that, we definitely should turn to higher–order algorithms.

Notice that all these computations were performed on a 32 bit machine in double precision, thus, the roundoff error is negligible in comparison with the truncation error. Just for fun, we repeated the same computations in simulated single precision by chopping eight digits off the state vector at the end of each integration step using the MATLAB statement:

$$\mathbf{x} = \mathrm{chop}(\mathbf{x}, 8)$$

The results of this effort are shown in Fig.3.8.

If the accuracy requirements are low, the simulation can use large step sizes, and therefore, roundoff is not a problem. Consequently, the algorithms behave in the same way as before. However, for higher accuracy requirements, roundoff sets in (due to small step sizes), and accordingly, a further reduction of the step size will not help to meet the required accuracy. No algorithm of any order will get us a global accuracy of better than 10^{-5} in this case.

To summarize this discussion, problems with roundoff make simulation in single precision on a 32 bit machine quite problematic for *any* model using *any* integration algorithm. Accuracy requirements put a lower bound on the approximation order of the integration algorithm. A local error tolerance of $\varepsilon_{\mathrm{local}} = 10^{-k}$ calls for at least a kth–order integration algorithm. Lower–order algorithms aren't necessarily cheaper even if the accuracy requirements aren't stringent. Accuracy domains aren't quite as handy as one could hope for due to their heavy dependence on the chosen initial conditions.

Let us see whether we can come up with yet another tool to describe the accuracy of an integration algorithm, a tool that isn't plagued by the same flaw as the accuracy domain.

Let us look once more at our standard linear test problem of Eq.3.59 with the analytical solution of Eq.3.62. Obviously, the analytical solution

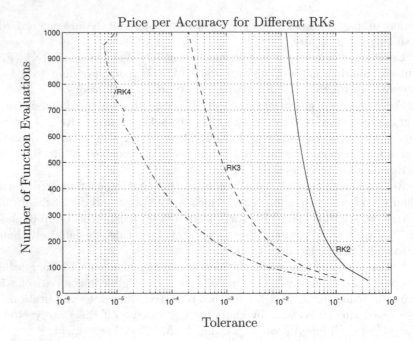

FIGURE 3.8. Cost for accuracy for different RKs in single precision.

is correct for any value of t, t_0, and $\mathbf{x_0}$, and in particular, it is true for $t_0 = t_k$, $\mathbf{x_0} = \mathbf{x_k}$ and $t = t_{k+1}$. With this substitution, we find:

$$\mathbf{x_{k+1}} = \exp(\mathbf{A} \cdot h) \cdot \mathbf{x_k} \tag{3.67}$$

Thus, the analytical $\mathbf{F}$–matrix of this system is:

$$\mathbf{F_{anal}} = \exp(\mathbf{A} \cdot h) \tag{3.68}$$

For a scalar problem, we can specialize the general solution of Eq.3.62 into:

$$x(t) = c_1 \cdot \exp(-\sigma \cdot t) \cdot \cos(\omega \cdot t) + c_2 \cdot \exp(-\sigma \cdot t) \cdot \sin(\omega \cdot t) \tag{3.69}$$

where σ is the distance of the eigenvalue from the imaginary axis and is called the *damping* of the eigenvalue, whereas ω is the distance of the eigenvalue from the real axis and is called the *eigenfrequency* of the eigenvalue.

It is easy to show that the eigenvalues of the analytical $\mathbf{F_{anal}}$–matrix are related to those of the $\mathbf{A}$–matrix through:

$$\text{Eig}\{\mathbf{F_{anal}}\} = \exp(\text{Eig}\{\mathbf{A}\} \cdot h) \tag{3.70}$$

or:

$$\lambda_{disc} = \exp(\lambda_{cont} \cdot h) = \exp((-\sigma + j \cdot \omega) \cdot h) = \exp(-\sigma \cdot h) \cdot \exp(j \cdot \omega \cdot h) \tag{3.71}$$

Consequently, the damping, i.e., the distance of an eigenvalue from the imaginary axis in the λ–plane maps into a distance from the origin in the $\exp(\lambda \cdot h)$–plane, whereas the eigenfrequency, i.e., the distance of an eigenvalue from the real axis in the λ–plane maps into an angle away from the positive real axis in the $\exp(\lambda \cdot h)$–plane.

Notice that we just introduced a new plane: the $\exp(\lambda \cdot h)$–plane. Control engineers call this plane the z–domain, where:

$$z = \exp(\lambda \cdot h) \tag{3.72}$$

Since even the analytical $\mathbf{F_{anal}}$–matrix depends on the step size, h, it makes sense to introduce a *discrete damping*, $\sigma_d = h \cdot \sigma$, and a *discrete frequency*, $\omega_d = h \cdot \omega$. Obviously, we can write:

$$
\begin{aligned}
|z| &= \exp(-\sigma_d) & \text{(3.73a)} \\
\angle z &= \omega_d & \text{(3.73b)}
\end{aligned}
$$

Now, let us replace the analytical $\mathbf{F_{anal}}$–matrix by the one that belongs to the numerical integration routine, $\mathbf{F_{simul}}$. The numerical $\mathbf{F_{simul}}$–matrix is either a rational (for implicit integration algorithms) or a polynomial (for explicit integration algorithms) approximation of the analytical $\mathbf{F_{anal}}$–matrix. We define:

$$\hat{z} = \exp(\hat{\lambda}_d) \tag{3.74}$$

with:

$$\hat{\lambda}_d = -\hat{\sigma}_d + j \cdot \hat{\omega}_d \tag{3.75}$$

Therefore:

$$
\begin{aligned}
|\hat{z}| &= \exp(-\hat{\sigma}_d) & \text{(3.76a)} \\
\angle \hat{z} &= \hat{\omega}_d & \text{(3.76b)}
\end{aligned}
$$

As $\hat{z}$ approximates z, so must $\hat{\sigma}_d$ be an approximation of σ_d, and $\hat{\omega}_d$ must approximate ω_d. It makes sense to study the relationship between the analytical discrete damping, σ_d, on the one hand, and the numerical discrete damping, $\hat{\sigma}_d$, on the other. Similarly, we can study the relationship between the analytical discrete frequency, ω_d, and the numerical discrete frequency, $\hat{\omega}_d$.

We can define:

$$
\begin{aligned}
\varepsilon_\sigma &= \sigma_d - \hat{\sigma}_d & \text{(3.77a)} \\
\varepsilon_\omega &= \omega_d - \hat{\omega}_d & \text{(3.77b)}
\end{aligned}
$$

where ε_σ denotes the *damping error*, and ε_ω denotes the *frequency error* committed by the numerical integration algorithm.

Since the case of all continuous eigenvalues being negative and real occurs so frequently (e.g. *all* thermal systems are of that nature), it is worthwhile to study the damping error when moving an eigenvalue left or right along the negative real axis. We can plot σ_d and $\hat{\sigma}_d$ as functions of σ_d itself. The following program will compute $\hat{\sigma}_d$ for any single–step integration algorithm.

```
function [sdhat] = damp(sd, algor)
    %
    f = ff(-sd, 1, algor);
    sdhat = -log(f);
return
```

We can then sweep across a range of σ_d–values, and plot both $-\sigma_d$ and $-\hat{\sigma}_d$ against $-\sigma_d$. Such a graph is called a *damping plot*. Figure 3.9 shows the damping plot of RK4.

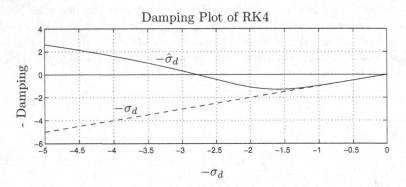

FIGURE 3.9. Damping plot of RK4.

In the vicinity of the origin, the numerical damping value, $\hat{\sigma}_d$, follows the analytical damping, σ_d, very well. However, already for very moderate eigenvalue locations, the numerical damping behavior deviates drastically from the analytical one, and somewhere around $\sigma_d = 2.8$, the numerical damping becomes negative, which coincides with the border of numerical stability. The area where the approximation of σ_d by $\hat{\sigma}_d$ is accurate, is called the *asymptotic region* of the integration algorithm.

Let us now look at the damping plot of BI4. This plot is shown in Fig.3.10.

Since BI4 is A–stable, the numerical damping stays positive for all values of λ. Since BI4 is F–stable, the numerical damping approaches zero as λ approaches $-\infty$. Figure 3.11 shows the damping plot of the ϑ–method with $\vartheta = 0.4$.

Now, the damping no longer approaches zero, but it doesn't go to infinity either. From Eq.3.57d, we can conclude that:

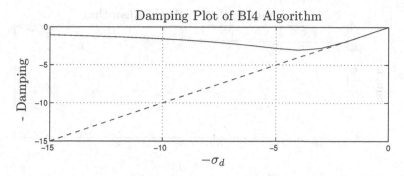

FIGURE 3.10. Damping plot of BI4.

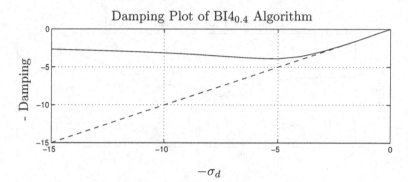

FIGURE 3.11. Damping plot of BI4 with $\vartheta = 0.4$.

$$\hat{\sigma}_d(-\infty) = -4 \cdot \log(\frac{\vartheta}{1-\vartheta}) \tag{3.78}$$

Consequently, for $\vartheta = 0.5$, we find that $\hat{\sigma}_d(-\infty) = 0.0$, a fact that we already knew (F–stability), and for $\vartheta = 0.4$, we find that $\hat{\sigma}_d(-\infty) = 1.6219$. The algorithm with $\vartheta = 0.0$, i.e., BRK4, is L–stable, since $\hat{\sigma}_d(-\infty) \rightarrow -\infty$. Unfortunately, the true power of L–stability is not as glamorous as one might think, as the BRK4 damping plot of Fig.3.12 demonstrates. Although BRK4 is L–stable, the increase in damping when moving the pole to the left is despairingly slow as a result of the logarithm function in Eq.(3.78). For $\sigma_d = -10^{-9}$, we find that $\hat{\sigma}_d \approx -80$. L–stability is thus somewhat overrated.

Just for completeness, let us draw the damping plot of IEX4 as well. It is shown inf Fig.3.13. Now, this is interesting. Somewhere around $\sigma_d = -6.7$, the numerical damping, $\hat{\sigma}_d$, intersects with the analytical damping, σ_d, thus the damping error is exactly equal to zero.

Let us extend our search to the entire complex $(\lambda \cdot h)$–plane. Figure 3.14 plots $\hat{\sigma}_d - \sigma_d$ over the complex plane.

Points on Fig.3.14 with positive amplitude represent a surplus in numer-

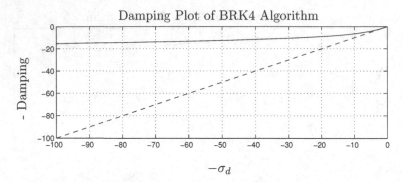

FIGURE 3.12. Damping plot of BRK4.

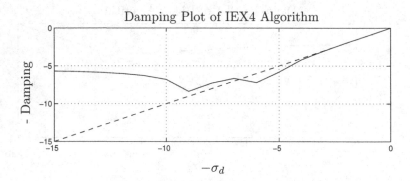

FIGURE 3.13. Damping plot of IEX4.

ical damping, whereas points with negative amplitude represent a lack in numerical damping. We can strip away the magnitude information, and only display the sign of the damping error. This is shown in Fig.3.15. '+' means that there is surplus damping at this point, whereas '−' indicates that there is not enough damping.

There obviously exists a locus of at least partially connected points of the $\lambda \cdot h$–plane, where the damping error is zero. This locus has been plotted in Fig.3.16 for the IEX4 method. Such a locus is called an *order star* [3.13, 3.16, 3.23].

Figure 3.14 shows that there exist points where the simulated damping $\hat{\sigma}_d$ is infinite. Contrary to $\mathbf{F}_{\mathbf{anal}}$, which is a very smooth function, any rational function approximation $\mathbf{F}_{\mathbf{simul}}$ has poles, i.e., points with infinite numerical damping.

The $\mathbf{F}$–matrix of IEX4 can be written as follows:

$$
\begin{aligned}
\mathbf{F} = & -\frac{1}{6} \cdot [\mathbf{I}^{(n)} - \mathbf{A} \cdot h]^{-1} + 4 \cdot [\mathbf{I}^{(n)} - \frac{\mathbf{A} \cdot h}{2}]^{-2} \\
& -\frac{27}{2} \cdot [\mathbf{I}^{(n)} - \frac{\mathbf{A} \cdot h}{3}]^{-3} + \frac{32}{3} \cdot [\mathbf{I}^{(n)} - \frac{\mathbf{A} \cdot h}{4}]^{-4}
\end{aligned}
\tag{3.79}
$$

3D Damping Plot for IEX4

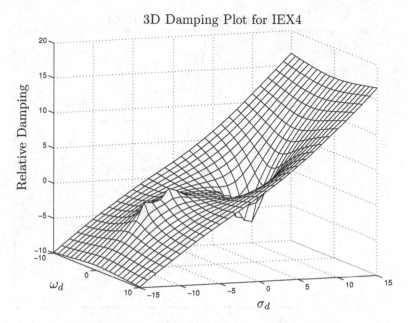

FIGURE 3.14. 3D–plot of damping error for IEX4.

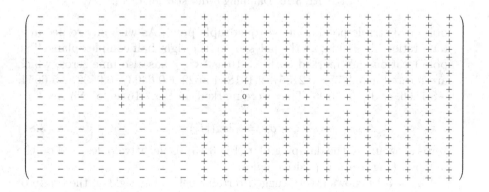

FIGURE 3.15. Damping error sign for IEX4.

Let us analyze the scalar case with:

$$q = \lambda \cdot h$$

We can write:

$$f = -\frac{1}{6} \cdot \frac{1}{1-q} + 4 \cdot \frac{1}{(1-q/2)^2} - \frac{27}{2} \cdot \frac{1}{(1-q/3)^3} + \frac{32}{3} \cdot \frac{1}{(1-q/4)^4} \quad (3.80)$$

f is a rational function with 10 poles located at $+1$ (single pole), $+2$ (double

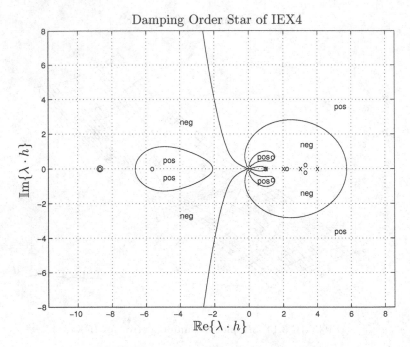

FIGURE 3.16. Damping order star for IEX4.

pole), +3 (triple pole), and +4 (quadruple pole). As with any respectable integration algorithm, all the poles are in the right half complex plane. The pole locations are marked on Fig.3.16 as '×'. f has also nine zeros, which are located at $3.277 \pm 0.2155j$, 2.1918, $1.3622 \pm 0.6493j$, 0.9562, -5.6654, -8.7506, and -65.0105. They were marked on Fig.3.16 as 'o'. Since:

$$\hat{\sigma}_d = -\log(|f|) \tag{3.81}$$

zeros show up on the damping plot (Fig.3.13) as negative poles, and on the 3D–plot (Fig.3.14) as positive poles.

Figure 3.14 shows a very rugged terrain just to the right of the origin of the complex plain. For this reason, extrapolation techniques surely aren't suitable for the integration of unstable systems.

In stiff system integration, we requested that $\hat{\sigma}_d \to -\infty$ as $\lambda \to \infty$. From Eq.(3.81), we conclude that $f \to 0$ as $q \to \infty$. This is obviously only possible if f is a strictly proper rational function. This is the reason why explicit integration algorithms can never be L–stable.

IEX4 also has many zeros, some of which are even in the left half complex plain. This can pose a problem. The BRK algorithms don't have any zeros. This may sometimes be beneficial.

Let us now look at the damping order star of a backinterpolation technique. It is shown for BI4 on Fig.3.17.

The terrain in the vicinity of the origin (the asymptotic region) is much

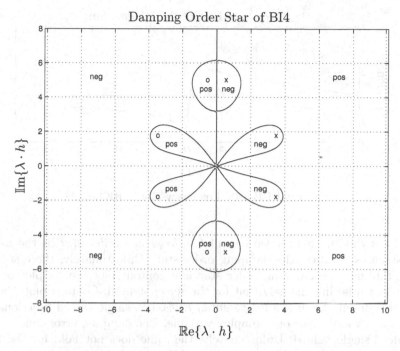

FIGURE 3.17. Damping order star for BI4.

smoother than in the case of IEX4. BI4 has four poles and four zeros that
are marked on Fig.3.17. As in the case of IEX4, BI4 has zeros in the left
half complex plane, but at least none on the negative real axis.

Let us now discuss the *frequency error*. It may be worthwhile to study the
frequency error when moving an eigenvalue up or down along the positive
imaginary axis. We can plot ω_d and $\hat{\omega}_d$ as functions of ω_d itself. The following program will compute $\hat{\omega}_d$ for any single–step integration algorithm.

```
function [wdhat] = freq(wd, algor)
%
f  = ff(wd, 1, algor);
wdhat  = atan2(imag(f),real(f));
return
```

We can then sweep across a range of ω_d–values, and plot both ω_d and $\hat{\omega}_d$
against ω_d. Such a graph is called a *frequency plot*. Figure 3.18 shows the
frequency plot of RK4.

The frequency plot of RK4 seems to exhibit a discontinuity around $\omega_d =$
2.5. Yet, the plot is misleading. The discrete frequency ω_d is 2π–periodic.
The "discontinuity" simply represents a jump of $\hat{\omega}_d$ from $+\pi$ to $-\pi$. We
could easily compensate for the jump, and get a $\hat{\omega}_d$ curve that is totally
smooth, as long as the chosen path for ω_d doesn't lead through either a
pole or a zero.

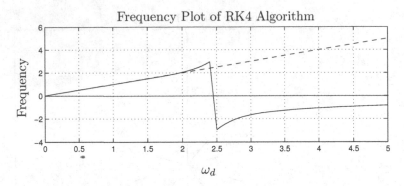

FIGURE 3.18. Frequency plot of RK4.

Does this mean that we can plot a *frequency order star* in the same fashion as we drew the damping order star? Unfortunately, there still is a problem. The damping order star is a contour plot, i.e., a plot of an equipotential line, namely that for the zero potential. Contour plots, however, can only be drawn for *potential fields*, i.e., single–valued functions in two real variables or one complex variable. The damping error function is indeed single–valued. Unfortunately, the same does not hold for the frequency error function. For each value of the complex independent variable $s_d = \sigma_d + j \cdot \omega_d$, the dependent variable $\hat{\omega}_d$ assumes infinitely many values, as $\hat{\omega}_d$ is 2π–periodic.

Can we fix the problem by eliminating the artificial discontinuities, as proposed above? Unfortunately, this does not solve the problem. If we choose a closed path in s_d that encircles either a pole or a zero, the total frequency contribution around the pole or zero is $\pm 2\pi$. The terrain of the $\hat{\omega}_d$ function in the vicinity of any pole or zero looks like an infinitely long spiral staircase. Consequently, the $\hat{\omega}_d$–function is not a potential field.

The $\hat{\omega}_d$–function can be turned into a potential field by limiting its range to e.g. $(-\pi, +\pi]$, in which case we can indeed plot a frequency order star just as easily as in the case of the damping order star.

Figure 3.19 shows the frequency order star of BI4, and Fig.3.20 exhibits the frequency order star of IEX4.

Frequency order stars aren't depicted often, although they should be, and were it only for their exquisite beauty.

What can we do with these tools? We have seen that both the damping plot and the frequency plot exhibit asymptotic regions, i.e., regions, in which $\mathbf{F}_{\mathbf{simul}}$ deviates only little from $\mathbf{F}_{\mathbf{anal}}$ both in terms of the absolute value (damping) and in terms of the phase value (frequency). The asymptotic region surrounds the origin. Figure 3.21 shows the asymptotic regions of the RK4 algorithm. The top graph shows the asymptotic region for $s_d = -\sigma_d$, whereas the bottom graph depicts the asymptotic region for $s_d = j \cdot \omega_d$.

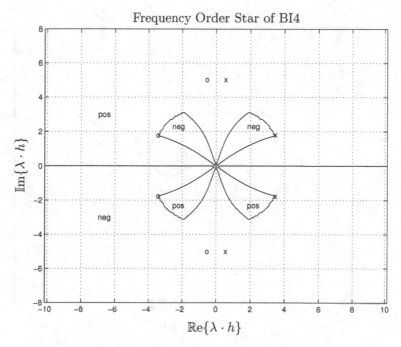

FIGURE 3.19. Frequency order star of BI4.

Hence it may make sense to define an error function that accounts for both types of errors, the damping error and the frequency error, simultaneously, e.g.:

$$os_{err} = |\sigma_d - \hat{\sigma}_d| + |\omega_d - \hat{\omega}_d| \tag{3.82}$$

If both the damping error and the frequency error are defined such that they form potential fields, then the *order star error* function, os_{err}, must also form a potential field. Consequently, we can draw equipotential lines of $os_{err} = 10^{-4}$, $os_{err} = 10^{-3}$, and $os_{err} = 10^{-2}$ as contour plots, and superpose them onto the same graph. Figure 3.22 depicts the order star accuracy domain of the RK4 algorithm.

The *order star accuracy domain* has an important advantage over the previously introduced *accuracy domain*. It is totally independent of the problem to be solved or the initial conditions being used. It only depends on the algorithm itself. It is a metric that is as "pure" as the stability domain.

Notice that the order star accuracy domain is not asymmetric w.r.t. the imaginary axis. The reason is simple. The order star accuracy domain compares the *analytical* solution of the original continuous–time problem with the equally *analytical* solution of the derived discrete–time problem. Consequently, it only accounts for the truncation error. It does not consider

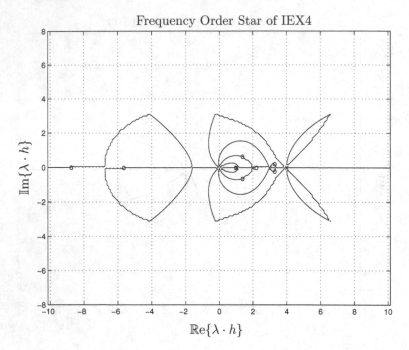

FIGURE 3.20. Frequency order star of IEX4.

either the roundoff or the accumulation errors. The roundoff error is relatively harmless, as it can be easily controlled by the length of the mantissa
used in the numerical computation. The accumulation error, on the other
hand, is anything but harmless. It is responsible for the narrow region of
accurate simulations in the right half plane of the accuracy domain. As the
order star accuracy domain doesn't account for accumulation errors, it allows for an equally large region of accurate computations in the right–half
$\lambda \cdot h$–plane as in the left–half $\lambda \cdot h$–plane.

Hence and in spite of its other shortcomings, the previously introduced
accuracy domain is a considerably more conservative measure of the ability
of a code to perform accurate simulations than the newly introduced order
star accuracy domain.

Figure 3.23 shows the order star accuracy domain of the IEX4 algorithm.
This time, the order star accuracy domain is indeed asymmetrical to the
imaginary axis. However, the reason here is not related to the accumulation
errors, but rather to the poles and zeros of this algorithm that are located
close to the origin in the right–half $\lambda \cdot h$–plane, leading to a very rugged
terrain of the order star in this region. Hence the IEX4 algorithm has no
chance of simulating accurately unstable systems, even irrespective of the
problem of error accumulation.

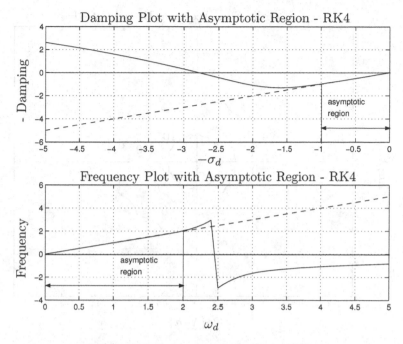

FIGURE 3.21. Asymptotic regions of RK4.

3.9   Step–size and Order Control

Although we have talked about the various sources of errors that can corrupt our numerical integration results, we have done nothing so far to contain them. We know that smaller step sizes will, in general, lead to smaller integration errors at a higher computational cost, whereas larger step sizes will lead to larger errors at a smaller cost. The right step size is a compromise between containment of error and cost. However, we don't know how to choose the most appropriate step size. As engineers, we certainly know how accurate we need our results to be, and we also know how much money we are willing to spend in order to get them, but while this knowledge indirectly determines the step size, we don't have a good algorithm yet that would translate the error/cost knowledge into an adequate value for the step size. This problem will be discussed next.

Since the relationship between error/cost on the one hand and the step size on the other depends heavily on the numerical properties of the system to be integrated, it is not clear that a fixed step size will lead to the same integration error throughout the simulation. It may well be that a variation of the step size during the simulation period in order to keep the error at a constant level close to the maximum allowed error tolerance can reduce the overall cost of the simulation. This observation leads to the demand for *variable–step integration algorithms*, which in turn call for a *step–size*

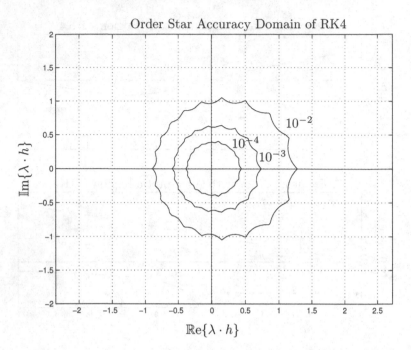

FIGURE 3.22. Order star accuracy domain of RK4.

control algorithm.

The following step–size control algorithm may work. Take any two arbitrary Runge–Kutta algorithms, and repeat the same step twice, once with each of the algorithms. The results of these two algorithms will differ by ε. If ε is larger than the tolerated error tol_{abs}, the step is rejected and repeated with half the step size. If ε is smaller than $0.5 \cdot tol_{abs}$ during four steps in a row, the next step will be computed with $1.5 \cdot h$.

This algorithm is *very* heuristic and somewhat unsatisfactory on several counts, but the reader certainly has no problems in understanding how the algorithm is supposed to work. The difference between the two solutions, ε, is taken as an estimate for the local integration error and is compared against the tolerated error. If the estimate is larger than the tolerated error, the step size needs to be reduced, but if it is smaller, the step size can be increased.

The first objection that comes to mind is the use of the absolute error as a performance measure. Intuitively, if the state variable (i.e., the output of the integrator) has a value of the order of 10^6, we can tolerate a much larger absolute error than if the state variable has a value of 10^{-6}. It therefore may make sense to replace the absolute error by a relative error. The modified algorithm looks as follows:

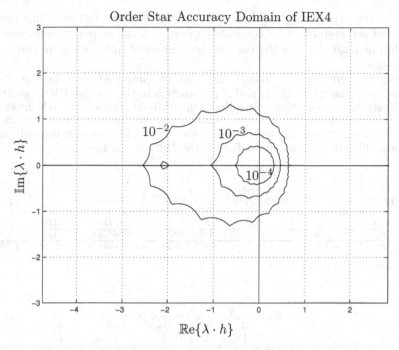

FIGURE 3.23. Order star accuracy domain of IEX4.

$$\varepsilon_{\mathrm{rel}} = \frac{|x_1 - x_2|}{|x_1|} \tag{3.83a}$$

$$\textbf{if } \varepsilon_{\mathrm{rel}} > tol_{\mathrm{rel}} \Rightarrow h_{\mathrm{new}} = 0.5 \cdot h \tag{3.83b}$$

$$\textbf{if } \varepsilon_{\mathrm{rel}} < 0.5 \cdot tol_{\mathrm{rel}} \text{ during four steps} \Rightarrow h_{\mathrm{new}} = 1.5 \cdot h \tag{3.83c}$$

where x_1 is the value of the state variable obtained by one of the two algorithms, whereas x_2 is the value obtained by the other algorithm.

Also this algorithm has its flaws. What if, by accident, $x_1 = 0.0$ at some point in time? This problem can be countered by modifying Eq.(3.83a) in the following fashion:

$$\varepsilon_{\mathrm{rel}} = \frac{|x_1 - x_2|}{\max(|x_1|, |x_2|, \delta)} \tag{3.84}$$

where δ is a fudge factor, e.g., $\delta = 10^{-10}$.

If an entire state vector needs to be integrated, the user may specify different relative error tolerances for each of the state variables separately. The above algorithm could then be applied to each of the state variables separately, resulting in different suggestions for the next step size, h_{new}, to be taken. The smallest of those values would then be applied.

The only remaining problem is the price tag associated with this procedure. Each step must be computed twice, i.e., the step–size control algorithm at least doubles the cost of the numerical integration. Is this truly necessary?

Edwin Fehlberg [3.9] didn't think so. He proposed to make use of the freedom in assigning the α and β parameters of the regular RK algorithms in designing a step–size controlled algorithm in which both RK methods share the early stages of the scheme, and vary only in the later stages. He created pairs of algorithms one order apart. The most commonly used among his methods is RKF4/5 with the Butcher tableau:

0	0	0	0	0	0	0
1/4	1/4	0	0	0	0	0
3/8	3/32	9/32	0	0	0	0
12/13	1932/2197	−7200/2197	7296/2197	0	0	0
1	439/216	−8	3680/513	−845/4104	0	0
1/2	−8/27	2	−3544/2565	1859/4104	−11/40	0
x_1	25/216	0	1408/2565	2197/4104	−1/5	0
x_2	16/135	0	6656/12825	28561/56430	−9/50	2/55

where:

$$f_1(q) = 1 + q + \frac{1}{2}q^2 + \frac{1}{6}q^3 + \frac{1}{24}q^4 + \frac{1}{104}q^5 \qquad (3.85a)$$

$$f_2(q) = 1 + q + \frac{1}{2}q^2 + \frac{1}{6}q^3 + \frac{1}{24}q^4 + \frac{1}{120}q^5 + \frac{1}{2080}q^6 \quad (3.85b)$$

Thus, x_1 is a five–stage RK4, and x_2 is a six–stage RK5. However, the RK4 and RK5 algorithms have the first five stages in common. Therefore, the step–size controlled algorithm is overall still a six–stage RK5, and the only additional cost associated with step–size control is the computation of the corrector of RK4. Step–size control comes almost for free.

How about the heuristic algorithm for modifying the step size? Luckily, we can do better than that. Since we have explicit expressions for $f_1(q)$ and $f_2(q)$, we can provide an explicit formula for the error estimate:

$$\varepsilon(q) = f_1(q) - f_2(q) = \frac{1}{780}q^5 - \frac{1}{2080}q^6 \qquad (3.86)$$

In a first approximation, we can write:

$$\varepsilon \sim h^5 \qquad (3.87)$$

or:

$$h \sim \sqrt[5]{\varepsilon} \qquad (3.88)$$

It makes sense to use the following step–size control algorithm:

$$h_{\text{new}} = \sqrt[5]{\frac{tol_{\text{rel}} \cdot \max(|x_1|, |x_2|, \delta)}{|x_1 - x_2|}} \cdot h_{\text{old}} \qquad (3.89)$$

The rational behind this algorithm is very simple. As long as:

$$\frac{|x_1 - x_2|}{\max(|x_1|, |x_2|, \delta)} = tol_{\text{rel}} \tag{3.90}$$

we got the right step size, and the step size won't change. However, as soon as $|x_1 - x_2|$ becomes too large, the step size will be reduced. On the other hand, if $|x_1 - x_2|$ becomes too small, the step size will be enlarged.

Contrary to the previously proposed algorithm, no step will ever be repeated. The algorithm accepts too large errors in a single step and just tries to prevent mishap from repeating itself. Algorithms that operate in this fashion are called *optimistic algorithms*, whereas algorithms that repeat steps exhibiting errors that are too large are called *conservative algorithms*.

The above procedure might work very well indeed if only we could trust that one of the algorithms always *overestimates* the true value, while the other always *underestimates* it, with the additional constraint that both algorithms smoothly approximate the true value as $h \rightarrow 0$.

Unfortunately, such a guarantee cannot be given. It is entirely feasible that both algorithms agree, by accident, on the same incorrect result. Thus, the algorithm can be fooled. Luckily, this doesn't happen too often in practice, and consequently, engineers are usually quite happy with this algorithm.

Kjell Gustafsson [3.11] had an even better idea. Kjell had a background in control engineering, and therefore viewed the step–size control problem from a control engineering perspective. This view is shown in Fig.3.24.

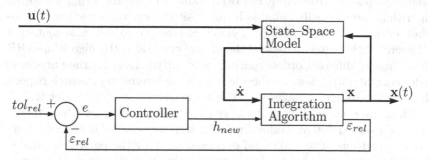

FIGURE 3.24. Step–size control viewed as a control problem.

The integration algorithm interacts with the state space model in a closed loop. We have seen that loop before. However, it also generates another output, namely the estimate of the relative integration error, ε_{rel}. This quantity is fed back and compared with the desired relative error, tol_{rel}. The resulting error signal, e, is then fed into a controller box that computes the next value of the step size, h_{new}.

The controller can be designed using standard control theory. It turns out that the previously proposed step–size adaptation rule of Eq.3.89 corresponds to a discrete proportional controller (P–controller). However, we can just as well implement either a discrete PI–controller or a discrete PID–controller. All we need to do is to modify Eq.3.89 accordingly.

Gustafsson found that a PI–controller can be implemented using the following modified step–size control algorithm:

$$ h_{\text{new}} = \left(\frac{0.8 \cdot tol_{\text{rel}}}{\varepsilon_{\text{rel}_{\text{now}}}} \right)^{\frac{0.3}{n}} \cdot \left(\frac{\varepsilon_{\text{rel}_{\text{last}}}}{\varepsilon_{\text{rel}_{\text{now}}}} \right)^{\frac{0.4}{n}} \cdot h_{\text{old}} \qquad (3.91) $$

where:

$$ \varepsilon_{\text{rel}_{\text{now}}} = \frac{\|\mathbf{x_1} - \mathbf{x_2}\|_\infty}{\max(\|\mathbf{x_1}\|_2, \|\mathbf{x_2}\|_2, \delta)} \qquad (3.92a) $$

$$ \varepsilon_{\text{rel}_{\text{last}}} = \text{same quantity one time step back} \qquad (3.92b) $$

and n is the approximation order of the integration algorithm, in the case of RKF4/5, $n = 5$.

In Fig.3.24, it was assumed that the same relative error applies to all state variables, i.e., we can operate on norms of the two state vectors rather than on individual state variables. Therefore, ε_{rel} is a scalar rather than a vector. However, the vector case could have been treated in exactly the same fashion.

In Chapter 2, we have seen that we always have a choice between a high–order algorithm with a larger step size and a low–order algorithm with a smaller step size. Although it has been shown in Chapter 3 that low–order algorithms aren't really suitable in most situations, the question remains whether order control might be a viable alternative to step–size control.

Several authors have extended the idea of embedded RK algorithms (RK algorithms of different orders sharing their early stages) for the purpose of order control [3.1]. However, beside from some interesting research papers, these efforts didn't go anywhere. The reason is simple. Low–order RK algorithms are dubious anyway. Step–size controlled low–order RKs are even worse, since the relative overhead paid for step–size control is larger for low–order algorithms. The additional overhead paid for the embedding makes these algorithms non–economic for practically all applications. Order control is a fashionable subject in multi–step integration, algorithms that will be discussed in the next chapter of this book. However, the order–control issue is, in our opinion, overrated even in the context of those algorithms.

3.10   Summary

This chapter has extended the general observations on numerical ODE integration made in Chapter 2 to a large class of higher–order integration algorithms, namely the so–called Runge–Kutta methods. Both explicit [3.3, 3.12] and implicit versions [3.13] of these algorithms were discussed, and their stability as well as accuracy properties were analyzed.

Most engineering applications call for fourth–order algorithms. The step–size controlled RKF4/5 algorithm [3.9] is easily the most popular among the explicit RK techniques. More recently, DOPRI4/5 [3.7, 3.11] became also fairly popular due to its somewhat smaller error coefficient ($97/120,000$ in comparison with $1/780$ of RKF4/5), which is paid for by one additional stage, i.e., one additional function evaluation per step.

The most popular implicit RK algorithms for stiff system simulation are the fully–implicit Radau algorithms [3.13]. These algorithms are highly efficient, and good (robust) implementations are available. They shall be discussed in greater detail in Chapter 8 of this book. Backinterpolation techniques, and in particular BI4/$5_{0.45}$, are a new and viable alternative that, in our opinion, will receive more attention in the future.

Implicit extrapolation techniques have been advocated for use on parallel processor architectures [3.6]. Parallelization of these algorithms is trivial, since the predictor stages can be computed in parallel. In this way, the number of consecutive stages of a fourth-order extrapolation method can be reduced from 10 to four. These algorithms have been successfully applied to the simulation of complex chemical processing plants, such as distillation columns with 50 trays.

F–stable algorithms for the simulation of marginally stable systems are a more specialized breed of animals in the zoo of numerical ODE integration algorithms. They will never become as popular as either the non–stiff explicit algorithms or the stiffly–stable implicit algorithms, since the number of suitable applications for these algorithms is more limited. However, these algorithms should be looked at more closely in the context of method–of–lines solutions to hyperbolic PDE problems.

3.11 References

[3.1] Dale G. Bettis. Efficient Embedded Runge–Kutta Methods. In Roland Bulirsch, Rolf Dieter Grigorieff, and Johann Schröder, editors, *Numerical Treatment of Differential Equations*, volume 631 of *Lecture Notes in Mathematics*, pages 9–18. Springer–Verlag, New York, 1976.

[3.2] Kevin Burrage. Efficiently Implementable Algebraically Stable Runge–Kutta Methods. *SIAM J. Numerical Analysis*, 19:245–258, 1982.

[3.3] John C. Butcher. *The Numerical Analysis of Ordinary Differential Equations: Runge–Kutta and General Linear Methods*. John Wiley, Chichester, United Kingdom, 1987. 512p.

[3.4] Alan R. Curtis. An Eighth Order Runge–Kutta Process With Eleven Function Evaluations Per Step. *Numerische Mathematik*, 16:268–277, 1970.

[3.5] Germund G. Dahlquist, Werner Liniger, and Olavi Nevanlinna. Stability of Two–Step Methods for Variable Integration Steps. *SIAM J. Numerical Analysis*, 20(5):1071–1085, 1983.

[3.6] Peter Deuflhard. Extrapolation Integrators for Quasilinear Implicit ODEs. In Peter Deuflhard and Björn Enquist, editors, *Large Scale Scientific Computing*, volume 7 of *Progress in Scientific Computing*, pages 37–50, Birkhäuser, Boston, Mass., 1987.

[3.7] John R. Dormand and Peter J. Prince. A Family of Embedded Runge–Kutta Formulae. *J. of Computational and Applied Mathematics*, 6(1):19–26, 1980.

[3.8] Leonhard Euler. De integratione æquationum differentialium per approximationem. In *Opera Omnia*, volume 11 of *first series*, pages 424–434. Institutiones Calculi Integralis, Teubner Verlag, Leipzig, Germany, 1913.

[3.9] Edwin Fehlberg. Classical 5th–, 6th–, 7th–, and 8th–Order Runge–Kutta Formulas. Technical Report NASA TR R–287, NASA Johnson Space Center, Houston, Texas, 1968.

[3.10] C. William Gear. *Numerical Initial Value Problems in Ordinary Differential Equations*. Series in Automatic Computation. Prentice–Hall, Englewood Cliffs, N.J., 1971. 253p.

[3.11] Kjell Gustafsson. *Control of Error and Convergence in ODE Solvers*. PhD thesis, Dept. of Automatic Control, Lund Institute of Technology, Lund, Sweden, 1992.

[3.12] Ernst Hairer, Syvert P. Nørsett, and Gerhard Wanner. *Solving Ordinary Differential Equations I: Nonstiff Problems*, volume 8 of *Series in Computational Mathematics*. Springer–Verlag, Berlin, Germany, 2nd edition, 2000. 528p.

[3.13] Ernst Hairer and Gerhard Wanner. *Solving Ordinary Differential Equations II: Stiff and Differential–Algebraic Problems*, volume 14 of *Series in Computational Mathematics*. Springer–Verlag, Berlin, Germany, 2nd edition, 1996. 632p.

[3.14] Karl Heun. Neue Methoden zur approximativen Integration der Differentialgleichungen einer unabhängigen Veränderlichen. *Zeitschrift für Mathematische Physik*, 45:23–38, 1900.

[3.15] Anton Huťa. Une amélioration de la méthode de Runge–Kutta–Nyström pour la résolution numérique des équations différentielles du premier ordre. *Acta Fac. Nat. Univ. Comenian. Math.*, 1:201–224, 1956.

[3.16] Arieh Iserles and Syvert P. Nørsett. *Order Stars*, volume 2 of *Applied Mathematics and Mathematical Computation*. Chapman & Hall, London, United Kingdom, 1991. 248p.

[3.17] Wilhelm Kutta. Beitrag zur näherungsweisen Integration totaler Differentialgleichungen. *Zeitschrift für Mathematische Physik*, 46:435–453, 1901.

[3.18] John D. Lambert. *Numerical Methods for Ordinary Differential Systems: The Initial Value Problem*. John Wiley, New York, 1991. 304p.

[3.19] Cleve Moler and Charles van Loan. Nineteen Dubious Ways to Compute the Exponential of a Matrix. *SIAM Review*, 20(4):801–836, 1978.

[3.20] Evert Johannes Nyström. Über die numerische Integration von Differentialgleichungen. *Acta Socialum Scientarum Fennicæ*, 50(13):1–55, 1925.

[3.21] Carl Runge. Über die numerische Auflösung von Differentialgleichungen. *Mathematische Annalen*, 46:167–178, 1895.

[3.22] Baylis Shanks. Solutions of Differential Equations by Evaluations of Functions. *Mathematics of Computation*, 20:21–38, 1966.

[3.23] Gerhard Wanner, Ernst Hairer, and Syvert Nørsett. Order Stars and Stability Theorems. *BIT*, 18:475–489, 1978.

[3.24] Wei Xie. Backinterpolation Methods for the Numerical Solution of Ordinary Differential Equations and Applications. Master's thesis, Dept. of Electrical & Computer Engineering, University of Arizona, Tucson, Ariz., 1995.

3.12   Homework Problems

[H3.1] Family of Explicit RK2 Algorithms

Verify that Eq.(3.10) is indeed the correct Taylor–Series expansion describing the parameterized family of all two–stage explicit RK2 algorithms.

[H3.2] Family of Explicit RK3 Algorithms

Derive the constraint equations in the α_i and β_{ij} parameters that characterize the family of all three–stage explicit RK3 algorithms.

To this end, the Taylor–series of $f(x+\Delta x, y+\Delta y)$ must now be expanded
up to the quadratic terms:

$$f(x + \Delta x, y + \Delta y) \approx f(x,y) + \frac{\partial f(x,y)}{\partial x} \cdot \Delta x + \frac{\partial f(x,y)}{\partial y} \cdot \Delta y +$$

$$\frac{\partial^2 f(x,y)}{\partial x^2} \cdot \frac{\Delta x^2}{2} + \frac{\partial^2 f(x,y)}{\partial x \cdot \partial y} \cdot \Delta x \cdot \Delta y + \frac{\partial^2 f(x,y)}{\partial y^2} \cdot \frac{\Delta y^2}{2} \quad \text{(H3.2a)}$$

[H3.3] Runge–Kutta–Simpson Algorithm

Given the four–stage Runge–Kutta algorithm characterized by the Butcher
tableau:

0	0	0	0	0
1/3	1/3	0	0	0
2/3	-1/3	1	0	0
1	1	-1	1	0
x	1/8	3/8	3/8	1/8

Write down the stages of this algorithm. Determine the linear order of
approximation accuracy of this method. How would you judge this method
in comparison with other Runge–Kutta algorithms discussed in this chapter?

[H3.4] RK Order Increase by Blending

Given two separate n^{th}–order accurate RK algorithms in at least $(n + 1)$
stages:

$$f_1(q) \;=\; 1 + q + \frac{q^2}{2!} + \cdots + \frac{q^n}{n!} + c_1 \cdot q^{n+1} \qquad \text{(H3.4a)}$$

$$f_2(q) \;=\; 1 + q + \frac{q^2}{2!} + \cdots + \frac{q^n}{n!} + c_2 \cdot q^{n+1} \qquad \text{(H3.4b)}$$

where $c_2 \neq c_1$.

Show that it is always possible to use blending:

$$\mathbf{x}^{\text{blended}} = \vartheta \cdot \mathbf{x^1} + (1 - \vartheta) \cdot \mathbf{x^2} \qquad \text{(H3.4c)}$$

where $\mathbf{x^1}$ is the solution found using method $f_1(q)$ and $\mathbf{x^2}$ is the solution
found using method $f_2(q)$, such that $\mathbf{x}^{\text{blended}}$ is of order $(n + 1)$.

Find a formula for ϑ that will make the blended algorithm accurate to
the order $(n + 1)$.

[H3.5] Stability Domains of RKF4/5

Find the stability domains of the two algorithms used in RKF4/5. Interpret the results. Is it better to use the fourth–order approximation to continue with the next step, or should the fifth–order approximation be used?

[H3.6] Runge–Kutta Integration

Given the following linear time–invariant continuous–time system:

$$
\dot{\mathbf{x}} = \begin{pmatrix}
1250 & -25113 & -60050 & -42647 & -23999 \\
500 & -10068 & -24057 & -17092 & -9613 \\
250 & -5060 & -12079 & -8586 & -4826 \\
-750 & 15101 & 36086 & 25637 & 14420 \\
250 & -4963 & -11896 & -8438 & -4756
\end{pmatrix} \cdot \mathbf{x} + \begin{pmatrix} 5 \\ 2 \\ 1 \\ -3 \\ 1 \end{pmatrix} \cdot u
$$

$$
\mathbf{y} = \begin{pmatrix} -1 & 26 & 59 & 43 & 23 \end{pmatrix} \cdot \mathbf{x} \tag{H3.6a}
$$

with initial conditions:

$$
\mathbf{x_0} = \begin{pmatrix} 1 & -2 & 3 & -4 & 5 \end{pmatrix}^T \tag{H3.6b}
$$

This is the same system that was used in Hw.[H2.1] Simulate the system across 10 seconds of simulated time with step input using the RK4 algorithm with the α–vector and β–matrix of Eq.(3.15). The following fixed step sizes should be tried:

1. $h = 0.32$,

2. $h = 0.032$,

3. $h = 0.0032$.

Plot the three trajectories on top of each other. What can you conclude about the accuracy of the results?

[H3.7] Implicit Extrapolation

Derive the α_1 and α_2 coefficients of the IEX2 method using the two approaches demonstrated in the chapter. Show that you indeed obtain the same coefficients using either of the two methods.

[H3.8] Implicit Extrapolation

Repeat Hw.[H3.6], this time using the IEX4 algorithm. Since the system to be simulated is linear, the implicit algorithm can be implemented by matrix inversion rather than by Newton iteration.

Which algorithm is more accurate for the same step size: RK4 or IEX4?

[H3.9] Explicit Integration Methods and Their Stability Domains

Prove that all explicit RK algorithms have stability domains that look qualitatively like that of FE, i.e., bend into the left–half $\lambda \cdot h$ plane. To this end, show that all explicit RK algorithms are characterized by a polynomial rather than rational $f(q)$, and analyze $f(q)$ for large values of $q \to \infty$.

[H3.10] Implicit Integration Methods and Their Stability Domains

Prove that all implicit RK algorithms with strictly proper rational $f(q)$ functions have stability domains that look qualitatively like that of BE, i.e., bend into the right–half $\lambda \cdot h$ plane.

Show furthermore that no integration algorithm with a non–strictly proper $f(q)$ function can exhibit infinite damping far away from the origin of the complex $\lambda \cdot h$ plane.

[H3.11] Stability Domains of BI4/5$_\vartheta$

Find the stability domain of BI4/5$_\vartheta$ using the approximation of Eq.(3.85a) for the explicit semi–step, and the approximation of Eq.(3.85b) for the implicit semi–step using the following ϑ values:

$$\vartheta = \{0.4, 0.45, 0.475, 0.48, 0.5\} \tag{H3.11a}$$

For this problem, it may be easier to use MATLAB's *contour* plot, than your own stability domain tracking routine.

[H3.12] BI4/5$_{0.45}$ for Linear Systems

Repeat Hw.[H3.6] this time using BI4/5$_{0.45}$. The explicit semi–step uses the fourth–order approximation of RKF4/5. There is no need to compute the fifth–order corrector. The implicit semi–step uses the fifth–order corrector. There is no need to compute the fourth–order corrector. Since the system to be simulated is linear, the implicit semi–step can be implemented using matrix inversion. No step–size control is attempted.

Compare the accuracy of this algorithm with that of RK4 and IEX4.

[H3.13] Stability Domain and Newton Iteration

Show that Newton iteration indeed does not modify the stability domain of BI2.

[H3.14] BI4/5$_{0.45}$ for Nonlinear Systems

Repeat Hw.[H3.12]. This time, we want to replace the matrix inversion by Newton iteration. Of course, since the problem is linear and time–invariant, Newton iteration and modified Newton iteration are identical. Iterate until

$\delta_{\text{rel}} \leq 10^{-5}$, where:

$$\delta_{\text{rel}} = \frac{\|\mathbf{x}_{k+\frac{1}{2}}^{\text{right}} - \mathbf{x}_{k+\frac{1}{2}}^{\text{left}}\|_\infty}{\max(\|\mathbf{x}_{k+\frac{1}{2}}^{\text{left}}\|_2, \|\mathbf{x}_{k+\frac{1}{2}}^{\text{right}}\|_2, \delta)} \tag{H3.14a}$$

Compare the results obtained with those found in Hw.[H3.12].

[H3.15] Backinterpolation With Step–Size Control

We want to repeat Hw.[H3.14] once more, this time using a step–size controlled algorithm. The step–size control to be used is the following. On the *explicit* semi–step, compute now both correctors, and find ε_{rel} according to the formula:

$$\varepsilon_{\text{rel}} = \frac{\|\mathbf{x_1} - \mathbf{x_2}\|_\infty}{\max(\|\mathbf{x_1}\|_2, \|\mathbf{x_2}\|_2, \delta)} \tag{H3.15a}$$

If $\varepsilon_{\text{rel}} \leq 10^{-4}$, use the Gustafsson algorithm to compute the step size to be used in the next step:

$$h_{\text{new}} = \left(\frac{0.8 \cdot 10^{-4}}{\varepsilon_{\text{rel}_{\text{now}}}}\right)^{0.06} \cdot \left(\frac{\varepsilon_{\text{rel}_{\text{last}}}}{\varepsilon_{\text{rel}_{\text{now}}}}\right)^{0.08} \cdot h_{\text{old}} \tag{H3.15b}$$

except during the first step, when we use:

$$h_{\text{new}} = \left(\frac{0.8 \cdot 10^{-4}}{\varepsilon_{\text{rel}_{\text{now}}}}\right)^{0.2} \cdot h_{\text{old}} \tag{H3.15c}$$

However, if $\varepsilon_{\text{rel}} > 10^{-4}$, we reject the step at once, i.e., we never even proceed to the implicit semi–step, and compute a new step size in accordance with Eq.(H3.15c).

If a step was repeated, the step size for the immediately following next step is also computed according to Eq.(H3.15c) rather than using Eq.(H3.15b).

Apply this step–size control algorithm to the same problem as before, and determine the largest global relative error by comparing the solution with the analytical solution of this linear time–invariant system.

Compute also largest global relative error of the three solutions of Hw.[H3.14].

Compute the number of floating–point operations of the step–size controlled algorithm as well as the numbers of floating–point operations of the fixed–step algorithms of Hw.[H3.14].

Use the product of accuracy and cost:

$$Q = \text{\# of floating point operations} \times \text{largest global relative error} \tag{H3.15d}$$

as a performance measure, and rank the four solutions accordingly. Is the step–size controlled algorithm economical when using this performance measure?

[H3.16] CSMP–III

The most widely used simulation software in the 70s was a program from IBM called *Continuous System Modeling Program III (CSMP–III)*. That software offered, as its default integration algorithm, the classical 4th–order Runge–Kutta algorithm presented in Eq.(3.15). For step–size control, the algorithm used an implementation of *Simpson's rule*, also known under the name of 4th–order Milne algorithm, an implicit linear multi–step method that we shall meet again in Chapter 4 of this book. The algorithm is usually written as:

$$x_{k+1} = x_{k-1} + \frac{h}{3} \cdot (f_{k+1} + 4 \cdot f_k + f_{k-1}) \qquad (H3.16a)$$

However, by shrinking the step size by a factor of two, it can also be written as:

$$x_{k+1} = x_k + \frac{h}{6} \cdot (f_{k+1} + 4 \cdot f_{k+\frac{1}{2}} + f_k) \qquad (H3.16b)$$

CSMP–III implemented this formula as a predictor–corrector technique, using two semi–steps of FE to estimate the unknown derivative values, $f_{k+\frac{1}{2}}$ and f_{k+1}.

Write down the Butcher tableau of the combined RK4/Simpson algorithm, sharing as many stages between the two algorithms as possible.

Find the linear order of approximation accuracy of this implementation of Simpson's rule. How would you characterize this method?

What can you conclude about the usefulness of this technique for step–size control of the RK4 algorithm?

[H3.17] Embedded RK Algorithms

Given the two embedded RK algorithms characterized by the following Butcher tableau:

0	0	0	0
1	1	0	0
1/2	1/4	1/4	0
x_1	1/2	1/2	0
x_2	1/6	1/6	2/3

Write down the stages of these two algorithms. Determine the linear order of approximation accuracy for each of them.

[H3.18] Accuracy Domains

Determine the accuracy domains (left–half plane only) of IEX4 and BI4/$5_{0.45}$ for $tol = 10^{-4}$, and compare them to the accuracy domain of RK4. What do you conclude?

[H3.19] Order Star

Find the damping order star for $BI4/5_{0.45}$, and plot it together with the pole and zero locations. Compare with the order star of Fig.3.17. Find the frequency order star for $BI4/5_{0.45}$, and plot it together with the pole and zero locations. Compare with the order star of Fig.3.19. Finally, compute and plot the order star accuracy domain of this method.

[H3.20] Lie-series Integration, Algebraic Differentiation

The Van–der–Pol oscillator can be described by the following 2nd–order differential equation:

$$\ddot{x} - \mu \cdot (1 - x^2) \cdot \dot{x} + x = 0 \qquad (H3.20a)$$

Write down a state–space model of the Van–der–Pol oscillator with $x_1 = x$, and $x_2 = \dot{x}$.

Create a MATLAB function:

$$[\mathbf{f}, \dot{\mathbf{f}}, \ddot{\mathbf{f}}] = vdp(\mathbf{x}) \qquad (H3.20b)$$

that computes the first, second, and third state derivative vectors. Use *algebraic differentiation* to symbolically find expressions for the higher derivatives.

We wish to simulate the Van–der–Pol oscillator with $\mu = 2.0$ during 20 seconds using a step size of $h = 0.1$ by means of *Lie–series integration*, i.e., by making direct use of the Taylor–series expansion of the exponential function:

$$\mathbf{x_{k+1}} = \mathbf{x_k} + h \cdot \mathbf{f_k} + \frac{h^2}{2} \cdot \dot{\mathbf{f}}_\mathbf{k} + \frac{h^3}{6} \cdot \ddot{\mathbf{f}}_\mathbf{k} \qquad (H3.20c)$$

Use six different sets of initial conditions:

1. $x_{10} = 0.1$, $x_{20} = 0.1$,

2. $x_{10} = 0.1$, $x_{20} = -0.1$,

3. $x_{10} = -0.1$, $x_{20} = 0.1$,

4. $x_{10} = -0.1$, $x_{20} = -0.1$,

5. $x_{10} = -2.0$, $x_{20} = 2.0$,

6. $x_{10} = 2.0$, $x_{20} = -2.0$,

and plot $x_2(t)$ as a function of $x_1(t)$ in the phase plane, superposing the six solutions onto the same graph.

3.13   Projects

[P3.1] Accuracy Domain *vs.* Order–star Accuracy Domain

Draw the accuracy domains and order–star accuracy domains for different integration algorithms, and determine the relationship between these two approaches to characterizing the accuracy of an integration algorithm.

3.14   Research

[R3.1] ϑ–Method

Study the relationship between the locations of the eigenvalues of the Jacobian matrix of the system to be simulated, and the choice of the *theta*–parameter in BI4/5$_\vartheta$.

Design a general–purpose control algorithm to modify *theta* as a function of the (usually time–dependent) eigenvalue locations of the Jacobian matrix of the system to be simulated.

[R3.2] L–Stability

Analyze the effects of the shape of the damping plot on the accuracy of a stiff system integrator. Quantify the importance of L–stability. Determine a method to systematically find L–stable algorithms that minimize the distance between σ_d and $\hat{\sigma}_d$ in a least–square sense within a reasonable range of the negative real axis of the damping plot.

4

Multi–step Integration Methods

Preview

In this chapter, we shall look at several families of integration algorithms that all have in common the fact that only a single function evaluation needs to be performed in every integration step, irrespective of the order of the algorithm. Both explicit and implicit varieties of this kind of algorithms exist and shall be discussed. As in the last chapter, we shall spend some time discussing the stability and accuracy properties of these families of integration algorithms.

Whereas step–size and order control were easily accomplished in the case of the single–step techniques, these issues are much more difficult to tackle in the case of the multi–step algorithms. Consequently, their discussion must occupy a significant portion of this chapter.

The chapter starts out with mathematical preliminaries that shall simplify considerably the subsequent derivation of the multi–step methods.

4.1 Introduction

In the last chapter, we have looked at integration algorithms that, in one way or other, all try to approximate Taylor–Series expansions of the unknown solution around the current time instant. The trick was to never compute the higher derivatives explicitly, but to replace these higher derivatives by additional function evaluations to be taken at various time instants inside the integration step.

One disadvantage of this approach is that, with every new step, we start out again with an empty slate, i.e., in each new step, we have to build up the higher–order algorithms from scratch. Isn't it a pity that, at the end of every step, all the higher–order information is thrown away again? Isn't that wasteful? Wouldn't it be possible to preserve some of this information so that, in the subsequent step, the number of function evaluations can be kept smaller? The answer to this question is a definite yes. In fact, it is possible to find entire classes of integration algorithms of arbitrary order of approximation accuracy that require only a single function evaluation in every new step, because they preserve the complete information from the previous steps. That is the topic of our discussion in this chapter.

There are many ways how these families of algorithms can be derived. However, in order to make their introduction and derivation easy, we need some additional mathematical apparatus that we shall introduce first. To this end, we shall initially not talk about numerical integration at all. Instead, we shall focus our interest on higher–order interpolation (extrapolation) polynomials.

4.2   Newton–Gregory Polynomials

Given a function of time, $f(t)$. We shall denote the values of this function at various points in time, t_0, t_1, t_2, etc. as f_0, f_1, f_2, etc. We shall introduce Δ as a *forward difference operator*, thus, $\Delta f_0 = f_1 - f_0$, $\Delta f_1 = f_2 - f_1$,   etc.
Higher–order forward difference operators can be defined accordingly:

$$\Delta^2 f_0 = \Delta(\Delta f_0) = \Delta(f_1 - f_0) = \Delta f_1 - \Delta f_0 = f_2 - 2f_1 + f_0 \qquad (4.1a)$$

$$\Delta^3 f_0 = \Delta(\Delta^2 f_0) = f_3 - 3f_2 + 3f_1 - f_0 \qquad (4.1b)$$

etc.

In general:

$$\Delta^n f_i = f_{i+n} - n \cdot f_{i+n-1} + \frac{n(n-1)}{2!} \cdot f_{i+n-2} - \frac{n(n-1)(n-2)}{3!} \cdot f_{i+n-3} + \cdots \qquad (4.2)$$

or:

$$\Delta^n f_i = \binom{n}{0} f_{i+n} - \binom{n}{1} f_{i+n-1} + \binom{n}{2} f_{i+n-2} - \binom{n}{3} f_{i+n-3} + \cdots \pm \binom{n}{n} f_i \qquad (4.3)$$

Let us now assume that the time points at which the f_i values are given are a fixed distance h apart from each other. We wish to find an interpolation (extrapolation) polynomial of nth order that passes through the $(n+1)$ given function values f_0, f_1, f_2, $\ldots$, f_n at the given time instants t_0, $t_1 = t_0 + h$, $t_2 = t_0 + 2h$, $\ldots$, $t_n = t_0 + n \cdot h$.
Let us introduce an auxiliary variable s defined as follows:

$$s = \frac{t - t_0}{h} \qquad (4.4)$$

Thus, for $t = t_0 \leftrightarrow s = 0.0$, for $t = t_1 \leftrightarrow s = 1.0$, etc. The real–valued variable s assumes integer values at the sampling points. At those points, the value of s corresponds to the index of the sampling point.
The desired interpolation polynomial can be written as a function of s:

$$f(s) \approx \binom{s}{0} f_0 + \binom{s}{1} \Delta f_0 + \binom{s}{2} \Delta^2 f_0 + \cdots + \binom{s}{n} \Delta^n f_0 \qquad (4.5)$$

This is called a *Newton–Gregory forward polynomial*. It is easy to prove that this polynomial indeed possesses the desired qualities. First of all, it is clearly an nth–order polynomial in s. Since s is linear in t, it is also an nth–order polynomial in t. By plugging in integer values of s in the range 0 to n, we can verify easily that the polynomial indeed passes through f_0 to f_n. Since there exists exactly one n^{th}–order polynomial that passes through any given set of $(n+1)$ points, the assertion has been proven.

Sometimes, it is more useful to have an nth–order polynomial that passes through $(n+1)$ time points in the past. The *Newton–Gregory backward polynomial* can be written as:

$$f(s) \approx f_0 + \binom{s}{1} \Delta f_{-1} + \binom{s+1}{2} \Delta^2 f_{-2} + \binom{s+2}{3} \Delta^3 f_{-3} + \ldots$$
$$+ \binom{s+n-1}{n} \Delta^n f_{-n} \qquad (4.6)$$

It is equally easy to show that this n^{th}–order polynomial passes through the $(n+1)$ points f_0, f_{-1}, f_{-2}, $\ldots$, f_{-n} at the time instants t_0, t_{-1}, t_{-2}, $\ldots$, t_{-n} by plugging in values of $s = 0.0$, $s = -1.0$, $s = -2.0$, $\ldots$, $s = -n$.

It is common practice to also introduce a *backward difference operator*, ∇, defined as:

$$\nabla f_i = f_i - f_{i-1} \qquad (4.7)$$

with the higher–order operators:

$$\nabla^2 f_i = \nabla(\nabla f_i) = \nabla(f_i - f_{i-1}) = \nabla f_i - \nabla f_{i-1}$$
$$= f_i - 2 f_{i-1} + f_{i-2} \qquad (4.8a)$$
$$\nabla^3 f_i = \nabla(\nabla^2 f_i) = f_i - 3f_{i-1} + 3f_{i-2} - f_{i-3} \qquad (4.8b)$$
etc.

or, in general:

$$\nabla^n f_i = \binom{n}{0} f_i - \binom{n}{1} f_{i-1} + \binom{n}{2} f_{i-2} - \binom{n}{3} f_{i-3} + \cdots \pm \binom{n}{n} f_{i-n} \qquad (4.9)$$

The Newton–Gregory backward polynomial can be expressed in terms of the ∇–operator as:

$$f(s) \approx f_0 + \binom{s}{1}\nabla f_0 + \binom{s+1}{2}\nabla^2 f_0 + \binom{s+2}{3}\nabla^3 f_0 + \cdots + \binom{s+n-1}{n}\nabla^n f_0$$
$$(4.10)$$

It is also quite common to introduce yet another operator, namely the *shift operator*, $\mathcal{E}$. It is defined as:

$$\mathcal{E} f_i = f_{i+1} \qquad (4.11)$$

with the higher–order operators:

$$\mathcal{E}^2 f_i = \mathcal{E}(\mathcal{E} f_i) = \mathcal{E}(f_{i+1}) = f_{i+2} \qquad (4.12a)$$
$$\mathcal{E}^3 f_i = \mathcal{E}(\mathcal{E}^2 f_i) = \mathcal{E}(f_{i+2}) = f_{i+3} \qquad (4.12b)$$

etc.

It is obviously true that:

$$\Delta f_i = \mathcal{E} f_i - f_i = (\mathcal{E} - 1) f_i \qquad (4.13a)$$
$$\nabla f_i = f_i - \mathcal{E}^{-1} f_i = (1 - \mathcal{E}^{-1}) f_i \qquad (4.13b)$$
$$\mathcal{E}(\nabla f_i) = \mathcal{E}(f_i - f_{i-1}) = f_{i+1} - f_i = \Delta f_i \qquad (4.13c)$$

By abstraction:

$$\Delta = \mathcal{E} - 1 \qquad (4.14a)$$
$$\nabla = 1 - \mathcal{E}^{-1} \qquad (4.14b)$$
$$\Delta = \mathcal{E}\nabla \qquad (4.14c)$$

Since these are all linear operators, we can formally calculate with them as with other algebraic quantities. In particular:

$$\Delta^n = (\mathcal{E} - 1)^n = \mathcal{E}^n - n\mathcal{E}^{n-1} + \binom{n}{2}\mathcal{E}^{n-2} - + \cdots \pm \binom{n}{n-1}\mathcal{E} \mp 1 \quad (4.15)$$

Using this calculus, the derivation of the two Newton–Gregory polynomials becomes trivial.

$$f(s) \approx \mathcal{E}^s f_0 = (1 + \Delta)^s f_0 = \left[1 + \binom{s}{1}\Delta + \binom{s}{2}\Delta^2 + \binom{s}{3}\Delta^3 + \cdots \right] f_0$$
$$(4.16)$$

is the Newton–Gregory forward polynomial, and:

$$f(s) \approx (1 - \nabla)^{-s} f_0 = \left[1 + \binom{s}{1} \nabla + \binom{s+1}{2} \nabla^2 + \binom{s+2}{3} \nabla^3 + \dots \right] f_0$$

$$(4.17)$$

is the Newton–Gregory backward polynomial.

Since differentiation is also a linear operation, we can find the first time derivative of $f(t)$ in the following manner:

$$\dot{f}(t) = \frac{d}{dt} f(t) = \frac{\partial}{\partial s} f(s) \cdot \frac{ds}{dt}$$

$$\approx \frac{1}{h} \cdot \frac{\partial}{\partial s} \left(f_0 + s \Delta f_0 + \frac{s(s-1)}{2!} \Delta^2 f_0 + \dots \right) \qquad (4.18)$$

and in particular:

$$\dot{f}(t_0) \approx \frac{1}{h} \cdot \left(\Delta f_0 - \frac{1}{2} \Delta^2 f_0 + \frac{1}{3} \Delta^3 f_0 - \dots \pm \frac{1}{n} \Delta^n f_0 \right) \qquad (4.19)$$

We introduce a new operator, the *differentiation operator*, $\mathcal{D}$, as:

$$\mathcal{D} = \frac{1}{h} \cdot \left(\Delta - \frac{1}{2} \Delta^2 + \frac{1}{3} \Delta^3 - \dots \pm \frac{1}{n} \Delta^n \right) \qquad (4.20)$$

Consequently, we can compute the second derivative as:

$$\mathcal{D}^2 = \frac{1}{h^2} \cdot \left(\Delta - \frac{1}{2} \Delta^2 + \frac{1}{3} \Delta^3 - \dots \pm \frac{1}{n} \Delta^n \right)^2$$

$$= \frac{1}{h^2} \cdot \left(\Delta^2 - \Delta^3 + \frac{11}{12} \Delta^4 - \frac{5}{6} \Delta^5 + \dots \right) \qquad (4.21)$$

etc.

A more thorough discussion of these and other interpolation polynomials can be found in [4.7]. However, for our purpose, the material presented here will suffice.

4.3 Numerical Integration Through Polynomial Extrapolation

The idea behind multi–step integration is straightforward. We can employ a Newton–Gregory backward polynomial setting $t_k = t_0$ and evaluating for $s = 1.0$. This should give us an estimate of $x(t_{k+1}) = f_1$. The back values f_0, f_{-1}, f_{-2}, etc. are the previously computed solutions $x(t_k)$, $x(t_{k-1})$, $x(t_{k-2})$, etc. Until here, we have written the Newton–Gregory polynomials for the scalar case, but the concept extends without complications also to the vector case.

The trick is to somehow modify the notation of the Newton–Gregory backward polynomial such that values of $\dot{f}$ are used beside from the values of f in order to incorporate the knowledge available through the state–space model, but such that higher derivatives, as $\ddot{f}$, are avoided, since they are difficult to compute accurately.

4.4 Explicit Adams–Bashforth Formulae

Let us write a Newton–Gregory backward polynomial for the state derivative vector $\dot{\mathbf{x}}(t)$ around the time point t_k:

$$\dot{\mathbf{x}}(t) = \mathbf{f}_k + \binom{s}{1}\nabla\mathbf{f}_k + \binom{s+1}{2}\nabla^2\mathbf{f}_k + \binom{s+2}{3}\nabla^3\mathbf{f}_k + \ldots \qquad (4.22)$$

where:

$$\mathbf{f}_k = \dot{\mathbf{x}}(t_k) = \mathbf{f}(\mathbf{x}(t_k), t_k) \qquad (4.23)$$

is the state derivative vector at time t_k. We wish to find an expression for $\mathbf{x}(t_{k+1})$. Therefore, we need to integrate Eq.(4.22) in the interval $[t_k, t_{k+1}]$:

$$\int_{t_k}^{t_{k+1}} \dot{\mathbf{x}}(t)dt = \mathbf{x}(t_{k+1}) - \mathbf{x}(t_k)$$

$$= \int_{t_k}^{t_{k+1}} \left[\mathbf{f}_k + \binom{s}{1}\nabla\mathbf{f}_k + \binom{s+1}{2}\nabla^2\mathbf{f}_k + \binom{s+2}{3}\nabla^3\mathbf{f}_k + \ldots \right] dt$$

$$= \int_{0.0}^{1.0} \left[\mathbf{f}_k + \binom{s}{1}\nabla\mathbf{f}_k + \binom{s+1}{2}\nabla^2\mathbf{f}_k + \binom{s+2}{3}\nabla^3\mathbf{f}_k + \ldots \right] \cdot \frac{dt}{ds} \cdot ds$$

$$(4.24)$$

Thus:

$$\mathbf{x}(t_{k+1}) = \mathbf{x}(t_k) + h \int_0^1 \left[\mathbf{f}_k + s\nabla\mathbf{f}_k + \left(\frac{s^2}{2} + \frac{s}{2}\right)\nabla^2\mathbf{f}_k \right.$$

$$\left. + \left(\frac{s^3}{6} + \frac{s^2}{2} + \frac{s}{3}\right)\nabla^3\mathbf{f}_k + \ldots \right] ds \qquad (4.25)$$

and therefore:

$$\mathbf{x}(t_{k+1}) = \mathbf{x}(t_k) + h\left(\mathbf{f}_k + \frac{1}{2}\nabla\mathbf{f}_k + \frac{5}{12}\nabla^2\mathbf{f}_k + \frac{3}{8}\nabla^3\mathbf{f}_k + \dots\right) \qquad (4.26)$$

If we truncate Eq.(4.26) after the quadratic term and expand the ∇–operators, we obtain:

$$\mathbf{x}(t_{k+1}) = \mathbf{x}(t_k) + \frac{h}{12}\left(23\mathbf{f}_k - 16\mathbf{f}_{k-1} + 5\mathbf{f}_{k-2}\right) \qquad (4.27)$$

which is the well–known third–order Adams–Bashforth algorithm, abbreviated as AB3. Since the expressions on the right are multiplied by h, we obtain a third–order approximation by truncating the infinite series after the quadratic term.

If we truncate Eq.(4.26) only after the cubic term, we obtain:

$$\mathbf{x}(t_{k+1}) = \mathbf{x}(t_k) + \frac{h}{24}\left(55\mathbf{f}_k - 59\mathbf{f}_{k-1} + 37\mathbf{f}_{k-2} - 9\mathbf{f}_{k-3}\right) \qquad (4.28)$$

which is the AB4 algorithm.

Also these algorithms can be represented through an α–vector and a β–matrix. These are:

$$\alpha = \begin{pmatrix} 1 & 2 & 12 & 24 & 720 & 1440 \end{pmatrix}^T \qquad (4.29a)$$

$$\beta = \begin{pmatrix} 1 & 0 & 0 & 0 & 0 & 0 \\ 3 & -1 & 0 & 0 & 0 & 0 \\ 23 & -16 & 5 & 0 & 0 & 0 \\ 55 & -59 & 37 & -9 & 0 & 0 \\ 1901 & -2774 & 2616 & -1274 & 251 & 0 \\ 4277 & -7923 & 9982 & -7298 & 2877 & -475 \end{pmatrix} \qquad (4.29b)$$

Here, the ith row contains the coefficients of the ABi algorithm, i.e., the coefficients of the ith order Adams–Bashforth algorithm. The ith row of the β–matrix contains the multipliers of the $\mathbf{f}$–vectors at different time points, and the ith row of the α–vector contains the common denominator, i.e., the divider of h.

All algorithms within the class of ABi algorithms are *explicit* algorithms. Of course, AB1 is:

$$\mathbf{x}(t_{k+1}) = \mathbf{x}(t_k) + \frac{h}{1}\left(1\mathbf{f}_k\right) \qquad (4.30)$$

which is immediately recognized as the FE–algorithm. There exists only one explicit first–order algorithm, namely the *Forward Euler* algorithm.

Let us now look at the stability domains of the ABi algorithms. However, before we can do so, we must find the $\mathbf{F}$–matrices of the ABi algorithms. Let us look at AB3 for example. Applying Eq.(4.27) to the linear homogeneous problem, we find:

$$\mathbf{x}(t_{k+1}) = \left[\mathbf{I}^{(n)} + \frac{23}{12}\mathbf{A}h\right]\mathbf{x}(t_k) - \frac{4}{3}\mathbf{A}h\ \mathbf{x}(t_{k-1}) + \frac{5}{12}\mathbf{A}h\ \mathbf{x}(t_{k-2}) \quad (4.31)$$

We can transform the third–order vector differential equation into three first–order vector differential equations with the substitutions:

$$\mathbf{z_1}(t_k) = \mathbf{x}(t_{k-2}) \quad\quad (4.32a)$$
$$\mathbf{z_2}(t_k) = \mathbf{x}(t_{k-1}) \quad\quad (4.32b)$$
$$\mathbf{z_3}(t_k) = \mathbf{x}(t_k) \quad\quad (4.32c)$$

With these substitutions, we find:

$$\mathbf{z_1}(t_{k+1}) = \mathbf{z_2}(t_k) \quad\quad\quad\quad (4.33a)$$
$$\mathbf{z_2}(t_{k+1}) = \mathbf{z_3}(t_k) \quad\quad\quad\quad (4.33b)$$
$$\mathbf{z_3}(t_{k+1}) = \frac{5}{12}\mathbf{A}h\ \mathbf{z_1}(t_k) - \frac{4}{3}\mathbf{A}h\ \mathbf{z_2}(t_k) + \left[\mathbf{I}^{(n)} + \frac{23}{12}\mathbf{A}h\right]\mathbf{z_3}(t_k) \quad (4.33c)$$

or, in a matrix form:

$$\mathbf{z}(t_{k+1}) = \begin{pmatrix} \mathbf{O}^{(n)} & \mathbf{I}^{(n)} & \mathbf{O}^{(n)} \\ \mathbf{O}^{(n)} & \mathbf{O}^{(n)} & \mathbf{I}^{(n)} \\ \frac{5}{12}\mathbf{A}h & -\frac{4}{3}\mathbf{A}h & (\mathbf{I}^{(n)} + \frac{23}{12}\mathbf{A}h) \end{pmatrix} \cdot \mathbf{z}(t_k) \quad (4.34)$$

Thus, for a 2×2 $\mathbf{A}$–matrix, we obtain a $2i \times 2i$ $\mathbf{F}$–matrix for ABi. The stability domains that result when plugging these $\mathbf{F}$–matrices into the general routine of Chapter 2 are shown in Fig.4.1.

As the ABi methods are explicit algorithms, their borders of stability must loop into the left–half complex $\lambda \cdot h$–plane. This was to be expected. Unfortunately, the stability domains of the ABi algorithms look very disappointing. We proceed to higher orders of approximation accuracy, in order to use *larger* step sizes . . . yet, the stability domains *shrink*! AB7 is even totally unstable.

As we proceed to higher orders, the step size will very soon be limited by the stability domain rather than by the accuracy requirements. In comparison with the RK algorithms, it is true that we need only one function evaluation per step, yet, we probably will have to use considerably smaller step sizes due to the disappointingly small stable regions in the left–half $\lambda \cdot h$–plane.

The reasons for these unfortunate results are easy to understand. It is not true that higher–order polynomials necessarily lead to a more accurate interpolation everywhere. They only allow us to fit more points precisely. In between these points, higher–order polynomials have a tendency

Stability Domains of AB

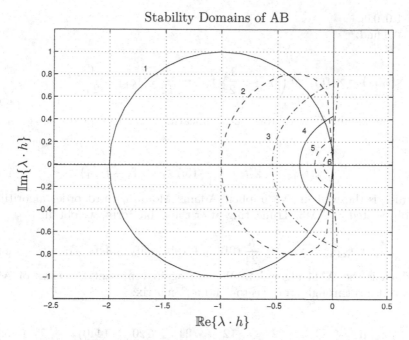

FIGURE 4.1. Stability domains of explicit AB algorithms.

to oscillate. Worse, while higher–order polynomial *interpolation* may still be acceptable, higher–order polynomial *extrapolation* is a disaster. These polynomials have a tendency to deviate quickly from the approximated curve outside the interpolation interval. Unfortunately, extrapolation is what numerical integration is all about.

The previous paragraph indicates that the discovered shortcoming of this class of algorithms will not be limited to the explicit Adams–Bashforth methods, but is an inherent disease of *all* multi–step integration algorithms.

4.5   Implicit Adams–Moulton Formulae

Let us check whether we have more luck with implicit multi–step algorithms. To this end, we again develop $\dot{\mathbf{x}}(t)$ into a Newton–Gregory backward polynomial, however this time, we shall develop the polynomial around the point t_{k+1}.

$$\dot{\mathbf{x}}(t) = \mathbf{f}_{k+1} + \binom{s}{1}\nabla\mathbf{f}_{k+1} + \binom{s+1}{2}\nabla^2\mathbf{f}_{k+1} + \binom{s+2}{3}\nabla^3\mathbf{f}_{k+1} + \dots \quad (4.35)$$

We integrate again from time t_k to time t_{k+1}. However, this time, $s = 0.0$ corresponds to $t = t_{k+1}$, thus, we need to integrate across the range $s \in$

$[-1.0, 0.0]$.

We find:

$$\mathbf{x}(t_{k+1}) = \mathbf{x}(t_k) + h \left(\mathbf{f}_{k+1} - \frac{1}{2}\nabla\mathbf{f}_{k+1} - \frac{1}{12}\nabla^2\mathbf{f}_{k+1} - \frac{1}{24}\nabla^3\mathbf{f}_{k+1} + \dots \right) \tag{4.36}$$

Expanding the ∇–operators and truncating after the quadratic term, we find:

$$\mathbf{x}(t_{k+1}) = \mathbf{x}(t_k) + \frac{h}{12} \left(5\mathbf{f}_{k+1} + 8\mathbf{f}_k - \mathbf{f}_{k-1} \right) \tag{4.37}$$

which is the well–known implicit Adams–Moulton third–order algorithm, abbreviated as AM3. Truncating after the cubic term, we obtain:

$$\mathbf{x}(t_{k+1}) = \mathbf{x}(t_k) + \frac{h}{24} \left(9\mathbf{f}_{k+1} + 19\mathbf{f}_k - 5\mathbf{f}_{k-1} + \mathbf{f}_{k-2} \right) \tag{4.38}$$

which is the AM4 algorithm. We can again represent the class of AMi algorithms through an α–vector and a β–matrix:

$$\alpha = \begin{pmatrix} 1 & 2 & 12 & 24 & 720 & 1440 \end{pmatrix} \tag{4.39a}$$

$$\beta = \begin{pmatrix} 1 & 0 & 0 & 0 & 0 & 0 \\ 1 & 1 & 0 & 0 & 0 & 0 \\ 5 & 8 & -1 & 0 & 0 & 0 \\ 9 & 19 & -5 & 1 & 0 & 0 \\ 251 & 646 & -264 & 106 & -19 & 0 \\ 475 & 1427 & -798 & 482 & -173 & 27 \end{pmatrix} \tag{4.39b}$$

Clearly, AM1 is the same as BE. This was to be expected since there can exist only one implicit first–order integration algorithm. AM2 is the trapezoidal rule, thus while AM1 is L–stable, AM2 is F–stable.

Let us now look at the stability domains of the higher–order AMi formulae. Plugging the linear homogeneous system into AM3, we find:

$$\left[\mathbf{I}^{(n)} - \frac{5}{12}\mathbf{A}h \right] \mathbf{x}(t_{k+1}) = \left[\mathbf{I}^{(n)} + \frac{2}{3}\mathbf{A}h \right] \mathbf{x}(t_k) - \frac{1}{12}\mathbf{A}h\, \mathbf{x}(t_{k-1}) \tag{4.40}$$

Using the same substitution as in the case of the ABi formulae, we find:

$$\mathbf{F} = \begin{pmatrix} \mathbf{O}^{(n)} & \mathbf{I}^{(n)} \\ -[\mathbf{I}^{(n)} - \frac{5}{12}\mathbf{A}h]^{-1} \cdot [\frac{1}{12}\mathbf{A}h] & [\mathbf{I}^{(n)} - \frac{5}{12}\mathbf{A}h]^{-1} \cdot [\mathbf{I}^{(n)} + \frac{2}{3}\mathbf{A}h] \end{pmatrix} \tag{4.41}$$

Thus, for a 2×2 $\mathbf{A}$–matrix, we obtain a $2(i-1) \times 2(i-1)$ $\mathbf{F}$–matrix for AMi. The stability domains that result when plugging these $\mathbf{F}$–matrices into the general routine of Chapter 2 are shown in Fig.4.2.

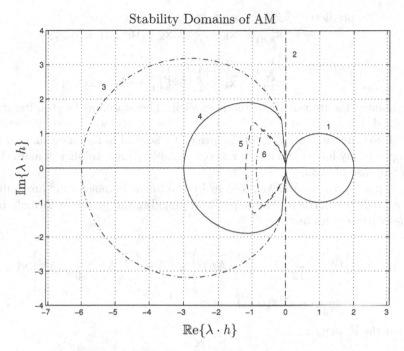

FIGURE 4.2. Stability domains of implicit AM algorithms.

As in the case of the ABi algorithms, the results are disappointing. AM1 and AM2 are useful algorithms ... but they were already known to us under different names. Starting from the third–order, the stability domains of the AMi algorithms loop again into the left–half $\lambda \cdot h$–plane. It is unclear to us why we should want to pay the high price of Newton iteration, if we don't get a stiffly–stable technique after all.

4.6 Adams–Bashforth–Moulton Predictor–Corrector Formulae

The ABi algorithms were rejected due to their miserably small stable regions in the left–half $\lambda \cdot h$–plane. The AMi algorithms, on the other hand, were rejected because they are implicit, yet not stiffly–stable. Maybe all is not lost yet. We can try a compromise between ABi and AMi. Let us construct a predictor–corrector method with one step of ABi as a predictor, and one step of AMi as a corrector. For example, the (third–order accurate) ABM3 algorithm would look as follows:

predictor: $\dot{\mathbf{x}}_k = \mathbf{f}(\mathbf{x}_k, t_k)$

$\mathbf{x}_{k+1}^P = \mathbf{x}_k + \frac{h}{12}(23\dot{\mathbf{x}}_k - 16\dot{\mathbf{x}}_{k-1} + 5\dot{\mathbf{x}}_{k-2})$

corrector: $\dot{\mathbf{x}}_{k+1}^P = \mathbf{f}(\mathbf{x}_{k+1}^P, t_{k+1})$

$\mathbf{x}_{k+1}^C = \mathbf{x}_k + \frac{h}{12}(5\dot{\mathbf{x}}_{k+1}^P + 8\dot{\mathbf{x}}_k - \dot{\mathbf{x}}_{k-1})$

Evidently, the overall algorithm is explicit. Therefore, no Newton iteration is needed, and consequently, the fact that the method won't be stiffly–stable is of no concern. However, for the price of a second function evaluation per step, we may have bargained for a considerably larger stability domain than in the case of AB3.

Replacing the nonlinear problem by the linear homogeneous problem in the predictor–corrector technique, and plugging the predictor formula into the corrector, we find:

$$\mathbf{x}(t_{k+1}) = \left[\mathbf{I}^{(n)} + \frac{13}{12}\mathbf{A}h + \frac{115}{144}(\mathbf{A}h)^2\right]\mathbf{x}(t_k) - \left[\frac{1}{12}\mathbf{A}h + \frac{5}{9}(\mathbf{A}h)^2\right]\mathbf{x}(t_{k-1})$$

$$+ \frac{25}{144}(\mathbf{A}h)^2\mathbf{x}(t_{k-2}) \tag{4.42}$$

with the $\mathbf{F}$–matrix:

$$\mathbf{F} = \begin{pmatrix} \mathbf{O}^{(n)} & \mathbf{I}^{(n)} & \mathbf{O}^{(n)} \\ \mathbf{O}^{(n)} & \mathbf{O}^{(n)} & \mathbf{I}^{(n)} \\ \frac{25}{144}(\mathbf{A}h)^2 & -\left[\frac{1}{12}\mathbf{A}h + \frac{5}{9}(\mathbf{A}h)^2\right] & \left[\mathbf{I}^{(n)} + \frac{13}{12}\mathbf{A}h + \frac{115}{144}(\mathbf{A}h)^2\right] \end{pmatrix} \tag{4.43}$$

The stability domains of some ABMi algorithms are shown in Fig.4.3.

Indeed, the approach worked. The stability domains of the ABMi algorithms are considerably larger than those of the ABi algorithms, although they are still considerably smaller than those of the AMi algorithms — especially for orders three and four. Since Newton iteration takes usually about three iterations per step, i.e., three additional function evaluations in the case of these multi–step algorithms, ABMi is about twice as expensive as ABi, and AMi is about twice as expensive as ABMi. Thus, if ABMi allows us to use a step size that is at least twice as large as the step size we could employ when using ABi, the predictor–corrector method becomes economical. If AMi allows us to use a step size that is at least four times as large as the step size we could employ when using ABi, the implicit algorithm becomes economical in spite of the need for Newton iteration.

4.7   Backward Difference Formulae

Let us check whether we can find a set of multi–step formulae whose stability domains loop into the right–half $\lambda \cdot h$–plane. This time, we write the

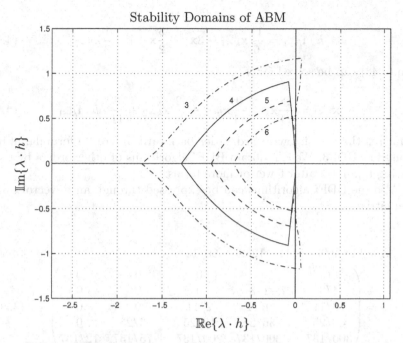

FIGURE 4.3. Stability domains of predictor–corrector ABM algorithms.

Newton–Gregory backward polynomial in $\mathbf{x}(t)$ rather than in $\dot{\mathbf{x}}(t)$ around the time instant t_{k+1}. Thus:

$$\mathbf{x}(t) = \mathbf{x}_{k+1} + \binom{s}{1}\nabla\mathbf{x}_{k+1} + \binom{s+1}{2}\nabla^2\mathbf{x}_{k+1} + \binom{s+2}{3}\nabla^3\mathbf{x}_{k+1} + \dots \quad (4.44)$$

or:

$$\mathbf{x}(t) = \mathbf{x}_{k+1} + s\nabla\mathbf{x}_{k+1} + \left(\frac{s^2}{2} + \frac{s}{2}\right)\nabla^2\mathbf{x}_{k+1} + \left(\frac{s^3}{6} + \frac{s^2}{2} + \frac{s}{3}\right)\nabla^3\mathbf{x}_{k+1} + \dots$$
$$(4.45)$$

We now compute the derivative of Eq.(4.45)) with respect to time:

$$\dot{\mathbf{x}}(t) = \frac{1}{h}\left[\nabla\mathbf{x}_{k+1} + \left(s + \frac{1}{2}\right)\nabla^2\mathbf{x}_{k+1} + \left(\frac{s^2}{2} + s + \frac{1}{3}\right)\nabla^3\mathbf{x}_{k+1} + \dots\right]$$
$$(4.46)$$

We evaluate Eq.(4.46) for $s = 0.0$, and obtain:

$$\dot{\mathbf{x}}(t_{k+1}) = \frac{1}{h}\left[\nabla\mathbf{x}_{k+1} + \frac{1}{2}\nabla^2\mathbf{x}_{k+1} + \frac{1}{3}\nabla^3\mathbf{x}_{k+1} + \dots\right] \quad (4.47)$$

Multiplying Eq.4.47 with h, truncating after the cubic term, and expanding the ∇–operators, we obtain:

$$h \cdot \mathbf{f}_{k+1} = \frac{11}{6}\mathbf{x}_{k+1} - 3\mathbf{x}_k + \frac{3}{2}\mathbf{x}_{k-1} - \frac{1}{3}\mathbf{x}_{k-2} \qquad (4.48)$$

Eq.(4.48) can be solved for $\mathbf{x}_{k+1}$:

$$\mathbf{x}_{k+1} = \frac{18}{11}\mathbf{x}_k - \frac{9}{11}\mathbf{x}_{k-1} + \frac{2}{11}\mathbf{x}_{k-2} + \frac{6}{11} \cdot h \cdot \mathbf{f}_{k+1} \qquad (4.49)$$

which is the well–known third–order backward difference formula, abbreviated as BDF3. We can obtain BDFi algorithms of other orders by truncating Eq.(4.47) after fewer or more terms.

Also the BDFi algorithms can be expressed through an α–vector and a β–matrix:

$$\alpha = \begin{pmatrix} 1 & 2/3 & 6/11 & 12/25 & 60/137 \end{pmatrix}^T \qquad (4.50a)$$

$$\beta = \begin{pmatrix} 1 & 0 & 0 & 0 & 0 \\ 4/3 & -1/3 & 0 & 0 & 0 \\ 18/11 & -9/11 & 2/11 & 0 & 0 \\ 48/25 & -36/25 & 16/25 & -3/25 & 0 \\ 300/137 & -300/137 & 200/137 & -75/137 & 12/137 \end{pmatrix} \qquad (4.50b)$$

where the ith row represents the BDFi algorithm. The coefficients of the β–matrix are here the multipliers of past values of the state vector $\mathbf{x}$, whereas the coefficients of the α–vector are the multipliers of the state derivative vector $\dot{\mathbf{x}}$ at time t_{k+1}. The BDF techniques are implicit algorithms, thus clearly, BDF1 is the same as BE.

The stability domains of the BDFi algorithms are presented in Fig.4.4.

It becomes evident at once that, finally, we have found a set of stiffly–stable multi–step algorithms. Unfortunately, they (not unexpectedly) also suffer from the high–order polynomial extrapolation disease. As the order of the extrapolation polynomials grows, the methods become less and less stable. Although the stability domains loop into the right–half $\lambda \cdot h$–plane, they are pulled over more and more into the left–half plane. BDF6 (not shown on Fig.4.4) has only a very narrow band of stable area to the left of the origin. BDF6 is thus only useful for simulation of problems with all eigenvalues strictly on the negative real axis, such as method–of–lines solutions to parabolic PDEs. BDF7, is unstable in the entire $\lambda \cdot h$–plane.

Yet, due to the simplicity of these techniques, the BDFi algorithms are today easily the most widely used stiff system solvers on the market. In the engineering literature, these algorithms are often called *Gear algorithms*, after Bill Gear who discovered their stiffly–stable properties [4.5]. The most widely used code based o the BDF formulae is DASSL. DASSL is the default simulation algorithm used in Dymola. We shall talk more about DASSL in Chapter 8 of this book.

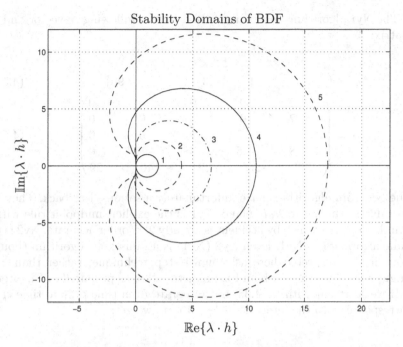

FIGURE 4.4. Stability domains of implicit BDF algorithms.

By evaluating Eq.(4.46) for $s = -1.0$, we can obtain a series of explicit BDFi algorithms. Unfortunately, they are not useful, since they are all unstable in the entire $\lambda \cdot h$–plane.

4.8 Nyström and Milne Algorithms

There exist two more classes of multi–step techniques that are sometimes talked about in the numerical ODE literature, the explicit *Nyström techniques* [4.10], and the implicit *Milne methods* [4.10]. Let us derive them and look at their stability behavior.

We start again out with Eq.(4.22). However this time, we integrate from t_{k-1} to t_{k+1}, thus, from $s = -1.0$ to $s = +1.0$. We find:

$$\mathbf{x}(t_{k+1}) = \mathbf{x}(t_{k-1}) + h \left(2\mathbf{f}_k + \frac{1}{3}\nabla^2 \mathbf{f}_k + \frac{1}{3}\nabla^3 \mathbf{f}_k + \dots \right) \qquad (4.51)$$

The term in $\nabla \mathbf{f}_k$ drops out. Truncating Eq.(4.51) after the cubic term and expanding the ∇–operators, we obtain:

$$\mathbf{x}(t_{k+1}) = \mathbf{x}(t_{k-1}) + \frac{h}{3} \left(8\mathbf{f}_k - 5\mathbf{f}_{k-1} + 4\mathbf{f}_{k-2} - \mathbf{f}_{k-3} \right) \qquad (4.52)$$

which is the fourth–order Nyström algorithm, abbreviated as Ny4.

The Nyi algorithms are characterized by the following α–vector and β–matrix:

$$\alpha = \begin{pmatrix} 1 & 1 & 3 & 3 & 90 \end{pmatrix}^T \tag{4.53a}$$

$$\beta = \begin{pmatrix} 2 & 0 & 0 & 0 & 0 \\ 2 & 0 & 0 & 0 & 0 \\ 7 & -2 & 1 & 0 & 0 \\ 8 & -5 & 4 & -1 & 0 \\ 269 & -266 & 294 & -146 & 29 \end{pmatrix} \tag{4.53b}$$

The Nyström algorithms have unfortunately a serious drawback. They are unstable in the entire $\lambda \cdot h$–plane. Ny1 is the explicit midpoint rule with a double step size, which by accident is already 2nd–order accurate. Ny2 is the same algorithm as Ny1. Even Ny2 (Ny1) is an unstable algorithm though, since it is interpreted here as a multi–step technique, rather than as a single–step algorithm with an FE predictor step, as proposed in Chapter 3.

If we start out with Eq.4.35, but integrate from time t_{k-1} to time t_{k+1}, corresponding to the interval $s \in [-2.0, 0.0]$, we get:

$$\mathbf{x}(t_{k+1}) = \mathbf{x}(t_{k-1}) + h \left(2\mathbf{f}_{k+1} - 2\nabla\mathbf{f}_{k+1} + \frac{1}{3}\nabla^2\mathbf{f}_{k+1} + 0\nabla^3\mathbf{f}_{k+1} + \ldots \right) \tag{4.54}$$

This time around, the term in $\nabla^3\mathbf{f}_{k+1}$ drops out. Truncating Eq.4.54 after the cubic term (the quadratic term really) and expanding the ∇–operators, we find:

$$\mathbf{x}(t_{k+1}) = \mathbf{x}(t_{k-1}) + \frac{h}{3} \left(\mathbf{f}_{k+1} + 4\mathbf{f}_k + \mathbf{f}_{k-1} \right) \tag{4.55}$$

which is the implicit fourth–order *Milne algorithm*, abbreviated as Mi4. The same algorithm is also known under the name of *Simpson's rule*.

The α–vector and β–matrix for the Mii algorithms are as follows:

$$\alpha = \begin{pmatrix} 1 & 1 & 3 & 3 & 90 \end{pmatrix}^T \tag{4.56a}$$

$$\beta = \begin{pmatrix} 2 & 0 & 0 & 0 & 0 \\ 0 & 2 & 0 & 0 & 0 \\ 1 & 4 & 1 & 0 & 0 \\ 1 & 4 & 1 & 0 & 0 \\ 29 & 124 & 24 & 4 & -1 \end{pmatrix} \tag{4.56b}$$

Mi1 is recognizable as backward Euler with a double step size. Mi2 is by accident explicit, since the coefficient of $\mathbf{f}_{k+1}$ drops out. Mi2 is the same as Ny2, i.e., the explicit midpoint rule with a double step size.

Mi4 is truly remarkable. Due to a combination of "lucky" circumstances, a lot of terms dropped away, leading to a fourth–order accurate multi–step methods with only two memory elements (two past values are used in the algorithm). It is truly regrettable that our "good fortune" comes at a high price. The stability domain of Mi4 is extremely small — it consists of the origin only (!) Therefore, while Simpson's rule is very fashionable for quadrature problems (to numerically determine the integral of a function), it is entirely useless for solving differential equations. The higher–order Milne formulae are all unstable as well.

Nyström and Milne formulae are sometimes useful as partners within predictor–corrector methods. The fact that these formulae are unstable by themselves does not preclude the possibility that they may be combined with other formulae either in a predictor-corrector scheme, or in a blended method, or in a cyclic method, thereby leading to perfectly usable algorithms with appropriate stability properties.

4.9   In Search for Stiffly–stable Methods

Until now, we used the Newton–Gregory polynomials to derive multi–step algorithms. This technique has its advantages. In particular, it generates the integration algorithms using the ∇–operator, which is useful. However, the technique called for lots of symbolic or at least semi–symbolic operations that are hard to implement in MATLAB in search for new algorithms with pre–specified stability and/or accuracy properties.

To this end, let us introduce another technique that can alternatively be used to derive the coefficients of multi–step integration algorithms. Let us derive once again the BDF3 algorithm.

We know that every n^{th}–order multi–step algorithm is defined through an nth–order polynomial fitted through $(n + 1)$ points, a mixture of state values and state derivative values, at the time points t_{k+1}, t_k, t_{k-1}, etc. Let us write this polynomial once again in the variable s, assuming that $s = 1.0$ corresponds to $t = t_{k+1}$, $s = 0.0$ corresponds to $t = t_k$, etc. The polynomial can be written as:

$$p(s) = a_0 + a_1 \, s + a_2 \, s^2 + a_3 \, s^3 + \cdots + a_n \, s^n \qquad (4.57)$$

in the yet unknown coefficients a_i. The time derivative of $p(s)$ can be written as:

$$h \cdot \dot{p}(s) = a_1 + 2a_2 \, s + 3a_3 \, s^2 + \cdots + n \, a_n \, s^{n-1} \qquad (4.58)$$

In the case of BDF3, we know that $p(0) = x_k, p(-1) = x_{k-1}, p(-2) = x_{k-2}$, and $h \cdot \dot{p}(+1) = h \cdot f_{k+1}$. This gives us four equations in the four unknowns a_0, a_1, a_2, and a_3. These are:

$$h \cdot f_{k+1} = a_1 + 2a_2 + 3a_3 \tag{4.59a}$$

$$x_k = a_0 \tag{4.59b}$$

$$x_{k-1} = a_0 - a_1 + a_2 - a_3 \tag{4.59c}$$

$$x_{k-2} = a_0 - 2a_1 + 4a_2 - 8a_3 \tag{4.59d}$$

or, in a matrix form:

$$\begin{pmatrix} h \cdot f_{k+1} \\ x_k \\ x_{k-1} \\ x_{k-2} \end{pmatrix} = \begin{pmatrix} 0 & 1 & 2 & 3 \\ 1 & 0 & 0 & 0 \\ 1 & -1 & 1 & -1 \\ 1 & -2 & 4 & -8 \end{pmatrix} \cdot \begin{pmatrix} a_0 \\ a_1 \\ a_2 \\ a_3 \end{pmatrix} \tag{4.60}$$

which can be solved for the unknown parameter vector by matrix inversion. We wish to evaluate:

$$x_{k+1} = p(+1) = a_0 + a_1 + a_2 + a_3 \tag{4.61}$$

Thus, we simply add up the elements in each column of the inverse matrix, and receive directly the desired coefficients of BDF3:

$$x_{k+1} = \frac{6}{11}h \cdot f_{k+1} + \frac{18}{11}x_k - \frac{9}{11}x_{k-1} + \frac{2}{11}x_{k-2} \tag{4.62}$$

This procedure can easily be automated in a MATLAB function.

You had been told that BDF6 is not a very useful technique due to its narrow corridor of stable territory to the left of the origin. The method is (A,α)–stable, but only with $\alpha = 19°$. Let us ascertain whether the above outlined procedure allows us to find a better sixth–order stiffly–stable algorithm than BDF6.

Obviously, any sixth–order linear multi–step method can be written as:

$$p(s) = a_0 + a_1\, s + a_2\, s^2 + a_3\, s^3 + a_4\, s^4 + a_5\, s^5 + a_6\, s^6 \tag{4.63}$$

in seven unknown parameters. Consequently, we must provide seven solution points through which the polynomial will be fitted. In the past, we talked about the high–order extrapolation disease. Maybe, it will work better if we shorten the tail of the algorithm by providing both values for x and for f at $s = 0$, at $s = -1$, and at $s = -2$. Clearly, the list of data points *must* contain $f(t_{k+1})$ in order for the method to be implicit. The rationale for this idea is that the interpolated curve may look more like a polynomial over a shorter time span.

Thus:

$$
\begin{pmatrix}
h \cdot f_{k+1} \\
x_k \\
h \cdot f_k \\
x_{k-1} \\
h \cdot f_{k-1} \\
x_{k-2} \\
h \cdot f_{k-2}
\end{pmatrix}
= \mathbf{M} \cdot \mathbf{a}
\tag{4.64}
$$

where:

$$
\mathbf{M} =
\begin{pmatrix}
0 & 1 & 2 & 3 & 4 & 5 & 6 \\
1 & 0 & 0 & 0 & 0 & 0 & 0 \\
0 & 1 & 0 & 0 & 0 & 0 & 0 \\
1 & -1 & 1 & -1 & 1 & -1 & 1 \\
0 & 1 & -2 & 3 & -4 & 5 & -6 \\
1 & -2 & 4 & -8 & 16 & -32 & 64 \\
0 & 1 & -4 & 12 & -32 & 80 & -192
\end{pmatrix}
\tag{4.65}
$$

Computing the inverse of $\mathbf{M}$ and adding up columns, we find:

$$
\mathbf{x_{k+1}} = \frac{3}{11} h \cdot \mathbf{f_{k+1}} - \frac{27}{11} \mathbf{x_k} + \frac{27}{11} h \cdot \mathbf{f_k} + \frac{27}{11} \mathbf{x_{k-1}} + \frac{27}{11} h \cdot \mathbf{f_{k-1}} + \mathbf{x_{k-2}} + \frac{3}{11} h \cdot \mathbf{f_{k-2}}
\tag{4.66}
$$

This is a beautiful new method, it is clearly sixth–order accurate, and it has only one single drawback . . . it is unfortunately unstable in the entire $\lambda \cdot h$–plane (!)

Thus, we need to expand our search. Let us allow values to be included as far back as $t = t_{k-5}$, corresponding to $s = -5$. Since we definitely want $\mathbf{f_{k+1}}$ to be included and since we can pick either $\mathbf{x}$–values or $\mathbf{f}$–values otherwise, we need to pick six items out of 12. This gives us 924 different methods to try.

I quickly programmed this problem in MATLAB, calculating the $\mathbf{F}$–matrices for all 924 techniques, and checking for each of them whether or not its stability domain intersects with the positive real axis, making it a potential candidate for a stiffly–stable algorithm.

Most of the 924 methods are entirely unstable. Others behave like Adams–Moulton. Only six out of the 924 methods have an intersection of their respective stability domains with the positive real axis. These are:

$$
\mathbf{x_{k+1}} = \frac{20}{49} h \cdot \mathbf{f_{k+1}} + \frac{120}{49} \mathbf{x_k} - \frac{150}{49} \mathbf{x_{k-1}} + \frac{400}{147} \mathbf{x_{k-2}} - \frac{75}{49} \mathbf{x_{k-3}}
$$
$$
+ \frac{24}{49} \mathbf{x_{k-4}} - \frac{10}{147} \mathbf{x_{k-5}}
\tag{4.67a}
$$

$$x_{k+1} = \frac{308}{745} h \cdot f_{k+1} + \frac{1776}{745} x_k - \frac{414}{149} x_{k-1} + \frac{944}{447} x_{k-2} - \frac{87}{149} x_{k-3}$$
$$- \frac{288}{745} h \cdot f_{k-4} - \frac{2}{15} x_{k-5} \qquad (4.67b)$$

$$x_{k+1} = \frac{8820}{21509} h \cdot f_{k+1} + \frac{52200}{21509} x_k - \frac{63900}{21509} x_{k-1} + \frac{400}{157} x_{k-2}$$
$$- \frac{28575}{21509} x_{k-3} + \frac{6984}{21509} x_{k-4} + \frac{600}{21509} h \cdot f_{k-5} \qquad (4.67c)$$

$$x_{k+1} = \frac{179028}{432845} h \cdot f_{k+1} + \frac{206352}{86569} x_k - \frac{34452}{12367} x_{k-1} + \frac{26704}{12367} x_{k-2}$$
$$- \frac{65547}{86569} x_{k-3} - \frac{83808}{432845} h \cdot f_{k-4} + \frac{24}{581} h \cdot f_{k-5} \qquad (4.67d)$$

$$x_{k+1} = \frac{12}{29} h \cdot f_{k+1} + \frac{1728}{725} x_k - \frac{81}{29} x_{k-1} + \frac{64}{29} x_{k-2} - \frac{27}{29} x_{k-3}$$
$$+ \frac{97}{725} x_{k-5} + \frac{12}{145} h \cdot f_{k-5} \qquad (4.67e)$$

$$x_{k+1} = \frac{30}{71} h \cdot f_{k+1} + \frac{162}{71} x_k - \frac{675}{284} x_{k-1} + \frac{100}{71} x_{k-2} - \frac{54}{71} x_{k-4}$$
$$+ \frac{127}{284} x_{k-5} + \frac{15}{71} h \cdot f_{k-5} \qquad (4.67f)$$

Among those six methods, Eq.(4.67a) is the well-known BDF6 technique. The methods of Eq.(4.67b) and Eq.(4.67f) are useless, since their stability domains also intersect with the negative real axis. The stability domains of the survivors are shown in Fig.4.5.

How can we evaluate the relative merits of these four algorithms against each other? One useful criterium is the angle α of the $A(\alpha)$ stability. Gear [4.5] proposed an alternate method to judge the stability of a stiffly–stable method consisting of two parameters, the distance a away from the imaginary axis that the stability domain reaches into the left–half $\lambda \cdot h$–plane, and the distance c away from the negative real axis, which defines the closest distance of the stability domain to the negative real axis. These three parameters are shown in Fig.4.6.

Yet, we shall need to look at accuracy as well. It has become customary [4.10] to judge the accuracy of a multi–step method in the following way. Any multi–step method can be written in the form:

$$x_{k+1} = \sum_{i=0}^{\ell} a_i \cdot x_{k-i} + \sum_{i=-1}^{\ell} b_i \cdot h \cdot f_{k-i} \qquad (4.68)$$

where ℓ is a suitably large index to include the entire history needed for the method. We can take all terms to the left–hand side of the equal sign, and shift the equation by ℓ steps into the future. By doing so, we obtain:

$$\sum_{i=0}^{m} \alpha_i \cdot x_{k+i} + h \cdot \sum_{i=0}^{m} \beta_i \cdot f_{k+i} = 0 \qquad (4.69)$$

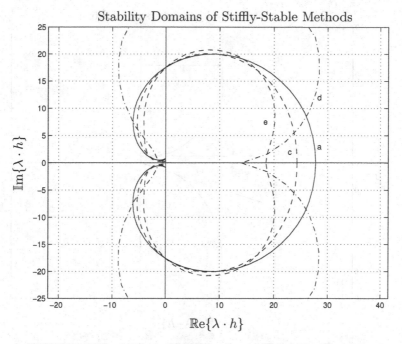

FIGURE 4.5. Stability domains of some stiffly–stable algorithms.

where α_i, β_i, and m can be easily expressed in terms of the previously used parameters a_i, b_i, and ℓ. Assuming that our system is *smooth*, i.e., the solution is continuous and continuously differentiable at least n times, where n is the order of the integration algorithm, we can develop $\mathbf{x_{k+i}}$ and $\mathbf{f_{k+i}}$ into Taylor series around $\mathbf{x_k}$ and $\mathbf{f_k}$, and come up with an expression in $\mathbf{x_k}$ and its derivatives:

$$c_0 \cdot \mathbf{x_k} + c_1 \cdot h \cdot \mathbf{\dot{x}_k} + \cdots + c_q \cdot h^q \cdot \mathbf{x_k^{(q)}} + \ldots \qquad (4.70)$$

where $\mathbf{x_k^{(q)}}$ is the qth time derivative of $\mathbf{x_k}$. The coefficients can be expressed in terms of the previously used parameters α_i, β_i, and m as follows:

$$c_0 = \sum_{i=0}^{m} \alpha_i \qquad (4.71)$$

$$c_1 = \sum_{i=0}^{m} (i \cdot \alpha_i - \beta_i) \qquad (4.72)$$

$$\vdots \qquad (4.73)$$

$$c_q = \sum_{i=0}^{m} \left(\frac{1}{q!} \cdot i^q \cdot \alpha_i - \frac{1}{(q-1)!} \cdot i^{q-1} \cdot \beta_i \right), \quad q = 2, 3, \ldots \ (4.74)$$

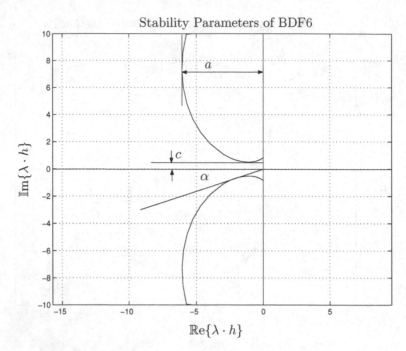

FIGURE 4.6. Stability parameters of a stiffly–stable algorithm.

Since the function that has been developed into a Taylor series is the zero function, all of these coefficients ought to be equal to zero. However, since the approximation is only n^{th}–order accurate, the coefficients for $q > n$ may be different from zero. Hence we can define the dominant of those coefficients as the *error coefficient* of the multi–step integration algorithm:

$$c_{err} = \sum_{i=0}^{m} \left(\frac{1}{(n+1)!} \cdot i^{n+1} \cdot \alpha_i - \frac{1}{n!} \cdot i^n \cdot \beta_i \right) \qquad (4.75)$$

A small value of the error coefficient is indicative of a good n^{th}–order multi–step formula.

We may also wish to look at the damping properties of the algorithm. The *damping plot*, that had been introduced in Chapter 3 of this book, can easily be extended to multi-step methods by redefining the *discrete damping* as:

$$\sigma_d = -\log(\max(\text{abs}(\text{eig}(\mathbf{F})))) \qquad (4.76)$$

The damping plots of BDF6 and the other three surviving algorithms are shown in Fig.4.7. The top graph shows the damping plots as depicted in the past. The bottom graph shows the same plots using a semi–logarithmic scale for the independent variable.

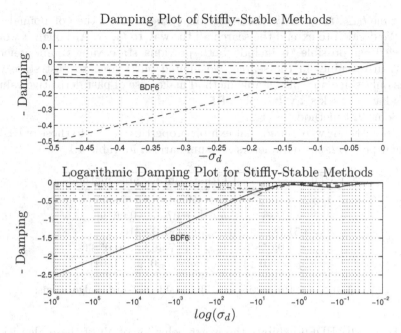

FIGURE 4.7. Damping plots of BDF6 and other 6th–order stiffly–stable algorithms.

The top graph of Fig.4.7 shows that the *asymptotic region* of BDF6 is clearly larger than that of its three competitors. The graph shows furthermore that the discrete damping of the method exhibits a sharp bend at the place where it starts deviating from the analytical damping, i.e., at approximately $\sigma_d = -0.125$. This bend is caused by a *spurious eigenvalue* taking over at that point. This is a new phenomenon that we didn't observe in the case of the single–step algorithms. Since the **F**–matrix is of a larger size than the **A**–matrix, it has more eigenvalues, which may become dominant eventually.

The bottom graph shows that BDF6 is *L–stable*, whereas its three contenders are not. Although the concept of L–stability is somewhat overrated in the numerical ODE literature (at $\sigma_d = 10^6$, the discrete damping of the BDF6 method assumes only a value of $\hat{\sigma}_d = 2.5$), this is still a nice property for a stiffly–stable method to possess.

Hence, and in spite of all our efforts, we haven't hit a mark yet. BDF6 is still the winner.

Looking at the surviving algorithms, we notice at once that all of them make use of the entire interpolation span. Quite obviously, it was a lousy idea to try to shorten the tail of the algorithm. The reason for this result is also understandable. By extending the interpolation range, the relative distance of extrapolation necessary to predict $\mathbf{x_{k+1}}$ becomes shorter. This is beneficial. Thus, let us extend this idea, and allow also interpolation points at time t_{k-6}, t_{k-7}, etc without increasing the order of the polynomial.

We decided to extend the search all the way to t_{k-11}. Although it would have been possible to include both previous state values and previous derivative values in the search, we limited our search to past state values only, since no stiffly–stable method makes use of past derivative values. We have to pick six values out of 12.

Some 924 methods later ...

314 methods were found that exhibit properties similar to those of BDF6. Their performance parameters are summarized in Table 4.1.

BDF6	Other stiffly–stable methods
$\alpha = 19°$	$\alpha \in [19°, 48°]$
$a = -6.0736$	$a \in [-6.0736, -0.6619]$
$c = 0.5107$	$c \in [0.2250, 0.8316]$
$c_{err} = -0.0583$	$c_{err} \in [-7.4636, -0.0583]$
$as.reg. = -0.14$	$as.reg. \in [-0.30, -0.01]$

TABLE 4.1. Properties of stiffly–stable 6th–order algorithms.

Evidently, BDF6 exhibits the worst behavior of all of these algorithms w.r.t. its α and a values. Yet, BDF6 is characterized by the smallest error coefficient. Unfortunately, as the length of the tail of the algorithm grows, so does the error coefficient. The c parameter is somewhere in the middle range, and so is the asymptotic region, which we defined as the value of σ_d, where $|\hat{\sigma}_d - \sigma_d| = 0.01$.

How can these 314 algorithms be rank–ordered? To this end, we shall need to define a performance index, something along the lines of:

$$P.I._{\cdot i} = \frac{|\alpha_i|}{\|\alpha\|} - \frac{|a_i|}{\|a\|} + \frac{|c_i|}{\|c\|} - k \cdot \frac{|c_{err_i}|}{\|c_{err}\|} + \frac{|as.reg._i|}{\|as.reg.\|} = max! \qquad (4.77)$$

where each of the five parameters is normalized to make them comparable with each other. A k–factor was assigned to the error coefficient to be able to vary the importance of the error coefficient within the overall performance index. We chose a value of $k = 20$.

The best three methods are now compared with BDF6. Their coefficients are given by:

$$\mathbf{x_{k+1}} = \frac{72}{167} h \cdot \mathbf{f_{k+1}} + \frac{2592}{1169} \mathbf{x_k} - \frac{2592}{1169} \mathbf{x_{k-1}} + \frac{1152}{835} \mathbf{x_{k-2}}$$
$$- \frac{324}{835} \mathbf{x_{k-3}} + \frac{81}{5845} \mathbf{x_{k-7}} - \frac{32}{5845} \mathbf{x_{k-8}} \qquad (4.78a)$$
$$\mathbf{x_{k+1}} = \frac{420}{977} h \cdot \mathbf{f_{k+1}} + \frac{19600}{8793} \mathbf{x_k} - \frac{2205}{977} \mathbf{x_{k-1}} + \frac{1400}{977} \mathbf{x_{k-2}}$$

$$-\frac{1225}{2931}x_{k-3} + \frac{40}{2931}x_{k-6} - \frac{7}{8793}x_{k-9} \qquad (4.78b)$$

$$x_{k+1} = \frac{44}{103}h \cdot f_{k+1} + \frac{5808}{2575}x_k - \frac{242}{103}x_{k-1} + \frac{484}{309}x_{k-2}$$

$$-\frac{363}{721}x_{k-3} + \frac{242}{7725}x_{k-5} - \frac{4}{18025}x_{k-10} \qquad (4.78c)$$

Table 4.2 summarizes the five performance parameters of these methods.

BDF6	SS6a	SS6b	SS6c
$\alpha = 19°$	$\alpha = 45°$	$\alpha = 44°$	$\alpha = 43°$
$a = -6.0736$	$a = -2.6095$	$a = -2.7700$	$a = -3.0839$
$c = 0.5107$	$c = 0.7994$	$c = 0.8048$	$c = 0.8156$
$c_{err} = -0.0583$	$c_{err} = -0.1478$	$c_{err} = -0.1433$	$c_{err} = -0.1343$
$as.reg. = -0.14$	$as.reg. = -0.21$	$as.reg. = -0.21$	$as.reg. = -0.21$

TABLE 4.2. Properties of stiffly–stable 6th–order algorithms.

The stability domains of the four methods are presented in Fig.4.8. The damping plots are shown in Fig.4.9.

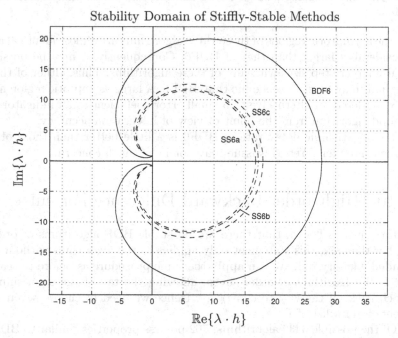

FIGURE 4.8. Stability domains of BDF6 and other 6th–order stiffly–stable algorithms.

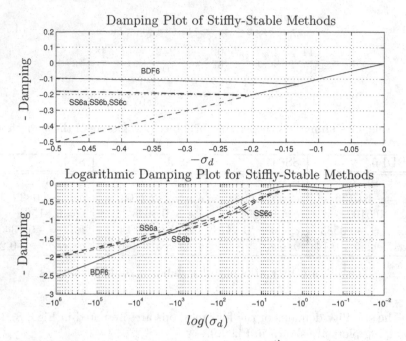

FIGURE 4.9. Damping plots of BDF6 and other 6th–order stiffly–stable algorithms.

The asymptotic regions of these three algorithms are almost identical and considerably larger than that of BDF6. Consequently, it may be possible to use larger step sizes with any of these algorithms. Thus, either of these methods may be more economic than BDF6. A large asymptotic region may be considered an alternative to a small error coefficient as an indicator for a good algorithm from the point of view of integration accuracy.

The logarithmic decay rate of BDF6 is a little better than those of its three contenders. Yet, this issue may not be of much concern.

4.10   High–order Backward Difference Formulae

Although it is known that there are no stable BDF algorithms of orders higher than six, this statement only applies to the algorithms without extended memory tail. We can apply the same procedure as before to search for BDF algorithms of order seven, allowing the tail of the algorithm to reach back all the way to e.g. t_{k-13}. Hence we need to choose seven elements from a list of 14.

Of the possible 3432 algorithms, 762 possess properties similar to BDF6, i.e., they are $A(\alpha)$–stable and also L–stable. The search was limited to algorithms with $\alpha \geq 10°$. Their performance parameters are tabulated in Table 4.3.

BDF6	7th–order stiffly–stable methods
$\alpha = 19°$	$\alpha \in [10°, 48°]$
$a = -6.0736$	$a \in [-6.1261, -0.9729]$
$c = 0.5107$	$c \in [0.0811, 0.7429]$
$c_{err} = -0.0583$	$c_{err} \in [-6.6498, -0.1409]$
$as.reg. = -0.14$	$as.reg. \in [-0.23, -0.01]$

TABLE 4.3. Properties of stiffly–stable 7th–order algorithms.

The smallest error coefficient is now almost three times larger than in the case of the 6th–order algorithms. The other parameters are comparable in their ranges with those of the 6th–order algorithms.

This time, we used a value of $k = 15$ in Eq.(4.77). The best three algorithms are characterized by the following sets of coefficients:

$$x_{k+1} = \frac{5148}{12161} h \cdot f_{k+1} + \frac{552123}{243220} x_k - \frac{200772}{85127} x_{k-1} + \frac{184041}{121610} x_{k-2}$$
$$- \frac{184041}{425635} x_{k-3} + \frac{20449}{1702540} x_{k-8} - \frac{4563}{851270} x_{k-10} + \frac{99}{121610} x_{k-12}$$
$$(4.79a)$$

$$x_{k+1} = \frac{234}{551} h \cdot f_{k+1} + \frac{13689}{6061} x_k - \frac{492804}{212135} x_{k-1} + \frac{4056}{2755} x_{k-2}$$
$$- \frac{4563}{11020} x_{k-3} + \frac{169}{19285} x_{k-8} - \frac{507}{121220} x_{k-11} + \frac{54}{30305} x_{k-12}$$
$$(4.79b)$$

$$x_{k+1} = \frac{3276}{7675} h \cdot f_{k+1} + \frac{17199}{7675} x_k - \frac{191646}{84425} x_{k-1} + \frac{596232}{422125} x_{k-2}$$
$$- \frac{74529}{191875} x_{k-3} + \frac{1183}{191875} x_{k-8} - \frac{882}{422125} x_{k-12} + \frac{2106}{2110625} x_{k-13}$$
$$(4.79c)$$

Their performance parameters are tabulated in Table 4.4.

BDF6	SS7a	SS7b	SS7c
$\alpha = 19°$	$\alpha = 37°$	$\alpha = 39°$	$\alpha = 35°$
$a = -6.0736$	$a = -3.0594$	$a = -2.9517$	$a = -3.2146$
$c = 0.5107$	$c = 0.6352$	$c = 0.6664$	$c = 0.6331$
$c_{err} = -0.0583$	$c_{err} = -0.3243$	$c_{err} = -0.3549$	$c_{err} = -0.3136$
$as.reg. = -0.14$	$as.reg. = -0.15$	$as.reg. = -0.16$	$as.reg. = -0.15$

TABLE 4.4. Properties of stiffly–stable 7th–order algorithms.

The error coefficients of these methods have grown quite a bit, but luckily, the asymptotic regions haven't shrunk yet significantly.

The stability domains of the three methods are presented in Fig.4.10, where they are also compared to that of BDF6. The damping plots are shown in Fig.4.11.

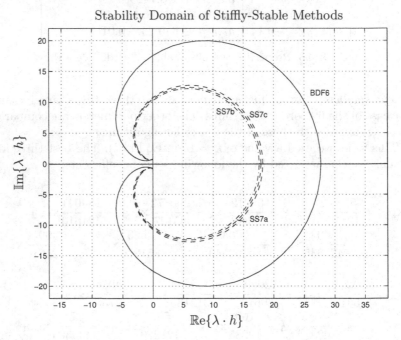

FIGURE 4.10. Stability domains of BDF6 and a set of 7th–order stiffly–stable algorithms.

The three algorithms are very similar indeed in almost every respect, and they should also perform quite similarly in simulations.

We can now proceed to algorithms of 8th order. We searched for algorithms with tails reaching all the way back to t_{k-15}. Hence we had to choose 8 elements out of a list of 16 candidates. Of the possible 12870 algorithms, 493 exhibit properties similar to those of BDF6.

Their performance parameters are tabulated in Table 4.5.

BDF6	8th–order stiffly–stable methods
$\alpha = 19°$	$\alpha \in [10°, 48°]$
$a = -6.0736$	$a \in [-5.3881, -1.4382]$
$c = 0.5107$	$c \in [0.0859, 0.6485]$
$c_{err} = -0.0583$	$c_{err} \in [-6.4014, -0.4416]$
$as.reg. = -0.14$	$as.reg. \in [-0.16, -0.01]$

TABLE 4.5. Properties of stiffly–stable 8th–order algorithms.

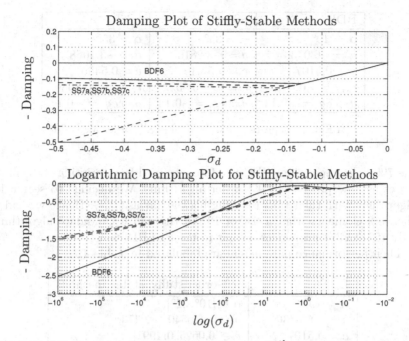

FIGURE 4.11. Damping plots of BDF6 and a set of 7th order stiffly stable algorithms.

The smallest error coefficient has unfortunately again grown by about a factor of three, and this time, also the largest asymptotic region has begun to shrink.

This time around, we used a factor of $k = 10$ in Eq.(4.77). Two among the best of these algorithms are characterized by the following sets of coefficients:

$$
\begin{aligned}
x_{k+1} ={}& \frac{112}{267}h \cdot f_{k+1} + \frac{71680}{31239}x_k - \frac{2800}{1157}x_{k-1} + \frac{179200}{114543}x_{k-2} \\
& - \frac{3920}{8811}x_{k-3} + \frac{112}{12015}x_{k-9} - \frac{160}{12727}x_{k-13} + \frac{7168}{572715}x_{k-14} \\
& - \frac{35}{10413}x_{k-15}
\end{aligned}
$$
(4.80a)

$$
\begin{aligned}
x_{k+1} ={}& \frac{208}{497}h \cdot f_{k+1} + \frac{216320}{93933}x_k - \frac{93600}{38269}x_{k-1} + \frac{16640}{10437}x_{k-2} \\
& - \frac{67600}{147609}x_{k-3} + \frac{5408}{469665}x_{k-9} - \frac{1280}{147609}x_{k-12} + \frac{3328}{574035}x_{k-14} \\
& - \frac{65}{31311}x_{k-15}
\end{aligned}
$$
(4.80b)

Their performance parameters are tabulated in Table 4.6.

Their stability domains and damping plots look almost the same as for

BDF6	SS8a	SS8b
$\alpha = 19°$	$\alpha = 35°$	$\alpha = 35°$
$a = -6.0736$	$a = -3.2816$	$a = -3.4068$
$c = 0.5107$	$c = 0.5779$	$c = 0.5456$
$c_{err} = -0.0583$	$c_{err} = -0.9322$	$c_{err} = -0.8636$
$as.reg. = -0.14$	$as.reg. = -0.14$	$as.reg. = -0.13$

TABLE 4.6. Properties of stiffly–stable 8th–order algorithms.

the 7th–order algorithms. We therefore refrained from printing these plots.

Let us now proceed to 9th–order algorithms. We decided to search for algorithms with their tails reaching back as far as t_{k-17}. Hence we had to choose 9 elements from a list of 18. Of the 48620 candidate algorithms, only 152 exhibit properties similar to those of BDF6.

Their performance parameters are tabulated in Table 4.7.

BDF6	9th–order stiffly–stable methods
$\alpha = 19°$	$\alpha \in [10°, 32°]$
$a = -6.0736$	$a \in [-5.0540, -2.4730]$
$c = 0.5107$	$c \in [0.0625, 0.4991]$
$c_{err} = -0.0583$	$c_{err} \in [-5.9825, -1.2492]$
$as.reg. = -0.14$	$as.reg. \in [-0.10, -0.02]$

TABLE 4.7. Properties of stiffly–stable 9th–order algorithms.

The smallest error coefficient has once again grown by about a factor of three, and also the largest asymptotic region has now shrunk significantly.

This time, we used a factor of $k = 5$ in Eq.(4.77). Two among the best of these algorithms are characterized by the following sets of coefficients:

$$
\begin{aligned}
x_{k+1} =& \frac{4080}{9947}h \cdot f_{k+1} + \frac{165240}{69629}x_k - \frac{16854480}{6336239}x_{k-1} + \frac{1664640}{905177}x_{k-2} \\
&- \frac{5618160}{9956947}x_{k-3} + \frac{23120}{1462209}x_{k-8} - \frac{332928}{9956947}x_{k-14} + \frac{351135}{6336239}x_{k-15} \\
&- \frac{29160}{905177}x_{k-16} + \frac{1360}{208887}x_{k-17}
\end{aligned}
$$

(4.81a)

$$
\begin{aligned}
x_{k+1} =& \frac{1904}{4651}h \cdot f_{k+1} + \frac{719712}{302315}x_k - \frac{62424}{23255}x_{k-1} + \frac{6214656}{3325465}x_{k-2} \\
&- \frac{873936}{1511575}x_{k-3} + \frac{18496}{1046475}x_{k-8} - \frac{249696}{16627325}x_{k-13} + \frac{7803}{302315}x_{k-15} \\
&- \frac{6048}{302315}x_{k-16} + \frac{952}{209295}x_{k-17}
\end{aligned}
$$

(4.81b)

Their performance parameters are tabulated in Table 4.8.

BDF6	SS9a	SS9b
$\alpha = 19°$	$\alpha = 18°$	$\alpha = 18°$
$a = -6.0736$	$a = -4.3280$	$a = -4.3321$
$c = 0.5107$	$c = 0.3957$	$c = 0.3447$
$c_{err} = -0.0583$	$c_{err} = -1.7930$	$c_{err} = -1.6702$
$as.reg. = -0.14$	$as.reg. = -0.10$	$as.reg. = -0.08$

TABLE 4.8. Properties of stiffly–stable 9th–order algorithms.

Their stability domains and damping plots look almost the same as for the 7th and 8th–order algorithms.

In this section, a number of new algorithms have been developed and presented that extend the concept of BDF algorithms to higher orders.

4.11   Newton Iteration

As we have seen, many of the truly interesting multi–step algorithms are implicit. Let us look once more at BDF3.

$$\mathbf{x_{k+1}} = \frac{6}{11}h \cdot \mathbf{f_{k+1}} + \frac{18}{11}\mathbf{x_k} - \frac{9}{11}\mathbf{x_{k-1}} + \frac{2}{11}\mathbf{x_{k-2}} \qquad (4.82)$$

Plugging in the α–vector and the β–matrix of the BDF3 method, we find:

$$\mathbf{x_{k+1}} = \alpha_3 \, h \cdot \mathbf{f_{k+1}} + \beta_{31} \, \mathbf{x_k} + \beta_{32} \, \mathbf{x_{k-1}} + \beta_{33} \, \mathbf{x_{k-2}} \qquad (4.83)$$

Thus, the ith–order BDFi algorithm can be written as:

$$\mathbf{x_{k+1}} = \alpha_i \, h \cdot \mathbf{f_{k+1}} + \sum_{j=1}^{i} \beta_{ij} \, \mathbf{x_{k-j+1}} \qquad (4.84)$$

Plugging in the linear homogeneous problem and solving for $\mathbf{x_{k+1}}$, we find:

$$\mathbf{x_{k+1}} = - \left[\alpha_i \cdot (\mathbf{A} \cdot h) - \mathbf{I^{(n)}} \right]^{-1} \sum_{j=1}^{i} \beta_{ij} \, \mathbf{x_{k-j+1}} \qquad (4.85)$$

On a nonlinear problem, we cannot apply matrix inversion. We can rewrite Eq.(4.84) as:

$$\mathcal{F}(\mathbf{x_{k+1}}) = \alpha_i \, h \cdot \mathbf{f}(\mathbf{x_{k+1}}, t_{k+1}) - \mathbf{x_{k+1}} + \sum_{j=1}^{i} \beta_{ij} \, \mathbf{x_{k-j+1}} = 0.0 \qquad (4.86)$$

and use Newton iteration on Eq.(4.86):

$$\mathbf{x}_{k+1}^{\ell+1} = \mathbf{x}_{k+1}^{\ell} - [\mathcal{H}^{\ell}]^{-1} \cdot [\mathcal{F}^{\ell}] \qquad (4.87)$$

where the Hessian $\mathcal{H}$ can be computed as:

$$\mathcal{H} = \alpha_i \cdot (\mathcal{J} \cdot h) - \mathbf{I}^{(\mathbf{n})} \tag{4.88}$$

and $\mathcal{J}$ is the Jacobian of the system. By plugging the linear homogeneous system into Eq.(4.87), it is easy to show that we get Eq.(4.85) back, i.e., Newton iteration doesn't change the stability domain of the method.

Most of the professional multi–step codes use modified Newton iteration, i.e., they do not reevaluate the Jacobian during the iteration. They usually don't evaluate the Jacobian even once every step. Instead, they use the error estimate of the method as an indicator when the Jacobian needs to be reevaluated. As long as the error estimate remains approximately constant, the Jacobian is still acceptable. However, as soon as the absolute gradient of the error estimate starts to grow, indicating the need for a change in step size, this is a clear indication that a new Jacobian computation is in order, and only if a reevaluation of the Jacobian doesn't get the error estimate back to where it was before, will the step size of the method be adjusted.

Even the Hessian is not reevaluated frequently. The Hessian needs to be recomputed either if the Jacobian has been reevaluated, or if the step size has just been modified.

Most professional codes offer several options for how to treat the Jacobian. The user can choose between (i) providing an analytical expression for the Jacobian, (ii) having the full Jacobian evaluated by means of a numerical approximation, and (iii) having only the diagonal elements of the Jacobian evaluated by means of a numerical approximation ignoring the off–diagonal elements altogether.

Both the convergence speed and the convergence range of the Newton iteration scheme are strongly influenced by the quality of the Jacobian. A diagonal approximation is cheap, but leads to a heavy increase in the number of iterations necessary for the algorithm to converge, and necessitates more frequent Jacobian evaluations as well. In our experience, it hardly ever pays off to consider this option.

The full Jacobian is usually determined by first–order approximations. The ith state variable, x_i, is modified by a small value, Δx_i. The state derivative vector is then reevaluated using the modified state value. We find:

$$\frac{\partial \mathbf{f}(\mathbf{x}, t)}{\partial x_i} \approx \frac{\mathbf{f}_{\mathbf{pert}} - \mathbf{f}}{\Delta x_i} \tag{4.89}$$

where $\mathbf{f}$ is the state derivative vector evaluated at the current nominal values of all state variables, whereas $\mathbf{f}_{\mathbf{pert}}$ is the perturbed state derivative vector evaluated at $x_i + \Delta x_i$ with all other state variables being kept at their currently nominal values. Thus, an nth–order system calls for n additional function evaluations in order to obtain one full approximation of the Jacobian.

Yet, even by spending these additional n function evaluations, we gain only a first–order approximation of the Jacobian. Any linear model can be converged in precisely one step of Newton iteration with the correct Jacobian being used, irrespective of the location of its eigenvalues or the size of the system. Using the first–order approximation, however, we may need three to four iterations in order to get the *iteration error* down to a value below the *integration error*. On a sufficiently nasty nonlinear problem, the ratio of the numbers of iterations needed to converge may be even worse.

For these reasons, we strongly advocate the analytical option. In the past, this option has rarely been used ... because we engineers are a lazy lot. An nth–order model calls for n^2 additional equations in order to analytically describe its Jacobian. Moreover, these equations may be longer and more involved than the original n equations due to the analytical differentiation. Thus, deriving the Jacobian equations by hand is a tedious and error–prone process.

However, there is really no good reason why these equations should have to be derived by hand. As you already know if you read the companion book to this text *Continuous System Modeling* [4.2], engineers anyway don't usually write down their models by hand in a form that the numerical integration software could use directly. They employ a *modeling language*, such as Dymola [4.3], from which, by means of compilation, a simulation program is generated.

There is no good reason why the analytical Jacobian equations could not be generated automatically in this process, i.e., the Jacobian can be generated once and for all at compile time by means of symbolic differentiation. Indeed, Dymola [4.3] already offers such a feature. Symbolic differentiation is very useful also for other purposes that we shall talk about in due course [4.1].

We are fully convinced that mixed symbolic/numerical algorithms are the way of the future. Many problems can be tackled either numerically or symbolically. However in some cases, the numerical solution is more efficient, whereas in others, the symbolic approach is clearly superior. By merging these two approaches into one integrated software environment, we can preserve the best of both worlds. In continuous system modeling and simulation, models specified by the user in a form most convenient to him or her are symbolically preconditioned for optimal use by the subsequent numerical algorithms, such as the numerical integration software.

Even in this chapter, we have already made use of symbolic algorithms without explicitly mentioning it. We explained in Eqs.(4.60) and (4.61), how the coefficients of high–order stiffly–stable integration algorithms can be found. Yet, if this is done numerically in MATLAB, e.g. using the statement:

$$\mathbf{coef} = \text{sum}(\text{inv}(\mathbf{M})); \tag{4.90}$$

the resulting coefficient vector will be generated in a real–valued format, rather than as rational expressions. Numerical mathematicians prefer to be provided with rational expressions for these coefficients, so that the mantissa of the machine on which the integration algorithm is to be implemented can be fully exploited without leading to additional and unnecessary roundoff errors.

We could have tried to obtain rational expressions making use of the fact that both the determinant and the adjugate of an integer–valued matrix are integer–valued, i.e.:

$$
\begin{aligned}
Mdet &= \text{round}(\det(\mathbf{M})); \\
\mathbf{Madj} &= \text{round}(Mdet * \text{inv}(\mathbf{M})); \\
\mathbf{coef\_num} &= \text{sum}(\mathbf{Madj});
\end{aligned}
\tag{4.91}
$$

In this way, the numerators of the coefficients, **coef_num**, can be computed as an integer–valued vector, whereas the common denominator is the equally integer–valued determinant. We could then use any one among a number of well-known algorithms to determine the common dividers between the numerators and denominators of each coefficient.

This approach works well for BDF algorithms of orders three or four, but fails, when dealing with 8th– or 9th–order algorithms. The reason is that the determinant of $\mathbf{M}$ grows so rapidly with the size of $\mathbf{M}$ that the mantissa of a 32–bit machine is exhausted quickly, in spite of the fact that MATLAB computes everything in double precision.

For this reason, we generated the coefficient vectors of e.g. Eqs.(4.81) by means of the MATLAB statement:

$$
\mathbf{coef} = \text{sum}(\text{inv}(\text{sym}(\mathbf{M})));
\tag{4.92}
$$

i.e., making use of MATLAB's *symbolic toolbox*. The symbolic toolbox represents integers as character strings, and is thereby not limited by the length of the mantissa. The *sym*–operator converts the numeric integer–valued matrix $\mathbf{M}$ into a symbolic representation. The *inv*– and *sum*–functions are overloaded MATLAB functions that make use of different algorithms depending on the type declaration of their operand.

Using similar techniques, Gander and Gruntz [4.4] recently corrected a number of errors in well-known and frequently used numerical algorithms that had gone unnoticed for several decades.

4.12   Step–size and Order Control

We have seen in the previous chapter that the appropriate order of an RK algorithm is determined by the accuracy requirements. Therefore, since the relative accuracy requirements usually are the same throughout the entire simulation run, order control makes little sense.

Let us ascertain whether the same is true in the case of the multi–step algorithms. To this end, we shall simulate our fifth–order linear test problem of Eq.(H4.8a) across 10 seconds, using zero input and applying an initial value of 1.0 to all five state variables. For different global relative error requirements, we find the largest step sizes that keep the numerical error just below the required error bounds, and plot the number of function evaluations needed to simulate the system across 10 seconds as a function of the required accuracy. The process is repeated for AB2, AB3, and AB4. The same quantity for RK4 is plotted on the same graph for comparison. The results of this analysis are shown on Fig.4.12.

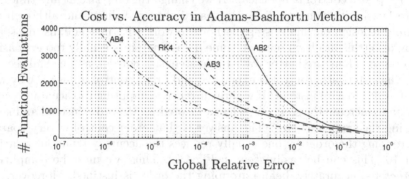

FIGURE 4.12. Cost versus accuracy for different ABi algorithms.

The results are a little deceiving, since only the number of function evaluations (one per step for all ABi algorithms) is plotted, not taking into account the higher cost associated with data management within the higher–order ABi algorithms.

If we decide that we are willing to spend about 500 function evaluations on this simulation, we can get a global relative accuracy of about 10% with AB2, we can get about 1% global accuracy with AB3, and we can get about 0.1% accuracy with AB4. Thus, just as in the case of the RKi algorithms, the accuracy requirements determine the minimum order of the algorithm that is economical to employ. Since the order of the algorithm is dictated by the (constant) accuracy requirements specified by the user, order control doesn't make too much sense.

As a little caveat: AB4 is about 25% cheaper than RK4 in this example. Notice that this is a linear time–invariant non–stiff problem, i.e., a problem where the ABi algorithms perform at their best. Although RK4 takes four function evaluations per step, whereas AB4 takes only one function evaluation per step, we never gain a factor of four in efficiency, since the asymptotic regions of the RKi algorithms are considerably larger than those of the ABi algorithms, forcing us to use a smaller step size in the latter case. The situation gets worse with higher orders of approximation accuracy due to the detrimental influence of spurious eigenvalues.

Why is order control fashionable in multi–step algorithms? The answer is simple: because order control is *cheap*. Remember how multi–step algorithms work. At all times, we keep a record of back storage values of states and/or state derivatives. When we proceed from time t_k to time t_{k+1}, we simply throw the oldest values (the rear end of the tail) away, shift all the vectors by one into the past, and add the newest state information to the front end of the queue. If we decide to increase the order by one, we simply don't throw away anything. On the other hand, if the decide to decrease the order by one, we simply throw away the two oldest values. Thus, order control is trivial.

Step–size control is *not* cheap. If we change the step size at any time, we are faced with non–equidistantly spaced storage values, and although we could redesign our multi–step methods to work with non–equidistant spacing also (this has been done on some occasions), it is usually too expensive to do so. There are better ways to do step–size control, as we shall see.

Since step–size control is expensive and order control is cheap, why not use order control instead? If we are currently computing too accurately and wish to increase the step size, why can't we drop the order instead and keep using the same step size? The answer is that order control is very coarse. Dropping the order by one usually reduces the accuracy by about a factor of 10. This can be easily seen on Fig.4.12. Thus, we must be computing *much* too accurately, before dropping the order is justified. Moreover, we don't even save that much by doing so. After all, the number of function evaluations remains the same. The fact that everybody does order control, doesn't mean, it's the smart thing to do (!)

How then can we do step–size control efficiently? The trick is actually quite simple. Let us reconsider the Newton–Gregory backward polynomials. We start out with:

$$\mathbf{x}(t) = \mathbf{x_k} + s\nabla\mathbf{x}_k + \left(\frac{s^2}{2} + \frac{s}{2}\right)\nabla^2\mathbf{x}_k + \left(\frac{s^3}{6} + \frac{s^2}{2} + \frac{s}{3}\right)\nabla^3\mathbf{x}_k + \dots \quad (4.93)$$

Differentiation with respect to time yields:

$$\dot{\mathbf{x}}(t) = \frac{1}{h}\left[\nabla\mathbf{x}_k + \left(s + \frac{1}{2}\right)\nabla^2\mathbf{x}_k + \left(\frac{s^2}{2} + s + \frac{1}{3}\right)\nabla^3\mathbf{x}_k + \dots\right] \quad (4.94)$$

The second derivative becomes:

$$\ddot{\mathbf{x}}(t) = \frac{1}{h^2}\left[\nabla^2\mathbf{x}_k + (s+1)\nabla^3\mathbf{x}_k + \dots\right] \quad (4.95)$$

etc.

Truncating Eqs.(4.93)–(4.95) after the cubic term, expanding the ∇–operators, and evaluating for $t = t_k$ $(s = 0.0)$, we obtain:

$$
\begin{pmatrix} x_k \\ h \cdot \dot{x}_k \\ \frac{h^2}{2} \cdot \ddot{x}_k \\ \frac{h^3}{6} \cdot x_k^{(iii)} \end{pmatrix} = \frac{1}{6} \cdot \begin{pmatrix} 6 & 0 & 0 & 0 \\ 11 & -18 & 9 & -2 \\ 6 & -15 & 12 & -3 \\ 1 & -3 & 3 & -1 \end{pmatrix} \cdot \begin{pmatrix} x_k \\ x_{k-1} \\ x_{k-2} \\ x_{k-3} \end{pmatrix} \qquad (4.96)
$$

The vector to the left of the equal sign is called the *Nordsieck vector*, here written of third order. Using this trick, it has become possible to translate state information stored at different points in time by means of a simple multiplication with a constant matrix into state– and state derivative information stored at one time point only. The transformation was written here for a single state variable. The vector case works in exactly the same fashion, but the notation is less convenient.

This discovery allows us to solve the step–size control problem. If we wish to change the step size at time t_k, we simply multiply the vector containing the past state values with the transformation matrix, thereby transforming the state history vector to an equivalent Nordsieck vector. In this new form, the step size can be modified easily, e.g. by multiplying the Nordsieck vector from the left with the diagonal matrix:

$$
\mathbf{H} = \begin{pmatrix} 1 & 0 & 0 & 0 \\ 0 & \frac{h_{new}}{h_{old}} & 0 & 0 \\ 0 & 0 & \left(\frac{h_{new}}{h_{old}}\right)^2 & 0 \\ 0 & 0 & 0 & \left(\frac{h_{new}}{h_{old}}\right)^3 \end{pmatrix} \qquad (4.97)
$$

We now have the Nordsieck vector expressed in the modified time step, h_{new}. We then multiply this modified Nordsieck vector from the left with the inverse transformation matrix. This operation results in a modified state history vector, where the "stored" $\mathbf{x}$ values represent the solution at the modified "sampling points."

This is today the preferred method for step–size control in multi–step integration algorithms. Step–size control is still fairly expensive. In the case of implicit algorithms, we need to also evaluate a new Hessian after modifying the step size using the above matrix multiplications. Since we are already in a "spending mood," we might just as well use the opportunity to get a new Jacobian also.

For this reason, the Gustafsson algorithm [4.8] is not practical for use in multi–step integration. We don't want to change the step size after every step. If we need to reduce the step size, we shall reduce it at once to at least one half of its former value to prevent the danger of having to reduce the step size immediately again. We don't want to increase the step size either until we are fairly certain that we can at least double it without violating the accuracy constraints. How do we know that? The next section will tell.

4.13   The Startup Problem

One problem we have not discussed yet is how the integration process is started in the first place. Usually, the initial state vector is given at time t_0, but no back values are available. How can we compute estimates for these values such that the multi–step formulae become applicable?

Traditionally, applied mathematicians have chosen the easy route: applying order control. They employ a first–order method during the first integration step. This provides them with a second data point at time $t_1 = t + h$. Since, by now, they have two data points, they can raise the order by one, and perform the second integration step using a second–order formula, etc. After a suitable number of steps, the algorithm has acquired the desired state history vector in order to make an nth–order multi–step method applicable.

This method "works," in the sense that it can be programmed into an algorithm. However, it is not acceptable on any other grounds. The problem is accuracy. In order to satisfy our accuracy requirements, we must use a very small step size during the first low–order steps. Even after we have built up the order, we should not immediately increase the step size by use of the Nordsieck transformation, since some of the back values currently in the storage area of the algorithm are low–order accurate. In the transformation, we may pick up bad numerical errors. It is better to wait for at least another n steps, before we even dream of changing the step size to a more decent value. This is utterly wasteful.

Bill Gear has shown another way [4.6]. We can use Runge–Kutta algorithms for startup purposes. In this way, also the early steps are of the correct order, and a more decent step size can be used from the beginning.

Let us explain our own version of this general idea. If we decide that we want to employ an ith–order algorithm, we start out performing $(i - 1)$ steps of RKi using a fixed–step algorithm. The step size doesn't matter too much as long as it is chosen sufficiently small to ensure that the accuracy requirements are met. For example, we can use step–size control during the first step to determine the right step size, and then disable the step–size control algorithm.

By now, we have i equidistantly–spaced storage values of the state vector, and we are able to start using the multi–step algorithm. However, what step size should we use? In order to determine the answer to this question, let us look once more at Fig.4.12. This time, we plotted the same curves as before using a double–logarithmic scale.

We notice that the logarithm of the step size is, for all practical purposes, linear in the logarithm of the accuracy. Thus, we can perform one step of the multi–step technique using the step size h_1 from the RK starter, and obtain an error estimate ε_1. We then reduce the step size to $h_2 = h_1/2$, and repeat the step. We obtain a new error estimate ε_2.

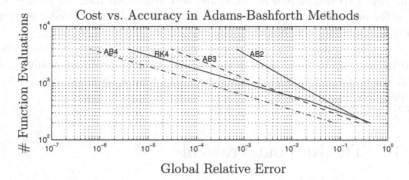

FIGURE 4.13. Cost versus accuracy for different ABi algorithms.

We now place a linear curve through the two points, and interpolate (extrapolate) with the desired accuracy ε_{des} to obtain a good value for the true step size to be used by the multi–step algorithm:

$$\begin{pmatrix} \ln(h_1) \\ \ln(h_2) \end{pmatrix} = \begin{pmatrix} \ln(\varepsilon_1) & 1 \\ \ln(\varepsilon_2) & 1 \end{pmatrix} \cdot \begin{pmatrix} a_1 \\ a_2 \end{pmatrix} \tag{4.98a}$$

$$\ln(h_{\text{des}}) = a_1 \cdot \ln(\varepsilon_{\text{des}}) + a_2 \tag{4.98b}$$

$$h_{\text{new}} = 0.8 \cdot h_{\text{des}} \tag{4.98c}$$

Eq.(4.98a) solves a linear set of two equations for the unknown parameters a_1 and a_2, Eq.(4.98b) solves the interpolation (extrapolation) problem, and Eq.(4.98c) computes the new step size using a safety margin of 20%.

Runge–Kutta starters work very well in the case of non–stiff problems, i.e., as start–up algorithms for Adams–Bashforth or Adams–Bashforth–Moulton algorithms. They are more problematic in the case of the Backward Difference Formulae. The reason for this observation is simple: The RK start–up algorithm may need to use exceedingly small steps because of the stiffness of the problem to be solved. Once we switch over to the BDF algorithm, we may thus wish to increase the step size dramatically. The Nordsieck approach allows us to do so, but in the process, the sampling points get extended over a wide range, which corresponds to heavy extrapolation, a process that is invariably associated with sources of inaccuracy. We may thus prefer to limit the allowed step size enlargement to maybe a factor of 10, then perform n steps of BDF with that intermediate step size to gain a new accurate history vector, before increasing the step size once more by a factor of 10, and repeat this process, until the appropriate step size has been reached.

In step–size control, we can use the same algorithm. We don't want to change the step size unless it needs to be changed by a large value. Therefore, if we have decided that a step–size change is truly in order, we can afford to calculate one additional test step with half the former step size (or double the former step size) in order to obtain a decent estimate of where the step size ought to be.

A yet better approach might have been to use a number of $BI4/5_{0.45}$ steps during startup to avoid both the numerical stability problems haunting the RK starters and the numerical accuracy problems associated with the low–order BDF starters.

4.14   The Readout Problem

The last problem to be discussed in this chapter is the readout problem. If we simulate a system, we want to obtain results, i.e., we wish to obtain the values of one or several output variables at pre–specified points in time, the so–called *communication points*.

Often, the communication points are equidistantly spaced. In this case, the distance between neighboring communication points is referred to as the *communication interval*.

In single–step integration, we simply reduce the step size when approaching the next communication point in order to hit the communication point accurately. In multi–step integration, this approach is too expensive. We cannot afford to modify the step size for no other purpose than to deliver some output values.

The solution is simple. We integrate past the communication point using the actual step size. We then interpolate back to calculate an estimation of the state vector at the communication point. In multi–step integration, interpolation with the same order of approximation accuracy that the currently employed integration method uses is cheap. All we need to do is to convert the state history vector at the end of the integration step to the Nordsieck form, then correct the step size such that the end of the "previous step" coincides with the communication point, then record the new "immediate past value" as the readout value.

After the results have been recorded, the algorithm returns to the end of the integration step, and proceeds as if no interruption had taken place.

4.15   Summary

20 years ago, the chase for new integration algorithms was still on. In numerical integration workshops around the globe, new integration meth-

ods were presented by the dozen. Hardly any of them survived the test of time. The reason for this surprising fact is simple. To come up with a new algorithm is the easy part. To incorporate that algorithm in a robust general–purpose production code is an entirely different matter.

Most engineering users of simulation software use the numerical integration software as a black box. They don't have the foggiest idea of how the code works, and frankly, they couldn't care less. All they are interested in is that the code *reliably* and *efficiently* generates accurate estimates of the output variables at the communication points.

In a mature production code of a multi–step integration algorithm, the actual algorithm occupies certainly less than 5% of the code. All the rest is boiler plate: code for initializing the coefficient matrices; code for starting up the integration algorithm, e.g., using an RK starter; code for update, maintenance, and disposal of the state history information; code for interpolating the results at communication points; code for step–size and (alas!) order control; and finally, code for handling of discontinuities — a topic to be discussed in a separate chapter of this book.

Software engineers may be inclined to believe that the answer to this problem is software modularization. Let us structure the software in such a way that e.g. the step–size control is handled by one routine, interpolation is handled by another, etc., in such a fashion that these routines can be modularly plugged together. Unfortunately, even this doesn't work. A step–size control algorithm that is efficient for an RK algorithm, such as the Gustafsson method, could theoretically also be used for a multi–step algorithm, but it would be terribly inefficient.

What *has* happened is a certain standardization of the interfaces of numerical integration software (the parameter calling sequences), such that a user can fairly easily replace one entire code by another to check which one works better. In this context, it is important to mention the efforts of Alan Hindmarsh whose various LSODE codes are clearly among the survivors.

More could certainly be done. Today's multi–step codes are unnecessarily unreadable. What we would need is an efficient *MATLAB compiler* that would allow us to develop new production codes in MATLAB, making them easily readable, and once they are fully debugged, generate automatically efficient C–code for use as stand–alone programs. Availability of such a software design tool would make the life of applied mathematicians much easier. Although a C–compiler for MATLAB has been developed, the generated code is unfortunately anything but efficient, since it makes use of the generic data management tools of MATLAB. Consequently, C–compiled MATLAB code doesn't execute much faster than the interpreted MATLAB code itself.

You, the reader, may have noticed that we are somewhat critical vis-à-vis the multi–step integration methods. Runge–Kutta methods are, in our opinion, considerably more robust, and it is easier to design production codes for them. Multi–step techniques are fashionable because the algorithms themselves are so much easier to design, but the price to be paid is dear. Mature multi–step codes are very difficult to design, and even with the best of all such codes, it still happens that it breaks down in the face of a nasty nonlinear simulation problem, and if it does, it may be very difficult for even knowledgeable users to determine what precisely it was that the algorithm didn't like, and which parameter to fiddle around with in order to get around the problem.

Single–step codes are much simpler to develop and maintain, and they offer a smaller number of tuning parameters that the user might need to worry about. They are considerably more robust. Whereas non–stiff RK codes are available and are being widely used, stiff implicit RK production codes are slow in coming. BDF codes are still far more frequently used in practice than IRK codes. However, this is true not because of the superiority of these algorithms, but due to the wider distribution of good production codes.

4.16 References

[4.1] François E. Cellier and Hilding Elmqvist. Automated Formula Manipulation in Object–Oriented Continuous–System Modeling. *IEEE Control Systems*, 13(2):28–38, 1993.

[4.2] François E. Cellier. *Continuous System Modeling*. Springer Verlag, New York, 1991. 755p.

[4.3] Hilding Elmqvist. *A Structured Model Language for Large Continuous Systems*. PhD thesis, Dept. of Automatic Control, Lund Institute of Technology, Lund, Sweden, 1978.

[4.4] Walter Gander and Dominik Gruntz. Derivation of Numerical Methods Using Computer Algebra. *SIAM Review*, 41(3):577–593, 1999.

[4.5] C. William Gear. *Numerical Initial Value Problems in Ordinary Differential Equations*. Series in Automatic Computation. Prentice–Hall, Englewood Cliffs, N.J., 1971. 253p.

[4.6] C. William Gear. Runge–Kutta Starters for Multistep Methods. *ACM Trans. Math. Software*, 6(3):263–279, 1980.

[4.7] Curtis F. Gerald and Patrick O. Wheatley. *Applied Numerical Analysis*. Addison–Wesley, Reading, Mass., 6th edition, 1999. 768p.

[4.8] Kjell Gustafsson. *Control of Error and Convergence in ODE Solvers*. PhD thesis, Dept. of Automatic Control, Lund Institute of Technology, Lund, Sweden, 1992.

[4.9] Klaus Hermann. *Solution of Stiff Systems Described by Ordinary Differential Equations Using Regression Backward Difference Formulae.* Master's thesis, Dept. of Electrical & Computer Engineering, University of Arizona, Tucson, Ariz., 1995.

[4.10] John D. Lambert. *Numerical Methods for Ordinary Differential Systems: The Initial Value Problem.* John Wiley, New York, 1991. 304p.

[4.11] William E. Milne. *Numerical Solution of Differential Equations.* John Wiley, New York, 1953. 275p.

[4.12] Cleve Moler and Charles van Loan. Nineteen Dubious Ways to Compute the Exponential of a Matrix. *SIAM Review,* 20(4):801–836, 1978.

4.17 Homework Problems

[H4.1] The Differentiation Operator

Rewrite Eq.(4.20) and Eq.(4.21) in terms of the ∇–operator. Develop also a formula for $\mathcal{D}^3$.

[H4.2] Nyström–Milne Predictor–Corrector Techniques

Follow the reasoning of the Adams–Bashforth–Moulton predictor–corrector techniques, and develop similar pairs of algorithms using a Nyström predictor stage followed by a Milne corrector stage.

Plot the stability domains for NyMi3 and NyMi4. What do you conclude?

[H4.3] New Methods

Using the Gregory–Newton backward polynomial approach, design a set of algorithms of the type:

$$\mathbf{x_{k+1}} = \mathbf{x_{k-2}} + \frac{h}{\alpha_i} \cdot \left[\sum_{j=1}^{i} \beta_{ij}\, \mathbf{x}_{k-j+1} \right] \tag{H4.3a}$$

Plot their stability domains. Compare them with those of the Adams–Bashforth techniques and those of the Nyström techniques. What do you conclude?

[H4.4] Milne Integration

Usually, the term "Milne integration algorithm," when used in the literature, denotes a specific predictor–corrector technique, namely:

predictor: $\dot{\mathbf{x}}_k = \mathbf{f}(\mathbf{x}_k, t_k)$
$\mathbf{x}_{k+1}^P = \mathbf{x}_{k-3} + \frac{h}{3}(8\dot{\mathbf{x}}_k - 4\dot{\mathbf{x}}_{k-1} + 8\dot{\mathbf{x}}_{k-2})$

corrector: $\dot{\mathbf{x}}_{k+1}^P = \mathbf{f}(\mathbf{x}_{k+1}^P, t_{k+1})$
$\mathbf{x}_{k+1}^C = \mathbf{x}_{k-1} + \frac{h}{3}(\dot{\mathbf{x}}_{k+1}^P + 4\dot{\mathbf{x}}_k + \dot{\mathbf{x}}_{k-1})$

The corrector is clearly Simpson's rule. However, the predictor is something new that we haven't seen yet.

Derive the order of approximation accuracy of the predictor. To this end, use the Newton–Gregory backward polynomial in order to derive a set of formulae with a distance of four steps apart between their two state values.

Plot the stability domain of the predictor–corrector method, and compare it with that of NyMi4. What do you conclude? Why did William E. Milne [4.11] propose to use this particular predictor?

[H4.5] Damping Plots of Adams–Bashforth Techniques

Find the damping plots of AB2, AB3, and AB4 for σ_d in the range [-1.0,0.0]. Compare them with the corresponding damping plots of RK2, RK3, and RK4. What do you conclude about the size of the asymptotic regions of these algorithms?

[H4.6] Damping Plots of Adams–Moulton Techniques

Find the damping plots of AM2, AM3, and AM4 for σ_d in the range $[-1.0, 0.0]$. Compare them with those of AB2, AB3, and AB4. What can you say about the comparative sizes of the asymptotic regions of these algorithms?

[H4.7] Damping Plots of Backward Difference Formulae

Find the damping plots of BDF2, BDF3, and BDF4 for σ_d logarithmically spaced between 10^{-1} and 10^6, and plot them on a logarithmic scale like in Fig.4.7. What do you conclude about the damping properties of these algorithms at large values of σ_d? Do all these methods share the desirable properties of BDF6, or was this a happy accident?

[H4.8] Cost Versus Accuracy

Compute the cost–versus–accuracy plots of AB4, ABM4, and AM4 for the linear non–stiff test problem:

$$\dot{\mathbf{x}} = \begin{pmatrix} 1250 & -25113 & -60050 & -42647 & -23999 \\ 500 & -10068 & -24057 & -17092 & -9613 \\ 250 & -5060 & -12079 & -8586 & -4826 \\ -750 & 15101 & 36086 & 25637 & 14420 \\ 250 & -4963 & -11896 & -8438 & -4756 \end{pmatrix} \cdot \mathbf{x} + \begin{pmatrix} 5 \\ 2 \\ 1 \\ -3 \\ 1 \end{pmatrix} \cdot u$$

(H4.8a)

with zero input and with the initial condition $\mathbf{x_0} = \text{ones}(5, 1)$, and plot them together on one graph. ABM4 consumes always two function evaluations per step. In the case of AM4, the situation is more involved, but for simplicity, we want to assume that AM4 needs, on the average, four function evaluations per step.

Although we plot the number of steps versus the accuracy, it is more efficient to vary the number of steps and check what accuracy we obtain in each case. We suggest that you select a set of values of steps in the range $[200, 4000]$, e.g. $nbrstp = [200, 500, 1000, 2000, 4000]$. You then need to compute the step sizes. In the case of ABi, they would be $h = 10/nbrstp$, since we want to integrate across 10 seconds of simulated time using $nbrstp$ steps in total. In the case of AMi, we would use the formula $h = 40/nbrstp$.

Since the test problem is linear, you can simulate the system using the $\mathbf{F}$–matrices. In the case of the AMi algorithms, you can use matrix inversion rather than Newton iteration (we indirectly accounted for the iteration by allowing four function evaluations per step).

Since these methods are not self–starting, you need to start out with $(i - 1)$ steps of RKi using the same step sizes. Of course, the RKi steps are also simulated using their respective $\mathbf{F}$–matrices.

As a gauge, we need the analytical solution of the test problem. Since the input is constant between sampling points (in fact, it is zero), we can find the *exact* solution by converting the differential equations into a set of equivalent difference equations using MATLAB's *c2d*–function. This generates the analytical $\mathbf{F}$–matrix. Theoretically:

$$\mathbf{F} = \exp(\mathbf{A} \cdot h) \tag{H4.8b}$$

but doesn't use MATLAB's *expm*–function. For sufficiently large step sizes, you'll get an overflow error. It would be asked a little too much to explain here why this happens. If you are interested to know more about this numerical problem, we refer you to Cleve Moler's excellently written paper on this subject [4.12].

The local relative error is computed using the formula:

$$\varepsilon_{\text{local}}(t_k) = \frac{\|\mathbf{x}_{\text{anal}}(t_k) - \mathbf{x}_{\text{simul}}(t_k)\|}{\max(\|\mathbf{x}_{\text{anal}}(t_k)\|, \text{eps})} \tag{H4.8c}$$

where *eps* is MATLAB's machine constant.

The global relative error is computed using the formula:

$$\varepsilon_{\text{global}} = \max_k(\varepsilon_{\text{local}}(t_k)) \tag{H4.8d}$$

What do you conclude about the relative efficiency of these three algorithms to solve the test problem?

[H4.9] Cost Versus Accuracy

We wish to repeat the same analysis as before, but this time for the stiff linear test problem:

$$\dot{x} = \begin{pmatrix} 0 & 1 & 0 \\ 0 & 0 & 1 \\ -10001 & -10201 & -201 \end{pmatrix} \cdot x + \begin{pmatrix} 0 \\ 0 \\ 1 \end{pmatrix} \cdot u \qquad (H4.9a)$$

We wish to compute the step response of this system across ten seconds of simulated time.

This time, we are going to use BDF2, BDF3, and BDF4, in order to compare their relative efficiency at solving this stiff test problem. As in the case of the previous homework, we are going to simulate the system using the F–matrices. As with the AMi algorithms, we use matrix inversion, and simply assume that each step consumes, on the average, four function evaluations.

As a reference, also compute the cost–versus–accuracy plot of the BI4/$5_{0.45}$ algorithm using RKF4/5 for its semi–steps. For reasons of fairness, we shall assume that the implicit semi–step uses four iterations. Together with the single explicit semi–step, one entire step of BI4/$5_{0.45}$ consumes five semi–steps with six function evaluations each, thus: $h = 300/nbrstp$.

Which technique is more efficient, BDF4 or BI4/5, to solve this stiff test problem?

[H4.10] The Nordsieck Form

Equation (4.96) showed the transformation matrix that converts the state history vector into an equivalent Nordsieck vector. Since, at the time of conversion, we also have the current state derivative information available, it is more common to drop the oldest state information in the state history vector, and replace it by the current state derivative information. Consequently, we are looking for a transformation matrix T of the form:

$$\begin{pmatrix} x_k \\ h \cdot \dot{x}_k \\ \frac{h^2}{2} \cdot \ddot{x}_k \\ \frac{h^3}{6} \cdot x_k^{(iii)} \end{pmatrix} = T \cdot \begin{pmatrix} x_k \\ h \cdot \dot{x}_k \\ x_{k-1} \\ x_{k-2} \end{pmatrix} \qquad (H4.10a)$$

The matrix T can easily be found by manipulating the individual equations of Eq.(4.96).

Find corresponding T–matrices of dimensions 3×3 and 5×5.

[H4.11] Backward Difference Formulae

We wish to simulate the stiff test problem of Eq.(H4.9a) once more using BDF4. However this time around, we no longer want to make use of the knowledge that the system is linear.

Implement BDF4 with Newton iteration in MATLAB. Use three steps of RK4 for startup. Use the outlined procedure for step–size control. We wish to record the values of the three state variables once every second. Solve the readout problem using the algorithm outlined in this chapter.

4.18   Projects

[P4.1] Stiffly–Stable Methods

Extend one of the widely used variable–order, variable–step size stiff system solvers to include methods of orders seven, eight, and nine, as developed in this chapter.

Compare the efficiency of the so modified code with that of the original code when solving a stiff system with high accuracy requirements.

[P4.2] Stiffly–Stable Methods

Study the effect of the start–up algorithm on a stiff system solver by comparing an order buildup approach with a single–step start–up approach.

Compare RK starters with BI starters.

[P4.3] Stiffly–Stable Methods

Study the importance of a small error coefficient *vs.* a large asymptotic region on the efficiency of a stiff system solver.

Compare different BDF algorithms of the same order using the same start–up and step–size control strategies against each other. The codes are supposed to differ only in the formulae being used. Choose some formulae with small error coefficients, and compare them with formulae with large asymptotic regions.

Draw cost *vs.* accuracy plots to compare their relative economy when solving identical stiff systems.

4.19   Research

[R4.1] Regression Backward Difference Algorithms

In the development of the stiffly–stable algorithms, we always made use of $n + 1$ terms to define an nth–order algorithm. It may be beneficial to allow more terms in the algorithm without increasing its order.

For example, we could allow a 3rd–order accurate BDF algorithm to make use of the term x_{k-3} as well. In that case, Eq.(4.60) needs to be modified as follows:

$$
\begin{pmatrix} h \cdot f_{k+1} \\ x_k \\ x_{k-1} \\ x_{k-2} \\ x_{k-3} \end{pmatrix} = \begin{pmatrix} 0 & 1 & 2 & 3 \\ 1 & 0 & 0 & 0 \\ 1 & -1 & 1 & -1 \\ 1 & -2 & 4 & -8 \\ 1 & -3 & 9 & -27 \end{pmatrix} \cdot \begin{pmatrix} a_0 \\ a_1 \\ a_2 \\ a_3 \end{pmatrix} \tag{R4.1a}
$$

The **M**–matrix in this case is no longer square. The above equation can thus only be solved for the unknown parameter vector in a least square sense. We thus call these algorithms *Regression Backward Difference Formulae* [4.9].

We can multiply the equation:

$$
\mathbf{z} = \mathbf{M} \cdot \mathbf{a} \tag{R4.1b}
$$

from the left with **M**′:

$$
\mathbf{M}' \cdot \mathbf{z} = (\mathbf{M}' \cdot \mathbf{M}) \cdot \mathbf{a} \tag{R4.1c}
$$

where $\mathbf{M}' \cdot \mathbf{M}$ is a square matrix of full rank. Hence, we can multiply the equation with its inverse:

$$
\mathbf{a} = (\mathbf{M}' \cdot \mathbf{M})^{-1} \mathbf{M}' \cdot \mathbf{z} \tag{R4.1d}
$$

where $(\mathbf{M}' \cdot \mathbf{M})^{-1} \mathbf{M}'$ is the *Penrose–Moore pseudoinverse* of the rectangular matrix **M**. It solves the over–determined linear system in a least square sense.

Extend the search for high–order stiffly–stable methods by allowing extra terms in the algorithm. The hope is that the added flexibility may enable us to either reduce the error coefficient or (even better!) enlarge the asymptotic region.

5

Second Derivative Systems

Preview

In this chapter, we shall look at integration algorithms designed to deal with system descriptions containing second–order derivatives in time. Such system descriptions occur naturally in the mathematical modeling of mechanical systems, as well as in the mathematical modeling of distributed parameter systems leading to hyperbolic partial differential equations.

In this chapter, we shall concentrate on mechanical systems. The discussion of partial differential equations is postponed to the next chapter.

Whereas it is always possible to convert second derivative systems to state–space form, integration algorithms that deal with the second derivatives directly may, in some cases, offer a numerical advantage.

5.1 Introduction

Let us start the discussion by modeling a human body riding in a car. Some people with weak muscles in their neck region suffer from a so–called *cervical syndrome* [5.8]. When riding in a car for extended periods of time, they suffer awful headache attacks, because their head vibrates (oscillates) with the vibrations of the car on the road, since their head is not attached stiffly enough to the shoulders.

Since the passengers in a car are seated, their legs can be eliminated from the model, as they won't affect the motion of the neck at all. The body can be modeled in a first approximation as a mechanical system. A very simple mechanical model of a sitting human body is depicted in Fig.5.1.

The model is decomposed into four masses that can move separately from each other, representing the head, the upper torso, the two arms, and the lower body. Due to linearity, the two arms can be treated as a single mass. The connections between the four masses are modeled as damped springs. The bumping of the road is modeled by a time–dependent force attached to the lower body. To be determined is the distance between head and shoulders as a function of the driving force.

Only vertical motions are to be considered. Hence each mass represents a single mechanical degree of freedom. The overall model is obtained by applying Newton's law to each of the four masses separately:

$$M_1 \cdot \ddot{x}_1 = k_1 \cdot (x_2 - x_1) + B_1 \cdot (\dot{x}_2 - \dot{x}_1) \qquad (5.1)$$

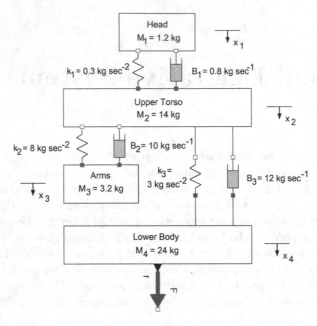

FIGURE 5.1. Mechanical model of a sitting human body.

$$
\begin{aligned}
M_2 \cdot \ddot{x}_2 \;=\;& k_2 \cdot (x_3 - x_2) + B_2 \cdot (\dot{x}_3 - \dot{x}_2) + k_3 \cdot (x_4 - x_2) \\
& + B_3 \cdot (\dot{x}_4 - \dot{x}_2) - k_1 \cdot (x_2 - x_1) - B_1 \cdot (\dot{x}_2 - \dot{x}_1) \quad (5.2) \\
M_3 \cdot \ddot{x}_3 \;=\;& -k_2 \cdot (x_3 - x_2) - B_2 \cdot (\dot{x}_3 - \dot{x}_2) \quad\quad\quad\quad\quad\quad (5.3) \\
M_4 \cdot \ddot{x}_4 \;=\;& F - k_3 \cdot (x_4 - x_2) - B_3 \cdot (\dot{x}_4 - \dot{x}_2) \quad\quad\quad\quad (5.4)
\end{aligned}
$$

Let:

$$
\mathbf{x} = \begin{pmatrix} x_1 \\ x_2 \\ x_3 \\ x_4 \end{pmatrix} \quad\quad\quad (5.5)
$$

be the partial state vector consisting of the four mass positions. Using the partial state vector, the model can be written in matrix/vector form as follows:

$$
\mathbf{M} \cdot \ddot{\mathbf{x}} + \mathbf{C} \cdot \dot{\mathbf{x}} + \mathbf{K} \cdot \mathbf{x} = \mathbf{f} \quad\quad\quad (5.6)
$$

where:

$$
\mathbf{M} = \begin{pmatrix} M_1 & 0 & 0 & 0 \\ 0 & M_2 & 0 & 0 \\ 0 & 0 & M_3 & 0 \\ 0 & 0 & 0 & M_4 \end{pmatrix} \quad\quad\quad (5.7)
$$

is the *mass matrix*,

$$C = \begin{pmatrix} B_1 & -B_1 & 0 & 0 \\ -B_1 & (B_1 + B_2 + B_3) & 0 & 0 \\ 0 & -B_2 & B_2 & 0 \\ 0 & -B_3 & 0 & B_3 \end{pmatrix} \qquad (5.8)$$

is the *damping matrix*,

$$K = \begin{pmatrix} k_1 & -k_1 & 0 & 0 \\ -k_1 & (k_1 + k_2 + k_3) & 0 & 0 \\ 0 & -k_2 & k_2 & 0 \\ 0 & -k_3 & 0 & k_3 \end{pmatrix} \qquad (5.9)$$

is the *stiffness matrix*, and

$$f = \begin{pmatrix} 0 \\ 0 \\ 0 \\ F \end{pmatrix} \qquad (5.10)$$

is the vector of (generalized) forces.

The mass matrix turned out to be a diagonal matrix in this example, but this is only true, because no rotational motions were considered in the given example. Generally, this will not be the case.

Assuming that the mass matrix is non–singular, i.e., there are as many mechanical degrees of freedom in the system as were formulated into second–order differential equations, i.e., there are no *structural singularities* in the model [5.3], the model can be solved for the highest derivatives:

$$\ddot{x} = A^2 \cdot x + B \cdot \dot{x} + u \qquad (5.11)$$

where:

$$A = \sqrt{-M^{-1} \cdot K} \qquad (5.12)$$
$$B = -M^{-1} \cdot C \qquad (5.13)$$
$$u = M^{-1} \cdot f \qquad (5.14)$$

or, more generally, in a nonlinear case:

$$\ddot{x} = f(x, \dot{x}, u, t) \qquad (5.15)$$

Of special interest is the case of the *conservative*, i.e., friction–less system with the second–derivative form:

$$\ddot{x} = A^2 \cdot x + u \qquad (5.16)$$

or, more generally:

$$\ddot{\mathbf{x}} = \mathbf{f}(\mathbf{x}, \mathbf{u}, t) \tag{5.17}$$

and especially, we may want to look at the homogeneous, conservative, linear system with the second–derivative model:

$$\ddot{\mathbf{x}} = \mathbf{A}^2 \cdot \mathbf{x} \tag{5.18}$$

5.2   Conversion of Second–derivative Models to State–space Form

It is always possible to convert a second–derivative model to state–space form. To this end, we introduce the velocity vector, $\mathbf{v}$:

$$
\begin{aligned}
\dot{\mathbf{x}} &= \mathbf{v} & (5.19)\\
\dot{\mathbf{v}} &= \ddot{\mathbf{x}} = \mathbf{A}^2 \cdot \mathbf{x} + \mathbf{B} \cdot \mathbf{v} + \mathbf{u} & (5.20)
\end{aligned}
$$

We introduce the state vector:

$$\xi = \begin{pmatrix} \mathbf{x} \\ \mathbf{v} \end{pmatrix} \tag{5.21}$$

which leads us to the state–space form:

$$\dot{\xi} = \begin{pmatrix} \mathbf{Z}^{(n)} & \mathbf{I}^{(n)} \\ \mathbf{A}^2 & \mathbf{B} \end{pmatrix} \cdot \xi + \begin{pmatrix} \mathbf{0}^{(n)} \\ \mathbf{u} \end{pmatrix} \tag{5.22}$$

where $\mathbf{Z}^{(n)}$ is a zero matrix of dimensions $n \times n$, $\mathbf{I}^{(n)}$ is an identity matrix of dimensions $n \times n$, and $\mathbf{0}^{(n)}$ is a zero vector of length n.

After the conversion, the state–space model can be simulated using any of the integration algorithms introduced in the previous chapters of this book.

Unfortunately, the resulting state vector is of length $2 \cdot n$, which makes simulation methods (integration algorithms) that can deal with the second–derivative model directly potentially interesting.

We might conclude at this point in time that direct methods should be of particular interest for the simulation of *conservative systems*, as those systems do not require computing the velocity vector at all, i.e., half of the state variables can simply be thrown away.

5.3   Velocity–free Models

We shall define a velocity–free model as one that satisfies, in the linear case, the differential vector equation:

$$\ddot{x} = A^2 \cdot x + u \qquad (5.23)$$

and, in the nonlinear case, the differential vector equation:

$$\ddot{x} = f(x, u, t) \qquad (5.24)$$

Notice that every conservative system leads to a velocity–free second–derivative model. Yet, not every velocity–free second–derivative model is conservative. This can be recognized easily.

Given a linear, time–invariant, homogeneous state–space model of the form:

$$\dot{x} = A \cdot x \qquad (5.25)$$

Depending on the eigenvalues of A, the system is either damped or undamped, stable or unstable. We can differentiate the state–space model, leading to:

$$\ddot{x} = A \cdot \dot{x} = A^2 \cdot x \qquad (5.26)$$

Thus, any linear, time–invariant, homogeneous state–space model can also be written in the form of a velocity–free second–derivative model, irrespective of where its eigenvalues are located. Yet, a conservative linear system has its eigenvalues spread up and down along the imaginary axis of the complex plane.

How can the special structure of a velocity–free second–derivative model be exploited by a simulation algorithm?

Let us start by developing the solution vector at time $(t+h)$ into a Taylor series around time t:

$$x_{k+1} = x_k + h \cdot \dot{x}_k + \frac{h^2}{2} \cdot \ddot{x}_k + \frac{h^3}{6} \cdot x_k^{(iii)} + \frac{h^4}{24} \cdot x_k^{(iv)} + \dots \qquad (5.27)$$

We shall also need to develop the solution vector at time $(t - h)$ into a Taylor series around time t:

$$x_{k-1} = x_k - h \cdot \dot{x}_k + \frac{h^2}{2} \cdot \ddot{x}_k - \frac{h^3}{6} \cdot x_k^{(iii)} + \frac{h^4}{24} \cdot x_k^{(iv)} \mp \dots \qquad (5.28)$$

Adding these two equations together, we obtain:

$$x_{k+1} + x_{k-1} = 2 \cdot x_k + h^2 \cdot \ddot{x}_k + \frac{h^4}{12} \cdot x_k^{(iv)} + \dots \qquad (5.29)$$

thus:

$$x_{k+1} = 2 \cdot x_k - x_{k-1} + h^2 \cdot \ddot{x}_k + o(h^4) \qquad (5.30)$$

We just found a 3rd–order accurate explicit linear multi–step method that makes use of the second derivative directly. In some references, the method is referred to as Godunov's method [5.2], in spite of the fact that Sergei Konstantinovic Godunov, a famous Russian applied mathematician of the middle of the 20th century, was much more interested in *conservation laws*, i.e., in partial differential equations of the hyperbolic type, where he used this technique to approximate spatial derivatives.

The second derivative can be plugged in from the homogeneous linear second–derivative model of Eq.(5.18):

$$\mathbf{x_{k+1}} \approx 2 \cdot \mathbf{x_k} - \mathbf{x_{k-1}} + (\mathbf{A} \cdot h)^2 \cdot \mathbf{x_k} \tag{5.31}$$

We find the $\mathbf{F}$–matrix of the algorithm in the usual fashion. Let

$$\xi_k = \begin{pmatrix} \mathbf{x_{k-1}} \\ \mathbf{x_k} \end{pmatrix} \tag{5.32}$$

Then:

$$\xi_{k+1} \approx \mathbf{F} \cdot \xi_k \tag{5.33}$$

where:

$$\mathbf{F} = \begin{pmatrix} \mathbf{Z}^{(n)} & \mathbf{I}^{(n)} \\ -\mathbf{I}^{(n)} & [2 \cdot \mathbf{I}^{(n)} + (\mathbf{A} \cdot h)^2] \end{pmatrix} \tag{5.34}$$

Consequently, we should be able to plot the stability domain of this algorithm as a function of the eigenvalues of $\mathbf{A} \cdot h$, exactly as we did in the case of the algorithms presented in the previous two chapters. We shall attempt to do so in due course.

5.4   Linear Velocity Models

We shall next look at the case of a possibly nonlinear second derivative model with a linear velocity term, i.e., a model of the type:

$$\ddot{\mathbf{x}} + \mathbf{B} \cdot \dot{\mathbf{x}} = \mathbf{f}(\mathbf{x}, \mathbf{u}, t) \tag{5.35}$$

We shall demonstrate that this problem can be reduced to the case of the velocity–free model.

To this end, we apply the variable transformation:

$$\xi = \exp\left(\frac{\mathbf{B} \cdot t}{2}\right) \cdot \mathbf{x} \tag{5.36}$$

Therefore:

$$\mathbf{x} = \exp\left(\frac{-\mathbf{B} \cdot t}{2}\right) \cdot \xi \tag{5.37}$$

$$\dot{\mathbf{x}} = -\frac{\mathbf{B}}{2} \cdot \exp\left(\frac{-\mathbf{B} \cdot t}{2}\right) \cdot \xi + \exp\left(\frac{-\mathbf{B} \cdot t}{2}\right) \cdot \dot{\xi} \tag{5.38}$$

$$\ddot{\mathbf{x}} = \frac{\mathbf{B}^2}{4} \cdot \exp\left(\frac{-\mathbf{B} \cdot t}{2}\right) \cdot \xi - \mathbf{B} \cdot \exp\left(\frac{-\mathbf{B} \cdot t}{2}\right) \cdot \dot{\xi}$$

$$+ \exp\left(\frac{-\mathbf{B} \cdot t}{2}\right) \cdot \ddot{\xi} \tag{5.39}$$

We introduce the abbreviation:

$$\mathbf{E} = \exp\left(\frac{-\mathbf{B} \cdot t}{2}\right) \tag{5.40}$$

Thus:

$$\mathbf{x} = \mathbf{E} \cdot \xi \tag{5.41}$$

$$\dot{\mathbf{x}} = -\frac{\mathbf{B}}{2} \cdot \mathbf{E} \cdot \xi + \mathbf{E} \cdot \dot{\xi} \tag{5.42}$$

$$\ddot{\mathbf{x}} = \frac{\mathbf{B}^2}{4} \cdot \mathbf{E} \cdot \xi - \mathbf{B} \cdot \mathbf{E} \cdot \dot{\xi} + \mathbf{E} \cdot \ddot{\xi} \tag{5.43}$$

Plugging these expressions into the original second–derivative model, we obtain:

$$\ddot{\xi} = \mathbf{E}^{-1} \cdot \frac{\mathbf{B}^2}{4} \cdot \mathbf{E} \cdot \xi + \mathbf{E}^{-1} \cdot \mathbf{f}(\mathbf{E} \cdot \xi, \mathbf{u}, t) \tag{5.44}$$

which has taken on the form of a velocity–free second–derivative model.

5.5   Nonlinear Velocity Models

If the second–derivative model assumes the general form:

$$\ddot{\mathbf{x}} = \mathbf{f}(\mathbf{x}, \dot{\mathbf{x}}, \mathbf{u}, t) \tag{5.45}$$

the velocity vector $\dot{\mathbf{x}}$ needs to be computed as well. However, since we have access to the second derivative, we can compute the velocity vector using any one of the integration algorithms proposed in the previous two chapters.

It seems reasonable to employ an explicit linear multi–step algorithm. Furthermore, since the derivative vector always gets multiplied by the step size, h, it suffices to use a second–order accurate formula. A reasonable choice might be AB2:

$$\dot{\mathbf{x}}_{k+1} = \dot{\mathbf{x}}_k + \frac{h}{2} \cdot (3 \cdot \ddot{\mathbf{x}}_k - \ddot{\mathbf{x}}_{k-1}) \tag{5.46}$$

For the computation of the stability domain, we are dealing with the following set of three equations:

$$\mathbf{x}_{k+1} = 2 \cdot \mathbf{x}_k - \mathbf{x}_{k-1} + h^2 \cdot \dot{\mathbf{v}}_k \tag{5.47}$$

$$\mathbf{v}_{k+1} = \mathbf{v}_k + \frac{h}{2} \cdot (3 \cdot \dot{\mathbf{v}}_k - \dot{\mathbf{v}}_{k-1}) \tag{5.48}$$

$$\dot{\mathbf{v}}_k = \mathbf{A}^2 \cdot \mathbf{x}_k + \mathbf{B} \cdot \mathbf{v}_k \tag{5.49}$$

Plugging the linear second–derivative model of Eq.(5.49) into the two integrator equations, Eq.(5.47) and Eq.(5.48), we obtain:

$$\mathbf{x}_{k+1} = 2 \cdot \mathbf{x}_k - \mathbf{x}_{k-1} + (\mathbf{A} \cdot h)^2 \cdot \mathbf{x}_k + (\mathbf{B} \cdot h) \cdot (h \cdot \mathbf{v}_k) \tag{5.50}$$

$$(h \cdot \mathbf{v}_{k+1}) = (h \cdot \mathbf{v}_k) + \frac{3}{2} \cdot (\mathbf{A} \cdot h)^2 \cdot \mathbf{x}_k + \frac{3}{2} \cdot (\mathbf{B} \cdot h) \cdot (h \cdot \mathbf{v}_k)$$
$$- \frac{1}{2} \cdot (\mathbf{A} \cdot h)^2 \cdot \mathbf{x}_{k-1} - \frac{1}{2} \cdot (\mathbf{B} \cdot h) \cdot (h \cdot \mathbf{v}_{k-1}) \tag{5.51}$$

which can be rewritten in a matrix/vector form as follows:

$$\begin{pmatrix} \mathbf{x}_k \\ h \cdot \mathbf{v}_k \\ \mathbf{x}_{k+1} \\ h \cdot \mathbf{v}_{k+1} \end{pmatrix} = \mathbf{F} \cdot \begin{pmatrix} \mathbf{x}_{k-1} \\ h \cdot \mathbf{v}_{k-1} \\ \mathbf{x}_k \\ h \cdot \mathbf{v}_k \end{pmatrix} \tag{5.52}$$

where:

$$\mathbf{F} = \begin{pmatrix} \mathbf{Z}^{(n)} & \mathbf{Z}^{(n)} & \mathbf{I}^{(n)} & \mathbf{Z}^{(n)} \\ \mathbf{Z}^{(n)} & \mathbf{Z}^{(n)} & \mathbf{Z}^{(n)} & \mathbf{I}^{(n)} \\ -\mathbf{I}^{(n)} & \mathbf{Z}^{(n)} & \left[2 \cdot \mathbf{I}^{(n)} + (\mathbf{A} \cdot h)^2 \right] & \mathbf{B} \cdot h \\ -\frac{1}{2} \cdot (\mathbf{A} \cdot h)^2 & -\frac{1}{2} \cdot (\mathbf{B} \cdot h) & \frac{3}{2} \cdot (\mathbf{A} \cdot h)^2 & \left[\mathbf{I}^{(n)} + \frac{3}{2} \cdot (\mathbf{B} \cdot h) \right] \end{pmatrix} \tag{5.53}$$

5.6   Stability and Damping of Godunov Scheme

As with all other integration techniques, we shall now introduce a classification code for the Godunov scheme. Since the algorithm is explicit and third–order accurate, we shall call this scheme GE3. We shall call the enhanced scheme, that also computes the velocity vector, GE3/AB2.

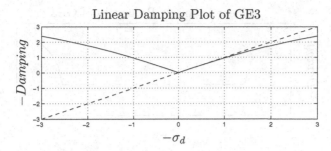

FIGURE 5.2. Linear damping plot of GE3 algorithm.

Before we attempt drawing a stability domain of GE3, we shall draw the linear damping plot. It is shown in Fig. 5.2.

How very disappointing! The scheme is unstable in the left half plane. The result should not surprise us too much. Since the **F**–matrix of Eq.(5.34) is an even function in $\mathbf{A} \cdot h$, the damping properties must be symmetric to the imaginary axis. Thus there cannot exist an asymptotic region around the origin, as we would expect of any well–behaved integration algorithm.

To gain a better understanding of the damping properties of the algorithm, let us plot the damping order star.

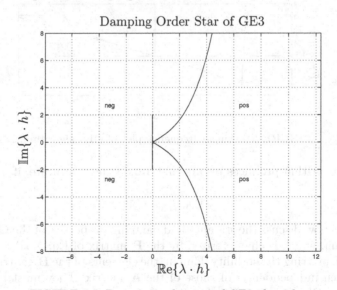

FIGURE 5.3. Damping order star of GE3 algorithm.

Interesting is the line segment stretching from $-2j$ to $+2j$ along the imaginary axis. Evidently, there is zero damping along this line segment, which is exactly, what it should be. To verify the results, let us plot the linear damping properties once more, but this time along the imaginary rather than the real axis.

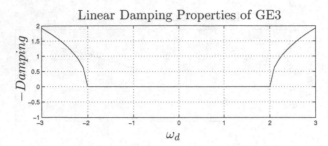

FIGURE 5.4. Linear damping properties of GE3 along imaginary axis.

The GE3 algorithm is only useful for strictly conservative systems. In order to obtain marginally stable results, the largest absolute eigenvalue multiplied by the step size must be smaller than or equal to 2:

$$|\lambda|_{\max} \cdot h \leq 2 \qquad (5.54)$$

Let us now plot the linear frequency properties of GE3 along the imaginary axis.

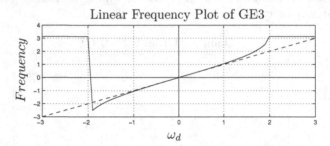

FIGURE 5.5. Linear frequency plot of GE3 algorithm.

The algorithm produces results that are decently accurate, if

$$|\lambda|_{\max} \cdot h \leq 1 \qquad (5.55)$$

Let us now discuss the stability and damping properties of the GE3/AB2 algorithm, which is characterized by the $\mathbf{F}$–matrix of Eq.(5.53).

When plotting the stability domain, the elements of the $\mathbf{B}$–matrix cannot be chosen independently of those of the $\mathbf{A}$–matrix. They must be chosen such that the overall system has its eigenvalues located on the unit circle. The procedure is explained in detail in Hw.[H5.3].

Since this is a third–order accurate linear explicit multi–step method similar in scope to AB3, we decided to plot the stability domain of AB3 on top of the stability domain of GE3/AB2.

GE3/AB2 performs similar to AB3 for systems with eigenvalues located in the vicinity of the negative real axis, but it outperforms AB3 when

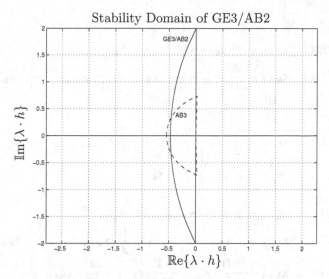

FIGURE 5.6. Stability domain of GE3/AB2 algorithm.

employed to simulating systems with their eigenvalues located close to the imaginary axis.

5.7 Explicit and Implicit Godunov Algorithms of Different Orders

Classes of both explicit and implicit integration algorithms of different orders of approximation accuracy for second–derivative systems can be derived using Newton–Gregory polynomials.

To this end, we shall develop $\mathbf{x}(t)$ into a Newton–Gregory backward polynomial around t_{k+1}. We then compute the second derivative of the Newton–Gregory polynomial. Evaluating this second derivative polynomial for $s = -1$, we obtain the class of explicit Godunov schemes. Evaluating the second derivative polynomial for $s = 0$, we obtain the class of implicit Godunov methods.

We shall denote the explicit Godunov scheme of order n as GE_n, and the implicit Godunov algorithm of the same order as GI_n. The enhanced algorithms, that also compute the velocity vector, are denoted as GE_n/AB_{n-1} and GI_n/BDF_{n-1}, respectively.

Since the approach is very similar to those presented in Chapter 4 for the derivation of the AB, AM, and BDF algorithms, we shall refrain from repeating this derivation here once more. After all, some problems need to remain to be dealt with in the homework section.

The resulting algorithms are summarized below:

$$GE3: \quad \mathbf{x_{k+1}} = 2 \cdot \mathbf{x_k} - \mathbf{x_{k-1}} + h^2 \cdot \ddot{\mathbf{x}}_k \tag{5.56}$$

$$GE4: \quad \mathbf{x_{k+1}} = \frac{20}{11} \cdot \mathbf{x_k} - \frac{6}{11} \cdot \mathbf{x_{k-1}} - \frac{4}{11} \cdot \mathbf{x_{k-2}} + \frac{1}{11} \cdot \mathbf{x_{k-3}}$$
$$+ \frac{12}{11} \cdot h^2 \cdot \ddot{\mathbf{x}}_k \tag{5.57}$$

$$GE5: \quad \mathbf{x_{k+1}} = \frac{3}{2} \cdot \mathbf{x_k} + \frac{2}{5} \cdot \mathbf{x_{k-1}} - \frac{7}{5} \cdot \mathbf{x_{k-2}} + \frac{3}{5} \cdot \mathbf{x_{k-3}}$$
$$- \frac{1}{10} \cdot \mathbf{x_{k-4}} + \frac{6}{5} \cdot h^2 \cdot \ddot{\mathbf{x}}_k \tag{5.58}$$

$$GI2: \quad \mathbf{x_{k+1}} = 2 \cdot \mathbf{x_k} - \mathbf{x_{k-1}} + h^2 \cdot \ddot{\mathbf{x}}_{k+1} \tag{5.59}$$

$$GI3: \quad \mathbf{x_{k+1}} = \frac{5}{2} \cdot \mathbf{x_k} - 2 \cdot \mathbf{x_{k-1}} + \frac{1}{2} \cdot \mathbf{x_{k-2}} + \frac{1}{2} \cdot h^2 \cdot \ddot{\mathbf{x}}_{k+1} \tag{5.60}$$

$$GI4: \quad \mathbf{x_{k+1}} = \frac{104}{35} \cdot \mathbf{x_k} - \frac{114}{35} \cdot \mathbf{x_{k-1}} + \frac{56}{35} \cdot \mathbf{x_{k-2}} - \frac{11}{35} \cdot \mathbf{x_{k-3}}$$
$$+ \frac{12}{35} \cdot h^2 \cdot \ddot{\mathbf{x}}_{k+1} \tag{5.61}$$

$$GI5: \quad \mathbf{x_{k+1}} = \frac{154}{45} \cdot \mathbf{x_k} - \frac{214}{45} \cdot \mathbf{x_{k-1}} + \frac{52}{15} \cdot \mathbf{x_{k-2}} - \frac{61}{45} \cdot \mathbf{x_{k-3}}$$
$$+ \frac{2}{9} \cdot \mathbf{x_{k-4}} + \frac{12}{45} \cdot h^2 \cdot \ddot{\mathbf{x}}_{k+1} \tag{5.62}$$

The algorithms can be summarized using the following α– and β–matrices:

$$\alpha_{\text{GE}} = \begin{pmatrix} 0 \\ 1 \\ 1 \\ \frac{12}{11} \\ \frac{6}{5} \end{pmatrix} \quad ; \quad \beta_{\text{GE}} = \begin{pmatrix} 0 & 0 & 0 & 0 & 0 \\ 2 & -1 & 0 & 0 & 0 \\ 2 & -1 & 0 & 0 & 0 \\ \frac{20}{11} & -\frac{6}{11} & -\frac{4}{11} & \frac{1}{11} & 0 \\ \frac{3}{2} & \frac{2}{5} & -\frac{7}{5} & \frac{3}{5} & -\frac{1}{10} \end{pmatrix} \tag{5.63}$$

$$\alpha_{\text{GI}} = \begin{pmatrix} 0 \\ 1 \\ \frac{1}{2} \\ \frac{12}{35} \\ \frac{12}{45} \end{pmatrix} \quad ; \quad \beta_{\text{GI}} = \begin{pmatrix} 0 & 0 & 0 & 0 & 0 \\ 2 & -1 & 0 & 0 & 0 \\ \frac{5}{2} & -2 & \frac{1}{2} & 0 & 0 \\ \frac{104}{35} & -\frac{114}{35} & \frac{56}{35} & -\frac{11}{35} & 0 \\ \frac{154}{45} & -\frac{214}{45} & \frac{52}{15} & -\frac{61}{45} & \frac{2}{9} \end{pmatrix} \tag{5.64}$$

Unfortunately, all of these methods have $\mathbf{F}$–matrices that are even functions in $\mathbf{A} \cdot h$. Thus, none of these methods can be expected to offer an asymptotic region for eigenvalues located along the real axis. In fact, all of the above techniques are unstable everywhere in the vicinity of the origin, with the exception of the imaginary axis itself, where they exhibit marginal stability, as they should.

Let us plot the damping properties of these algorithms up and down along the imaginary axis. The damping properties of GE4 and GE5 are shown in Fig. 5.7.

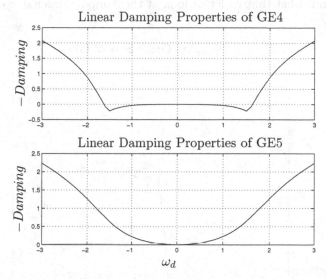

FIGURE 5.7. Damping properties of GE4 and GE5 along the imaginary axis.

Both of these algorithms may be used. Unfortunately, all of the explicit Godunov schemes, when used as stand–alone algorithms, are only applicable to the simulation of *linear conservation laws*. Engineers will likely shrug them off, because there aren't many real–life engineering applications that call for the simulation of linear conservation laws.

Yet, these algorithms may be perfectly suitable as part of blending algorithms.

Let us now discuss the implicit Godunov schemes. Their damping properties along the imaginary axis are shown in Fig. 5.8.

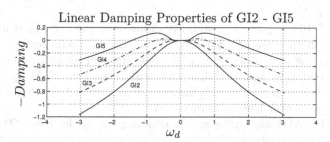

FIGURE 5.8. Damping properties of GI2 ... GI5 along the imaginary axis.

All of these algorithms could be used as well for the simulation of linear conservation laws, but there is no good reason, why we would ever want to do so. These algorithms have no advantages over their explicit brethren. They are only less efficient.

Furthermore, there is something else wrong with these algorithms. To understand, what that is, let us look at the damping order star of GI5.

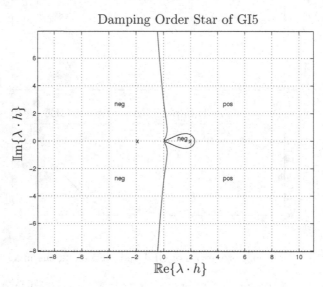

FIGURE 5.9. Damping order star of GI5 algorithm.

The locus of correct damping approximates the imaginary axis, which is nice. However, all of the algorithms of the GI class have a pole pair symmetric to the imaginary axis. Thus, all of these algorithms have a pole in the left half plane, which is something we should shun away from.

Let us now discuss the enhanced algorithms, i.e., the ones that also compute the velocity vector. We begin with GE4/AB3, and compare this algorithm with AB4.

Figure 5.10 compares the stability domains of the two algorithms.

Just as in the case of GE3/AB2, also GE4/AB3 will perform similarly to AB4 in the case of systems with their eigenvalues located close to the negative real axis, but the algorithm will outperform AB4 by leaps and bounds when employed to simulating systems with their dominant eigenvalues located near the imaginary axis.

The GE5/AB4 algorithm is unfortunately unstable. Even GE5/ABM4 turns out to be unstable.

It makes sense to try enhancing the implicit Godunov schemes by computing the velocity vector using a BDF formula of one order below that of the Godunov scheme itself.

For example, the GI3/BDF2 algorithm may be written as:

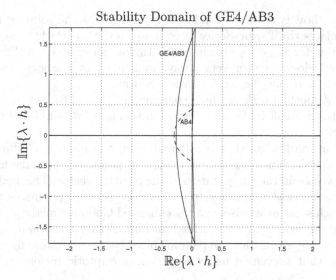

FIGURE 5.10. Stability domain of GE4/AB3.

$$x_{k+1} = \frac{5}{2} \cdot x_k - 2 \cdot x_{k-1} + \frac{1}{2} \cdot x_{k-2} + \frac{h^2}{2} \cdot \ddot{x}_{k+1} \qquad (5.65)$$

$$h \cdot \dot{x}_{k+1} = \frac{4 \cdot h}{3} \cdot \dot{x}_k - \frac{h}{3} \cdot \dot{x}_{k-1} + \frac{2 \cdot h^2}{3} \cdot \ddot{x}_{k+1} \qquad (5.66)$$

Unfortunately, it didn't work. Some of the resulting algorithms are indeed A–stable. They have nice unstable regions in the right half complex $\lambda \cdot h$ plane, but unfortunately, the damping is exactly equal to zero everywhere else. None of these algorithms produces an asymptotic region everywhere around the origin.

5.8   The Newmark Algorithm

An integration algorithm for the direct numerical solution of second derivative systems that has seen quite a bit of publicity in the mechanical engineering literature is the integration algorithm by Newmark [5.7]. Although the algorithm had originally been proposed for the solution of problems in structural dynamics, such as the simulation of earthquakes, it can be used for the simulation of other mechanical systems as well [5.1, 5.5].

The method can be described as follows:

$$x_{k+1} = x_k + h \cdot \dot{x}_k + \frac{h^2}{2} \cdot [(1 - \vartheta_1) \cdot \ddot{x}_k + \vartheta_1 \cdot \ddot{x}_{k+1}] \qquad (5.67)$$

$$h \cdot \dot{x}_{k+1} = h \cdot \dot{x}_k + h^2 \cdot [(1 - \vartheta_2) \cdot \ddot{x}_k + \vartheta_2 \cdot \ddot{x}_{k+1}] \qquad (5.68)$$

The method is clearly second–order accurate, as the solution for $\mathbf{x_{k+1}}$ approximates the Taylor–Series directly up to the quadratic term, whereas the solution for $\dot{\mathbf{x}}_{k+1}$ approximates the Taylor–Series up to the linear term. Since the velocity vector gets always multiplied by the step size, h, the overall method must be second–order accurate.

It is a ϑ–method with two fudge parameters, ϑ_1 and ϑ_2. For $\vartheta_1 = \vartheta_2 = 0$, the method is explicit; for all other combinations of ϑ_1 and ϑ_2, the method is implicit.

The Newmark algorithm is different from the implicit Godunov techniques. The Godunov algorithms made it a point, not to make use of the velocity vector in the computation of the position vector. This makes sense in the special case of velocity–free second–derivative systems, but doesn't make much sense otherwise, i.e., the enhanced Godunov methods, that also compute the velocity vector, added an unnecessary constraint on the design of the algorithm that we paid for bitterly, since we were fighting even functions that prevented us from getting asymptotic regions around the origin.

Let us plot the stability domains of the Newmark algorithm for $\vartheta_1 = \vartheta_2 = \{0.0, 0.25, 0.5, 0.75, 1.0\}$.

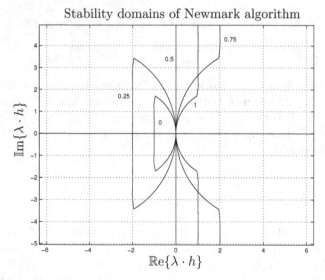

FIGURE 5.11. Stability domains of Newmark algorithm as a function of ϑ_1 and ϑ_2.

These are interesting looking and quite unusual stability domains. The algorithms are symmetric to $\vartheta_1 = \vartheta_2 = 0.5$. For $\vartheta_1 = \vartheta_2 = 0$, the algorithm exhibits a stable region in the left half plane that looks like an ascending half moon. The stable region is limited by $\mathbb{Re}\{\lambda \cdot h\} \geq -1.0$. As ϑ_1 and ϑ_2 increase their values, the half moon grows in size, and the stable region extends further to the left. For $\vartheta_1 = \vartheta_2 = 0.25$, the region is limited by $\text{Re}\{\lambda \cdot h\} \geq -2.0$. As ϑ_1 and ϑ_2 approach values of 0.5, the entire left half plane is covered by the stable region. For $\vartheta_1 = \vartheta_2 = 0.5$, the resulting algorithm is F–stable. At $\vartheta_1 = \vartheta_2 = 0.5$, the stable region wraps around infinity. For $\vartheta_1 = \vartheta_2 = 0.75$, the entire left half plane is stable, but also the region limited by $\text{Re}\{\lambda \cdot h\} \geq 2.0$.

¿From a practical perspective, the algorithms with $\vartheta_1 = \vartheta_2 \in (0, 0.5)$ are probably not of much interest. $\vartheta_1 = \vartheta_2 = 0$ could be interesting, as this is an explicit algorithm. Also the algorithms with $\vartheta_1 = \vartheta_2 \geq 0.5$ are of interest, since they are all A–stable.

Let us look at some damping plots. Figure 5.12 shows the linear and logarithmic damping plots of the F–stable Newmark algorithm for $\vartheta_1 = \vartheta_2 = 0.5$.

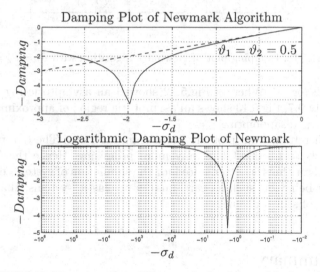

FIGURE 5.12. Damping plot of Newmark algorithm with $\vartheta_1 = \vartheta_2 = 0.5$.

As was to be expected, the damping is zero everywhere at infinity, and in fact, the damping assumes very small values in large portions of the left half plane, which could potentially become a problem when dealing with stiff systems.

Let us check, whether the Newmark algorithms with larger ϑ–values fare any better. Figure 5.13 shows the damping plots of the A–stable Newmark algorithm for $\vartheta_1 = \vartheta_2 = 0.75$.

It did not help. The damping still approaches zero at infinity. We should have been able to predict this result from the stability domain alone. Since the border of stability reaches all the way to infinity at some place, the numerical scheme must exhibit marginal stability at infinity, irrespective of how infinity is being approached.

Furthermore, the algorithm with $\vartheta_1 = \vartheta_2 = 0.75$ exhibits a much smaller

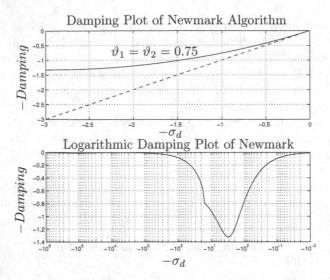

FIGURE 5.13. Damping plot of Newmark algorithm with $\vartheta_1 = \vartheta_2 = 0.75$.

asymptotic region. Whereas Fig.5.12 showed an asymptotic region of approximately 0.7, Fig.5.13 shows an asymptotic region of approximately 0.2. This result is disappointing.

We shall refrain from plotting more damping plots. The damping plots for $\vartheta_1 = \vartheta_2 = 1.0$ don't look any more promising. Hence the Newmark family of algorithms is better suited for the simulation of systems exhibiting oscillatory behavior, such as earthquakes or elastic systems, than for the simulation of stiff problems.

5.9   Summary

In this chapter, we have looked at new classes of simulation algorithms that can deal with second derivative systems directly, i.e., without first converting them to state–space form.

Our original goal had been to design algorithms for velocity–free problems that would avoid the seemingly unnecessary computation of the velocity vector altogether. To this end, we developed two families of algorithms, the explicit and implicit Godunov schemes, using our old friends, the Newton–Gregory polynomials.

Unfortunately, this approach was unsuccessful. None of the resulting algorithms possess any asymptotic regions around their origins, as the $\mathbf{F}$–matrices associated with these algorithms are even functions of $\mathbf{A} \cdot h$. Consequently, these algorithms, when used in a stand–alone mode, can only be employed for the simulation of linear conservation laws, a result that is not overly exciting.

We then enhanced these algorithms to make them suitable for the simulation of systems with damping, and found two explicit algorithms, GE3/AB2 and GE4/AB3, that can be considered serious contenders of AB3 and AB4 for the simulation of mechanical systems with strong oscillatory behavior. This is certainly a nice, albeit still not a truly exciting, result.

We then analyzed an algorithm that has seen quite a bit of publicity in the mechanical engineering literature: the algorithm by Newmark, an algorithm that is still being used quite frequently for the simulation of mechanical systems exhibiting oscillatory behavior. Unfortunately, the algorithm is only second–order accurate, which makes it unsuitable for general purpose simulation, as most engineering applications call for third and fourth–order accurate integration algorithms, in order to be simulated efficiently.

The Newmark family of algorithms can be, and have often been, used in *real–time applications*, because most real–time applications call for small step sizes to track external input signals, and therefore, are often simulated using low–order integration algorithms. Since real–time applications cannot deal with implicit algorithms very well, we should probably use the linearly–implicit variant of the Newmark algorithm in those cases. We shall deal intensively and extensively with real–time simulation in Chapter 10 of this book.

... And here comes the most exciting aspect of this chapter. Nathan Newmark, in 1959, accomplished for second derivative systems, what Leonhard Euler accomplished for first derivative systems (i.e., state–space models) in 1768 [5.4], and yet, researchers still write articles about this algorithm today, and the algorithm is still being employed frequently in engineering applications.

In the previous two chapters, we had to work very hard to make a mark. The field of numerical ODE solvers for state–space models is a very mature research topic, and consequently, it is difficult to come up with something new that no–one else has thought about before.

In contrast, the field of numerical ODE solvers for second derivative systems is still in its infancy, largely ignored by the community of applied mathematicians so far, and is therefore a fruitful research area for young and aspiring researchers of numerical methods.

5.10 References

[5.1] Klaus-Jürgen Bathe. *Finite Element Procedures in Engineering Analysis*. Prentice–Hall, 1982.

[5.2] Christopher Paul Beamis. Solution of Second Order Differential Equations Using the Godunov Integration Method. Master's thesis, Dept. of Electrical & Computer Engineering, University of Arizona, Tucson, Ariz., 1990.

[5.3] François E. Cellier. *Continuous System Modeling*. Springer Verlag, New York, 1991. 755p.

[5.4] Leonhard Euler. De integratione æquationum differentialium per approximationem. In *Opera Omnia*, volume 11 of *first series*, pages 424–434. Institutiones Calculi Integralis, Teubner Verlag, Leipzig, Germany, 1913.

[5.5] Javier Garcia de Jalón and Eduardo Bayo. *Kinematic and Dynamic Simulation of Multibody Systems –The Real–Time Challenge–*. Wiley, 1994.

[5.6] John C. Houbolt. A Recurrence Matrix Solution for the Dynamic Response of Elastic Aircraft. *Journal of Aeronautical Science*, 17:540–550, 1950.

[5.7] Nathan M. Newmark. A Method of Computation for Structural Dynamics. *ASCE Journal of the Engineering Mechanics Division*, pages 67–94, 1959.

[5.8] Raymond T. Stefani, Clement J. Savant Jr., Bahram Shahian, and Gene H. Hostetter. *Design of Feedback Control Systems*. Saunders College Publishing, Orlando, Florida, 1994. 819p.

[5.9] Edward L. Wilson. A Computer Program for the Dynamic Stress Analysis of Underground Structures. Technical Report SESM Report, 68-1, University of California, Berkeley, Division of Structural Engineering and Structural Mechanics, 1968.

5.11   Bibliography

[B5.1] Edda Eich-Söllner and Claus Führer. *Numerical Methods in Multibody Dynamics*. Teubner–Verlag, Stuttgart, Germany, 1998.

[B5.2] Michel Géradin and Alberto Gardona. *Flexible Multibody Dynamics: A Finite Element Approach*. John Wiley & Sons, Chichester, New York, 2001.

[B5.3] Parviz E. Nikravesh. *Computer–aided Analysis of Mechanical Systems*. Prentice Hall, Englewood Cliffs, New Jersey, 1988. 370p.

5.12   Homework Problems

[H5.1] Explicit Godunov Algorithms

We wish to generate the class of GE algorithms introduced in this chapter. These linear multi–step algorithms compute x_{k+1} as a function of previous values of $\mathbf{x}$ and of $\ddot{\mathbf{x}}_{\mathbf{k}}$.

Develop $\mathbf{x}(t)$ into a Newton–Gregory backward polynomial around the time instant t_{k+1}. Compute the second derivative:

$$\ddot{\mathbf{x}}(t) = \frac{1}{h^2} \cdot \mathbf{f}(s) \qquad \text{(H5.1a)}$$

Evaluate this expression for $s = -1$. Truncating this expression after the quadratic ∇^2 term, you obtain a second–order accurate formula for $\ddot{\mathbf{x}}_{\mathbf{k}}$, which can be solved for $\mathbf{x}_{k+1}$. Truncating after the cubic ∇^3 term, you obtain a third–order accurate formula, etc.

There is no first–order accurate GE scheme. From (H5.1a), one can recognize by inspection that the second–order accurate GE2 scheme is already third–order accurate. Explain.

[H5.2] Implicit Godunov Algorithms

We wish to generate the class of GI algorithms introduced in this chapter. These linear multi–step algorithms compute $\mathbf{x}_{k+1}$ as a function of previous values of $\mathbf{x}$ and of $\ddot{\mathbf{x}}_{k+1}$.

Develop $\mathbf{x}(t)$ into a Newton–Gregory backward polynomial around the time instant t_{k+1}. Compute the second derivative:

$$\ddot{\mathbf{x}}(t) = \frac{1}{h^2} \cdot \mathbf{f}(s) \qquad \text{(H5.2a)}$$

Evaluate this expression for $s = 0$. Truncating this expression after the quadratic ∇^2 term, you obtain a second–order accurate formula for $\ddot{\mathbf{x}}_{k+1}$, which can be solved for $\mathbf{x}_{k+1}$. Truncating after the cubic ∇^3 term, you obtain a third–order accurate formula, etc.

[H5.3] Stability Domain of GE4/AB3

The method introduced in earlier chapters for drawing stability domains was geared towards linear time–invariant homogeneous multi–variable state–space models:

$$\dot{\mathbf{x}} = \mathbf{A} \cdot \mathbf{x} \qquad \text{(H5.3a)}$$

We generated real–valued $\mathbf{A}$–matrices $\in \mathbb{R}^{2 \times 2}$ with their eigenvalues located on the unit circle, at an angle α away from the negative real axis. We then computed the $\mathbf{F}$–matrix corresponding to that $\mathbf{A}$–matrix for the given algorithm, and found the largest value of the step size h, for which all eigenvalues of $\mathbf{F}$ remained inside the unit circle. This gave us one point on the stability domain. We repeated this procedure for all suitable values of the angle α.

The algorithm needs to be modified for dealing with second derivative systems described by the linear time–invariant homogeneous multi–variable second–derivative model:

$$\ddot{\mathbf{x}} = \mathbf{A}^2 \cdot \mathbf{x} + \mathbf{B} \cdot \dot{\mathbf{x}} \qquad \text{(H5.3b)}$$

We need to find real–valued $\mathbf{A}$– and $\mathbf{B}$–matrices such that the model of (H5.3b) has its eigenvalues located on the unit circle.

This can be accomplished using the scalar model:

$$\ddot{x} = a^2 \cdot x + b \cdot \dot{x} \qquad \text{(H5.3c)}$$

where:

$$a = \sqrt{a_{21}} \qquad \text{(H5.3d)}$$
$$b = a_{22} \qquad \text{(H5.3e)}$$

of the formerly used $\mathbf{A}$–matrix.

Write the GE4/AB3 algorithm as follows:

$$x_{k+1} = \frac{20}{11} \cdot x_k - \frac{6}{11} \cdot x_{k-1} - \frac{4}{11} \cdot x_{k-2} + \frac{1}{11} \cdot x_{k-3}$$
$$+ \frac{12 \cdot h^2}{11} \cdot \ddot{x}_k \qquad \text{(H5.3f)}$$

$$h \cdot \dot{x}_{k+1} = h \cdot \dot{x}_k + \frac{23 \cdot h^2}{12} \cdot \ddot{x}_k - \frac{4 \cdot h^2}{3} \cdot \ddot{x}_{k-1}$$
$$+ \frac{5 \cdot h^2}{12} \cdot \ddot{x}_{k-2} \qquad \text{(H5.3g)}$$

$$\ddot{x} = a^2 \cdot x + b \cdot \dot{x} \qquad \text{(H5.3h)}$$

Substitute Eq.(H5.3h) into Eq.(H5.3f) and Eq.(H5.3g), and rewrite the resulting equations in a state–space form:

$$\xi_{\mathbf{k+1}} = \mathbf{F} \cdot \xi_{\mathbf{k}} \qquad \text{(H5.3i)}$$

whereby the state vector ξ is chosen as:

$$\xi_{\mathbf{k}} = \begin{pmatrix} x_{k-3} \\ h \cdot \dot{x}_{k-3} \\ x_{k-2} \\ h \cdot \dot{x}_{k-2} \\ x_{k-1} \\ h \cdot \dot{x}_{k-1} \\ x_k \\ h \cdot \dot{x}_k \end{pmatrix} \qquad \text{(H5.3j)}$$

The $\mathbf{F}$–matrix turns out to be a function of $(a \cdot h)^2$ and of $b \cdot h$.

The remainder of the algorithm remains the same as before. Draw the stability domain of GE4/AB3 using this approach.

[H5.4] Stability Domain of GI3/BDF2

We want to draw the stability domain of GI3/BDF2. However, we shall use a different approach from that advocated in Hw.[H5.3], an approach that is a bit slower in execution, but more general.

We start by writing the algorithm as follows:

$$x_{k+1} = \frac{5}{2} \cdot x_k - 2 \cdot x_{k-1} + \frac{1}{2} \cdot x_{k-2} + \frac{h^2}{2} \cdot \ddot{x}_{k+1} \qquad \text{(H5.4a)}$$

$$h \cdot \dot{x}_{k+1} = \frac{4 \cdot h}{3} \cdot \dot{x}_k - \frac{h}{3} \cdot \dot{x}_{k-1} + \frac{2 \cdot h^2}{3} \cdot \ddot{x}_{k+1} \qquad \text{(H5.4b)}$$

$$\ddot{x} = a^2 \cdot x + b \cdot \dot{x} \qquad \text{(H5.4c)}$$

We substitute Eq.(H5.4c) into Eq.(H5.4a) and Eq.(H5.4b), and rewrite the resulting equations in a state–space form:

$$\xi_{k+1} = \mathbf{F} \cdot \xi_k \qquad \text{(H5.4d)}$$

whereby the state vector ξ is chosen as:

$$\xi_k = \begin{pmatrix} x_{k-2} \\ h \cdot \dot{x}_{k-2} \\ x_{k-1} \\ h \cdot \dot{x}_{k-1} \\ x_k \\ h \cdot \dot{x}_k \end{pmatrix} \qquad \text{(H5.4e)}$$

The $\mathbf{F}$–matrix turns out to be a function of $(a \cdot h)^2$ and of $b \cdot h$.

We choose a and b such that the two eigenvalues of the second–order differential equation are located in the position:

$$\lambda_{1,2} = \sigma \pm j \cdot \omega \qquad \text{(H5.4f)}$$

We perform a double loop over σ and ω to cover an entire area of the complex plane with eigenvalue locations.

For each eigenvalue location, we compute the corresponding values of α and h, compute the corresponding $\mathbf{F}$–matrix, and determine the discrete damping value using the equation:

$$damp = -\log(\max(abs(eig(\mathbf{F})))) \qquad \text{(H5.4g)}$$

The damping value can be interpreted as a real-valued function of the complex–valued argument $\sigma + j \cdot \omega$.

Use MATLAB's contour plot to draw the locus of all points of zero damping. This will be the stability domain of the method.

Interpret the results that you get.

[H5.5] Cervical Syndrome

Given the model of a sitting human body presented in Fig. 5.1. We exert this model by a sinusoidal force of 1.5 Hz. Simulate this model during 10 seconds, once using the GE4/AB3 algorithm, and once using the AB4 algorithm. Choose a fixed step size of $h = 0.1$ seconds.

Determine, which of the two algorithms is faster, i.e., requires a smaller number of floating point operations.

Assume that MATLAB's solution using the ODE45 algorithm of variable step size with a relative tolerance of 0.0001 is accurate, and compare the two solutions of your own simulation algorithms against MATLAB's solution. Determine, which of the algorithms produced more accurate results.

5.13   Projects

[P5.1] Houbolt's Integration Algorithm

John Houbolt proposed already in 1950 a second–derivative integration algorithm [5.6] that is very similar to the GI3/BDF2 method introduced in this chapter. Houbolt's algorithm can be written as follows:

$$x_{k+1} = \frac{5}{2} \cdot x_k - 2 \cdot x_{k-1} + \frac{1}{2} \cdot x_{k-2} + \frac{h^2}{2} \cdot \ddot{x}_{k+1} \quad \text{(P5.1a)}$$

$$h \cdot \dot{x}_{k+1} = \frac{11}{6} \cdot x_{k+1} - 3 \cdot x_k + \frac{3}{2} \cdot x_{k-1} - \frac{1}{3} \cdot x_{k-2} \quad \text{(P5.1b)}$$

The second derivative formula of Houbolt's algorithm can immediately be identified as GI3. The formula used for the velocity vector is BDF3; however, the formula was used differently from the way, it had been employed by us in the description of the GI3/BDF2 algorithm. Clearly, the Houbolt algorithm is third–order accurate.

Although it would have sufficed to use BDF2 for the velocity vector, nothing would have been gained computationally by choosing the reduced–order algorithm.

We can transform the Houbolt algorithm to the form that we meanwhile got used to by substituting Eq.(P5.1a) into Eq.(P5.1b). The so rewritten Houbolt algorithm assumes the form:

$$x_{k+1} = \frac{5}{2} \cdot x_k - 2 \cdot x_{k-1} + \frac{1}{2} \cdot x_{k-2} + \frac{h^2}{2} \cdot \ddot{x}_{k+1} \quad \text{(P5.1c)}$$

$$h \cdot \dot{x}_{k+1} = \frac{19}{12} \cdot x_k - \frac{13}{6} \cdot x_{k-1} + \frac{7}{12} \cdot x_{k-2}$$

$$+ \frac{11 \cdot h^2}{12} \cdot \ddot{x}_{k+1} \quad \text{(P5.1d)}$$

Find the stability domain and damping plot of Houbolt's algorithm, and discuss the properties of this algorithm in the same way, as Nemark's algorithm was discussed in this chapter.

Repeat the analysis, replacing the BDF3 formula by the BDF2 formula.

Analyze whether stable GI4/BDF3 and/or GI4/BDF4 algorithms can be constructed in the same fashion.

[P5.2] Wilson's Integration Algorithm

In 1968, Wilson proposed yet another second–derivative integration algorithm [5.9] that is quite similar to Newmark's method. A clean derivation of Wilson's algorithm can be found in Bathe [5.1].

The algorithm can be converted easily to the form that we embraced in this chapter:

$$x_{k+1} = x_k + h \cdot \dot{x}_k + \frac{h^2}{6} \cdot \ddot{x}_{k+1} + \frac{h^2}{3} \cdot \ddot{x}_k \tag{P5.2a}$$

$$h \cdot \dot{x}_{k+1} = h \cdot \dot{x}_k + \frac{h^2}{2} \cdot \ddot{x}_{k+1} + \frac{h^2}{2} \cdot \ddot{x}_k \tag{P5.2b}$$

Clearly, Eq.(P5.2a) is second–order accurate in x_k, whereas Eq.(P5.2b) is first–order accurate in $\dot{x}_k$. Hence the overall algorithm is second–order accurate.

Find the stability domain and damping plot of Wilson's algorithm, and discuss the properties of this algorithm in the same way, as Nemark's algorithm was discussed in this chapter.

Compare Wilson's and Newmark's algorithms with each other. Which of them would you use when and why?

5.14 Research

[R5.1] Second–derivative Runge–Kutta Algorithms

When designing explicit Runge–Kutta algorithms, we started out with a partial step of Euler, used the result obtained by that stage as a predictor, and added another stage as a corrector, in which we blended the solutions of the two stages. We continued in the same way to more and more stages, trying to approximate the Taylor–series expansion to increasingly higher orders. Contrary to the multi–step methods, we were able to determine also the nonlinear order of approximation accuracy in the case of the Runge–Kutta algorithms.

It must be possible to generalize the Runge–Kutta algorithms to second–derivative form. To this end, we postulate a first partial step using the algorithm:

$$\mathbf{x}^{P_1}(t + \alpha_1 \cdot h) = \mathbf{x_k} + \alpha_1 \cdot h \cdot \dot{\mathbf{x}}_k + \frac{(\alpha_1 \cdot h)^2}{2} \cdot \ddot{\mathbf{x}}_k \qquad (R5.1a)$$

We then proceed to building corrector stages in the same fashion, as we did in the case of the regular RK algorithms.

Whereas we left the first derivative alone in the case of the RK algorithms, and only expanded second and higher–order derivatives into Taylor series, we now must leave the first and second derivatives alone, and only develop third and higher–order derivatives into Taylor series.

We shall use a regular RK algorithm of one order lower than the second–derivative algorithm to compute the velocity vector. We use the freedom in the design of such algorithms to ensure that the individual stages of the second–derivative algorithm for computing the position vector and the first–derivative algorithm for computing the velocity vector are evaluated at the same instants of simulated time, i.e.:

$$\alpha_i(2^{\text{nd}} \text{ derivative algorithm}) = \alpha_i(1^{\text{st}} \text{ derivative algorithm}) \qquad (R5.1b)$$

Clearly, there don't exist first–order accurate generalized RK algorithms, and the second–order accurate generalized explicit RK algorithm, GERK2, can be written as follows:

$$\mathbf{x}_{k+1} \;=\; \mathbf{x_k} + h \cdot \dot{\mathbf{x}}_k + \frac{h^2}{2} \cdot \ddot{\mathbf{x}}_k \qquad (R5.1c)$$

$$h \cdot \dot{\mathbf{x}}_{k+1} \;=\; h \cdot \dot{\mathbf{x}}_k + h^2 \cdot \ddot{\mathbf{x}}_k \qquad (R5.1d)$$

which is identical to the Newmark algorithm with $\vartheta_1 = \vartheta 2 = 0.0$.

It should be possible to develop second–derivative (generalized) explicit Runge–Kutta algorithms of any order, and we would expect that these algorithms should outperform their regular RK cousins when dealing with nonlinear mechanical systems that exhibit highly oscillatory behavior.

6
Partial Differential Equations

Preview

In this chapter, we shall deal with method–of–lines solutions to models that are described by individual partial differential equations, by sets of coupled partial differential equations, or possibly by sets of mixed partial and ordinary differential equations.

Emphasis will be placed on the process of converting partial differential equations to equivalent sets of ordinary differential equations, and particular attention will be devoted to the problem of converting boundary conditions. To this end, we shall again consult our –meanwhile well–understood– Newton–Gregory polynomials.

We shall then spend some time analyzing the particular difficulties that await us when numerically solving the sets of resulting differential equations in the cases of parabolic, hyperbolic, and elliptic partial differential equations. It turns out that each class of partial differential equations exhibits its own particular and peculiar types of difficulties.

6.1 Introduction

Partial differential equation (PDE) modeling and simulation are certainly among the more difficult topics to deal with. PDE modeling is still in its infancy. You hardly ever encounter models of coupled PDEs that contain more than three or four PDEs at a time. This situation is comparable with ordinary differential equation (ODE) modeling some 30 years ago. At that time, researchers were content to analyze simple ODE models consisting of three or four coupled ODEs. No special software tools were needed to help the modeler organize his or her models. The modeling process was utterly trivial. What was difficult was the process of converting these ODEs to a form such that a numerical differential equation solver could tackle them, and then the process of simulation itself.

This way of looking at simulation still prevails in large portions of the simulation literature. However, reality of ODE modeling has changed drastically over the years. Today, continuous system modelers frequently deal with models containing hundreds or even thousands of coupled differential and algebraic equations, and the process of first deriving and then maintaining these ODE models has become the truly difficult part.

This was the focus point of the companion book to this text *Continuous*

System Modeling [6.5]. In that book, PDEs weren't mentioned with even one word. The reason for this is obvious. No special software tools or modeling methodologies are needed yet to derive or maintain PDE models, since PDE models are still very simple. You don't encounter models containing hundreds or even only tens of PDEs. It just isn't done. If you end up with three or four coupled PDEs, this is a lot. So, from a modeling perspective, PDE modeling is still a fairly trivial undertaking.

On the other hand, the numerical solution of PDE models is by no means trivial. Whereas we have learnt meanwhile pretty well how to numerically handle large classes of ODE models, the numerical solution of PDE models still presents a challenge.

Many different approaches to simulating PDE models have been described in the literature, partly purely numerical, such as the finite element methods used mostly to tackle elliptic PDE problems, and partly semi–analytical, such as the method–of–characteristics approach to solving hyperbolic systems of equations. It is not the aim of this chapter at all to duplicate or compete with that literature.

Among all the techniques that are known for tackling PDE models, only one specific technique shall be dealt with in this book, namely the method–of–lines (MOL) approach to numerically solving PDE models. The MOL methodology converts PDEs into (large) sets of (in some way equivalent) ODEs that are then solved by standard ODE solvers. Since this book deals explicitly and extensively with ODE solvers, the MOL approach to PDE solving fits well within the overall framework of this book methodologically. This is the only reason why this text focuses on MOL solutions. It is not our intention to convey the impression that MOL solutions are, in each and every case, the most suitable way of dealing with PDE problems. PDE problems are notoriously difficult to tackle, and the MOL approach is only one, among many, techniques that can provide a partial answer to these challenges.

6.2   The Method of Lines

The Method of Lines (MOL) is a technique that enables us to convert partial differential equations (PDEs) into sets of ordinary differential equations (ODEs) that, in some sense, are equivalent to the former PDEs.

The basic idea behind the MOL methodology is straightforward. Let us look at the simple *heat equation* or *diffusion equation* in a single space variable:

$$\frac{\partial u}{\partial t} = \sigma \cdot \frac{\partial^2 u}{\partial x^2} \tag{6.1}$$

Rather than looking at the solution $u(x, t)$ everywhere in the two–dimensional space spanned by the spatial variable x and the temporal variable t, we can discretize the spatial variable, and look at the solutions $u_i(t)$ where the index i denotes a particular point x_i in space. To this end, we replace the second–order partial derivative of u with respect to x by a finite difference, such as:

$$\frac{\partial^2 u}{\partial x^2}\bigg|_{x=x_i} \approx \frac{u_{i+1} - 2u_i + u_{i-1}}{\delta x^2} \tag{6.2}$$

where δx is the (here equidistantly chosen) distance between two neighboring discretization points in space, i.e., the so–called *grid width* of the discretization.

Plugging Eq.(6.2) into Eq.(6.1), we find:

$$\frac{du_i}{dt} \approx \sigma \cdot \frac{u_{i+1} - 2u_i + u_{i-1}}{\delta x^2} \tag{6.3}$$

and we have already converted the former PDE in u into a set of ODEs in u_i.

The principal idea behind the MOL methodology is thus utterly trivial. However, the devil is in the detail.

It is reasonable to use the same order of approximation accuracy for the discretization in space as for the discretization in time achieved by the numerical integration algorithm. Thus, if we plan to integrate the set of ODEs with a fourth–order method, we should better find a discretization formula for $\partial^2 u / \partial x^2$ that is also fourth–order accurate.

This can be accomplished by use of our old friends, the *Newton–Gregory polynomials*. A fourth–order polynomial needs to be fitted through five points. Since we prefer *central differences* over *biased differences*, we fit the polynomial through the five points x_{i-2}, x_{i-1}, x_i, x_{i+1}, and x_{i+2}. Using Newton–Gregory backward polynomials, we will have to write the polynomial around the point that is located most to the right, in our case, the point x_{i+2}. Thus, we write:

$$u(x) = u_{i+2} + s\nabla u_{i+2} + \left(\frac{s^2}{2} + \frac{s}{2}\right)\nabla^2 u_{i+2} + \left(\frac{s^3}{6} + \frac{s^2}{2} + \frac{s}{3}\right)\nabla^3 u_{i+2} + \dots \tag{6.4}$$

Notice that we write the approximation polynomial as $u(x)$ rather than as $u(t)$, since we want to discretize along the spatial axis.

Consequently, the second derivative can be written as:

$$\frac{\partial^2 u}{\partial x^2} = \frac{1}{\delta x^2}\left[\nabla^2 u_{i+2} + (s+1)\nabla^3 u_{i+2} + \left(\frac{s^2}{2} + \frac{3s}{2} + \frac{11}{12}\right)\nabla^4 u_{i+2} + \dots\right] \tag{6.5}$$

Eq.(6.5) needs to be evaluated at $x = x_i$, corresponding to $s = -2$. Truncating after the quartic term and expanding the ∇–operators, we find:

$$\frac{\partial^2 u}{\partial x^2}\bigg|_{x=x_i} \approx \frac{1}{12\delta x^2}\left(-u_{i+2} + 16u_{i+1} - 30u_i + 16u_{i-1} - u_{i-2}\right) \qquad (6.6)$$

which is the fourth–order central difference approximation to the second partial derivative of $u(x,t)$ with respect to x evaluated at $x = x_i$.

We could have obtained the same result using the Newton–Gregory forward polynomial written around the point x_{i-2}, evaluating it for $s = +2$.

Had we decided that we wish to integrate with a second–order algorithm, we would have developed the Newton–Gregory backward polynomial around the point x_{i+1}, truncating Eq.(6.5) after the quadratic term, and evaluating for $s = -1$. This would have led to:

$$\frac{\partial^2 u}{\partial x^2}\bigg|_{x=x_i} \approx \frac{1}{\delta x^2}\left(u_{i+1} - 2u_i + u_{i-1}\right) \qquad (6.7)$$

which is the second–order central difference formula for $\partial^2 u/\partial x^2$, the one that had been used in Eq.(6.2).

The third–order case is again a little different. For geometric reasons, it is obviously impossible to fit a central difference approximation of an odd order around x_i using only x_i and its nearest three neighbors. Thus, we can choose between a biased formula using the points x_{i-2} up to x_{i+1}, i.e., develop the Newton–Gregory backward polynomial around the point x_{i+1} and evaluate it for $s = -1$, and another biased formula using the points x_{i-1} up to x_{i+2}, i.e., develop the Newton–Gregory backward polynomial around the point x_{i+2} and evaluate it for $s = -2$.

It turns out that both cases lead to exactly the same formula, namely Eq.(6.7). Just by accident, a lot of terms drop out, and Eq.(6.7) turns out to be third–order accurate.

Looking more deeply into the matter, we find that the "lucky accident" is no accident at all, but has to do with the symmetry conditions. Every central difference approximation is one order more accurate than the number of points fitted by it would make us believe. Consequently, Eq.(6.6) is in fact fifth–order accurate.

The next difficulty arises as we approach the spatial domain boundary. Let us assume the heat equation applies to the temperature distribution along a rod of length $\ell = 1$ m. Let us assume we cut the rod into segments of a length of $\delta\ell = 10$ cm. Thus, we get 10 segments. If the left end of the rod corresponds to index $i = 1$, the right end corresponds to index $i = 11$. Let us further assume that we wish to integrate using a fourth–order algorithm. Thus, we shall apply Eq.(6.6) to the points x_3 up to x_9. However, for the remaining points, we need biased formulae, since we cannot use points outside the range where the solution $u(x,t)$ is defined.

In order to find a biased formula for x_2, we shall have to write the Newton–Gregory backward polynomial around the point u_5 and evaluate for $s = -3$, or alternatively, we can write a Newton–Gregory forward polynomial around the point u_1 and evaluate for $s = +1$. In order to find a biased formula for x_1, we shall have to write the Newton–Gregory backward polynomial around the point u_5 and evaluate for $s = -4$, or alternatively, we can write a Newton–Gregory forward polynomial around the point u_1 and evaluate for $s = 0$. Similarly for the points x_{10} and x_{11}.

Using the above example, we obtain the following biased approximation formulae:

$$\left.\frac{\partial^2 u}{\partial x^2}\right|_{x=x_1} = \frac{1}{12\delta x^2}\left(11u_5 - 56u_4 + 114u_3 - 104u_2 + 35u_1\right) \tag{6.8a}$$

$$\left.\frac{\partial^2 u}{\partial x^2}\right|_{x=x_2} = \frac{1}{12\delta x^2}\left(-u_5 + 4u_4 + 6u_3 - 20u_2 + 11u_1\right) \tag{6.8b}$$

$$\left.\frac{\partial^2 u}{\partial x^2}\right|_{x=x_{10}} = \frac{1}{12\delta x^2}\left(11u_{11} - 20u_{10} + 6u_9 + 4u_8 - u_7\right) \tag{6.8c}$$

$$\left.\frac{\partial^2 u}{\partial x^2}\right|_{x=x_{11}} = \frac{1}{12\delta x^2}\left(35u_{11} - 104u_{10} + 114u_9 - 56u_8 + 11u_7\right) \tag{6.8d}$$

In the MOL methodology, all derivatives w.r.t. spatial variables are discretized using either central or biased difference approximations, whereas derivatives w.r.t. the temporal variable are left unchanged. In this way, PDEs are converted into sets of ODEs that can, at least in theory, be solved just like any other ODE models by means of standard ODE solvers.

Next, we need to discuss what is to be done with the boundary conditions. Every PDE has beside from *initial conditions* in time *boundary conditions* in space. For example, the heat equation may have the two boundary conditions:

$$u(x = 0.0, t) = 100.0 \tag{6.9a}$$

$$\frac{\partial u}{\partial x}(x = 1.0, t) = 0.0 \tag{6.9b}$$

The boundary condition of Eq.(6.9a) is called *boundary value condition*. This is the simplest case. All we need to do is to eliminate the differential equation for $u_1(t)$, and replace it by an algebraic equation, in our case:

$$u_1 = 100.0 \tag{6.10}$$

The boundary condition of Eq.(6.9b) is also a special case. It is called a *boundary symmetry condition*. It is handled in the following way. Imagine that there is a mirror at $x = 1.0$. This mirror maps the solution $u(x, t)$ into the range $x \in [1.0, 2.0]$, such that $u(2.0 - x, t) = u(x, t)$. Obviously, the boundary condition at $x = 2.0$ is the same as that at $x = 0.0$. There is then no need at all to specify any boundary condition at $x = 1.0$, since, through symmetry, the desired boundary symmetry condition will be satisfied. Knowing this, we can replace Eqs.(6.8c–d) by:

$$\frac{\partial^2 u}{\partial x^2}\bigg|_{x=x_{10}} = \frac{1}{12\delta x^2}\left(-u_{12} + 16u_{11} - 30u_{10} + 16u_9 - u_8\right) \qquad (6.11a)$$

$$\frac{\partial^2 u}{\partial x^2}\bigg|_{x=x_{11}} = \frac{1}{12\delta x^2}\left(-u_{13} + 16u_{12} - 30u_{11} + 16u_{10} - u_9\right) \qquad (6.11b)$$

i.e., by central difference approximations. However, since (due to symmetry) $u_{12} = u_{10}$ and $u_{13} = u_9$, we can rewrite Eqs.(6.11a–b) as:

$$\frac{\partial^2 u}{\partial x^2}\bigg|_{x=x_{10}} = \frac{1}{12\delta x^2}\left(16u_{11} - 31u_{10} + 16u_9 - u_8\right) \qquad (6.12a)$$

$$\frac{\partial^2 u}{\partial x^2}\bigg|_{x=x_{11}} = \frac{1}{12\delta x^2}\left(-30u_{11} + 32u_{10} - 2u_9\right) \qquad (6.12b)$$

and having done this, we can happily forget our virtual mirror again. We don't need to bother to actually compute a solution for the range $x \in [1.0, 2.0]$, since we already know the solution ... it is the mirror image of the solution in the range $x \in [0.0, 1.0]$.

A third type of special boundary conditions is the so–called *temporal boundary condition* of the type:

$$\frac{\partial u}{\partial t}(x = 0.0, t) = f(t) \qquad (6.13)$$

In this case, the boundary condition of the PDE is itself described through an ODE. This case is also easy. We simply replace the ODE for u_1 by the boundary ODE:

$$\dot{u}_1 = f(t) \qquad (6.14)$$

The more general boundary condition of the type:

$$g\left(u(x = 1.0, t)\right) + h\left(\frac{\partial u}{\partial x}(x = 1.0, t)\right) = f(t) \qquad (6.15)$$

where f, g, and h are arbitrary functions, is more tricky. For example, we may have to deal with a boundary condition of the type:

$$\frac{\partial u}{\partial x}(x = 1.0, t) = -k \cdot \left(u(x = 1.0, t) - u_{\text{amb}}(t)\right) \qquad (6.16)$$

where $u_{\text{amb}}(t)$ is the ambient temperature. How would we handle such a general boundary condition? The answer is simple. We again replace all spatial derivatives by appropriate Newton–Gregory polynomials, e.g. in the above case:

$$\frac{\partial u}{\partial x}\bigg|_{x=x_{11}} = \frac{1}{12\delta x}\left(25u_{11} - 48u_{10} + 36u_9 - 16u_8 + 3u_7\right) \tag{6.17}$$

is the fourth–order biased difference approximation polynomial. Plugging Eq.(6.17) into Eq.(6.16), and solving for u_{11}, we find:

$$u_{11} = \frac{12k \cdot \delta x \cdot u_{amb} + 48u_{10} - 36u_9 + 16u_8 - 3u_7}{12k \cdot \delta x + 25} \tag{6.18}$$

By this process, the *general boundary condition* has been transformed into a *boundary value condition*, and the ODE defining u_{11} can be dropped.

Often we are faced with *nonlinear boundary conditions*, such as the radiation condition:

$$\frac{\partial u}{\partial x}(x = 1.0, t) = -k \cdot \left(u(x = 1.0, t)^4 - u_{amb}(t)^4\right) \tag{6.19}$$

which leads to:

$$\mathcal{F}(u_{11}) = 12k \cdot \delta x \cdot u_{11}^4 + 25u_{11} - 12k \cdot \delta x \cdot u_{amb}^4 - 48u_{10} + 36u_9$$
$$- 16u_8 + 3u_7 = 0.0 \tag{6.20}$$

i.e., an *implicit boundary value condition* that can be solved by Newton iteration. Convergence should be fast since we can always use the value of $u_{11}(t_k - h)$ as the starting value of the iteration.

Finally, let us consider diffusion of heat through a wall. Assume that the wall has two layers consisting of two different materials, one of 1 m thickness, the other of 10 cm thickness. In that case, the diffusion coefficient, σ, assumes a different value in the two materials. We can formulate this problem as follows:

$$\frac{\partial u}{\partial t} = \sigma_u \cdot \frac{\partial^2 u}{\partial x^2} \tag{6.21a}$$

$$\frac{\partial v}{\partial t} = \sigma_v \cdot \frac{\partial^2 v}{\partial x^2} \tag{6.21b}$$

where the PDE for $u(x,t)$ is valid in the region $x \in [0.0, 1.0]$, and the PDE for $v(x,t)$ is valid in the region $x \in [1.0, 1.1]$, with boundary conditions at the boundary between the two layers:

$$\frac{\partial u}{\partial x}(x = 1.0, t) = -k_u \cdot \left(u(x = 1.0, t) - v(x = 1.0, t)\right) \tag{6.22a}$$

$$\frac{\partial v}{\partial x}(x = 1.0, t) = -k_v \cdot \left(v(x = 1.0, t) - u(x = 1.0, t)\right) \tag{6.22b}$$

which leads to the following two equations:

$$(12k_u \cdot \delta x_u + 25)u_{11} - 12k_u \cdot \delta x_u \cdot v_1 = 48u_{10} - 36u_9 + 16u_8 - 3u_7$$
$$(6.23a)$$

$$-12k_v \cdot \delta x_v \cdot u_{11} + (12k_v \cdot \delta x_v + 3)v_1 = 16v_2 - 36v_3 + 48u_4 - 25v_5$$
$$(6.23b)$$

Eqs.(6.23a–b) constitute a *linear algebraic loop* in the unknown variables u_{11} and v_1 that can be solved either symbolically or numerically.

6.3   Parabolic PDEs

Some very simple types of PDEs are so common that they were given special names. Let us consider the following PDE in two variables x and y:

$$a\frac{\partial^2 u}{\partial x^2} + b\frac{\partial^2 u}{\partial x \partial y} + c\frac{\partial^2 u}{\partial y^2} = d \qquad (6.24)$$

which is characteristic of many field problems in physics. x and y can be either spatial or temporal variables, and a, b, c, and d can be arbitrary functions of x, y, u, $\partial u/\partial x$, and $\partial u/\partial y$. Such a PDE is called *quasi–linear*, since it is linear in the highest derivatives.

Depending on the numerical relationship between a, b, and c, Eq.(6.24) is classified as either being *parabolic*, *hyperbolic*, or *elliptic*. The classification is as follows:

$$b^2 - 4ac > 0 \implies \text{PDE is hyperbolic} \qquad (6.25a)$$
$$b^2 - 4ac = 0 \implies \text{PDE is parabolic} \qquad (6.25b)$$
$$b^2 - 4ac < 0 \implies \text{PDE is elliptic} \qquad (6.25c)$$

This classification makes sense, since the numerical methods most suitable for these three types of PDEs are vastly different. In this section, we shall deal with PDEs of the parabolic type exclusively.

Parabolic PDEs are very common. For example, all thermal field problems are of that nature. The simplest example of a parabolic PDE is the one–dimensional heat diffusion problem of Eq.(6.1). A complete example of such a problem is specified once more below.

$$\frac{\partial u}{\partial t} = \frac{1}{10\pi^2} \cdot \frac{\partial^2 u}{\partial x^2} \quad ; \quad x \in [0,1] \quad ; \quad t \in [0,\infty) \qquad (6.26a)$$

$$u(x, t = 0) = \cos(\pi \cdot x) \qquad (6.26b)$$

$$u(x = 0, t) = \exp(-t/10) \qquad (6.26c)$$

$$\frac{\partial u}{\partial x}(x = 1, t) = 0 \qquad (6.26d)$$

Equation (6.26a) is the one–dimensional heat equation, Eq.(6.26b) constitutes its single initial condition, and Eqs.(6.26c–d) describe its two boundary conditions.

Let us discretize this problem using the MOL approach. We split the spatial axis into n segments of length $\delta x = 1/n$. We shall apply the third–order accurate central difference formula of Eq.(6.7) for the approximation of the spatial derivatives. We furthermore use the symmetry boundary condition approach at the right end of the interval. This leads to the following set of ODEs:

$$u_1 = \exp(-t/10) \qquad (6.27a)$$

$$\dot{u}_2 = \frac{n^2}{10\pi^2} \cdot (u_3 - 2u_2 + u_1) \qquad (6.27b)$$

$$\dot{u}_3 = \frac{n^2}{10\pi^2} \cdot (u_4 - 2u_3 + u_2) \qquad (6.27c)$$

etc.

$$\dot{u}_n = \frac{n^2}{10\pi^2} \cdot (u_{n+1} - 2u_n + u_{n-1}) \qquad (6.27d)$$

$$\dot{u}_{n+1} = \frac{n^2}{5\pi^2} \cdot (-u_{n+1} + u_n) \qquad (6.27e)$$

with initial conditions:

$$u_2(0) = \cos\left(\frac{\pi}{n}\right) \qquad (6.28a)$$

$$u_3(0) = \cos\left(\frac{2\pi}{n}\right) \qquad (6.28b)$$

$$u_4(0) = \cos\left(\frac{3\pi}{n}\right) \qquad (6.28c)$$

etc.

$$u_n(0) = \cos\left(\frac{(n-1)\pi}{n}\right) \qquad (6.28d)$$

$$u_{n+1}(0) = \cos(\pi) \qquad (6.28e)$$

This is a linear, time–invariant, inhomogeneous, n^{th}–order, single–input system of the type:

$$\dot{\mathbf{x}} = \mathbf{A} \cdot \mathbf{x} + \mathbf{b} \cdot u \tag{6.29}$$

where:

$$\mathbf{A} = \frac{n^2}{10\pi^2} \cdot \begin{pmatrix} -2 & 1 & 0 & 0 & \cdots & 0 & 0 & 0 \\ 1 & -2 & 1 & 0 & \cdots & 0 & 0 & 0 \\ 0 & 1 & -2 & 1 & \cdots & 0 & 0 & 0 \\ \vdots & \vdots & \vdots & \vdots & \ddots & \vdots & \vdots & \vdots \\ 0 & 0 & 0 & 0 & \cdots & 1 & -2 & 1 \\ 0 & 0 & 0 & 0 & \cdots & 0 & 2 & -2 \end{pmatrix} \tag{6.30}$$

$\mathbf{A}$ is a band–structured matrix of dimensions $n \times n$. Let us calculate its eigenvalues. They are tabulated in Table 6.1.

$n = 3$	$n = 4$	$n = 5$	$n = 6$	$n = 7$
-0.0244	-0.0247	-0.0248	-0.0249	-0.0249
-0.1824	-0.2002	-0.2088	-0.2137	-0.2166
-0.3403	-0.4483	-0.5066	-0.5407	-0.5621
	-0.6238	-0.9884	-0.9183	-0.9929
		-0.8044	-1.2454	-1.4238
			-1.4342	-1.7693
				-1.9610

TABLE 6.1. Eigenvalue distribution for diffusion model.

All eigenvalues are strictly negative and real. This is characteristic of all thermal field problems and all parabolic PDEs converted to sets of ODEs by the MOL technique.

We notice at once that, whereas the damping properties of the system (determined by the location of the dominant pole) don't change significantly with the number of segments, the *stiffness ratio*, i.e., the ratio between the absolute largest real part and the absolute smallest real part of any eigenvalue depends heavily on the number of segments. Figure 6.1 shows the square root of the stiffness ratio plotted over the number of segments chosen.

It turns out that, for all practical purposes, the stiffness ratio grows quadratically with the number of segments chosen in the spatial discretization process. The more accurate we wish to solve the diffusion equation, the stiffer the corresponding ODE problem will become. Since diffusion problems are usually quite smooth, the BDF algorithms are optimally suited to simulate the resulting set of ODEs.

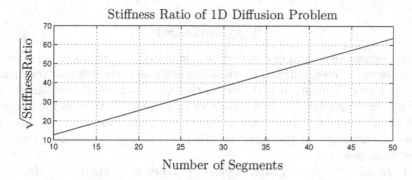

FIGURE 6.1. Dependence of stiffness ratio on discretization.

We chose a PDE problem, the analytical solution of which is known. It happens to be:

$$u_c(x,t) = \exp(-t/10) \cdot \cos(\pi \cdot x) \tag{6.31}$$

Hence we can compare the analytical solution of the original PDE problem with the equally analytical solution of the discretized ODE problem after applying the MOL discretization.

The analytical solution of the discretized ODE problem is a little harder to come by. We can create a system description of the continuous–time problem:

$$\dot{\mathbf{x}} = \mathbf{A} \cdot \mathbf{x} + \mathbf{b} \cdot u \tag{6.32a}$$
$$\mathbf{y} = \mathbf{C} \cdot \mathbf{x} + \mathbf{d} \cdot u \tag{6.32b}$$

where $\mathbf{C}$ is an identity matrix of suitable dimensions, and $\mathbf{d}$ is a zero vector using MATLAB's *control system toolbox*:

$$\mathbf{Sc} = \mathrm{ss}(\mathbf{A}, \mathbf{b}, \mathbf{C}, \mathbf{d}) \tag{6.33}$$

This continuous–time system can then be converted to an equivalent discrete–time system:

$$\mathbf{x_{k+1}} = \mathbf{F} \cdot \mathbf{x_k} + \mathbf{g} \cdot u_k \tag{6.34a}$$
$$\mathbf{y_k} = \mathbf{H} \cdot \mathbf{x_k} + \mathbf{i} \cdot u_k \tag{6.34b}$$

using the statement:

$$\mathbf{Sd} = \mathrm{c2d}(\mathbf{Sc}, h) \tag{6.35}$$

from which the $\mathbf{F}$–matrix and $\mathbf{g}$–vector of the discrete state equations can be extracted using the statement:

$$[\mathbf{F}, \mathbf{g}] = \text{ssdata}(\mathbf{Sd}) \qquad (6.36)$$

The discrete–time system can now be "simulated" by means of iteration of the discrete state equations. The solution of the discrete difference equation (ΔE) system is identical with that of the continuous ODE problem at the sampling points $k \cdot h$, where h is the step size (sampling rate) of the discrete problem, except for the discretization of the input function. The discrete system assumes that the input function $u(t)$ is kept constant in between sampling points.

Consequently, the step size, h, must be chosen small enough for the effect of the discretization of the input function to be negligible.

Let us look at the results of the experiment. The top left graph of Fig.6.2 shows the solution of the PDE problem, u_c, as a function of space and time, whereas the top right graph shows the solution of the discretized ODE problem, u_d, simulated using the approach discussed above. The two graphs look identical by visual inspection. The bottom left graph of Fig.6.2 displays the difference between the two functions, i.e.:

$$err = u_c - u_d \qquad (6.37)$$

and the bottom right graph of Fig.6.2 presents the maximum error, er_{max}, as a function of the number of segments used in the discretization. The maximum error was computed using the MATLAB statement:

$$er_{max} = \max(\max(\text{abs}(err))); \qquad (6.38)$$

The step size, h, was chosen small enough so that a further reduction of h would not visibly change the bottom right graph of Fig.6.2 any longer. In the given example, a step size of $h = 0.001$ had to be chosen to accomplish this goal.

We have just come across a new type of error. The *consistency error* describes the difference between the original PDE problem that we wish to solve, and the discretized ODE problem that we are actually solving.

Evidently, the consistency error cannot be overcome by either step–size or order control of the underlying ODE solver. Even the best ODE solver can only approximate the analytical solution, u_d, of the discretized ODE problem, but never the true analytical solution, u_c, of the original PDE problem.

Is the consistency error a *modeling error* or a *simulation error*? The answer to this question depends on the point of view. If we use a modeling environment that allows us to describe the PDE problem directly, we are inclined to call this a simulation error. However, it is an error that is incurred during the symbolic formulae manipulations that accompany the compilation of the model, rather than at run time. On the other hand, if we use a lower–level modeling environment that forces us to convert the

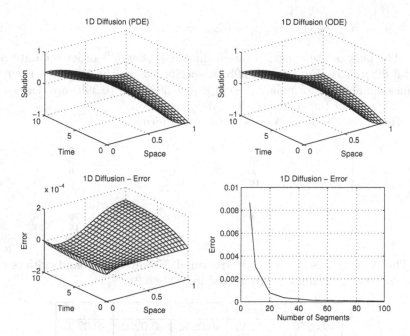

FIGURE 6.2. Solution of the 1D heat diffusion problem.

PDE manually into a set of ODEs, we would be more inclined to call this a modeling error.

Can the consistency error be overcome by choosing a more accurate scheme for the computation of the spacial derivatives? Let us use a 5th–order accurate central difference scheme together with an equally 5th–order accurate biased difference scheme for the discretization points near the two boundaries, hence:

$$u_1 = \exp(-t/10) \tag{6.39a}$$

$$\dot{u}_2 = \frac{n^2}{120\pi^2} \cdot (u_6 - 6u_5 + 14u_4 - 4u_3 - 15u_2 + 10u_1) \tag{6.39b}$$

$$\dot{u}_3 = \frac{n^2}{120\pi^2} \cdot (-u_5 + 16u_4 - 30u_3 + 16u_2 - u_1) \tag{6.39c}$$

$$\dot{u}_4 = \frac{n^2}{120\pi^2} \cdot (-u_6 + 16u_5 - 30u_4 + 16u_3 - u_2) \tag{6.39d}$$

$$etc.nonumber \tag{6.39e}$$

$$\dot{u}_{n-1} = \frac{n^2}{120\pi^2} \cdot (-u_{n+1} + 16u_n - 30u_{n-1} + 16u_{n-2} - u_{n-3}) \tag{6.39f}$$

$$\dot{u}_n = \frac{n^2}{120\pi^2} \cdot (16u_{n+1} - 31u_n + 16u_{n-1} - u_{n-2}) \tag{6.39g}$$

$$\dot{u}_{n+1} = \frac{n^2}{60\pi^2} \cdot (-15u_{n+1} + 16u_n - u_{n-1}) \tag{6.39h}$$

The bulk of the equations are formulated using 5th–order accurate central differences. Equation (6.39b) is specified using the 5th–order accurate biased difference formula, whereas Eqs.(6.39g) and (6.39h) are derived by making use of the symmetry boundary condition.

Hence the resulting **A**–matrix takes the form:

$$\mathbf{A} = \frac{n^2}{120\pi^2} \cdot \begin{pmatrix} -15 & -4 & 14 & -6 & 1 & \cdots & 0 & 0 & 0 & 0 \\ 16 & -30 & 16 & -1 & 0 & \cdots & 0 & 0 & 0 & 0 \\ -1 & 16 & -30 & 16 & -1 & \cdots & 0 & 0 & 0 & 0 \\ 0 & -1 & 16 & -30 & 16 & \cdots & 0 & 0 & 0 & 0 \\ \vdots & \vdots & \vdots & \vdots & \vdots & \ddots & \vdots & \vdots & \vdots & \vdots \\ 0 & 0 & 0 & 0 & 0 & \cdots & 16 & -30 & 16 & -1 \\ 0 & 0 & 0 & 0 & 0 & \cdots & -1 & 16 & -31 & 16 \\ 0 & 0 & 0 & 0 & 0 & \cdots & 0 & -2 & 32 & -30 \end{pmatrix} \tag{6.40}$$

The **A**–matrix is again band–structured. However, the bandwidth is now wider. Its eigenvalues are tabulated in Table 6.2.

$n = 5$	$n = 6$	$n = 7$	$n = 8$	$n = 9$
-0.0250	-0.0250	-0.0250	-0.0250	-0.0250
-0.2288	-0.2262	-0.2253	-0.2251	-0.2250
-0.5910	-0.6414	-0.6355	-0.6302	-0.6273
-0.7654	-0.9332	-1.1584	-1.2335	-1.2368
-1.2606	-1.3529	-1.4116	-1.6471	-1.9150
	-1.8671	-2.0761	-2.1507	-2.2614
		-2.5770	-2.9084	-3.0571
			-3.3925	-3.8460
				-4.3147

TABLE 6.2. Eigenvalue distribution for diffusion model.

The eigenvalue distribution has changed very little. In particular, all of them are still negative and real. Using this discretization scheme, the smallest number of segments is now five.

Figure 6.3 shows the square root of the stiffness ratio plotted as a function of the number of segments chosen. The corresponding stiffness ratio plot for the previously used **A**–matrix is presented also for comparison.

For the same number of segments, the stiffness ratio of the 5th–order scheme is slightly higher than that of the 3rd–order scheme. As the correct solution of the PDE problem corresponds to a discretization with infinitely many segments, i.e., an ODE problem with infinite stiffness, we may expect that the solution produced by the 5th–order scheme is indeed more accurate than that of the 3rd–order scheme.

Let us now perform the same experiment as before, this time using the 5th–order scheme. Figure 6.4 shows the consistency error as a function of

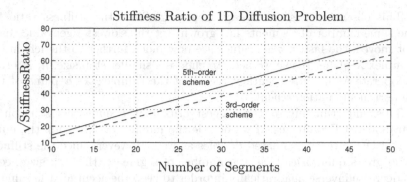

FIGURE 6.3. Dependence of stiffness ratio on discretization.

the number of segments used in the discretization scheme. The results of
using the 3rd–order accurate discretization scheme and those using the 5th–
order accurate discretization scheme are superposed on the same graph.

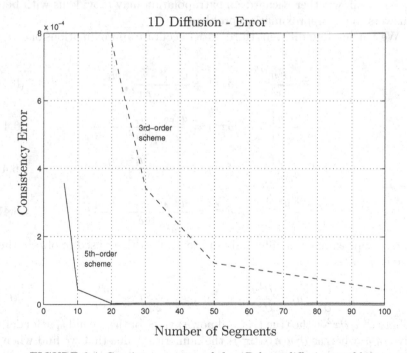

FIGURE 6.4. Consistency error of the 1D heat diffusion problem.

The improvement achieved by the more accurate discretization scheme
is quite dramatic. Yet, the "simulation" of the discretized problem is much
more expensive in this case. We had to choose a smaller step size of $h =$
0.0001 before the consistency error would no longer decrease by further
reducing the step size.

This observation is not overly surprising. Since the stiffness ratio for the same number of segments has grown, yet the slowest eigenvalues have not moved, the fastest eigenvalues are now much further to the left in the complex λ–plane. Hence we need to choose a smaller step size, h, in order to operate within the accuracy region of the complex $\lambda \cdot h$–plane of the numerical simulation scheme.

This, unfortunately, is the biggest crux in the numerical solution of parabolic PDE problems. If we double the number of segments, the number of ODEs to be simulated doubles as well. However, since the stiffness ratio grows quadratically in the number of segments, the step size needs to decrease inverse quadratically in order to keep the accuracy the same in the complex $\lambda \cdot h$–plane. Hence doubling the number of segments forces us to quadruple the number of time steps. Hence the *simulation effort* grows cubically in the number of segments.

Let us try another approach. You certainly remember the *Richardson extrapolation* technique that we talked about in Chapter 3 of this text. Let us ascertain whether Richardson extrapolation may provide us with better answers to our approximation problem.

We can find four different third–order accurate approximations of $\partial^2 u / \partial x^2$:

$$\frac{\partial^2 u}{\partial x^2}\bigg|_{x=x_i}^{P_1} (\delta x^2) = \frac{u_{i+1} - u_i + u_{i-1}}{\delta x^2} \tag{6.41a}$$

$$\frac{\partial^2 u}{\partial x^2}\bigg|_{x=x_i}^{P_2} (4\delta x^2) = \frac{u_{i+2} - u_i + u_{i-2}}{4\delta x^2} \tag{6.41b}$$

$$\frac{\partial^2 u}{\partial x^2}\bigg|_{x=x_i}^{P_3} (9\delta x^2) = \frac{u_{i+3} - u_i + u_{i-3}}{9\delta x^2} \tag{6.41c}$$

$$\frac{\partial^2 u}{\partial x^2}\bigg|_{x=x_i}^{P_4} (16\delta x^2) = \frac{u_{i+4} - u_i + u_{i-4}}{16\delta x^2} \tag{6.41d}$$

These approximations differ only in the grid width δx used to obtain them. We can write:

$$\frac{\partial^2 u}{\partial x^2}(\eta) = \frac{\partial^2 u}{\partial x^2} + e_1 \cdot \eta + e_2 \cdot \frac{\eta^2}{2!} + e_3 \cdot \frac{\eta^3}{3!} + \dots \tag{6.42}$$

where $\partial^2 u / \partial x^2$ is the true (yet unknown) value of the second spatial derivative of u, whereas $\partial^2 u(\eta)/\partial x^2$ is the numerical value that we find when we approximate the second spatial derivative using a grid width of η. Obviously, this value contains an error. Equation (6.42) is a Taylor–Series in η around the (unknown) correct value. The e_i variables are errors of the approximation.

We truncate the Taylor Series after the cubic term, and write Eq.(6.42) down for the same values of the grid width that had been used in Eqs.(6.41a–d). We find:

$$\frac{\partial^2 u}{\partial x^2}^{P_1}(\delta x^2) \approx \frac{\partial^2 u}{\partial x^2} + e_1 \cdot \delta x^2 + \frac{e_2}{2!} \cdot \delta x^4 + \frac{e_3}{3!} \cdot \delta x^6$$

$$\frac{\partial^2 u}{\partial x^2}^{P_2}(4\delta x^2) \approx \frac{\partial^2 u}{\partial x^2} + e_1 \cdot (4\delta x^2) + \frac{e_2}{2!} \cdot (4\delta x^2)^2 + \frac{e_3}{3!} \cdot (4\delta x^2)^3$$

$$\frac{\partial^2 u}{\partial x^2}^{P_3}(9\delta x^2) \approx \frac{\partial^2 u}{\partial x^2} + e_1 \cdot (9\delta x^2) + \frac{e_2}{2!} \cdot (9\delta x^2)^2 + \frac{e_3}{3!} \cdot (9\delta x^2)^3$$

$$\frac{\partial^2 u}{\partial x^2}^{P_4}(16\delta x^2) \approx \frac{\partial^2 u}{\partial x^2} + e_1 \cdot (16\delta x^2) + \frac{e_2}{2!} \cdot (16\delta x^2)^2 + \frac{e_3}{3!} \cdot (16\delta x^2)^3$$

$$(6.43)$$

or in a matrix notation:

$$
\begin{pmatrix} \frac{\partial^2 u}{\partial x^2}^{P_1} \\ \frac{\partial^2 u}{\partial x^2}^{P_2} \\ \frac{\partial^2 u}{\partial x^2}^{P_3} \\ \frac{\partial^2 u}{\partial x^2}^{P_4} \end{pmatrix} \approx
\begin{pmatrix}
(\delta x^2)^0 & (\delta x^2)^1 & (\delta x^2)^2 & (\delta x^2)^3 \\
(4\delta x^2)^0 & (4\delta x^2)^1 & (4\delta x^2)^2 & (4\delta x^2)^3 \\
(9\delta x^2)^0 & (9\delta x^2)^1 & (9\delta x^2)^2 & (9\delta x^2)^3 \\
(16\delta x^2)^0 & (16\delta x^2)^1 & (16\delta x^2)^2 & (16\delta x^2)^3
\end{pmatrix} \cdot
\begin{pmatrix} \frac{\partial^2 u}{\partial x^2} \\ e_1 \\ e_2/2 \\ e_3/6 \end{pmatrix}
$$

$$(6.44)$$

By inverting the Van–der–Monde matrix, we can solve for the unknown $\partial^2 u/\partial x^2$ and the three error variables. Since we aren't interested in the errors, we only look at the first row of the inverted Van–der–Monde matrix. It turns out that the values in this row don't depend at all on the grid width δx. We find:

$$\frac{\partial^2 u}{\partial x^2} \approx \begin{pmatrix} \frac{56}{35} & -\frac{28}{35} & \frac{8}{35} & -\frac{1}{35} \end{pmatrix} \cdot \begin{pmatrix} \frac{\partial^2 u}{\partial x^2}^{P_1} \\ \frac{\partial^2 u}{\partial x^2}^{P_2} \\ \frac{\partial^2 u}{\partial x^2}^{P_3} \\ \frac{\partial^2 u}{\partial x^2}^{P_4} \end{pmatrix} \qquad (6.45)$$

We can plug Eqs.(6.41) into Eq.(6.45), and find:

$$\left.\frac{\partial^2 u}{\partial x^2}\right|_{x=x_i} \approx \frac{1}{5040\delta x^2}(-9u_{i+4} + 128u_{i+3} - 1008u_{i+2} + 8064u_{i+1}$$

$$- 14350u_i + 8064u_{i-1} - 1008u_{i-2} + 128u_{i-3} - 9u_{i-4}) \quad (6.46)$$

which is exactly the central difference formula of order 9. Once again, the *Richardson extrapolation* has raised the approximation accuracy to the highest possible order.

Let us now look at a slightly different problem:

$$\frac{\partial u}{\partial t} = 4\frac{\partial^2 u}{\partial x^2} \ ; \quad x \in [0,1] \ ; \quad t \in [0,\infty) \tag{6.47a}$$

$$u(x, t = 0) = 20 \sin\left(\frac{\pi}{2}x\right) + 300 \tag{6.47b}$$

$$u(x = 0, t) = 20 \sin\left(\frac{\pi}{12}t\right) + 300 \tag{6.47c}$$

$$\frac{\partial u}{\partial x}(x = 1, t) = 0 \tag{6.47d}$$

We again solve a one–dimensional heat equation, but with a different time constant, and different initial and boundary conditions.

This time around, we don't know the analytical solution, hence we cannot compute the consistency error explicitly. What do we do? Similarly to the step–size control algorithms discussed in the previous chapters, we need an estimator of the spatial discretization error.

All numerical algorithms should have a second algorithm built in to them that reasons about the sanity of the first algorithm and starts screaming if it thinks that something is going awry. Without such a sanity check, numerical algorithms are never safe. It is precisely the availability of such alarm systems that constitutes one of the major distinctions between *production codes* and *experimental codes*.

We propose to compute all spatial derivatives twice, once with the grid size δx, and once with the grid size $2\delta x$ using central differences.

$$\left.\frac{\partial^2 u}{\partial x^2}\right|_{x=x_i}^{P_1} (\delta x^2) = \frac{u_{i+1} - u_i + u_{i-1}}{\delta x^2} \tag{6.48a}$$

$$\left.\frac{\partial^2 u}{\partial x^2}\right|_{x=x_i}^{P_2} (4\delta x^2) = \frac{u_{i+2} - u_i + u_{i-2}}{4\delta x^2} \tag{6.48b}$$

$$\tag{6.48c}$$

The two approximations form two separate partial derivative vectors, $\mathbf{u}_{xx}^{P_1}$ and $\mathbf{u}_{xx}^{P_2}$. Using these approximations, we can formulate a spatial error estimate:

$$\varepsilon_{rel} = \frac{|\mathbf{u}_{xx}^{P_1} - \mathbf{u}_{xx}^{P_2}|}{\max(|\mathbf{u}_{xx}^{P_1}|, |\mathbf{u}_{xx}^{P_2}|, \delta)} \tag{6.49}$$

where δ is a fudge factor, e.g., $\delta = 10^{-10}$.

If the estimated spatial discretization error is too big, we must either choose a more narrow grid, or alternatively, we must increase the approximation order of the spatial derivatives.

Is it wasteful to compute the entire vector of spatial derivatives twice? This question must clearly be answered in the negative. The two predictors can be used in a *Richardson corrector* step:

$$\mathbf{u}_{xx}^C = \frac{4}{3} \cdot \mathbf{u}_{xx}^{P_1} - \frac{1}{3} \cdot \mathbf{u}_{xx}^{P_2} \tag{6.50}$$

This is equivalent to having raised the approximation order of the spatial derivatives from three to five. However, by writing the 5th-order accurate spatial derivative formula in this way, we get an error estimator essentially for free.

Since the problem is stiff, a BDF formula may be appropriate for its integration. As we wish to obtain a global accuracy of 1%, we decided to simulate the system using BDF3. We chose $n_{seg} = 50$ in order to receive sufficiently many output points in space, and simulated across 10 seconds in time. The simulation results are shown in Fig.6.5.

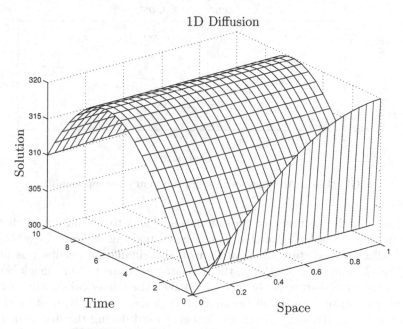

FIGURE 6.5. Solution of heat diffusion problem.

Figure 6.6 shows a slice through the solution at $x = 1.0$.

Unfortunately, the solution exhibits a fast transient precisely during the start–up period. The problem isn't truly stiff until the fast transients have died out. Initially, the solution is heavily controlled by accuracy requirements beside from the numerical stability constraints.

Assuming a fixed step size to be used throughout the solution, we repeated the simulation thrice, once using order buildup, i.e., a BDF starter, once using an RK3 starter, and once using an IEX3 starter. Figure 6.7 shows the step size required to achieve a desired level of accuracy using these three start–up algorithms.

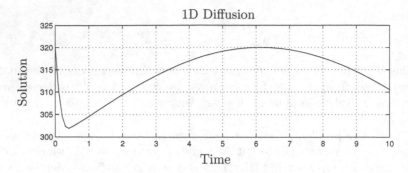

FIGURE 6.6. Solution of heat diffusion problem.

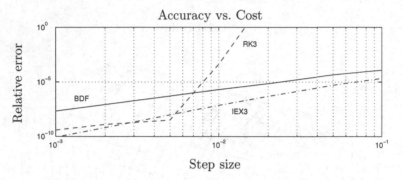

FIGURE 6.7. Accuracy *vs.* cost for different start–up algorithms.

Overall, the accuracy of the simulation seems to be quite a bit better than the 3rd–order algorithm would have made us believe. In addition, the effect of the start–up algorithm on the simulation accuracy is quite dramatic. For small step sizes, the RK3 starter seems to work much better than the BDF starter. However, at $h = 0.005$, the numerical stability is lost, and the overall accuracy of the simulation degrades rapidly, in spite of the fact that the RK3 algorithm is only being used during the first two steps of the simulation. Of course, an RK starter implemented in a production code would be expected to proceed with a smaller step size than during the remainder of the simulation, but we did not want to make use of any type of step–size control in this experiment, as this would make an interpretation of the obtained results much more difficult.

The IEX3 starter, implemented using BDF1 steps internally, performs similarly to the RK3 starter for small step sizes, but without being plagued by the numerical stability problems of the RK3 starter for larger step sizes.

We also tried a $BI4/5_{0.45}$ starter. It didn't work well at all in this application. The reason is the following. The backward RK semi–step is numerically highly unstable. It is only stabilized by the Newton iteration. In the given application, we ran into roundoff error problems. The unstable semi–step produced numbers so big that the Newton iteration could not stabilize them any longer due to roundoff.

Parabolic PDE problems discretized using the MOL approach always turn into very stiff ODE systems. The more accurate we wish to simulate, the stiffer the problem becomes. Yet, decent stiff system solvers, such as DASSL [6.1], are usually quite capable of dealing with such problems effectively and efficiently.

6.4 Hyperbolic PDEs

Let us now analyze the second class of PDE problems, the hyperbolic PDEs. The simplest specimen of this class of problems is the *wave equation* or *linear conservation law*:

$$\frac{\partial^2 u}{\partial t^2} = c^2 \cdot \frac{\partial^2 u}{\partial x^2} \tag{6.51}$$

We can easily transform this second–order PDE in time into two first order PDEs in time:

$$\frac{\partial u}{\partial t} = v \tag{6.52a}$$

$$\frac{\partial v}{\partial t} = c^2 \cdot \frac{\partial^2 u}{\partial x^2} \tag{6.52b}$$

At this point, we can replace the spatial derivatives again by finite difference approximations, and we seem to be in business.

Equations (6.53a–e) constitute a complete specification of such a model.

$$\frac{\partial^2 u}{\partial t^2} = \frac{\partial^2 u}{\partial x^2} \ ; \ \ x \in [0,1] \ ; \ \ t \in [0,\infty) \tag{6.53a}$$

$$u(x, t = 0) = \sin\left(\frac{\pi}{2}x\right) \tag{6.53b}$$

$$\frac{\partial u}{\partial t}(x, t = 0) = 0.0 \tag{6.53c}$$

$$u(x = 0, t) = 0.0 \tag{6.53d}$$

$$\frac{\partial u}{\partial x}(x = 1, t) = 0.0 \tag{6.53e}$$

Equation (6.53a) is the one–dimensional wave equation, Eqs.(6.53b–c) constitute its two initial conditions, and Eqs.(6.53d–e) describe its two boundary conditions.

Let us simulate this problem using the MOL approach. We decide to split the spatial axis into n segments of width $\delta x = 1/n$. If we work with the central difference formula of Eq.(6.7), and using the symmetry boundary condition approach at the right end of the interval, we obtain the following set of ODEs:

$$u_1 = 0.0 \tag{6.54a}$$
$$\dot{u}_2 = v_2 \tag{6.54b}$$

etc.

$$\dot{u}_{n+1} = v_{n+1} \tag{6.54c}$$
$$v_1 = 0.0 \tag{6.54d}$$
$$\dot{v}_2 = n^2 \left(u_3 - 2u_2 + u_1 \right) \tag{6.54e}$$
$$\dot{v}_3 = n^2 \left(u_4 - 2u_3 + u_2 \right) \tag{6.54f}$$

etc.

$$\dot{v}_n = n^2 \left(u_{n+1} - 2u_n + u_{n-1} \right) \tag{6.54g}$$
$$\dot{v}_{n+1} = 2n^2 \left(u_n - u_{n+1} \right) \tag{6.54h}$$

with the initial conditions:

$$u_2(0) = \sin\left(\frac{\pi}{2n} \right) \tag{6.55a}$$
$$u_3(0) = \sin\left(\frac{\pi}{n} \right) \tag{6.55b}$$
$$u_4(0) = \sin\left(\frac{3\pi}{2n} \right) \tag{6.55c}$$

etc.

$$u_n(0) = \sin\left(\frac{(n-1)\pi}{2n} \right) \tag{6.55d}$$
$$u_{n+1}(0) = \sin\left(\frac{\pi}{2} \right) \tag{6.55e}$$
$$v_2(0) = 0.0 \tag{6.55f}$$

etc.

$$v_{n+1}(0) = 0.0 \tag{6.55g}$$

This is a linear, time–invariant, inhomogeneous, $(2n)^{\text{th}}$–order, single–input system of the type specified in Eq.(6.29), where:

$$\mathbf{A} = \begin{pmatrix} \mathbf{0}^{(n)} & \mathbf{I}^{(n)} \\ \mathbf{A}_{21} & \mathbf{0}^{(n)} \end{pmatrix} \tag{6.56}$$

with:

$$\mathbf{A_{21}} = n^2 \begin{pmatrix} -2 & 1 & 0 & 0 & \cdots & 0 & 0 & 0 \\ 1 & -2 & 1 & 0 & \cdots & 0 & 0 & 0 \\ 0 & 1 & -2 & 1 & \cdots & 0 & 0 & 0 \\ \vdots & \vdots & \vdots & \vdots & \ddots & \vdots & \vdots & \vdots \\ 0 & 0 & 0 & 0 & \cdots & 1 & -2 & 1 \\ 0 & 0 & 0 & 0 & \cdots & 0 & 2 & -2 \end{pmatrix} \qquad (6.57)$$

$\mathbf{A}$ is a band–structured matrix of dimensions $2n \times 2n$ with two separate non–zero bands. Let us calculate its eigenvalues. They are tabulated in Table 6.3.

$n = 3$	$n = 4$	$n = 5$	$n = 6$
$\pm 1.5529j$	$\pm 1.5607j$	$\pm 1.5643j$	$\pm 1.5663j$
$\pm 4.2426j$	$\pm 4.4446j$	$\pm 4.5399j$	$\pm 4.5922j$
$\pm 5.7956j$	$\pm 6.6518j$	$\pm 7.0711j$	$\pm 7.3051j$
	$\pm 7.8463j$	$\pm 8.9101j$	$\pm 9.5202j$
		$\pm 9.8769j$	$\pm 11.0866j$
			$\pm 11.8973j$

TABLE 6.3. Eigenvalue distribution of linear conservation law.

All eigenvalues are strictly imaginary. All hyperbolic PDEs converted to sets of ODEs using the MOL technique show complex eigenvalues. Many of them have their eigenvalues spread up and down fairly close to the imaginary axis. The linear conservation law has all its eigenvalues exactly on the imaginary axis.

Figure 6.8 shows the *frequency ratio*, i.e., the ratio between the absolute largest and the absolute smallest imaginary parts of any eigenvalues plotted over the number of segments used in the discretization.

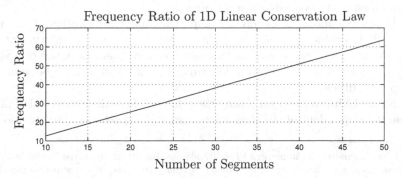

FIGURE 6.8. Frequency ratio of the 1D linear conservation law.

Evidently, the frequency ratio of the 1D linear conservation law grows linearly with the number of segments used in the discretization.

The numerical challenges are quite different from those in the parabolic case. The conservation law does not lead to a stiff set of ODEs. No "fast transients" appear that die out after some time, and consequently, the step size in the numerical integration must be kept small to account for all the eigenvalues of the discretized problem. The more narrow the grid width is chosen, the smaller the time steps will have to be in order to keep all eigenvalues within the asymptotic region of the numerical integration algorithm. Luckily, the spreading of the eigenvalues grows only linearly with the number of segments chosen.

We have seen that PDEs pose a new kind of challenge. In the case of ODE solutions, we only worried about stability and accuracy. In the case of PDE solution, we must concern ourselves with *stability*, *accuracy*, and *consistency*.

> *Definition:* "A discretization scheme is called *consistent* if the analytical solution of the discretized problem smoothly approaches the analytical solution of the original continuous problem as the grid width is being reduced to smaller and smaller values."

The *consistency error* is thus the deviation of the analytical solution of the discretized problem from the analytical solution of the continuous problem, whereas the *accuracy error* is the deviation of the numerical solution of the discretized problem from the analytical solution of the discretized problem.[1]

The example of Eqs.(6.53a–e) is so simple that an analytical solution of the continuous (field) problem can be given. It is:

$$u(x,t) = \frac{1}{2}\sin\left(\frac{\pi}{2}(x-t)\right) + \frac{1}{2}\sin\left(\frac{\pi}{2}(x+t)\right) \qquad (6.58)$$

Since the discretized problem is linear with constant input, we can use the method described in Hw.[H4.8] to derive its analytical solution. Thus, we can go after the consistency error directly.

Figure 6.9 shows in its top left graph the analytical solution of the original PDE problem, in its top right graph the analytical solution of the discretized ODE problem. The two solutions look identical when compared by the naked eye. The bottom left curve shows the difference between the top two curves.

Since the input function is zero, the solution of the discretized ODE problem is independent of the chosen step size, h, in time. The discretization in

---

[1]Traditionally, the numerical PDE literature talks about the three facets: *stability*, *consistency*, and *convergence*. It is then customary to prove that any two of the three imply the third one, i.e., it is sufficient to look at any selection of two of the three [6.13]. However, that way of reasoning is more conducive to fully discretized (finite difference or finite element) schemes, where the step size in time, h, is locked in a fixed relationship with the grid width in space, δx. Consequently, h and δx approach zero simultaneously. In the context of the MOL methodology, our approach may be more appealing.

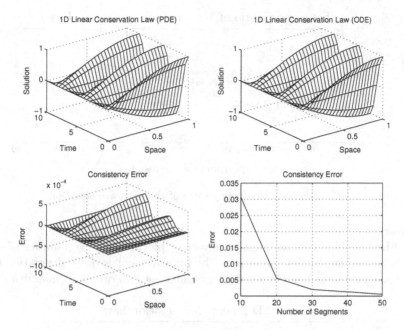

FIGURE 6.9. Analytical solutions of the 1D wave equation.

time serves here only for the purpose of generating sufficiently many output points. Hence the curve shown in the bottom left graph is the true *consistency error*. The only potential sources of numerical pollution could be due to roundoff and accumulation, but these are insignificant in magnitude in comparison with the analytical consistency error.

The bottom right graph shows the consistency error plotted against the number of segments chosen for the spatial discretization. The consistency error is here much larger than in the previous parabolic PDE examples. If we wish to obtain simulation results with a numerical accuracy of 1%, the consistency error itself ought to be at least one order of magnitude smaller. This means we should choose at least 40 segments for this simulation.

Just like in the case of the parabolic PDE problems, let us discuss what happens when we choose a higher–order discretization in space. Let us try first with 5th–order central differences.

Figure 6.10 shows the frequency ratio plotted against the number of segments chosen in the spatial discretization scheme. The frequency ratio of the 3rd–order scheme is plotted on the same graph for comparison.

The frequency ratio of the more accurate 5th–order scheme is consistently higher than that of the less accurate 3rd–order scheme for the same number of segments. Since the true PDE solution, corresponding to the solution with infinitely many infinitely dense discretization lines, has a frequency ratio that is infinitely large, we suspect that choosing a higher–order discretization scheme may indeed help with the reduction of the consistency

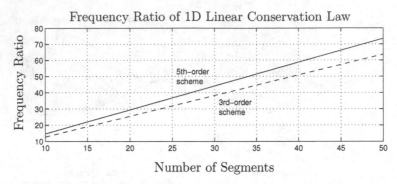

FIGURE 6.10. Frequency ratio of the 1D wave equation.

error.

Figure 6.11 shows the consistency error plotted over the number of segments used in the discretization. The improvement is quite dramatic. The consistency error has been reduced by at least two orders of magnitude.

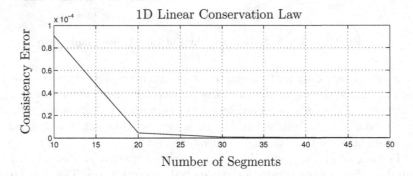

FIGURE 6.11. Consistency error of the 1D wave equation.

In a true simulation experiment, the 5th–order spatial discretization scheme should be implemented using the *Richardson predictor–corrector technique* presented earlier in this chapter.

Let us compute the cost–versus–accuracy plot for the above problem, comparing the various third–order algorithms to each other that we meanwhile know. We shall use 50 segments for the spatial discretization together with 5th–order central differences, in order to keep the consistency error sufficiently small, so that it won't affect the simulation results.

We computed the global accuracy of seven algorithms for simulating the discretized wave equation across 10 seconds of simulated time using a fixed step size of h, namely: RK3, IEX3, BI3, AB3, ABM3, AM3, and BDF3. We chose the step sizes: $h = 0.1$, $h = 0.05$, $h = 0.02$, $h = 0.01$, $h = 0.005$, $h = 0.002$, and $h = 0.001$ corresponding to 100, 200, 500, 1000, 2000, 5000, and 10000 steps, respectively. The results are tabulated in Table 6.4.

h	RK3	IEX3	BI3
0.1	unstable	0.6782e-4	0.4947e-6
0.05	unstable	0.8668e-5	0.2895e-7
0.02	unstable	0.5611e-6	0.1324e-8
0.01	0.7034e-7	0.7029e-7	0.2070e-8
0.005	0.8954e-8	0.8791e-8	0.2116e-8
0.002	0.2219e-8	0.2145e-8	0.2120e-8
0.001	0.2127e-8	0.2119e-8	0.2120e-8

h	AB3	ABM3	AM3	BDF3
0.1	unstable	unstable	unstable	garbage
0.05	unstable	unstable	unstable	garbage
0.02	unstable	unstable	unstable	garbage
0.01	unstable	0.6996e-7	unstable	garbage
0.005	0.7906e-7	0.8772e-8	0.8783e-8	0.9469e-2
0.002	0.5427e-8	0.2156e-8	0.2149e-8	0.1742e-6
0.001	0.2239e-8	0.2120e-8	0.2120e-8	0.4363e-7

TABLE 6.4. Comparison of accuracy of integration algorithms.

Using a step size of $h = 0.001$, all seven integration algorithms simulate the problem successfully. In fact, all of them with the exception of BDF3 are down to the level of the consistency error.

As the step size becomes smaller, the higher–order terms in the Taylor–series expansion become less and less important. For sufficiently small step sizes, all integration algorithms behave either like forward or backward Euler.

BDF3 performs a little poorer than the other algorithms, because its *error coefficient* is considerably larger than those of its competitors. BDF algorithms perform generally somewhat poor in terms of accuracy in comparison with their peers of equal order. The BDF algorithms had been known before they were made popular by Bill Gear in the early seventies [6.9]. However, they were considered "garbage algorithms" due to their poor accuracy properties.

It turns out that the problem is kind of "stiff," although it does not meet most of John Lambert's definitions of stiffness [6.11]. The problem is "stiff" in the sense that all the algorithms with stability domains looping into the left–half plane are unable to produce solutions with the desired accuracy of 1.0%, since they are numerically unstable when a step size is used that would produce the desired accuracy otherwise. BDF3 doesn't suffer the same fate, but it eventually succumbs to error accumulation problems. As the step sizes grow too big, the computations become so inaccurate that the simulation error exceeds the simulation output in magnitude. Hence

BDF3 starts accumulating numerical garbage.

Only IEX3 and BI3 are capable of solving the problem successfully for large step sizes. Between the two, BI3 seems to work a little better, which is no big surprise. Being an *F–stable algorithm*, BI3 is earmarked for these types of applications.

Figure 6.12 presents the same results graphically in a cost *vs.* accuracy plot.

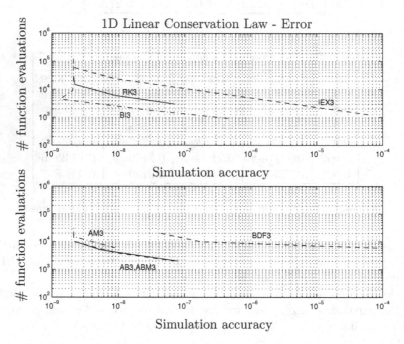

FIGURE 6.12. Cost *vs.* accuracy of the 1D wave equation.

These results are somewhat deceiving, since they do not take into account the effort spent in computing inverse Hessians. This decision was taken on purpose, since the number of function evaluations is the only objective measure available that depends on the algorithm alone, rather than on implementational details of the production code, as different codes vary a lot in how often and how accurately they compute inverse Hessians.

Of course, since the given problem is linear and since we don't vary the step size ever, it would suffice to compute one inverse Hessian at the beginning of the simulation. Yet, this fact is peculiar to the specific problem at hand. For nonlinear problems, the explicit algorithms, i.e., RK3, AB3, and ABM3, may be at least as attractive as BI3.

We would still argue in favor of the BI algorithms for these types of applications, not because of their superior cost–per–accuracy properties, but because of their better *robustness* characteristics. Using BI3, we can obtain a decent answer using any step size that we may try without having the algorithm blow up on us, and we get a meaningful accuracy in each and every case.

We could have included also GE3 in the comparison of this section. Since the problem to be solved is a linear conservation law, the stand–alone versions of the explicit Godunov schemes would have been excellently suited for the task at hand. However, we decided against doing so, because the comparison would have been quite unfair. All of the techniques compared against each other in this section are *general–purpose* numerical ODE solvers, whereas the stand–alone versions of the GE algorithms are limited to dealing with linear conservation laws only.

6.5 Shock Waves

Let us now study a more involved hyperbolic PDE problem. A thin tube of length 1 m is initially pressurized at $p_B = 1.1$ atm. The tube is located at sea level, i.e., the surrounding atmosphere has a pressure of $p_0 = 1.0$ atm $= 760.0$ Torr $= 1.0132 \cdot 10^5$ N m^{-2}. The current temperature is $T = 300.0$ K. At time zero, the tube is opened at one of its two ends. We wish to determine the pressure at various places inside the tube as functions of time.[1]

As the tube is opened, air rushes out of the tube, and a *rarefaction wave* enters the pipe. Had the initial pressure inside the pipe been smaller than the outside pressure, air would have rushed in, and a *compression wave* would have formed.

The problem can be mathematically described by a set of first–order hyperbolic PDEs:

$$\frac{\partial \rho}{\partial t} = -v \cdot \frac{\partial \rho}{\partial x} - \rho \cdot \frac{\partial v}{\partial x} \tag{6.59a}$$

$$\frac{\partial v}{\partial t} = -v \cdot \frac{\partial v}{\partial x} - \frac{a}{\rho} \tag{6.59b}$$

$$\frac{\partial p}{\partial t} = -v \cdot a - \gamma \cdot p \cdot \frac{\partial v}{\partial x} \tag{6.59c}$$

$$a = \frac{\partial p}{\partial x} + \frac{\partial q}{\partial x} + f \tag{6.59d}$$

$$q = \begin{cases} \beta \cdot \delta x^2 \cdot \rho \cdot \left(\frac{\partial v}{\partial x}\right)^2 & ; \; \frac{\partial v}{\partial x} < 0.0 \\ 0.0 & ; \; \frac{\partial v}{\partial x} \geq 0.0 \end{cases} \tag{6.59e}$$

$$f = \frac{\alpha \cdot \rho \cdot v \cdot |v|}{\delta x} \tag{6.59f}$$

---

[1]The problem can be found in a slightly modified form in the FORSIM–VI manual [6.4]. It is being reused here with the explicit permission by the author.

where $\rho(x,t)$ denotes the *gas density* inside the tube at position x and time t, $v(x,t)$ denotes the *gas velocity*, and $p(x,t)$ denotes the *gas pressure*. The quantity a was pulled out into a separate algebraic equation, since the same quantity is used in two places within the model. The two quantities computed in Eqs.(6.59e–f) are artificial, as their dependence on δx shows. Clearly, δx is not a physical quantity, but is introduced only in the process of converting the (small) set of PDEs into a (large) set of ODEs. q denotes the *pseudo viscous pressure*, and f denotes the *frictional resistance*. They were introduced by Richtmyer and Morton [6.14] in order to smoothen out numerical problems with the solution. We shall discuss this issue in due course. γ is the ratio of specific heat constants, a non–dimensional constant with a value of $\gamma = c_p/c_v = 1.4$. α and β are non–dimensional numerical fudge factors. We shall initially assign the following values to them: $\alpha = \beta = 0.1$. The "ideal" (i.e., undamped) problem has $\alpha = \beta = 0.0$.

Introduction of the two dissipative terms is not a bad idea, since the "ideal" solution does not represent a physical phenomenon in any true sense. Phenomena without any sort of dissipation belong allegedly in the world that we may enter after we die. They certainly don't form any part of *this* universe.

The initial conditions are:

$$\rho(x, t = 0.0) = \rho_B \tag{6.60a}$$
$$v(x, t = 0.0) = 0.0 \tag{6.60b}$$
$$p(x, t = 0.0) = p_B \tag{6.60c}$$

where ρ_B is determined by the *equation of state* for ideal gases (cf. Chapter 9 of the companion book *Continuous System Modeling* [6.5]):

$$\rho_B = \frac{p_B \cdot M_{\text{air}}}{R \cdot T} \tag{6.61}$$

where $T = 300.0$ is the absolute temperature (measured in Kelvin), $R = 8.314$ J K^{-1} mole^{-1} is the gas constant, and $M_{\text{air}} = 28.96$ g mole^{-1} is the average molar mass of air.[1] The boundary conditions are:

$$v(x = 0.0, t) = 0.0 \tag{6.62a}$$
$$\rho(x = 1.0, t) = \rho_0 \tag{6.62b}$$

$$\begin{cases} v(x = 1.0, t) = -\sqrt{\frac{2(p_0 - p(x=1.0,t))}{\rho(x=1.0,t)}} \; ; \; v(x = 1.0, t) < 0.0 \\ p(x = 1.0, t) = p_0 \; ; \; v(x = 1.0, t) \geq 0.0 \end{cases} \tag{6.62c}$$

---

[1] Air consists roughly to 78% of nitrogen (N$_2$) with a molar mass of 28 g mole^{-1}, to 21% of oxygen (O$_2$) with a molar mass of 32 g mole^{-1}, and to 1% of argon (Ar) with a molar mass of 40 g mole^{-1}.

As proposed in [6.4], we converted all spatial derivatives by means of second–order accurate central differences using the formula:

$$\left.\frac{\partial u}{\partial x}\right|_{x=x_i} \approx \frac{1}{2\delta x} \cdot (u_{i+1} - u_{i-1}) \tag{6.63}$$

except near the boundaries, where we used second–order accurate biased formulae:

$$\frac{\partial u}{\partial x}(x = x_1, t) \approx \frac{1}{2\delta x} \cdot (-u_3 + 4u_2 - 3u_1) \tag{6.64a}$$

$$\frac{\partial u}{\partial x}(x = x_{n+1}, t) \approx \frac{1}{2\delta x} \cdot (3u_{n+1} - 4u_n + u_{n-1}) \tag{6.64b}$$

where u can stand for either ρ, v, p, or q.

In order to keep the consistency error small, we chose 50 segments for each of the three PDEs. We created a MATLAB function:

$$\mathbf{u_x} = \text{partial}(\mathbf{u}, \delta x, bc, bctype) \tag{6.65}$$

which implements the above set of formulae with correction terms in the case of a *symmetry boundary condition*. The variable bc indicates whether the boundary condition is applied at the left end, $bc = -1$, or at the right end, $bc = +1$. The variable $bctype$ specifies the type of boundary condition. $bctype = 0$ indicates a symmetry boundary condition. $bctype = 1$ denotes a function value condition.

In the case of a symmetry boundary condition, the central formulae are used all the way to the boundary while folding the values that are outside the domain back into the domain, as explained earlier.

The correction formulae are:

$$\frac{\partial u}{\partial x}(x = x_1, t) \approx 0.0 \tag{6.66}$$

for a symmetry boundary condition at the left end, and:

$$\frac{\partial u}{\partial x}(x = x_{n+1}, t) \approx 0.0 \tag{6.67}$$

for a symmetry boundary condition at the right end.

The state–space model itself has been encoded in another MATLAB function:

```
function [xdot] = st_eq(x, t)
%
% State − space model of shock − tube problem
%
n  = round(length(x)/3);
n₁ = n + 1;
δx = 1/n;
```

```
%
% Constants
%
R = 8.314;
%
% Physical parameters
%
Temp = 300;
Mair = 0.02896;
p0 = 1.0132e5;
ρ0 = p0 * Mair/(R * Temp);
γ = 1.4;
%
% Fudge factors
%
global α β
%
% Unpack individual state vectors from total state vector
%
ρ = [ x(1 : n) ; ρ0 ];
v = [ 0 ; x(n1 : 2 * n) ];
p = x(n1 + n : n1 + 2 * n);
%
% Calculate nonlinear boundary condition
%
if v(n1) < 0,
    v(n1) = −sqrt(max([2 * (p0 − p(n1))/ρ(n1), 0]));
else
    p(n1) = p0;
end
%
% Calculate spatial derivatives
%
ρx = partial(ρ, δx, +1, +1);
vx = partial(v, δx, −1, +1);
px = partial(p, δx, +1, +1);
%
% Calculate algebraic quantities
%
f = α * (ρ .* v .* abs(v))/δx;
q = zeros(n1, 1);
for i = 1 : n1,
    if vx(i) < 0,
        q(i) = β * (δx^2) * ρ(i) * (vx(i)^2);
    end,
end
qx = partial(q, δx, −1, +1);
a = px + qx + f;
%
% Calculate temporal derivatives
%
ρt = −(v .* ρx) − (ρ .* vx);
vt = −(v .* vx) − (a ./ ρ);
```

$$p_t = -(v \mathrel{.*} a) - \gamma * (p \mathrel{.*} v_x);$$
```
%
% Pack individual state derivatives into total state derivative vector
%
```
$$xdot = [\, \rho_t(1:n)\, ;\; v_t(2:n1)\, ;\; p_t\,];$$
return

The resulting set of 151 nonlinear ODEs was simulated across 0.01 sec
using the RKF4/5 algorithm, as we learnt that RK algorithms are expected
to perform decently when faced with nonlinear hyperbolic PDE problems
converted to sets of ODEs by the MOL approach.

This time around, we used all the bells and whistles and included step–
size control in time. The results of this simulation are shown in Fig.6.13.

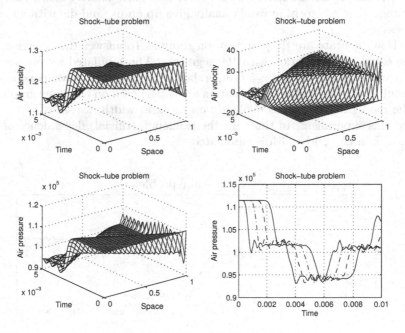

FIGURE 6.13. Shock tube simulation.

The first three graphs depict $\rho(x,t)$, $v(x,t)$, and $p(x,t)$. The solutions
look like the water falls of the Iguazu looked at from the Argentinean side
of the river. The bottom left parts of all three functions look dangerously
irregular in shape. Are the simulation results inaccurate?

The bottom right curve shows the air pressure as a function of time. The
solid curve depicts the pressure 20 cm away from the closed end, the dashed
line shows the pressure 40 cm away, the dot–dashed line 60 cm away, and
the dotted line 80 cm away.

As the end of tube opens, the point closest to the opening experiences the
rarefaction wave first. The points further into the tube experience the wave

later. From Fig.6.13, it can be concluded that the wave travels through the tube with a constant wave–front velocity of roughly 35 cm per 0.001 sec, or 350.0 m sec^{-1}. This is the correct value of the velocity of sound at sea level and at a temperature of $T = 300$ K. Thus, our simulation seems to be working fine. (There is nothing more healthy in simulation of physical systems than a little reality check once in a while!)

As the rarefaction wave reaches the closed end of the tube, the inertia of the flowing air creates a vacuum. The air flows further, but cannot be replaced by more air from the left. Consequently, the air pressure now sinks below that of the outside air.

As the vacuum reaches the open end of the tube, a new wave is created, this time a *compression wave*, that races back into the tube.

We ended the simulation at $t = 0.01$ sec, since shortly thereafter, the Runge–Kutta algorithm would finally give up on us, and die with an error message.

How accurate are these simulation results? To answer this question, we repeated the simulation with 100 segments. The simulated air pressure at the center of the tube, $x = 50$ cm, is shown in Fig.6.14. For comparison, the results of the 50–segment simulation are superposed on the same graph. As the model itself depends explicitly on the grid width, we set $\alpha = \beta = 0.0$ for this experiment. In this way, the explicit (artificial) dependence of the model on the grid width is eliminated.

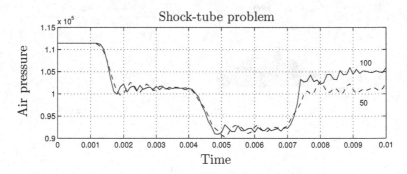

FIGURE 6.14. Consistency error for shock tube simulation.

The simulation results are visibly different. Moreover, the differences seem to grow over time. Is this a consistency error, or simply the result of an inaccurate simulation?

To answer this question, we repeated the same experiment, this time using a different integration algorithm. The F–stable Backinterpolation technique is supposed to work at least as well as the RK algorithm.

The simulation results are indistinguishable by naked eye. Whereas the largest relative distance between the air pressure with 50 and 100 segments:

$$err = \frac{\max(\max(\mathrm{abs}(p_{100} - p_{50})))}{\max([\|p_{100}\|, \|p_{50}\|])} \qquad (6.68)$$

is $err = 7.5726e - 4$, the largest relative distance between the air pressure with 50 segments comparing the two different integration algorithms is $err = 1.2374e - 7$, and with 100 segments, it is $err = 6.3448e - 7$.

Hence the *simulation error* is smaller than the *consistency error* by three orders of magnitude. Evidently, we are not faced with a *simulation problem* at all, but rather with a *modeling problem*. The simulation is as accurate as can be expected.

The BI4 algorithm is considerably less efficient than the RKF4/5 algorithm in simulating this problem. Its inefficiency is not caused by the step size. In fact, the step–size controlled BI4 algorithm can make use of step sizes that are quite a bit larger than those used by RKF4/5. The inefficiency is caused by the computation of the Jacobians and of the inverse Hessians.

Since the problem is nonlinear, the Jacobians need to be numerically estimated, using an algorithm such as:

```
function [J] = jacobian(x, t)
%
% Jacobian of shock − tube problem
%
n  = length(x);
J  = zeros(n, n);
xd_ref = st_eq(x, t);
for i = 1 : n,
    x_new  = x;
    if abs(x(i))  <  1.0e − 6,
        x_new(i)  =  0.05;
    else
        x_new(i)  =  1.05 ∗ x(i);
    end,
    xd_new  = st_eq(x, t);
    J(:, i)  =  (xd_new  −  xd_ref)/(x_new(i)  −  x(i));
 end
return
```

Thus, every single Jacobian, which is being computed once per integration step, requires 152 additional function evaluations in the case of a 50–segment simulation, and 302 additional function evaluations in the case of a 100–segment simulation. No wonder that production codes of implicit ODE solvers are frugal in the frequency of Jacobian evaluations.

The Hessian is of the same size as the Jacobian:

$$\mathbf{H} = \mathbf{I}^{(n)} + \mathbf{J} \cdot \bar{h} + \frac{1}{2!} \cdot (\mathbf{J} \cdot \bar{h})^2 + \frac{1}{3!} \cdot (\mathbf{J} \cdot \bar{h})^3 + \frac{1}{4!} \cdot (\mathbf{J} \cdot \bar{h})^4 \qquad (6.69)$$

where $\bar{h} = -h/2$ is the step size of the right half–step of the BI4 algorithm.

The Hessian is used in a Gauss elimination step once per iteration step:

```
while err_2 > 0.1 * tol,
    [x_right4, x_right5] = rkf45_step(x_new, t_new, -h/2);
    n_fct = n_fct + 6;
    x_new = x_new - H\(x_right4 - x_left4);
    err_2 = norm(x_right4 - x_left4, 'inf')/max([norm(x_left4), norm(x_right4), tol]);
end
```

The computational burden of these algorithms is atrocious. We shall have to do something about the size of these matrices. This problem shall be tackled in the next chapter of this book.

What can we do to reduce the consistency error? From our previous observation, we know the answer to this question. If we increase the approximation order of the spatial derivatives by two, the consistency error is expected to decrease by two orders of magnitude.

We modified the *partial* function to use fourth–order accurate central differences instead of the previously used second–order accurate central differences. To this end, the following formulae were now coded into the *partial* function:

$$\left.\frac{\partial u}{\partial x}\right|_{x=x_i} \approx \frac{1}{12\delta x} \cdot (-u_{i+2} + 8u_{i+1} - 8u_{i-1} + u_{i-2}) \tag{6.70}$$

except near the boundaries, where we used fourth–order accurate biased formulae:

$$\frac{\partial u}{\partial x}(x = x_1, t) \approx \frac{1}{12\delta x} \cdot (-3u_5 + 16u_4 - 36u_3 + 48u_2 - 25u_1) \tag{6.71a}$$

$$\frac{\partial u}{\partial x}(x = x_2, t) \approx \frac{1}{12\delta x} \cdot (u_5 - 6u_4 + 18u_3 - 10u_2 - 3u_1) \tag{6.71b}$$

$$\frac{\partial u}{\partial x}(x = x_n, t) \approx \frac{1}{12\delta x} \cdot (3u_{n+1} + 10u_n - 18u_{n-1} + 6u_{n-2} - u_{n-3}) \tag{6.71c}$$

$$\frac{\partial u}{\partial x}(x = x_{n+1}, t) \approx \frac{1}{12\delta x} \cdot (25u_{n+1} - 48 * u_n + 36u_{n-1} - 16u_{n-2}$$
$$+ 3u_{n-3}) \tag{6.71d}$$

In the case of a symmetry boundary condition, the central formulae are used all the way to the boundary while folding the values that are outside the domain back into the domain, as explained earlier.

The correction formulae are:

$$\frac{\partial u}{\partial x}(x = x_1, t) \approx 0.0 \tag{6.72a}$$

$$\frac{\partial u}{\partial x}(x = x_2, t) \approx \frac{1}{12\delta x} \cdot (-u_4 + 8u_3 + u_2 - 8u_1) \tag{6.72b}$$

$$\tag{6.72c}$$

for a symmetry boundary condition at the left end, and:

$$\frac{\partial u}{\partial x}(x = x_n, t) \approx \frac{1}{12\delta x} \cdot (8u_{n+1} - u_n - 8u_{n-1} + u_{n-2}) \tag{6.73a}$$

$$\frac{\partial u}{\partial x}(x = x_{n+1}, t) \approx 0.0 \tag{6.73b}$$

for a symmetry boundary condition at the right end.

We then simulated the system using RKF4/5. Unfortunately, the experiment failed miserably. The integration step size had to be reduced by three orders of magnitude to values around $h = 10^{-8}$, in order to obtain a numerically stable solution, and the results are still incorrect.

What happened? In the previous experiment, the global relative simulation error had been around $err = 10^{-7}$, which is small in comparison with the consistency error, but is still quite large, taking into account that MATLAB computes everything in double precision. With step sizes in the order of $h = 10^{-5}$, we had already sacrificed roughly nine digits to *shiftout*.

In the new experiment with step sizes smaller by three orders of magnitude, we lose at least another three digits to shiftout, i.e., the simulation error is now of the same order of magnitude as the former consistency error. Hence we have not gained anything.

In reality, the problem is even worse. With step sizes that small, the higher order terms of the Taylor–series expansion become irrelevant, and RKF4/5 behaves just like forward Euler. Consequently, also the stability domain of the method shrinks to that of forward Euler, which is totally useless with eigenvalues of the Jacobian spreading up and down along the imaginary axis of the complex $\lambda \cdot h$–plane.

How did BI4 fare in this endeavor? Unfortunately, its destiny is not much better than that of RKF4/5. Remember that BI4 consists of two semi–steps of RKF4/5. With larger step sizes, the left forward RKF4/5 semi–step produces highly unstable x_{left4} values, which the right backward RKF4/5 semi–step needs to stabilize in its Newton iteration.

Unfortunately, it cannot do so, because in the statement:

$$x_{new} = x_{new} - H \backslash (x_{right4} - x_{left4}); \tag{6.74}$$

we subtract a potentially very large number, x_{left4}, from another equally large number, x_{right4}, which again leads to an extreme case of *roundoff*.

With smaller step sizes, the BI4 algorithm degenerates to a forward Euler semi–step followed by a backward Euler semi–step, i.e., to an inefficient implementation of the trapezoidal rule. This is clearly superior to forward Euler alone, since also BI2 is still F–stable, but unfortunately, the semi–steps themselves still suffer from the shiftout problems of the RKF4/5 algorithm, i.e., the simulation error is still of the same order of magnitude as the former consistency error.

Why did all simulation attempts fail after a little more than 0.01 seconds of simulated time? In flow simulations (and in real flow phenomena), it can happen that the top of the wave travels faster than the bottom of the wave. When this happens, the wave will eventually topple over, and at this moment, the wave front becomes infinitely steep. The flow is no longer *laminar*, it has now become *turbulent*.

This is what happens in our shock–tube problem as subsequent versions of rarefaction and compression waves chase after each other back and forth through the tube at ever shorter time intervals. No wonder that the bottom of the three–dimensional plots of the shock–tube simulation look like the bottom of a water fall.

The MOL approach doesn't work for simulating turbulent flows. There exist other simulation techniques (such as the vortex methods [6.12]) that work well for very high Reynolds numbers (above 100 or 1000), and that don't work at all for laminar flows. Reynolds numbers between 1.0 (transition from laminar to turbulent flow) and 100, is where the real research in numerical solution of hyperbolic PDE problems is to be found. Until this day, we don't have any decent simulation methods that can deal appropriately with turbulent flows at low Reynolds numbers.

6.6   Upwind Discretization

In the previous section, we have recognized that hyperbolic PDEs, when converted to sets of ODEs using the MOL approach, lead to systems that share into some of the properties associated with stiff systems, although they do not meet most of the definitions of stiff systems. Yet, the step size had to be often reduced in order to obtain stable solutions when using explicit integration algorithms. In the case of the shock–tube example, the step size reduction was detrimental in that it led to a bad shiftout problem, before the consistency error could be reduced to an insignificantly small value.

How can we stabilize the RK algorithms when dealing with hyperbolic PDEs? One successful idea that was first proposed by Carver and Hinds is to bias the spatial discretization formulae of moving waves in the direction of the provenance of the wave [6.3].

Many wave propagation problems can be formulated in the following way:

$$\frac{\partial u}{\partial t} + v \cdot \frac{\partial u}{\partial x} = 0.0 \tag{6.75}$$

The velocity v determines the direction of flow of the wave. If $v > 0$, the wave moves from left to right. If $v < 0$, it moves from right to left.

The upwind discretization scheme can thus be implemented e.g. as follows:

$$\frac{\partial u}{\partial x}(x = x_i, t) \approx \begin{cases} (3u_i - 4u_{i-1} + u_{i-2})/(2\delta x) & , \quad v \gg 0 \\ (u_{i+1} - u_{i-1})/(2\delta x) & , \quad v \approx 0 \\ (-u_{i+2} + 4u_{i+1} - 3u_i)/(2\delta x) & , \quad v \ll 0 \end{cases} \tag{6.76}$$

if second–order accurate spatial differences are to be used.

Looking once more at the shock–tube problem with $\alpha = \beta = 0.0$:

$$\frac{\partial \rho}{\partial t} = -v \cdot \frac{\partial \rho}{\partial x} - \rho \cdot \frac{\partial v}{\partial x} \tag{6.77a}$$

$$\frac{\partial v}{\partial t} = -v \cdot \frac{\partial v}{\partial x} - \frac{1}{\rho} \cdot \frac{\partial p}{\partial x} \tag{6.77b}$$

$$\frac{\partial p}{\partial t} = -v \cdot \frac{\partial p}{\partial x} - \gamma \cdot p \cdot \frac{\partial v}{\partial x} \tag{6.77c}$$

we notice that all three of these PDEs look like Eq.(6.75), each with a correction term.

We thus encoded the fourth–order accurate upwind formulae in the function:

$$\mathbf{u_x} = \text{upwindv}(\mathbf{u}, \delta x, bc, bctype, \mathbf{fdirv}) \tag{6.78}$$

where **fdirv** is a vector of flow directions, and replaced each occurrence of *partial* in the state equations by *upwindv*, setting the argument *fdirv* as the velocity vector, **v**.

Unfortunately, it didn't work. The shock–tube model discretized using any fourth–order accurate spatial discretization scheme seems to be unstable beyond redemption.

Upwind discretization schemes have become quite fashionable in recent years and come in many different variations. They can be quite effective at times. We still like the original scheme [6.3] best for its simplicity. Yet, there doesn't seem to exist a clean recipe for when and how to use upwind discretization. Sometimes, it helps to only discretize one of several PDEs using an upwind scheme, while discretizing the remaining PDEs using a central difference scheme. What works best can often only be determined by trial and error.

6.7   Grid–width Control

How can we make the solution more accurate without paying too much for it? We already know that it is generally a bad idea to reduce the consistency error by decreasing the grid width. It is much more effective to increase the approximation order of the spatial discretization scheme, whenever possible. Yet, the shock–tube problem has demonstrated that this approach may not always work.

A more narrow grid may be needed in order to accurately compute a wave front. It seems intuitively evident that a more narrow grid width should be used where the absolute spatial gradient is large, thus:

$$\delta x_i(t) \propto \left| \frac{\partial u}{\partial x}(x = x_i, t) \right|^{-1} \tag{6.79}$$

When applied to hyperbolic PDEs, Eq.(6.79) unfortunately suggests use of an *adaptively moving grid*, since the narrowly spaced regions of the grid should follow the wave fronts through space and time.

As we mentioned earlier, naïvely implemented grid–width control is problematic, to say the least. However when implemented carefully, grid–width control can provide an answer to containing the consistency error without leading to either numerical stability problems or at least unacceptably expensive simulation runs. Mack Hyman published some very interesting results on this topic   [6.10]. The general gist of his algorithms is the following. We basically operate on a *fixed* grid as before. However, we want to make sure that:

$$\delta x_i(t) \cdot \left| \frac{\partial u}{\partial x}(x = x_i, t) \right| \leq k_{\max} \tag{6.80}$$

at all times. If the absolute spatial gradient grows at some point in space and time, we must reduce the local grid size in order to keep Eq.(6.80) satisfied. We do this by inserting a new auxiliary grid point in the middle between two existing points. We should do this *before* the consistency error grows too large. It thus makes sense to look at the quantity:

$$\frac{1}{h} \left(\left| \frac{\partial u}{\partial x}(x = x_i, t = t_k) \right| - \left| \frac{\partial u}{\partial x}(x = x_i, t = t_{k-1}) \right| \right) \approx \frac{d}{dt} \left(\left| \frac{\partial u}{\partial x}(x = x_i, t) \right| \right) \tag{6.81}$$

If Eq.(6.80) is in danger of not being satisfied any longer and if the temporal gradient of the absolute spatial gradient is positive, we insert a new grid point. On the other hand, if Eq.(6.80) shows a sufficiently small value and if furthermore the temporal gradient is negative, neighboring auxiliary grid points can be thrown out again.

The new grid point solutions are computed using spatial interpolation. These solutions are then used as initial conditions for the subsequent integration of the newly activated differential equations over time. When a grid point is thrown out again, so is the differential equation that accompanies it.

The entire process is completely transparent to the user. Only those solution points are reported for which a solution had been requested. The actually used basic grid width (determined using true grid–width control at time zero) and the auxiliary grid points that are introduced and removed during the simulation run are internal to the algorithm, and the casual user doesn't need to be made aware of their existence. This corresponds to the concept of *communication points* and a *communication interval* discussed in Chapter 4 of this book.

6.8   PDEs in Multiple Space Dimensions

In principle, the MOL methodology can be extended without modification to the case of PDEs in multiple space dimensions. For example, the two–dimensional heat flow problem:

$$\frac{\partial u}{\partial t} = \sigma \left(\frac{\partial^2 u}{\partial x^2} + \frac{\partial^2 u}{\partial y^2} \right) \tag{6.82}$$

discretized using third–order accurate finite difference formulae for both the discretization in the x– and in the y–directions leads to the following ODE at point $x = x_i$ and $y = y_j$:

$$\frac{du_{i,j}}{dt} \approx \sigma \left(\frac{u_{i+1,j} - 2u_{i,j} + u_{i-1,j}}{\delta x^2} + \frac{u_{i,j+1} - 2u_{i,j} + u_{i,j-1}}{\delta y^2} \right) \tag{6.83}$$

but the problems are formidable. The first, and most frightening, problem is concerned with the sheer numbers of resulting ODEs. Everything that we wrote about the consistency error still applies. Except for toy problems, we shall certainly need in the order of 50 segments in each space direction, in order to obtain sufficiently smooth output curves. In two space dimensions, this leads to $50 \times 50 = 2500$ ODEs. In the case of three space dimensions, we obtain $50 \times 50 \times 50 = 125,000$ ODEs. Let us assume the differential equation is linear, and we decided to write it in matrix form. The **A**–matrix of the three-dimensional problem consists of $125,000 \times 125,000 = 15,625,000,000$ elements. If you are interested in solving such problems, you better get yourself a *fast* computer and powerful *sparse matrix solvers*. This is the kind of problems for which supercomputers were invented.

The second problem has to do with the distribution of the non–zero elements in the **A**–matrix. Until now, it always happened that the **A**–matrix of a single linear PDE converted by use of finite differences was *band–structured* with a narrow band width. There exist special matrix routines for very efficient handling of band–structured matrices. Unfortunately, the same technique no longer applies to two– and three–dimensional PDEs. Figure 6.15 shows the distribution of non–zero elements in the two–dimensional and three–dimensional heat equations converted to ODEs by means of third–order accurate finite differences using 10 segments in each space dimension. The differential equations were numbered from left to right, from top to bottom, and from front to back, i.e., starting with the last of the three indices. We assumed function value boundary conditions along all edges of the solution cube.

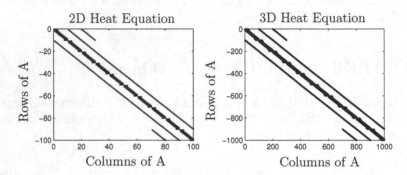

FIGURE 6.15. Distribution of non–zero elements in 2D and 3D heat equations.

Whereas the band width was five in the one–dimensional case, it is $4n+1$ in the two–dimensional case, and $4n^2 + 1$ in the three–dimensional case. Of course, the precise structure of the $\mathbf{A}$–matrix is application dependent. Unfortunately, this means that, when efficiency becomes truly an issue, we may no longer be able to apply the highly efficient algorithms for handling band–structured matrices. General sparse matrix techniques will still work, but they are considerably less efficient than the band–structured algorithms.

Special algorithms have been designed for renumbering a set of linear equations in such a manner as to minimize the band width of the resulting $\mathbf{A}$–matrix. For example the *red–black algorithm* often works well. These algorithms have been described in [6.15].

Unfortunately, we are not at the end of our misery yet. The next problem is illustrated in Fig.6.16.

Figure 6.16 shows a PDE that is defined on an irregularly shaped domain. Until now, we were always able to make the boundary condition coincide with one of the grid points. As Fig.6.16 shows, this may no longer be true in the multidimensional case.

Let us assume that four neighboring values on grid points in x–direction for $y = y_j$ are $u_{1,j}$, $u_{2,j}$, $u_{3,j}$, and $u_{4,j}$. Let us assume further that the boundary value is known at $x = x_{1.35}$ located between x_1 and x_2.

If we know the four solution values $u_{1,j}$, $u_{2,j}$, $u_{3,j}$, and $u_{4,j}$, we can use

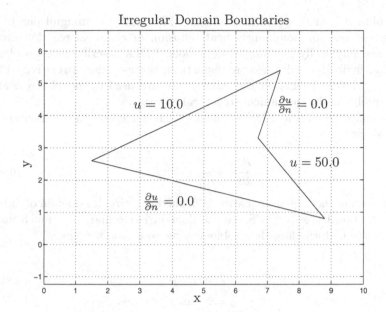

FIGURE 6.16. Irregular domain boundaries.

the Nordsieck vector approach presented in Chapter 4 to compute $u_{1.35,j}$. $u_{1.35,j}$ can be expressed as a weighted sum of $u_{1,j}$, $u_{2,j}$, $u_{3,j}$, and $u_{4,j}$. In reality, however, we know $u_{1.35,j}$ (boundary value), and $u_{2,j}$, $u_{3,j}$, and $u_{4,j}$ (through numerical integration). What is unknown is $u_{1,j}$. Thus, we need to solve the previously determined equation for the unknown $u_{1,j}$ instead for the known $u_{1.35,j}$.

To summarize this section: PDEs in one space dimension were still lots of fun. PDEs in multiple space dimensions are painful, to say the least. A large number of applied mathematicians devote their entire academic careers to nothing but solving these types of challenging numerical PDE problems. The purpose of the utterly brief description presented in this section is certainly not to add these specialists to the force of unemployed people, since you, by now, are able to solve all these problems on your own. The purpose of this section is to show you that there are still plenty of *very* challenging research topics around, and to possibly and hopefully wake your appetite for delving more deeply into one or the other of those areas.

6.9   Elliptic PDEs and Invariant Embedding

Equations (6.24) and (6.25) specified what elliptic PDEs are. However, this way of looking at the nature of PDEs is synthetic. People usually don't solve PDEs just for fun. They solve PDEs because they represent physical problems that they are interested in. Physically meaningful parabolic PDEs represent predominantly heat diffusion or chemical reaction problems, and physically meaningful hyperbolic PDEs describe field problems in either hydrodynamics, electromagnetism, optics, general relativity theory, etc. Elliptic PDEs, on the other hand, are used to model stress and strain problems in mechanical structural analysis.

The simplest elliptic PDE is the Laplace equation, e.g. in two space dimensions:

$$\frac{\partial^2 u}{\partial x^2} + \frac{\partial^2 u}{\partial y^2} = 0.0 \tag{6.84}$$

Let us assume the Laplace equation is defined in a circular domain of radius $r = 1.0$ around the origin. Since the domain is circular, it is much more appropriate to formulate the problem using *polar coordinates*.

$$x = r \cdot \cos\varphi \tag{6.85a}$$
$$y = r \cdot \sin\varphi \tag{6.85b}$$

or:

$$r = \sqrt{x^2 + y^2} \tag{6.86a}$$
$$\varphi = \arctan\left(\frac{y}{x}\right) \tag{6.86b}$$

We can express $u(x, y)$ as $\tilde{u}(r(x, y), \varphi(x, y))$. Thus,

$$\frac{\partial u}{\partial x} = \frac{\partial \tilde{u}}{\partial r} \cdot \frac{\partial r}{\partial x} + \frac{\partial \tilde{u}}{\partial \varphi} \cdot \frac{\partial \varphi}{\partial x} \tag{6.87}$$

or, in short–hand notation:

$$u_x = \tilde{u}_r \cdot r_x + \tilde{u}_\varphi \cdot \varphi_x \tag{6.88}$$

Using the chain rule and the multiplication rule, we find:

$$u_{xx} + u_{yy} = \left(r_x^2 + r_y^2\right)\tilde{u}_{rr} + 2\left(r_x\varphi_x + r_y\varphi_y\right)\tilde{u}_{r\varphi} + \left(\varphi_x^2 + \varphi_y^2\right)\tilde{u}_{\varphi\varphi}$$
$$+ \left(r_{xx} + r_{yy}\right)\tilde{u}_r + \left(\varphi_{xx} + \varphi_{yy}\right)\tilde{u}_\varphi \tag{6.89}$$

or finally:

$$\frac{\partial^2 \tilde{u}}{\partial r^2} + \frac{1}{r} \cdot \frac{\partial \tilde{u}}{\partial r} + \frac{1}{r^2} \cdot \frac{\partial^2 \tilde{u}}{\partial \varphi^2} = 0.0 \tag{6.90}$$

The boundary condition could be something like:

$$\frac{\partial \tilde{u}}{\partial r} = f(\varphi, t) \tag{6.91}$$

Notice that there is no need for any initial condition, since the PDE doesn't depend on time at all (except possibly through the boundary condition as in the above example). No numerical integration across time will take place at all. We are thus in trouble with our MOL methodology.

In some cases, we might still be able to apply the MOL approach by either differentiating along r and integrating along φ, or alternatively, by differentiating along φ and integrating along r. In both cases, however, we would be lacking one initial condition, and would instead have one final condition too many. This is therefore not an *initial value problem*, but rather a *boundary value problem*. We haven't discussed yet how those can be solved.

Does this mean that we have to give up for the time being, or is there a chance that we may turn this problem into one of our known initial value problems after all?

Let us simplify Eq.(6.91) a bit by assuming that the boundary condition does not depend on time. In this case, the problem is totally *static* in nature, i.e., the solution is not time–dependent at all. The solution consists simply of a set of u–values at the grid points.

We can now embed this problem within another problem as follows:

$$\frac{\partial \tilde{u}}{\partial t} = \frac{\partial^2 \tilde{u}}{\partial r^2} + \frac{1}{r} \cdot \frac{\partial \tilde{u}}{\partial r} + \frac{1}{r^2} \cdot \frac{\partial^2 \tilde{u}}{\partial \varphi^2} \tag{6.92}$$

with the boundary condition:

$$\frac{\partial \tilde{u}}{\partial r} = f(\varphi) \tag{6.93}$$

and with arbitrary initial conditions.

This is now clearly a parabolic initial value problem, which we already know how to solve. Since the PDE is analytically stable, and since the boundary condition is not a function of time, the solution will eventually settle into a *steady state*. However, once the steady state has been reached, the solution no longer changes with time, thus:

$$\frac{\partial \tilde{u}}{\partial t} = 0.0 \tag{6.94}$$

Therefore we conclude that the steady–state solution of the parabolic PDE is identical with the solution of the original elliptic PDE. This method of solving elliptic PDEs is called *invariant embedding*.[1] Of course, the price that we have to pay for this comfort is formidable. We were able to convert

---

[1] A majority of the references spell "imbedding" with an "i" rather than with an "e," probably because the inventor of the method didn't have a dictionary handy when he

a boundary value problem into an initial value problem at the expense of increasing the number of dimensions by one.

6.10 Finite Element Approximations

Those of you who read the companion book *Continuous System Modeling* [6.5] know our reservations against writing down mathematical formulae deprived of their physical meaning. Mathematics is no end in itself. *Mathematics is simply the language of physics.* Voltages and currents in an electronic circuit don't change their values as functions of time, because they observe some differential equations. They change their values in order to bring the system to a state of minimal energy. A differential equation is not the cause that makes physics tick, it is only one way of describing, in mathematical terms and after the fact, what happens in the process of energy exchange taking place in the physical system.

You may remember also that there are two ways of looking at energy conservation laws:

1. We can look at the energy itself. In the most general case, we write down a *Hamiltonian* or possibly a *Hamiltonian field* of the system (at least if the system is conservative), and from there, we can then derive a set of differential equations if we so choose.

2. Rather than looking at the stored energy itself, we can look at incremental energies, i.e., at *power flows*. This leads directly to the *bond graph approach* to modeling that was advocated in Chapters 7–9 of the companion book.

We strongly advocated the latter approach since power flow is a *local property* of the system, whereas energy is a *global property* of the system. Thus, power flow considerations lend themselves directly to an object–oriented approach to modeling.

In *distributed parameter system simulation*, the situation is a little different. As explained earlier, the PDE models that we are dealing with today are still structurally so simple that object orientation is of little concern. Also, especially if we are solving a boundary value problem anyway, as in the case of the elliptic PDEs, we need to solve a global optimization problem over the entire definition domain of the PDE, thus, the advantages of a local model description are gone.

Looking at the solution of the previously discussed Laplace equation, we know that the solution will minimize the amount of energy stored in

---

wrote his first paper about the method ... but we cannot bring ourselves to follow the trend — it looks so ugly (!)

the system. Consequently, we can write an energy function parameterized in the (unknown) solution values, and solve a minimization problem over the set of unknown parameters. This leads to a set of algebraic equations, possibly nonlinear, in the unknown solution vector.

Approaches that follow this line of reasoning are called *finite element methods.* They come in many shades and colors. The technique was originally developed by civil engineers trying to determine the static stress in bridges and other building structures. However, the method has a much broader range of possible applications. For all practical purposes, it can be viewed as an alternative to the finite difference approaches. Thus, it can conceptually also be used for other than elliptic PDEs.

The two approaches have their own particular advantages and disadvantages. Finite elements usually are less infected by problems with consistency errors than finite difference methods. Consequently, we can get by with a larger (and irregular) mesh, and thus, with a smaller number of equations. On the other hand, finite difference approximations always lead to sparse matrices. Finite element approximations do not share this property. As a consequence, although the number of equations is smaller in the finite element case, we may not be able to use sparse matrix techniques, and it is therefore not evident that the smaller system size truly leads to a more economical algorithm. Also, a finite difference formulation is usually easier to derive and harder to solve than a finite element formulation. However, it is easier to incorporate irregular and even non–convex domain boundaries into a finite element description.

Meanwhile, finite element methods have also been extended to the solution of non–stationary problems by means of a Galerkin formulation [6.17]. Thus, finite elements have suddenly become a contender to finite differences even in the context of the MOL methodology. However, more research in this area is still needed.

6.11 Summary

In this chapter, we have first and primarily discussed the numerical solution of PDEs in one space dimension. The method–of–lines approach lets us reduce such PDEs to large ODE systems that we can solve using regular ODE software.

Parabolic PDEs lead to sets of (artificially) stiff ODEs that can be treated appropriately using stiff system solvers such as the BDF algorithms. Since all of today's continuous–system modeling and simulation environments, such as Dymola [6.7, 6.8], offer stiff system solvers as part of their simulation run–time library, it became clear that they are perfectly capable of dealing with parabolic PDEs in one space dimension. The most cumbersome part in the conversion process was the derivation of the coefficients for the spatial finite difference approximations using Newton–Gregory polynomials, but this process can be easily automated.

Hyperbolic PDEs lead to large sets of marginally stable ODEs that can best be solved by F–stable integration algorithms, such as the backinterpolation techniques. However, explicit algorithms, such as AB3 or RKF4/5 may sometimes work just as well, as they avoid the need of computing expensive Jacobians and inverse Hessians. Hyperbolic PDEs are numerically more demanding than their parabolic cousins due to the occurrence of traveling shock waves. Adaptive moving mesh algorithms can provide a solution to this problem, but then call for special–purpose software, since these algorithms are non–trivial in their implementation. It would be too much of a burden to ask the user to implement such algorithms manually. Yet, powerful modeling environments can make also this process transparent to the modeler.

Elliptic PDEs in one space dimension are no PDEs at all. They are one class of boundary value ODEs, and we shall discuss later in this book how these can be tackled in general. However, one method was already provided here, namely the method of invariant embedding, a method that converts the boundary value ODE into a parabolic PDE in one space dimension, with which we can then proceed as elaborated above.

Multidimensional PDEs were discussed next. Although they can, in principle, be treated in exactly the same manner as their one–dimensional counterparts, the numerical problems are formidable, and efficiency considerations become here an issue of utmost importance.

Does there exist general–purpose PDE software? We had already mentioned the FORSIM–VI software [6.4]. FORSIM–VI is just a Fortran program. No preprocessor is involved at all. The user simply provides a Fortran subroutine describing his or her model. This makes FORSIM inappropriate for use in more complex ODE situations, since not even an equation sorter is offered, lest an object–oriented modeling facility. What makes FORSIM different from any other (simple–minded) ODE simulation system is that FORSIM provides built–in subroutines for converting spatial derivatives into finite difference approximations. These routines know how to compute the necessary coefficients, and consequently, the user doesn't need to worry about Newton–Gregory polynomials. FORSIM works with both equidistantly and non–equidistantly spaced grids. FORSIM also offers built–in routines for converting general and even nonlinear boundary conditions into boundary value conditions. Thus, FORSIM helps the user tremendously with the encoding of his or her PDEs. Routines are available for converting PDEs in one to three space dimensions, however, the two– and three–dimensional routines are not general since they work only on rectangular domains. FORSIM is strictly MOL–oriented. Spatial derivatives are discretized by means of finite difference approximations, whereas temporal derivatives are kept in the program for numerical integration across time. FORSIM offers a Gear (BDF) algorithm for the solution of parabolic problems, and an RKF4/5 algorithm for hyperbolic ones.

A fairly similar software system is DSS/2 [6.16]. The two systems, FORSIM–VI and DSS/2 are in fact so similar that a further discussion of DSS/2 can be skipped.

Other systems, such as PDEL [6.2], went another route. For the benefit of a more finely tuned numerical solution, they sacrificed generality for efficiency. These software systems allow the user to choose between a set of standard frequently occurring PDEs, and then employ different types of (not necessarily MOL) algorithms to solve the problem.

It may be noticed that all of these systems are fairly old. In the early seventies, it was hoped that PDE problems could be solved by general–purpose PDE software just as ODE problems are solved by general–purpose ODE software. This turned out to be an illusion. The ODE situation is *much* simpler. All we need to do is to provide a tool that allows to choose between a set of different numerical integration algorithms, and we are in business. Moreover, it often doesn't matter too much what algorithm we choose. One algorithm may be 30% faster or 20% slower than another, but who cares. Modern PCs have become so powerful that they can effectively and efficiently deal with the simulation of a large majority of lumped parameter models. In contrast, there exist many different techniques to solve PDE problems. Even if we limit our discussion to MOL–solutions, we must choose:

1. a numerical integration algorithm for integration across time,

2. a grid for discretization in space,

3. a numerical discretization scheme for differentiation across space,

4. an algorithm to translate boundary conditions specified at an arbitrary point in space to boundary conditions specified at the nearest grid point

5. an algorithm for converting general boundary conditions to boundary value conditions,

and this is only one among many approaches for numerically solving PDEs. Furthermore, the sensitivity of the solution to the selection of just the right combination of algorithms is much greater in the PDE case than in the ODE case. Selecting one method may mean that we have to wait for 50 hours until we obtain a (hopefully correct) answer, whereas the same problem may be solved by the best possible combination of algorithms in just a few seconds.

For these reasons, general–purpose PDE software hasn't lived up to its promise. The "casual" user of PDE software cannot be protected from having to understand the intricacies of the underlying numerical algorithms, and the numerical solution to all but toy PDE problems is so expensive that it is well worth spending some time on understanding what is going on before starting to crunch numbers. Getting coefficients out of Newton–Gregory polynomials may be but the least of our problems.

The situation is somewhat different in the case of elliptic PDEs. Elliptic PDEs are the simplest and most benign of all PDE problems. An extensive effort was undertaken by John Rice and his colleagues with large amounts of funding through the national agencies to solve that problem once and for all. They designed the ELLPACK software [6.15]. ELLPACK started out as a collection of useful algorithms to solve general–purpose elliptic PDEs in two and three space dimensions.

It turned out that the situation became soon too messy. Casual users no longer could learn to use these algorithms without help from the professional. To remedy the situation, a simple language was designed, and a compiler was written that would translate programs written in that language into a Fortran program that would then invoke the previously discussed algorithms that now form part of the run–time library. Thus, by this time, we are in the same situation as with the continuous–system simulation languages.

It turned out that it didn't work. The approach was too simple–minded. As a new algorithm became available, new keywords had to be added to the language in order to make this new algorithm accessible, and consequently, the compiler had to be updated frequently. This became too much of a hassle to the software designers. So they decided to parameterize the compiler. The compiler was generated out of a data template file that described both syntax and semantics of the ELLPACK language by means of a *compiler–compiler*. So, from now on, new features needed only to be incorporated into the data template file, and a new compiler for the so modified language could be generated at once.

Well, you may already have guessed . . . it didn't work. The researchers found the manual generation of the data template file much too cumbersome after all. That problem was taken care of easily. The precise details of the data template file were generated by a *data template compiler* out of a more abstract description of the data template file. Of course, also the data template compiler wasn't hand–coded. Why should it? Instead, the data template compiler was generated out of an abstract description of its duties by the same compiler–compiler that also generates the ELLPACK language compiler. This allows us to also update the data template compiler easily and readily.

At this point in time, only one question remains: Who wrote the compiler–compiler? We assume most of you read the story of Münchhausen who pulls himself out of the swamp by pulling on his own hair . . . the compiler–compiler wrote itself. A first (bootstrap) version of the compiler–compiler was hand–coded. This version was already able to read a language description in terms of its syntax and semantics. Well, the first language description it got to read was its own. So, by running the bootstrap compiler–

compiler through a description of itself, a second and much cleaner version of the compiler–compiler was obtained that could subsequently be used to generate new versions of the ELLPACK language compiler, the data template compiler, and –why not– itself.

Was it worth it? As an intellectual stimulus, most certainly. As an experimental toolbox for solving new kinds of elliptic PDEs, probably. As a general–purpose production tool for solving specific PDE problems posed in industry, not likely. We acquired the tool some years ago when we held a contract from the microelectronic industry to design a device simulator that could predict the breakdown behavior of bipolar power transistors (effectively, of any kind of reverse–biased p–n junction). The results that we obtained using ELLPACK were documented in [6.18]. ELLPACK allowed us to fairly quickly and easily go through a number of different algorithms and gain a feeling for which combination of algorithms might work decently well. However, the simulations obtained in this manner were painfully slow. A simple p–n junction milled for an hour or two on a VAX 11/780. More complex devices could not be handled at all within reasonable time limits. Therefore, we then designed our own special–purpose device simulator, ASEPS [6.19]. This program was able to simulate simple p–n junctions in a few seconds of CPU time on the same machine. ASEPS then enabled us to also study more complex device structures such as special geometric configurations of device termination structures for radiation–hardened power MOSFETs [6.6]. These simulation runs took a few minutes each, and optimization studies could be performed in batch mode over night.

Good special–purpose finite element software for structural analysis, such as NASTRAN, has been around for some time. This software doesn't attempt to solve general–purpose elliptic PDEs. Only one type of problem is solved, but the program is very flexible with respect to the specification of the domain on which the problem is to be solved and with respect to the selection of grid points (finite element programs aren't limited to using rectangular grids). Special–purpose numerical PDE solvers exist also for several other classes of applications, such as fluid dynamics.

It is disappointing to a generalist that the general–purpose approach to numerical PDE solution didn't work out. Unfortunately, we don't see any cure yet. Consequently, special–purpose solutions for specific PDE problems will be around for years to come.

6.12 References

[6.1] Kathryn E. Brenan, Stephen L. Campbell, and Linda R. Petzold. *Numerical Solution of Initial–Value Problems in Differential–Algebraic Equations*. North–Holland, New York, 1989. 256p.

[6.2] Alfonso F. Cárdenas and Walter J. Karplus. PDEL — A Language for Partial Differential Equations. *Comm. ACM*, 13:184–191, 1970.

[6.3] Michael B. Carver and H.W. Hinds. The Method of Lines and the Advective Equation. *Simulation*, 31:59–69, 1978.

[6.4] Michael B. Carver, D.G. Stewart, J.M. Blair, and W.M. Selander. The FORSIM VI Simulation Package for the Automated Solution of Arbitrarily Defined Partial and/or Ordinary Differential Equation Systems. Technical Report AECL–5821, Chalk River Nuclear Laboratories, Atomic Energy of Canada Limited, Chalk River, Ontario, Canada., 1978.

[6.5] François E. Cellier. *Continuous System Modeling*. Springer Verlag, New York, 1991. 755p.

[6.6] Kenneth R. Davis, Ronald D. Schrimpf, Kenneth F. Galloway, and François E. Cellier. The Effects of Ionizing Radiation on Power–MOSFET Termination Structures. *IEEE Trans. Nuclear Sci.*, 36(6):2104–2109, 1989.

[6.7] Hilding Elmqvist. *A Structured Model Language for Large Continuous Systems*. PhD thesis, Dept. of Automatic Control, Lund Institute of Technology, Lund, Sweden, 1978.

[6.8] Hilding Elmqvist. *Dymola — Dynamic Modeling Language, User's Manual, Version 5.3*. DynaSim AB, Research Park Ideon, Lund, Sweden, 2004.

[6.9] C. William Gear. *Numerical Initial Value Problems in Ordinary Differential Equations*. Series in Automatic Computation. Prentice–Hall, Englewood Cliffs, N.J., 1971. 253p.

[6.10] J. Mack Hyman. Moving Mesh Methods for Partial Differential Equations. In Jerome A. Goldstein, Steven Rosencrans, and Gary A. Sod, editors, *Mathematics Applied to Science: In Memoriam Edward D. Conway*, pages 129–153. Academic Press, Boston, Mass., 1988.

[6.11] John D. Lambert. *Numerical Methods for Ordinary Differential Systems: The Initial Value Problem*. John Wiley, New York, 1991. 304p.

[6.12] R. Ivan Lewis. *Vortex Element Methods for Fluid Dynamic Analysis of Engineering Systems*. Cambridge University Press, New York, 1991. 588p.

[6.13] Anthony Ralston and Herbert S. Wilf. *Mathematical Methods for Digital Computers*. John Wiley & Sons, New York, 1960. 287p.

[6.14] John R. Rice and Ronald F. Boisvert. *Solving Elliptic Problems Using Ellpack*. Springer–Verlag, New York, 1985. 497p.

[6.15] Robert D. Richtmyer and K. William Morton. *Difference Methods for Initial Value Problems.* Wiley Interscience, New York, 1967. 405p.

[6.16] William E. Schiesser. *The Numerical Method of Lines: Integration of Partial Differential Equations.* Academic Press, San Diego, Calif., 1991. 326p.

[6.17] V. Rao Vemuri and Walter J. Karplus. *Digital Computer Treatment of Partial Differential Equations.* Prentice–Hall, Englewood Cliffs, N.J., 1981. 449p.

[6.18] Qiming Wu and François E. Cellier. Simulation of High–Voltage Bipolar Devices in the Neighborhood of Breakdown. *Mathematics and Computers in Simulation,* 28:271–284, 1986.

[6.19] Qiming Wu, Chimin Yen, and François E. Cellier. Analysis of Breakdown Phenomena in High–Voltage Bipolar Devices. *Transactions of SCS,* 6(1):43–60, 1989.

6.13   Bibliography

[B6.1] Myron B. Allen, Ismael Herrera, and George F. Pinder. *Numerical Modeling in Science and Engineering.* John Wiley & Sons, New York, 1988. 418p.

[B6.2] William F. Ames. *Numerical Methods for Partial Differential Equations.* Academic Press, New York, 3rd edition, 1992. 433p.

[B6.3] T. J. Chung. *Computational Fluid Dynamics.* Cambridge University Press, Cambridge, United Kingdom, 2002. 800p.

[B6.4] Peter S. Huyakorn and George F. Pinder. *Computational Methods in Subsurface Flow.* Academic Press, New York, 1983. 473p.

[B6.5] Leon Lapidus and George F. Pinder. *Numerical Solution of Partial Differential Equations in Science and Engineering.* John Wiley & Sons, New York, 1999. 677p.

[B6.6] Robert Vichnevetsky and John B. Bowles. *Fourier Analysis of Numerical Approximations of Hyperbolic Equations.* SIAM Publishing, Philadelphia, Penn., 1982. 140p.

[B6.7] John Keith Wright. *Shock Tubes.* John Wiley & Sons, New York, 1961. 164p.

6.14   Homework Problems

[H6.1] Heat Diffusion in the Soil

Agricultural engineers are interested in knowing the temperature distribution in the soil as a function of the surface air temperature. As shown in Fig.H6.1a, we want to assume that we have a soil layer of 50 cm. Underneath the soil, there is a layer that acts as an ideal heat insulator.

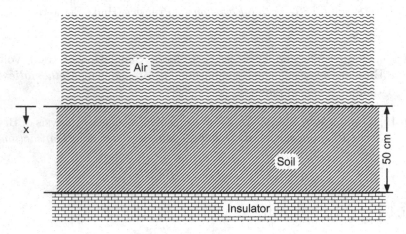

FIGURE H6.1a. Soil topology.

The heat flow problem can be written as:

$$\frac{\partial u}{\partial t} = \frac{\lambda}{\rho \cdot c} \cdot \frac{\partial^2 u}{\partial x^2} \tag{H6.1a}$$

where $\lambda = 0.004$ cal cm^{-1} sec^{-1} K^{-1} is the specific thermal conductance of soil, $\rho = 1.335$ g cm^{-3} is the density of soil, and $c = 0.2$ cal g^{-1} K^{-1} is the specific thermal capacitance of soil.

The surface air temperature has been recorded as a function of time. It is tabulated in Table H6.1a.

We want to assume that the surface soil temperature is identical with the surface air temperature at all times. We want to furthermore assume that the initial soil temperature is equal to the initial surface temperature everywhere.

Specify this problem using hours as units of time, and centimeters as units of space. Discretize the problem using third–order accurate finite differences everywhere. Simulate the resulting linear ODE system using MATLAB. Plot on one graph the soil temperature at the surface and at the insulator as functions of time. Generate also a three–dimensional plot showing the temperature distribution in the soil as a function of time and space.

t [hours]	u °C
0	6
6	16
12	28
18	21
24	18
36	34
48	18
60	25
66	15
72	4

TABLE H6.1a. Surface air temperature.

[H6.2] Electrically Heated Rod

We start out with a simple parabolic partial differential equation describing the temperature distribution in an electrically heated copper rod. This phenomenon can be modeled by the following equation:

$$\frac{\partial T}{\partial t} = \sigma \left(\frac{\partial^2 T}{\partial r^2} + \frac{1}{r} \cdot \frac{\partial T}{\partial r} + \frac{P_{electr}}{\lambda \cdot V} \right) \tag{H6.2a}$$

The first two terms represent the standard diffusion equation in polar coordinates as described previously in Eq.(6.92), and the constant term describes the electrically generated heat. It can be derived from Fig.8.13 of the companion book on *Continuous System Modeling* [6.5].

$$\sigma = \frac{\lambda}{\rho \cdot c} \tag{H6.2b}$$

is the diffusion coefficient, where $\lambda = 401.0$ J m^{-1} sec^{-1} K^{-1} is the specific thermal conductance of copper, $\rho = 8960.0$ kg m^{-3} is its density, and $c = 386.0$ J kg^{-1} K^{-1} is the specific thermal capacitance.

$$P_{electr} = u \cdot i \tag{H6.2c}$$

is the dissipated electrical power, and

$$V = \pi \cdot R^2 \cdot \ell \tag{H6.2d}$$

is the volume of the rod with the length $\ell = 1$ m and the radius $R = 0.01$ m. The rod is originally in an equilibrium state at room temperature $T_{room} = 298.0$ K.

The boundary conditions are:

$$\left.\frac{\partial T}{\partial r}\right|_{r=0.0} = 0.0 \tag{H6.2e}$$

$$\left.\frac{\partial T}{\partial r}\right|_{r=R} = -k_1\left(T(R)^4 - T_{\text{room}}^4\right) - k_2\left(T(R) - T_{\text{room}}\right) \tag{H6.2f}$$

where the quartic term models the heat radiation, whereas the linear term models convective heat flow away from the rod.

We want to simulate this system using the MOL approach with 20 spatial segments (in radial direction), and using second–order accurate finite difference approximations for the first–order spatial derivatives, and third–order accurate finite differences for the second–order spatial derivatives. We are going to treat the boundary condition at the center as a general boundary condition rather than as a symmetry boundary condition in order to circumvent the difficulties with computing the term $(\partial T/\partial r)/r$, which evaluates to $0/0$ at $r = 0.0$.

For internal segments, we obtain thus differential equations of the type:

$$\frac{dT_i}{dt} \approx \sigma\left(\frac{T_{i+1} - 2T_i + T_{i-1}}{\delta r^2} + \frac{1}{r}\cdot\frac{T_{i+1} - T_{i-1}}{2\delta r} + \frac{P_{\text{electr}}}{\lambda\cdot V}\right) \tag{H6.2g}$$

which are straightforward to implement. For the left–most segment, we have the condition:

$$\left.\frac{\partial T}{\partial r}\right|_{r=0.0} = 0.0 \approx \frac{1}{2\delta r}\left(-T_3 + 4T_2 - 3T_1\right) \tag{H6.2h}$$

and therefore:

$$T_1 \approx \frac{4}{3}T_2 - \frac{1}{3}T_3 \tag{H6.2i}$$

Consequently, we don't need to solve a differential equation at $r = 0.0$, and thereby, we skip the $0/0$ division.

At the right–most segment, we obtain:

$$\left.\frac{\partial T}{\partial r}\right|_{r=R} = -k_1\left(T_{21}^4 - T_{\text{room}}^4\right) - k_2\left(T_{21} - T_{\text{room}}\right) \approx \frac{1}{2\delta r}\left(-3T_{21} + 4T_{20} - T_{19}\right) \tag{H6.2j}$$

Thus, we obtain a nonlinear equation in the unknown T_{21}:

$$\mathcal{F}(T_{21}) = k_1\left(T_{21}^4 - T_{\text{room}}^4\right) + k_2\left(T_{21} - T_{\text{room}}\right) - \frac{1}{2\delta r}\left(-3T_{21} + 4T_{20} - T_{19}\right)$$

$$\approx 0.0 \tag{H6.2k}$$

which can be solved by Newton iteration:

$$T_{21}^0(t) = T_{21}(t - h) \tag{H6.2l}$$

$$T_{21}^1(t) = T_{21}^0(t) - \frac{\mathcal{F}(T_{21}^0)}{\mathcal{H}(T_{21}^0)} \tag{H6.2m}$$

$$T_{21}^2(t) = T_{21}^1(t) - \frac{\mathcal{F}(T_{21}^1)}{\mathcal{H}(T_{21}^1)} \tag{H6.2n}$$

until convergence

where:

$$\mathcal{H}(T_{21}) = \frac{\partial \mathcal{F}}{\partial T_{21}} = 4k_1 T_{21}^3 + k_2 - \frac{3}{2\delta r} \tag{H6.2o}$$

[H6.3] Wave Equation

The wave equation has been written as:

$$\frac{\partial^2 u}{\partial t^2} = c^2 \cdot \frac{\partial^2 u}{\partial x^2} \tag{H6.3a}$$

Let us rewrite $u(x, t)$ as $\tilde{u}(v, w)$, where:

$$v = t + x \tag{H6.3b}$$

$$w = t - x \tag{H6.3c}$$

What happens?

[H6.4] Shock Tube Simulation

We wish to analyze the influence of α and β on the accuracy of the simulation. Repeat the same 50 segment simulation with different values for α and β. What do you conclude about the relative influence of α and β in comparison with the consistency error. Assuming a small value of α, which is the largest value of β acceptable before the relative error exceeds 1%. Similarly, assuming a small value of β, which is the largest value of α acceptable before the relative error exceeds 1%.

Use one half the maximum values of α and β found above, and simulate across a longer period of time. Can you reach steady–state? Interpret the results.

[H6.5] River Bed Simulation

Hydrologists are interested in determining the movement of river beds with time. The dynamics of this system can be described through the PDE:

$$\frac{\partial v}{\partial t} + v \cdot \frac{\partial v}{\partial x} + g \cdot \frac{\partial h}{\partial x} + g \cdot \frac{\partial z}{\partial x} = w(v) \tag{H6.5a}$$

where $v(x, t)$ is the absolute value of the flow velocity of the water, $h(x, t)$ is the water depth, and $z(x, t)$ is the altitude of the river bed relative to an arbitrary constant level. $g = 9.81$ m sec^{-2} is the gravitational constant.

$w(v)$ is the friction of water at the river bed:

$$w(v) = -\frac{g \cdot v^2}{s_k^2 \cdot h^{4/3}} \tag{H6.5b}$$

where $s_k = 32.0$ m$^{3/4}$ sec^{-1} is the Strickler constant.

The continuity equation for the water can be written as:

$$\frac{\partial h}{\partial t} + v \cdot \frac{\partial h}{\partial x} + h \cdot \frac{\partial v}{\partial x} = 0.0 \tag{H6.5c}$$

and the continuity equation for the river bed can be expressed as:

$$\frac{\partial z}{\partial t} + \frac{df(v)}{dv} \cdot \frac{\partial v}{\partial x} = 0.0 \tag{H6.5d}$$

where $f(v)$ is the transport equation of Meyer–Peter simplified by means of regression analysis:

$$f(v) = f_0 + c_1(v - v_0)^{c_2} \tag{H6.5e}$$

where $c_1 = 1.272 \cdot 10^{-4}$ m, and $c_2 = 3.5$.

We want to study the Rhine river above Basel over a distance of 6.3 km. We want to simulate this system across 20 days of simulated time. The initial conditions are tabulated in Table H6.5a.

x [m]	v [m/s]	h [m]	z [m]
0.0	2.3630	3.039	59.82
630.0	2.3360	3.073	59.06
1260.0	2.2570	3.181	58.29
1890.0	1.6480	4.357	56.75
2520.0	1.1330	6.337	54.74
3150.0	1.1190	6.416	54.60
3780.0	1.1030	6.509	54.45
4410.0	1.0620	7.001	53.91
5040.0	0.8412	8.536	52.36
5670.0	0.7515	9.554	51.33
6300.0	0.8131	8.830	52.02

TABLE H6.5a. Initial data for river bed simulation.

We need three boundary conditions. We want to assume that the amount of water $q = h \cdot v$ entering the simulated river stretch is constant. For simplicity, we shall assume both h and v constant. At the lower end, there is a weir. Therefore, we can assume that the sum of z and h is constant at the lower end of the simulated stretch of river. Since the water moves much faster than the river bed, it doesn't make too much sense to apply boundary conditions to the river bed.

This system is pretty awful. The time constants of the water are measured in seconds, whereas those of the ground are measured in days. We are interested in the slow time constant, yet it is the fast time constant that dictates the integration step size. We can think of the first two PDEs as a nonlinear function generator for the third PDE. Let us therefore modify Eq.(H6.5d) as follows:

$$\frac{\partial z}{\partial t} + \beta \cdot \frac{df(v)}{dv} \cdot \frac{\partial v}{\partial x} = 0.0 \tag{H6.5f}$$

The larger we choose the tuning parameter, the faster will the river bed move. Select a value somewhere around $\beta = 100$ or even $\beta = 1000$. Later, we must analyze the damage that we did to the PDE system by introducing this tuning parameter. Maybe, we can extrapolate to the correct system behavior at $\beta = 1.0$.

The third boundary condition is analytically correct, but numerically not very effective since it is specified at the wrong end of the system. Since water always flows downhill, a boundary condition at the bottom is about as effective as commanding my dog to solve this homework problem. Let us therefore introduce yet another boundary condition at the top end:

$$\frac{\partial z}{\partial x} = constant \tag{H6.5g}$$

However, since we cannot specify a derivative boundary condition for a first–order equation, we reformulate Eq.(H6.5g) as:

$$z_1 = z_2 + constant \tag{H6.5h}$$

Plot the river bed altitude $z(x)$ measured at the end of every five day period superposed onto one graph.

Rerun the simulation for different values of β. Is it possible to extrapolate what the solution would look like for $\beta = 1.0$?

[H6.6] Boundary Value Conversion

A PDE in one space dimension is specified in the range $[0.0, 1.0]$ with $\delta x = 0.1$. Unfortunately, one of the boundary values if given as: $u(x = 0.98, t) = f(t)$.

We want to translate this boundary value to an equivalent boundary value at $u(x = 1.0, t)$. Use the Nordsieck vector approach to come up with a third–order accurate equation for $u(x = 1.0, t)$ as a function of $u(x = 0.98, t)$, $u(x = 0.9, t)$, $u(x = 0.8, t)$, and $u(x = 0.7, t)$.

[H6.7] Coordinate Transformation

Verify that Eq.(6.92) is indeed correct.

[H6.8] Coordinate Transformation

We wish to solve the Laplace equation for diffusion along the surface of a globe, assuming that no diffusion takes place in radial direction. To this end, we start out with the three–dimensional Laplace equation:

$$\left(\frac{\partial^2}{\partial x^2} + \frac{\partial^2}{\partial y^2} + \frac{\partial^2}{\partial z^2} \right) u(x, y, z) = 0.0 \qquad \text{(H6.8a)}$$

We want to rewrite this Laplace equation as a function of three different coordinates:

$$u(x, y, z) = \tilde{u}(\rho, \xi, \eta) \qquad \text{(H6.8b)}$$

where ρ is the radius of the globe, ξ is the longitude, and η is the latitude. We obtain a modified Laplace equation in these coordinates. We then specify that:

$$\frac{\partial^2 \tilde{u}}{\partial \rho^2} = \frac{\partial \tilde{u}}{\partial \rho} = 0.0 \qquad \text{(H6.8c)}$$

It is easy to make mistakes in such transformations. We therefore want to check whether the result is at least potentially correct. To this end, we let $\rho \to \infty$. Obviously, this must give us the original Laplace equation back, now expressed in ξ and η instead of x and y.

[H6.9] Poiseuille Flow Through a Pipe

The following equations describe the stationary flow of an incompressible fluid through a pipe:

$$\frac{d\hat{v}}{d\rho} = \frac{-\sqrt{2\Gamma}}{(\tau_M + 1)^2} \cdot \rho \cdot \tau^2 \qquad \text{(H6.9a)}$$

$$\frac{d}{d\rho} \left(\frac{\rho}{T} \cdot \frac{d\tau}{d\rho} \right) = \frac{-\Gamma}{(\tau_M + 1)^3} \cdot \rho^3 \cdot \tau^2 \qquad \text{(H6.9b)}$$

where:

$$\rho = \frac{r}{R} \qquad \text{(H6.9c)}$$

$$\tau = \frac{T(r)}{T_W} \qquad \text{(H6.9d)}$$

are two normalized coordinates. r is the distance from the center of the pipe, and R is the radius of the pipe. $T(r)$ is the temperature of the fluid at a distance r from the center, and T_W is the temperature of the pipe wall. T_W is assumed constant. $\hat{v} = k_1 * v$ is the normalized flow velocity, where

k_1 is a constant that depends on the viscosity, the thermal conductivity, and the average temperature of the fluid.

The boundary conditions are:

$$\frac{d\hat{v}}{d\rho}(\rho = 0.0) = 0.0 \qquad \text{(H6.9e)}$$

$$\frac{d\tau}{d\rho}(\rho = 0.0) = 0.0 \qquad \text{(H6.9f)}$$

$$\hat{v}(\rho = 1.0) = 0.0 \qquad \text{(H6.9g)}$$

$$\tau(\rho = 1.0) = 1.0 \qquad \text{(H6.9h)}$$

Thus, this is a boundary value problem. We could integrate this problem across ρ in the range $\rho = [0.0, 1.0]$ with unknown initial conditions $\hat{v}(\rho = 0.0) = \hat{v}_M$ and $\tau(\rho = 0.0) = \tau_M$.

However, in the light of what we learnt in this chapter, we shall try another approach. We embed this boundary value problem into a parabolic PDE, which we solve with arbitrary initial conditions until we reach steady-state.

Notice that the equations contain two yet unknown parameters. Γ is a constant that depends on the fluid. Let us assume that $\Gamma = 10.0$. τ_M is the value of the normalized temperature at the center of the pipe. We simply introduce the momentary value of that temperature into the equation, and modify that value as the simulation proceeds.

6.15   Projects

[P6.1] Grid–Width Control

Implement a moving grid algorithm for the shock tube problem using the ideas that were outlined in this chapter.

6.16   Research

[R6.1] Grid–Width Control

Generalize the idea of a moving grid algorithm to hyperbolic PDEs in two space dimensions.

Develop a general theory for assessment of the consistency error, and derive a grid–width control algorithm that contains the consistency error in a reliable and systematic fashion.

7

Differential Algebraic Equations

Preview

In this chapter, we shall analyze simulation problems that don't present themselves initially in an explicit state–space form. For many physical systems, it is quite easy to formulate a model where the state derivatives show up implicitly and possibly even in a nonlinear fashion anywhere within the equations. We call system descriptions that consist of a mixture of implicitly formulated algebraic and differential equations *Differential Algebraic Equations (DAEs)*. Since these cases constitute a substantial and important portion of the models encountered in science and engineering, they deserve our attention. In this chapter, we shall discuss the question, how sets of DAEs can be converted symbolically in an automated fashion to equivalent sets of ODEs.

7.1 Introduction

In the companion book *Continuous System Modeling* [7.5], we have demonstrated that object–oriented modeling of physical systems invariably leads to implicit DAE descriptions. Some of these can be converted to ODE descriptions quite easily by simple sorting algorithms, whereas others contain big *algebraic loops* or even *structural singularities*.

We shall now revisit these issues and present a set of symbolic formulae manipulation algorithms that allow us to convert implicit DAE descriptions to equivalent explicit ODE descriptions.

Let us once again begin with a simple electrical RLC circuit. Its schematic is shown in Fig.7.1.

As there are five circuit elements defining two variables each, namely the voltage across and the current through that element, we need 10 equations to describe the model, e.g. the five element equations, defining the relation between voltage across and current through the element, plus three mesh equations in the mesh voltages, plus two node equations in the node currents. A possible set of equations is:

$$u_0 = f(t) \tag{7.1a}$$
$$u_1 = R_1 \cdot i_1 \tag{7.1b}$$

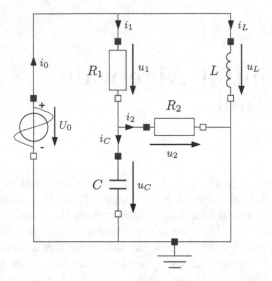

FIGURE 7.1. Schematic of electrical RLC circuit.

$$u_2 = R_2 \cdot i_2 \tag{7.1c}$$

$$u_L = L \cdot \frac{di_L}{dt} \tag{7.1d}$$

$$i_C = C \cdot \frac{du_C}{dt} \tag{7.1e}$$

$$u_0 = u_1 + u_C \tag{7.1f}$$

$$u_L = u_1 + u_2 \tag{7.1g}$$

$$u_C = u_2 \tag{7.1h}$$

$$i_0 = i_1 + i_L \tag{7.1i}$$

$$i_1 = i_2 + i_C \tag{7.1j}$$

As we wish to generate a state–space model, we define the outputs of the integrators, u_C and i_L, as our state variables. These can thus be considered known variables, for which no equations need to be found. In contrast, the inputs of the integrators, du_C/dt and di_L/dt, are unknowns, for which equations must be found. These are the state equations of the state–space description.

The structure of these equations can be captured in the so–called *structure incidence matrix*. The structure incidence matrix lists the equations in any order as rows, and the unknowns in any order as columns. If the ith equation contains the jth variable, the element $< i, j >$ of the structure incidence matrix assumes a value of 1, otherwise it is set to 0. The structure incidence matrix for the above set of equations could e.g. be written as:

$$\mathbf{S} = \begin{array}{c} \\ Eq.(7.1a) \\ Eq.(7.1b) \\ Eq.(7.1c) \\ Eq.(7.1d) \\ Eq.(7.1e) \\ Eq.(7.1f) \\ Eq.(7.1g) \\ Eq.(7.1h) \\ Eq.(7.1i) \\ Eq.(7.1j) \end{array} \begin{array}{c} \begin{matrix} u_0 & i_0 & u_1 & i_1 & u_2 & i_2 & u_L & \frac{di_L}{dt} & \frac{du_C}{dt} & i_C \end{matrix} \\ \begin{pmatrix} 1 & 0 & 0 & 0 & 0 & 0 & 0 & 0 & 0 & 0 \\ 0 & 0 & 1 & 1 & 0 & 0 & 0 & 0 & 0 & 0 \\ 0 & 0 & 0 & 0 & 1 & 1 & 0 & 0 & 0 & 0 \\ 0 & 0 & 0 & 0 & 0 & 0 & 1 & 1 & 0 & 0 \\ 0 & 0 & 0 & 0 & 0 & 0 & 0 & 0 & 1 & 1 \\ 1 & 0 & 1 & 0 & 0 & 0 & 0 & 0 & 0 & 0 \\ 0 & 0 & 1 & 0 & 1 & 0 & 1 & 0 & 0 & 0 \\ 0 & 0 & 0 & 0 & 1 & 0 & 0 & 0 & 0 & 0 \\ 0 & 1 & 0 & 1 & 0 & 0 & 0 & 0 & 0 & 0 \\ 0 & 0 & 0 & 1 & 0 & 1 & 0 & 0 & 0 & 1 \end{pmatrix} \end{array} \qquad (7.2)$$

Initially, all of these equations are *acausal*, meaning that the equal sign has to be interpreted in the sense of an equality, rather than in the sense of an assignment. For example, the above set of equations contains two equations that list u_0 to the left of the equal sign. Evidently, only one of those can be used to solve for u_0.

Two simple rules can be formulated that help us decide, which variables to solve for from which of the equations:

1. If an equation contains only a single unknown, i.e., one variable for which no solving equation has been found yet, we need to use that equation to solve for this variable. For example, Eq.(7.1a) contains only one unknown, u_0, hence that equation must be used to solve for u_0, and consequently, Eq.(7.1a) has now become a *causal equation*, and u_0 can henceforth be considered a known variable in all remaining equations.

2. If an unknown only appears in a single equation, that equation must be used to solve for it. For example, i_0 only appears in Eq.(7.1i). Hence we must use Eq.(7.1i) to solve for i_0.

These rules can be easily visualized in the structure incidence matrix. If a row contains a single element with a value of 1, that equation needs to be solved for the corresponding variable, and both the row and the column can be eliminated from the structure incidence matrix. If a column contains a single element with a value of 1, that variable must be solved for using the corresponding equation. Once again, both the column and the row can be eliminated from the structure incidence matrix. The algorithm proceeds iteratively, until no more rows and columns can be eliminated from the structure incidence matrix.

7.2   Causalization of Equations

Although the algorithm based on the structure incidence matrix will work, another algorithm has become more popular in recent years that is based on graph theory. It is an algorithm proposed first by Tarjan[1] [7.21]. Rather than capturing the structure of the set of DAEs in the form of a structure incidence matrix, it captures the same information in a graphical data structure, called the *structure digraph*.

The structure digraph depicts on the left hand side the equations as a column of nodes. On the right hand side, the unknowns are displayed also as a column of nodes. Since the number of equations must always equal the number of unknowns, the two column vectors are of equal length. A straight line connects an equation with an unknown, if that unknown appears in the equation.

The structure digraph of the above set of equations could be drawn e.g. as shown in Fig.7.2.

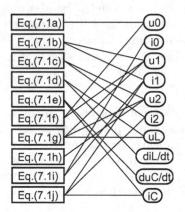

FIGURE 7.2. Structure digraph of electrical circuit.

We now implement our two rules for selecting which variable is to be solved for from which equation using the structure digraph.

When we select a variable to be solved for using a given equation, we color the connection between the equation and the variable in "red." Since this book is printed in black and white, we shall simulate the coloring by dashing it.

When we declare that a previously unknown variable is now known, because we already found an equation to solve for it, or because the equation,

---

[1]The algorithm presented in this section is not exactly the one originally proposed by Tarjan, but rather a somewhat modified algorithm, applied furthermore in a different context. Tarjan, in his original article, did not concern himself at all with the causalization of equations, but rather with detecting loops in a directed graph.

in which it occurs, is being used to solve for another variable, we color that connection in "blue." In this book, we shall simulate the coloring by dotting it.

A causal equation is an equation that has exactly one red (dashed) line attached to it. Acausal equations are equations that have only black (solid) and blue (dotted) lines attached to them. Known variables are variables that have exactly one red (dashed) line ending in them. An unknown variable has only black (solid) and blue (dotted) lines attached to it. No equation or variable has ever more than one red (dashed) line connecting to it.

We are now ready to implement our two rules.

1. For all acausal equations, if an equation has only one black (solid) line attached to it, color that line red (dash it), follow it to the variable it points at, and color all other connections ending in that variable in blue (dot the connections). Renumber the equation using the lowest free number starting from 1.

2. For all unknown variables, if a variable has only one black (solid) line attached to it, color that line red (dash it), follow it back to the equation it points at, and color all other connections emanating from that equation in blue (dot the connections). Renumber the equation using the highest free number starting from n, where n is the number of equations.

Figure 7.3 shows the structure digraph of the electrical circuit after one iteration through these two rules.

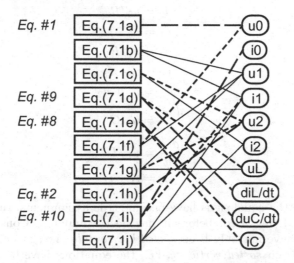

FIGURE 7.3. Structure digraph of electrical circuit after partial coloring.

In the first iteration, five of the 10 equations were made causal, two using
rule #1, and three using rule #2. However, the algorithm doesn't end here,
since these rules can be applied recursively.

Figure 7.4 shows the structure digraph of the electrical circuit after all
equations have been made causal.

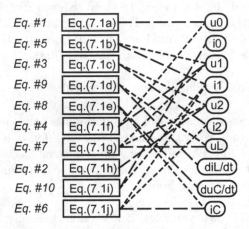

FIGURE 7.4. Structure digraph of electrical circuit after complete coloring.

We were able to complete the causalization of the equations. We can now
read out the 10 equations in their causal form.

$$u_0 = f(t) \tag{7.3a}$$

$$u_2 = u_C \tag{7.3b}$$

$$i_2 = u_2/R_2 \tag{7.3c}$$

$$u_1 = u_0 - u_C \tag{7.3d}$$

$$i_1 = u_1/R_1 \tag{7.3e}$$

$$i_C = i_1 - i_2 \tag{7.3f}$$

$$u_L = u_1 + u_2 \tag{7.3g}$$

$$\frac{du_C}{dt} = i_C/C \tag{7.3h}$$

$$\frac{di_L}{dt} = u_L/L \tag{7.3i}$$

$$i_0 = i_1 + i_L \tag{7.3j}$$

By now, the equal signs have become true assignments. In addition, no
variable is being used, before it has been computed. Consequently, the
equations have not only been *sorted horizontally*, i.e., made causal. They
have also been *sorted vertically*, i.e., the equations have been sorted into
an executable sequence.

Let us write down the structure incidence matrix of the horizontally and vertically sorted set of equations.

$$
\mathbf{S} =
\begin{array}{c}
\\
Eq.(7.3a) \\
Eq.(7.3b) \\
Eq.(7.3c) \\
Eq.(7.3d) \\
Eq.(7.3e) \\
Eq.(7.3f) \\
Eq.(7.3g) \\
Eq.(7.3h) \\
Eq.(7.3i) \\
Eq.(7.3j)
\end{array}
\begin{array}{c}
u_0 \; u_2 \; i_2 \; u_1 \; i_1 \; i_C \; u_L \; \frac{du_C}{dt} \; \frac{di_L}{dt} \; i_0 \\
\left(
\begin{array}{cccccccccc}
1 & 0 & 0 & 0 & 0 & 0 & 0 & 0 & 0 & 0 \\
0 & 1 & 0 & 0 & 0 & 0 & 0 & 0 & 0 & 0 \\
0 & 1 & 1 & 0 & 0 & 0 & 0 & 0 & 0 & 0 \\
1 & 0 & 0 & 1 & 0 & 0 & 0 & 0 & 0 & 0 \\
0 & 0 & 0 & 1 & 1 & 0 & 0 & 0 & 0 & 0 \\
0 & 0 & 1 & 0 & 1 & 1 & 0 & 0 & 0 & 0 \\
0 & 1 & 0 & 1 & 0 & 0 & 1 & 0 & 0 & 0 \\
0 & 0 & 0 & 0 & 0 & 1 & 0 & 1 & 0 & 0 \\
0 & 0 & 0 & 0 & 0 & 0 & 1 & 0 & 1 & 0 \\
0 & 0 & 0 & 0 & 1 & 0 & 0 & 0 & 0 & 1
\end{array}
\right)
\end{array}
\qquad (7.4)
$$

The structure incidence matrix of the sorted set of equations is now in lower–triangular form. Hence sorting the equations both horizontally and vertically is identical to finding permutation matrices that reduce the structure incidence matrix to lower–triangular form.

There exist algorithms to find these permutation matrices directly. This would be yet another approach to sorting the equations.

Variants of the Tarjan algorithm have become the most popular among all the available sorting algorithms for their efficiency, as their *computational effort* grows linearly with the size of the DAE system[1]. This is the best performance that can be expected of any algorithm.

A variant of the causalization algorithm, called *output set assignment*, can be found in a paper by Pantelides [7.20], who presented the algorithm once again in a somewhat different context. The Pantelides variant of the causalization algorithm has become the most popular of these algorithms, as it has the advantage that it can be implemented using a very compact and elegant recursive procedure.

7.3 Algebraic Loops

The previous section may leave the impression with you, the reader, that all DAE systems can be sorted as easily as the example system, by means of which the causalization algorithm has been demonstrated. Nothing could be farther from the truth.

Let us now look at a slightly modified circuit. Its schematic is shown in Fig.7.5. The capacitor has been replaced by a third resistor.

---

[1]It was shown in [7.6] that the computational complexity of the Tarjan algorithm grows in the worst case with $o(n \cdot m)$, where n is the number of equations, and m is the number of non–zero elements in the structure incidence matrix.

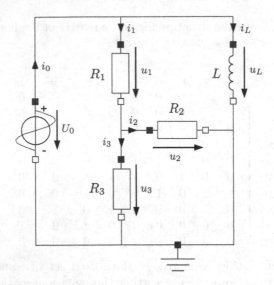

FIGURE 7.5. Schematic of modified electrical RLC circuit.

The resulting equations are almost the same as before. Only the element equation for the capacitor was replaced by a third element equation for a resistor.

$$u_0 = f(t) \tag{7.5a}$$
$$u_1 = R_1 \cdot i_1 \tag{7.5b}$$
$$u_2 = R_2 \cdot i_2 \tag{7.5c}$$
$$u_3 = R_3 \cdot i_3 \tag{7.5d}$$
$$u_L = L \cdot \frac{di_L}{dt} \tag{7.5e}$$
$$u_0 = u_1 + u_3 \tag{7.5f}$$
$$u_L = u_1 + u_2 \tag{7.5g}$$
$$u_3 = u_2 \tag{7.5h}$$
$$i_0 = i_1 + i_L \tag{7.5i}$$
$$i_1 = i_2 + i_3 \tag{7.5j}$$

The structure digraph for this new set of equations is presented in Fig.7.6.

Let us now apply the Tarjan algorithm to this structure digraph. Figure 7.7 shows the partially causalized structure digraph.

Unfortunately, the Tarjan algorithm stalls at this point. Every one of the remaining acausal equations and every one of the remaining unknowns has at least two black (solid) lines attached to it. Consequently, the DAE system cannot be sorted entirely.

Let us read out the partially sorted equations. We shall only list on the

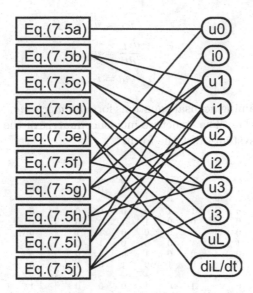

FIGURE 7.6. Structure digraph of modified electrical circuit.

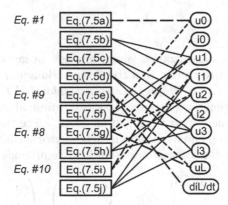

FIGURE 7.7. Structure digraph of partially causalized modified electrical circuit.

right side of the equal sign those variables that have already been computed.

$$u_0 = f(t) \tag{7.6a}$$
$$u_1 - R_1 \cdot i_1 = 0 \tag{7.6b}$$
$$u_2 - R_2 \cdot i_2 = 0 \tag{7.6c}$$
$$u_3 - R_3 \cdot i_3 = 0 \tag{7.6d}$$
$$u_1 + u_3 = u_0 \tag{7.6e}$$
$$u_2 - u_3 = 0 \tag{7.6f}$$
$$i_1 - i_2 - i_3 = 0 \tag{7.6g}$$

$$u_L = u_1 + u_2 \tag{7.6h}$$

$$\frac{di_L}{dt} = u_L/L \tag{7.6i}$$

$$i_0 = i_1 + i_L \tag{7.6j}$$

The six remaining acausal equations form an *algebraic loop*. They need to be solved together. The structure incidence matrix of the partially causalized equation system takes the form:

	u_0		u_1	i_1	u_2	i_2	u_3	i_3		u_L		$\frac{di_L}{dt}$		i_0
$Eq.(7.6a)$	1	$\vert$	0	0	0	0	0	0		0		0		0
$Eq.(7.6b)$	0	$\vert$	1	1	0	0	0	0	$\vert$	0		0		0
$Eq.(7.6c)$	0	$\vert$	0	0	1	1	0	0	$\vert$	0		0		0
$Eq.(7.6d)$	0	$\vert$	0	0	0	0	1	1	$\vert$	0		0		0
$Eq.(7.6e)$	1	$\vert$	1	0	0	0	1	0	$\vert$	0		0		0
$Eq.(7.6f)$	0	$\vert$	0	0	1	0	1	0	$\vert$	0		0		0
$Eq.(7.6g)$	0	$\vert$	0	1	0	1	0	1	$\vert$	0		0		0
$Eq.(7.6h)$	0		1	0	1	0	0	0	$\vert$	1	$\vert$	0		0
$Eq.(7.6i)$	0		0	0	0	1	0	0		1	$\vert$	1	$\vert$	0
$Eq.(7.6j)$	0		0	1	0	0	0	0		0		0	$\vert$	1

$$\mathbf{S} = \text{(above matrix)} \tag{7.7}$$

Although the causalization algorithm has been unable to convert the structure incidence matrix to a true lower–triangular form, it was at least able to reduce it to a *Block–Lower–Triangular (BLT)* form. Furthermore, the algorithm generates diagonal blocks of minimal sizes.

How can we deal with the algebraic loop? Since the model is linear, we can write the loop equations in a matrix–vector form, and solve for the six unknowns by a *Gaussian elimination* in six equations and six unknowns.

$$
\begin{pmatrix}
1 & -R_1 & 0 & 0 & 0 & 0 \\
0 & 0 & 1 & -R_2 & 0 & 0 \\
0 & 0 & 0 & 0 & 1 & -R_3 \\
1 & 0 & 0 & 0 & 1 & 0 \\
0 & 0 & 1 & 0 & -1 & 0 \\
0 & 1 & 0 & -1 & 0 & -1
\end{pmatrix}
\cdot
\begin{pmatrix}
u_1 \\ i_1 \\ u_2 \\ i_2 \\ u_3 \\ i_3
\end{pmatrix}
=
\begin{pmatrix}
0 \\ 0 \\ 0 \\ u_0 \\ 0 \\ 0
\end{pmatrix}
\tag{7.8}
$$

Had the model been nonlinear in the loop equations, we would have had to use a *Newton iteration*.

Are algebraic loops a rarity in physical system modeling? Unfortunately, DAE systems containing algebraic loops are much more common than those that can be sorted completely by the Tarjan algorithm. Furthermore, the algebraic loops can be of frightening dimensions. For example when modeling mechanical *Multi–Body Systems (MBS)* [7.16, 7.18] containing closed kinematic loops, there immediately result highly nonlinear algebraic loops in hundreds if not thousands of unknowns and equations.

7.4   The Tearing Algorithm

We have now been haunted by large algebraic equation systems long enough. It is time that we do something about them.

Let us look once again at the system of six algebraic equations in six unknowns that we had met in the last section.

$$u_1 - R_1 \cdot i_1 = 0 \tag{7.9a}$$

$$u_2 - R_2 \cdot i_2 = 0 \tag{7.9b}$$

$$u_3 - R_3 \cdot i_3 = 0 \tag{7.9c}$$

$$u_1 + u_3 = u_0 \tag{7.9d}$$

$$u_2 - u_3 = 0 \tag{7.9e}$$

$$i_1 - i_2 - i_3 = 0 \tag{7.9f}$$

Its structure digraph is shown in Fig.7.8

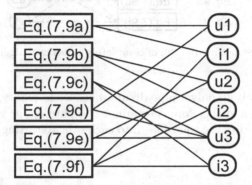

FIGURE 7.8. Structure digraph of algebraic equation system.

Clearly, every equation contains at least two unknowns, and every unknown appears in at least two equations. Yet, if only we could e.g. solve Eg.(7.9f) for the unknown i_3, as shown in Fig.7.9, then the entire set of equations could be made causal, as shown in Fig.7.10.

Now that the equation system has been causalized, we can write down the causal equations:

$$i_3 = i_1 - i_2 \tag{7.10a}$$

$$u_3 = R_3 \cdot i_3 \tag{7.10b}$$

$$u_1 = u_0 - u_3 \tag{7.10c}$$

$$i_1 = u_1/R_1 \tag{7.10d}$$

$$u_2 = u_3 \tag{7.10e}$$

$$i_2 = u_2/R_2 \tag{7.10f}$$

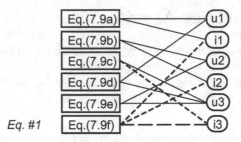

FIGURE 7.9. Structure digraph of partially causalized algebraic equation system.

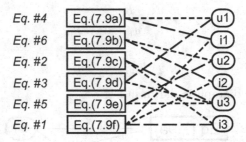

FIGURE 7.10. Structure digraph of completely causalized algebraic equation system.

Of course, it is all only a pipe dream, because in reality, we do not know either i_1 or i_2, and therefore, we cannot compute i_3. Or is it not?

Let us substitute the equations into each other, starting with Eq.(7.10a).

$$i_3 = i_1 - i_2 \tag{7.11a}$$

$$= \frac{1}{R_1} \cdot u_1 - \frac{1}{R_2} \cdot u_2 \tag{7.11b}$$

$$= \frac{1}{R_1} \cdot u_0 - \frac{1}{R_1} \cdot u_3 - \frac{1}{R_2} \cdot u_3 \tag{7.11c}$$

$$= \frac{1}{R_1} \cdot u_0 - \frac{R_1 + R_2}{R_1 \cdot R_2} \cdot u_3 \tag{7.11d}$$

$$= \frac{1}{R_1} \cdot u_0 - \frac{R_3 \cdot (R_1 + R_2)}{R_1 \cdot R_2} \cdot i_3 \tag{7.11e}$$

and thus:

$$\left[1 + \frac{R_3 \cdot (R_1 + R_2)}{R_1 \cdot R_2} \right] \cdot i_3 = \frac{1}{R_1} \cdot u_0 \tag{7.12}$$

or:

$$\frac{R_1 \cdot R_2 + R_1 \cdot R_3 + R_2 \cdot R_3}{R_2} \cdot i_3 = u_0 \tag{7.13}$$

Since the equation is linear in i_3, we can solve it explicitly for the unknown, and obtain:

$$i_3 = \frac{R_2}{R_1 \cdot R_2 + R_1 \cdot R_3 + R_2 \cdot R_3} \cdot u_0 \qquad (7.14)$$

Now, we can plug this equation back into the causalized equation system, replacing Eq.(7.10a) by it, and obtain the perfectly causal set of equations:

$$i_3 = \frac{R_2}{R_1 \cdot R_2 + R_1 \cdot R_3 + R_2 \cdot R_3} \cdot u_0 \qquad (7.15a)$$

$$u_3 = R_3 \cdot i_3 \qquad (7.15b)$$

$$u_1 = u_0 - u_3 \qquad (7.15c)$$

$$i_1 = u_1/R_1 \qquad (7.15d)$$

$$u_2 = u_3 \qquad (7.15e)$$

$$i_2 = u_2/R_2 \qquad (7.15f)$$

Evidently, it hadn't been a pipe dream after all.

After substituting the equations into each other in the proposed form, we end up with one equation in one unknown, instead of six equations in six unknowns. This is clearly much more economical.

Had the equations been nonlinear in the variable i_3, everything would have worked exactly the same way, except for the very last step, where we would have to involve a Newton iteration to solve for i_3, rather than solving for i_3 explicitly.

Substituting equations into each other may actually be a bad idea. The substituted equations may grow in size, and the same expressions may appear in them multiple times. It may be a better idea to iterate over the entire set of equations, but treat only i_3 as an iteration variable in the Newton iteration algorithm.

Given the set of equations:

$$u_3 = R_3 \cdot i_3 \qquad (7.16a)$$

$$u_1 = u_0 - u_3 \qquad (7.16b)$$

$$i_1 = u_1/R_1 \qquad (7.16c)$$

$$u_2 = u_3 \qquad (7.16d)$$

$$i_2 = u_2/R_2 \qquad (7.16e)$$

$$i_{3_{new}} = i_1 - i_2 \qquad (7.16f)$$

where i_3 is an initial guess, and $i_{3_{new}}$ is an improved version of that same variable, we can set up the following zero function:

$$\mathcal{F} = i_{3_{new}} - i_3 = 0.0 \qquad (7.17)$$

Since $\mathcal{F}$ is a scalar, also the Hessian is a scalar:

$$\mathcal{H} = \frac{\partial \mathcal{F}}{\partial i_3} \tag{7.18}$$

A convenient way to compute the Hessian $\mathcal{H}$ is by means of *algebraic differentiation* [7.11].

$$du_3 = R_3 \tag{7.19a}$$
$$du_1 = -du_3 \tag{7.19b}$$
$$di_1 = du_1/R_1 \tag{7.19c}$$
$$du_2 = du_3 \tag{7.19d}$$
$$di_2 = du_2/R_2 \tag{7.19e}$$
$$di_{3_{new}} = di_1 - di_2 \tag{7.19f}$$
$$\mathcal{H} = di_{3_{new}} - 1 \tag{7.19g}$$

We can then compute the next version of i_3 as:

$$i_3 = i_3 - \mathcal{H}\backslash\mathcal{F} \tag{7.20}$$

If the set of equations is linear, the Newton iteration converges in a single step. Hence it will not be terribly inefficient to employ Newton iteration even in the linear case.

The algorithm that we just described is a so–called *tearing algorithm*, as the set of equations is torn apart by making an assumption about one variable or possibly several variables to be known. The variables that are assumed known, such as i_3 in the given example, are called *tearing variables*, whereas the equations, from which the tearing variables are to be computed, such as Eq.(7.9f) in the given example, are called the *residual equations*.

Equation tearing is not exactly a new concept. The idea had originally been introduced by Gabriel Kron [7.12]. By now, many variations of different tearing algorithms have been reported in the literature. Some of the techniques are generally applicable, whereas others exploit particular matrix structures as they occur in special types of physical systems. Tearing has become most popular in chemical process engineering applications [7.13].

A version of tearing similar to the one described in this chapter has been implemented in Dymola [7.7] to accompany the Tarjan algorithm in the efficient solution of algebraic equation systems resulting from the automated symbolic conversion of DAE systems to ODE form [7.4].

How did we know to choose i_3 as tearing variable and Eq.(7.9f) as residual equation? What would have happened if we had chosen i_1 as the tearing variable and Eq.(7.9a) as the residual equation? The initial situation is depicted in Fig.7.11.

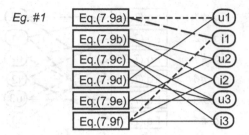

FIGURE 7.11. Structure digraph of partially causalized algebraic equation system.

We apply the Tarjan algorithm to the structure digraph. Unfortunately, the algorithm stalls once again after only one more step, as shown in Fig.7.12.

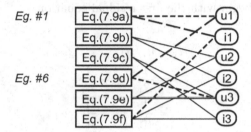

FIGURE 7.12. Structure digraph of partially causalized algebraic equation system.

We have been able to causalize only two of the six equations. Once again, we are faced with an algebraic loop in four equations and four unknowns, and therefore have to choose a second tearing variable and a second residual equation.

Let us proceed with the example to demonstrate, how the tearing algorithm can deal with multiple residual equations in multiple tearing variables. Let us select u_2 as the second tearing variable, and Eq.(7.9b) as the second residual equation. Now, we can complete the causalization of the equations. The completely colored structure digraph is shown in Fig.7.13.

We can read out the causal equations from the structure digraph of Fig.7.13.

$$i_1 = u_1/R_1 \tag{7.21a}$$
$$u_2 = R_2 \cdot i_2 \tag{7.21b}$$
$$u_3 = u_2 \tag{7.21c}$$
$$i_3 = u_3/R_3 \tag{7.21d}$$
$$i_2 = i_1 - i_3 \tag{7.21e}$$

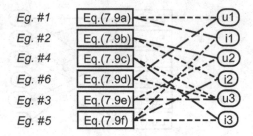

FIGURE 7.13. Structure digraph of completely causalized algebraic equation system.

$$u_1 = u_0 - u_3 \tag{7.21f}$$

Using the substitution technique, we can come up with two linearly independent equations in the two unknowns i_1 and u_2, i.e., in the two tearing variables. We begin with the first residual equation.

$$i_1 = u_1/R_1 \tag{7.22a}$$

$$= \frac{1}{R_1} \cdot u_0 - \frac{1}{R_1} \cdot u_3 \tag{7.22b}$$

$$= \frac{1}{R_1} \cdot u_0 - \frac{1}{R_1} \cdot u_2 \tag{7.22c}$$

Hence:

$$R_1 \cdot i_1 + u_2 = u_0 \tag{7.23}$$

We proceed with the second residual equation.

$$u_2 = R_2 \cdot i_2 \tag{7.24a}$$

$$= R_2 \cdot i_1 - R_2 \cdot i_3 \tag{7.24b}$$

$$= R_2 \cdot i_1 - \frac{R_2}{R_3} \cdot u_3 \tag{7.24c}$$

$$= R_2 \cdot i_1 - \frac{R_2}{R_3} \cdot u_2 \tag{7.24d}$$

$$\tag{7.24e}$$

Thus:

$$\left[1 + \frac{R_2}{R_3} \right] \cdot u_2 = R_2 \cdot i_1 \tag{7.25}$$

or:

$$R_2 \cdot R_3 \cdot i_1 - (R_2 + R_3) \cdot u_2 = 0 \tag{7.26}$$

We can write Eq.(7.23) and Eq.(7.26) in a matrix–vector form:

$$\begin{pmatrix} R_1 & 1 \\ R_2 \cdot R_3 & -(R_2 + R_3) \end{pmatrix} \cdot \begin{pmatrix} i_1 \\ u_2 \end{pmatrix} = \begin{pmatrix} u_0 \\ 0 \end{pmatrix} \tag{7.27}$$

which can be solved for the two unknowns i_1 and u_2. Instead of solving six linear equations in six unknowns, we have pushed the zeros out of the matrix, and ended up with two equations in two unknowns. In this sense, tearing can be considered a symbolic *sparse matrix technique*.

If we use Newton iteration instead of equation substitution, we need to place the residual equations at the end of each set, rather than at the beginning. The set of equations now takes the form:

$$u_3 = u_2 \tag{7.28a}$$
$$i_3 = u_3/R_3 \tag{7.28b}$$
$$i_2 = i_1 - i_3 \tag{7.28c}$$
$$u_{2_{new}} = R_2 \cdot i_2 \tag{7.28d}$$
$$u_1 = u_0 - u_3 \tag{7.28e}$$
$$i_{1_{new}} = u_1/R_1 \tag{7.28f}$$

We can formulate the following set of zero functions:

$$\mathcal{F} = \begin{pmatrix} f_1 \\ f_2 \end{pmatrix} = \begin{pmatrix} i_{1_{new}} - i_1 \\ u_{2_{new}} - u_2 \end{pmatrix} = \begin{pmatrix} 0 \\ 0 \end{pmatrix} \tag{7.29}$$

Hence the Hessian is a matrix of size 2×2:

$$\mathcal{H} = \begin{pmatrix} h_{11} & h_{12} \\ h_{21} & h_{22} \end{pmatrix} = \begin{pmatrix} \partial f_1/\partial i_1 & \partial f_1/\partial u_2 \\ \partial f_2/\partial i_1 & \partial f_2/\partial u_2 \end{pmatrix} \tag{7.30}$$

Using algebraic differentiation, we get:

$$d_1 u_3 = 0 \tag{7.31a}$$
$$d_1 i_3 = d_1 u_3/R_3 \tag{7.31b}$$
$$d_1 i_2 = 1 - d_1 i_3 \tag{7.31c}$$
$$d_1 u_{2_{new}} = R_2 \cdot d_1 i_2 \tag{7.31d}$$
$$d_1 u_1 = -d_1 u_3 \tag{7.31e}$$
$$d_1 i_{1_{new}} = d_1 u_1/R_1 \tag{7.31f}$$
$$d_2 u_3 = 1 \tag{7.31g}$$
$$d_2 i_3 = d_2 u_3/R_3 \tag{7.31h}$$
$$d_2 i_2 = -d_2 i_3 \tag{7.31i}$$
$$d_2 u_{2_{new}} = R_2 \cdot d_2 i_2 \tag{7.31j}$$

$$d_2 u_1 = -d_2 u_3 \tag{7.31k}$$

$$d_2 i_{1_{new}} = d_2 u_1 / R_1 \tag{7.31l}$$

$$h_{11} = d_1 i_{1_{new}} - 1 \tag{7.31m}$$

$$h_{12} = d_2 i_{1_{new}} \tag{7.31n}$$

$$h_{21} = d_1 u_{2_{new}} \tag{7.31o}$$

$$h_{22} = d_2 u_{2_{new}} - 1 \tag{7.31p}$$

where the prefix d_1 stands for the partial derivative with respect to i_1, and d_2 stands for the partial derivative with respect to u_2. Since i_1 and u_2 are mutually independent, the partial derivative of i_1 with respect to u_2 is zero, and vice–versa.

For each additional tearing variable, the causal model equations are repeated once in the computation of the Hessian. Hence given a system of n algebraic equations in $k < n$ tearing variables, we require $n \cdot k + k^2$ equations to explicitly compute the Hessian in symbolic form.

We have seen by now that the selection of tearing variables and residual equations is not arbitrary. Our first choice led to a single residual equation in a single tearing variable, whereas our second choice led to two residual equations in two tearing variables.

How can we determine the minimum number of tearing variables required? Unfortunately, this is a hard problem. It can be shown that this problem is *np–complete*, i.e., the computational effort grows exponentially in the number of equations forming the algebraic loop. Consequently, finding the minimal number of tearing variables is not practical.

Yet, it is possible to design a *heuristic procedure* that always results in a small number of tearing variables. It often results in the minimal number, but this cannot be guaranteed. The advantage of this heuristic procedure is that its computational effort grows quadratically rather than exponentially in the size of the algebraic system for most applications. The heuristic procedure is described in the sequel.

1. Using the structure digraph, determine the equations with the largest number of black (solid) lines attached to them.

2. For every one of these equations, follow its black (solid) lines, and determine those variables with the largest number of black (solid) lines attached to them.

3. For every one of these variables, determine how many additional equations can be made causal if that variable is assumed to be known.

4. Choose one of those variables as the next tearing variable that allows the largest number of additional equations to be made causal.

Looking at the structure digraph of Fig.7.7, we see that only Eq.(7.5j) has three black (solid) lines attached to it. All other acausal equations have only two black solid lines attached to them. Consequently, Eq.(7.5j) will be chosen as the first residual equation.

Following each of the three black (solid) lines to the right side, we notice that each of the variables, i_1, i_2, and i_3 has exactly two black (solid) lines attached to it. Consequently, we need to check for each one of them, what happens if it were chosen as the first tearing variable.

It turns out that each of these three variables could have been chosen as the first tearing variable, since all of them lead to a complete causalization of the equation system.

We shall see in the next chapter that the simple heuristic algorithm described in this section sometimes maneuvers itself into a corner. The heuristic algorithm implemented in Dymola has been refined in several respects. On the one hand, it never gets stuck. The algorithm may become slow at times, but it will always find a legal tearing structure. On the other hand, the tearing algorithm implemented in Dymola guarantees that the selection of tearing variables never leads to a division by zero at run time. This is a rather tricky demand, because parameter values can change after compilation.

The complete tearing algorithm, as implemented in Dymola, has not been published. It is a company secret, designed to give Dynasim a competitive edge over its competitors.

7.5   The Relaxation Algorithm

There is yet another symbolic algorithm for the solution of algebraic systems of equations to be discussed, which is called the *relaxation algorithm* [7.17].

Contrary to the tearing algorithm, which is a general algorithm that can be applied to all algebraic equation structures, the relaxation algorithm is limited to the solution of linear algebraic equation systems only.

Yet, linear algebraic systems assume a special role within the set of algebraic equation systems, and deserve special attention. One reason for this claim is the following. Within each Newton iteration of a nonlinear algebraic equation system, there is always a linear algebraic equation system to be solved. When we write the Newton iteration as:

$$\mathbf{x_{new}} = \mathbf{x_{old}} - \mathcal{H} \backslash \mathcal{F} \tag{7.32}$$

we are effectively saying that:

$$\mathbf{x_{new}} = \mathbf{x_{old}} - \mathbf{dx} \tag{7.33}$$

where $\mathbf{dx}$ is the solution of the linear algebraic equation system:

$$\mathcal{H} \cdot \mathbf{dx} = \mathcal{F} \tag{7.34}$$

Hence indeed, there is to be solved a linear algebraic equation system within each Newton iteration of the original nonlinear algebraic equation system.

Relaxation is a symbolic implementation of the Gaussian elimination algorithm without pivoting. Let us demonstrate how the relaxation algorithm works by means of the same example of a linear algebraic equation system in six equations and six unknowns that we had used in the previous section of this book.

We start out with the linear algebraic equation system in matrix–vector form, as presented in Eq.(7.8).

$$
\begin{pmatrix}
1 & -R_1 & 0 & 0 & 0 & 0 \\
0 & 0 & 1 & -R_2 & 0 & 0 \\
0 & 0 & 0 & 0 & 1 & -R_3 \\
1 & 0 & 0 & 0 & 1 & 0 \\
0 & 0 & 1 & 0 & -1 & 0 \\
0 & 1 & 0 & -1 & 0 & -1
\end{pmatrix}
\cdot
\begin{pmatrix}
u_1 \\ i_1 \\ u_2 \\ i_2 \\ u_3 \\ i_3
\end{pmatrix}
=
\begin{pmatrix}
0 \\ 0 \\ 0 \\ u_0 \\ 0 \\ 0
\end{pmatrix}
\tag{7.35}
$$

However, we wish to minimize the number of non–zero elements above the diagonal. To this end, we causalize the equations in the same way that we used in the tearing algorithm, but write the residual equation as the last equation in the set.

$$u_3 = R_3 \cdot i_3 \tag{7.36a}$$

$$u_1 = u_0 - u_3 \tag{7.36b}$$

$$i_1 = \frac{u_1}{R_1} \tag{7.36c}$$

$$u_2 = u_3 \tag{7.36d}$$

$$i_2 = \frac{u_2}{R_2} \tag{7.36e}$$

$$i_3 = i_1 - i_2 \tag{7.36f}$$

We now move all the unknowns to the left side of the equal sign and all the knows to the right side. At the same time, we eliminate the denominators.

$$u_3 - R_3 \cdot i_3 = 0 \tag{7.37a}$$

$$u_1 + u_3 = u_0 \tag{7.37b}$$

$$R_1 \cdot i_1 - u_1 = 0 \tag{7.37c}$$

$$u_2 - u_3 = 0 \tag{7.37d}$$

$$R_2 \cdot i_2 - u_2 = 0 \tag{7.37e}$$

$$i_3 - i_1 + i_2 = 0 \tag{7.37f}$$

We now rewrite these equations in a matrix–vector form, whereby we number the equations in the same order as above and list the variables in the same order as in the causal equations:

$$\begin{pmatrix} 1 & 0 & 0 & 0 & 0 & -R_3 \\ 1 & 1 & 0 & 0 & 0 & 0 \\ 0 & -1 & R_1 & 0 & 0 & 0 \\ -1 & 0 & 0 & 1 & 0 & 0 \\ 0 & 0 & 0 & -1 & R_2 & 0 \\ 0 & 0 & -1 & 0 & 1 & 1 \end{pmatrix} \cdot \begin{pmatrix} u_3 \\ u_1 \\ i_1 \\ u_2 \\ i_2 \\ i_3 \end{pmatrix} = \begin{pmatrix} 0 \\ u_0 \\ 0 \\ 0 \\ 0 \\ 0 \end{pmatrix} \tag{7.38}$$

There is now only a single non–zero element above the diagonal, and none of the diagonal elements are zero.

We can now apply Gaussian elimination without pivoting to this set of equations. Remember how Gaussian elimination works:

$$A_{ij}^{(n+1)} = A_{ij}^{(n)} - A_{ik}^{(n)} \cdot A_{kk}^{(n)^{-1}} \cdot A_{kj}^{(n)} \tag{7.39a}$$

$$b_i^{(n+1)} = b_i^{(n)} - A_{ik}^{(n)} \cdot A_{kk}^{(n)^{-1}} \cdot b_k^{(n)} \tag{7.39b}$$

We can apply this algorithm symbolically. After each step, we eliminate the first row and the first column, i.e., the pivot row and the pivot column. Rather than substituting expressions into the matrix, we introduce auxiliary variables where needed.

Since we constantly eliminate rows and columns, the index k in the above equations is always 1. Thus, in the n plus first iteration of the algorithm, the element in row $i - 1$ and column $j - 1$ of the matrix is equal to the element in the row i and column j of the nth iteration minus the product of the element at the very left end of the matrix (in row i) times the element at the very top end of the matrix (in column j) divided by the element in the top left corner.

For this reason, if an element in the top row is zero, the elements underneath it don't change at all during the iteration. Similarly, if an element in the leftmost column is zero, the elements to the right of it don't change.

Therefore, the only elements in the above equation system that change during the first iteration are the elements in the positions $< 2, 6 >$ and $< 4, 6 >$. Let us call the new elements c_1 and c_2. Thus, the second version of the equation system takes the form:

$$\begin{pmatrix} 1 & 0 & 0 & 0 & c_1 \\ -1 & R_1 & 0 & 0 & 0 \\ 0 & 0 & 1 & 0 & c_2 \\ 0 & 0 & -1 & R_2 & 0 \\ 0 & -1 & 0 & 1 & 1 \end{pmatrix} \cdot \begin{pmatrix} u_1 \\ i_1 \\ u_2 \\ i_2 \\ i_3 \end{pmatrix} = \begin{pmatrix} u_0 \\ 0 \\ 0 \\ 0 \\ 0 \end{pmatrix} \tag{7.40}$$

where $c_1 = R_3$, and $c_2 = -R_3$.

The only elements that can change in the next iteration are the element in the position $< 2, 5 >$ of the matrix, as well as the element in the position $< 2 >$ of the vector. Let us call those c_3 and c_4, respectively.

Thus, the third version of the equation system takes the form:

$$
\begin{pmatrix} R_1 & 0 & 0 & c_3 \\ 0 & 1 & 0 & c_2 \\ 0 & -1 & R_2 & 0 \\ -1 & 0 & 1 & 1 \end{pmatrix} \cdot \begin{pmatrix} i_1 \\ u_2 \\ i_2 \\ i_3 \end{pmatrix} = \begin{pmatrix} c_4 \\ 0 \\ 0 \\ 0 \end{pmatrix}
\tag{7.41}
$$

where $c_3 = c_1$, and $c_4 = u_0$.

In the next iteration, the only elements that can change are in the position $< 4, 4 >$ of the matrix, and in the position $< 4 >$ of the vector. Let us call these c_5, and c_6, respectively.

The fourth version of the equation system takes the form:

$$
\begin{pmatrix} 1 & 0 & c_2 \\ -1 & R_2 & 0 \\ 0 & 1 & c_5 \end{pmatrix} \cdot \begin{pmatrix} u_2 \\ i_2 \\ i_3 \end{pmatrix} = \begin{pmatrix} 0 \\ 0 \\ c_6 \end{pmatrix}
\tag{7.42}
$$

where $c_5 = 1 + c_3/R_1$, and $c_6 = c_4/R_1$.

In the next iteration, the only element that can change is in the position $< 2, 3 >$ of the matrix. Let us call the new element c_7.

The fifth version of the equation system takes the form:

$$
\begin{pmatrix} R_2 & c_7 \\ 1 & c_5 \end{pmatrix} \cdot \begin{pmatrix} i_2 \\ i_3 \end{pmatrix} = \begin{pmatrix} 0 \\ c_6 \end{pmatrix}
\tag{7.43}
$$

where $c_7 = c_2$.

In the final iteration, the only element that can change is in the position $< 2, 2 >$ of the matrix. Let us call the new element c_8.

The sixth and last version of the equation system takes the form:

$$
\begin{pmatrix} c_8 \end{pmatrix} \cdot \begin{pmatrix} i_3 \end{pmatrix} = \begin{pmatrix} c_6 \end{pmatrix}
\tag{7.44}
$$

where $c_8 = c_5 - c_7/R_2$.

This equation system can be solved at once for the unknown i_3:

$$
i_3 = \frac{c_6}{c_8}
\tag{7.45}
$$

¿From the previous set of equations, we can subsequently compute:

$$
i_2 = -\frac{c_7 \cdot i_3}{R_2}
\tag{7.46}
$$

and so forth.

Thus, the overall equation system can be replaced by the following set of symbolic scalar equations:

$$c_1 = R_3 \tag{7.47a}$$

$$c_2 = -R_3 \tag{7.47b}$$

$$c_3 = c_1 \tag{7.47c}$$

$$c_4 = u_0 \tag{7.47d}$$

$$c_5 = 1 + \frac{c_3}{R_1} \tag{7.47e}$$

$$c_6 = \frac{c_4}{R_1} \tag{7.47f}$$

$$c_7 = c_2 \tag{7.47g}$$

$$c_8 = c_5 - \frac{c_7}{R_2} \tag{7.47h}$$

$$i_3 = \frac{c_6}{c_8} \tag{7.47i}$$

$$i_2 = -\frac{c_7 \cdot i_3}{R_2} \tag{7.47j}$$

$$u_2 = -c_2 \cdot i_3 \tag{7.47k}$$

$$i_1 = \frac{c_4 - c_3 \cdot i_3}{R_1} \tag{7.47l}$$

$$u_1 = u_0 - c_1 \cdot i_3 \tag{7.47m}$$

$$u_3 = R_3 \cdot i_3 \tag{7.47n}$$

Of course, we can also combine the relaxation approach with tearing. Once an expression for the tearing variable i_3 has been found, the remaining variables can be computed from the original set of equations instead of using those from the back–substitution:

$$c_1 = R_3 \tag{7.48a}$$

$$c_2 = -R_3 \tag{7.48b}$$

$$c_3 = c_1 \tag{7.48c}$$

$$c_4 = u_0 \tag{7.48d}$$

$$c_5 = 1 + \frac{c_3}{R_1} \tag{7.48e}$$

$$c_6 = \frac{c_4}{R_1} \tag{7.48f}$$

$$c_7 = c_2 \tag{7.48g}$$

$$c_8 = c_5 - \frac{c_7}{R_2} \tag{7.48h}$$

$$i_3 = \frac{c_6}{c_8} \tag{7.48i}$$

$$u_3 = R_3 \cdot i_3 \tag{7.48j}$$

$$u_1 = u_0 - u_3 \qquad (7.48\text{k})$$

$$i_1 = \frac{u_1}{R_1} \qquad (7.48\text{l})$$

$$u_2 = u_3 \qquad (7.48\text{m})$$

$$i_2 = \frac{u_2}{R_2} \qquad (7.48\text{n})$$

What would have happened if we had started out with the second set of causal equations, i.e., the one derived involving two tearing variables. The causal equations present themselves as follows:

$$u_3 = u_2 \qquad (7.49\text{a})$$

$$i_3 = \frac{u_3}{R_3} \qquad (7.49\text{b})$$

$$i_2 = i_1 - i_3 \qquad (7.49\text{c})$$

$$u_2 = R_2 \cdot i_2 \qquad (7.49\text{d})$$

$$u_1 = u_0 - u_3 \qquad (7.49\text{e})$$

$$i_1 = \frac{u_1}{R_1} \qquad (7.49\text{f})$$

Moving all unknowns to the left side of the equal sign, we obtain:

$$u_3 - u_2 = 0 \qquad (7.50\text{a})$$

$$R_3 \cdot i_3 - u_3 = 0 \qquad (7.50\text{b})$$

$$i_2 - i_1 + i_3 = 0 \qquad (7.50\text{c})$$

$$u_2 - R_2 \cdot i_2 = 0 \qquad (7.50\text{d})$$

$$u_1 + u_3 = u_0 \qquad (7.50\text{e})$$

$$R_1 \cdot i_1 - u_1 = 0 \qquad (7.50\text{f})$$

This set of equations can be written in a matrix–vector form as follows:

$$
\begin{pmatrix}
1 & 0 & 0 & -1 & 0 & 0 \\
-1 & R_3 & 0 & 0 & 0 & 0 \\
0 & 1 & 1 & 0 & 0 & -1 \\
0 & 0 & -R_2 & 1 & 0 & 0 \\
1 & 0 & 0 & 0 & 1 & 0 \\
0 & 0 & 0 & 0 & -1 & R_1
\end{pmatrix}
\cdot
\begin{pmatrix}
u_3 \\ i_3 \\ i_2 \\ u_2 \\ u_1 \\ i_1
\end{pmatrix}
=
\begin{pmatrix}
0 \\ 0 \\ 0 \\ 0 \\ u_0 \\ 0
\end{pmatrix}
\qquad (7.51)
$$

Just as in the previous case, all the diagonal elements of the matrix are non–zero, allowing a Gaussian elimination without pivoting to be performed. However this time around, there are two non–zero elements above the diagonal, one involving the tearing variable u_2, the other involving the tearing variable i_1.

Finding a minimal set of non–zero elements above the diagonal of the matrix is thus identical to finding a minimal set of tearing variables. Hence also this problem is *np–complete*.

The same heuristic procedure that was proposed for tackling the problem of finding a small (though not necessarily the minimal) set of tearing variables can also be used to find a small (though not necessarily the smallest) set of non–zero elements above the diagonal of the linear equation matrix for the relaxation algorithm.

Why were we interested in minimizing the number of non–zero elements above the diagonal? Remember that we only need to introduce new auxiliary variables c_i in the symbolic Gaussian elimination, if both the elements in the top row and the elements in the leftmost column are non–zero. If an element in the top row is zero, then all the elements beneath it don't change in the next version of the system equations, i.e., in the next step of the Gaussian elimination algorithm. Thus by minimizing the number of non–zero elements above the diagonal, we also minimize the number of new auxiliary variables that need to be introduced, and for which expressions have to be evaluated.

Hence also the symbolic Gaussian elimination algorithm exploits the number and positions of the zero elements in the linear equation system, and therefore can be interpreted as a symbolic *sparse matrix technique*.

7.6   Structural Singularities

Let us now look at yet another circuit problem. This time, we shall exchange the capacitor and the inductor, as shown in Fig.7.14.

The set of differential and algebraic equations thus presents itself as:

$$u_0 = f(t) \tag{7.52a}$$

$$u_1 = R_1 \cdot i_1 \tag{7.52b}$$

$$u_2 = R_2 \cdot i_2 \tag{7.52c}$$

$$u_L = L \cdot \frac{di_L}{dt} \tag{7.52d}$$

$$i_C = C \cdot \frac{du_C}{dt} \tag{7.52e}$$

$$u_0 = u_1 + u_L \tag{7.52f}$$

$$u_C = u_1 + u_2 \tag{7.52g}$$

$$u_L = u_2 \tag{7.52h}$$

$$i_0 = i_1 + i_C \tag{7.52i}$$

$$i_1 = i_2 + i_L \tag{7.52j}$$

with the structure digraph as shown in Fig.7.15.

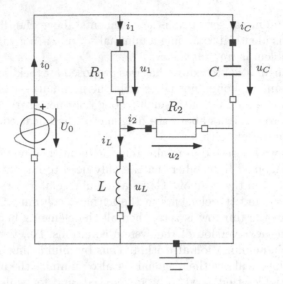

FIGURE 7.14. Schematic of once more modified electrical RLC circuit.

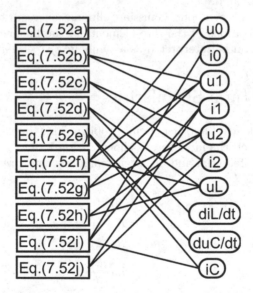

FIGURE 7.15. Structure digraph of once more modified electrical circuit.

Let us start to color the digraph. Figure 7.16 shows the partially colored digraph after the first iteration.

We notice that we are in trouble. The variable i_c has two blue (dotted) lines attached to it, and nothing else. Consequently, we no longer have an equation to compute the value of i_c.

Let us try another approach. We can introduce the node potentials, v_i, as additional variables, write down for each branch the relationship be-

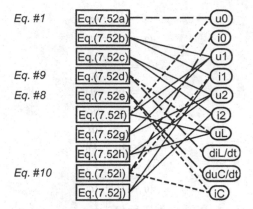

FIGURE 7.16. Structure digraph of partially causalized electrical circuit.

tween the branch voltage and the two neighboring node potentials, and eliminate the mesh equations instead. Figure 7.17 shows the schematic of the electrical circuit with node potentials.

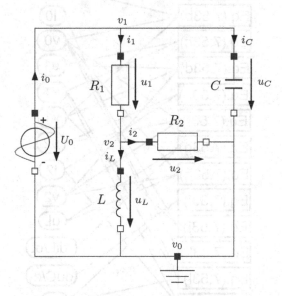

FIGURE 7.17. Schematic of electrical RLC circuit with node potentials.

The new set of equations can be written as follows:

$$u_0 = f(t) \tag{7.53a}$$
$$u_0 = v_1 - v_0 \tag{7.53b}$$
$$u_1 = R_1 \cdot i_1 \tag{7.53c}$$
$$u_1 = v_1 - v_2 \tag{7.53d}$$

$$u_2 = R_2 \cdot i_2 \tag{7.53e}$$

$$u_2 = v_2 - v_0 \tag{7.53f}$$

$$u_L = L \cdot \frac{di_L}{dt} \tag{7.53g}$$

$$u_L = v_2 - v_0 \tag{7.53h}$$

$$i_C = C \cdot \frac{du_C}{dt} \tag{7.53i}$$

$$u_C = v_1 - v_0 \tag{7.53j}$$

$$v_0 = 0 \tag{7.53k}$$

$$i_0 = i_1 + i_C \tag{7.53l}$$

$$i_1 = i_2 + i_L \tag{7.53m}$$

This time, we ended up with 13 equations in 13 unknowns. Its structure digraph is shown in Fig.7.18.

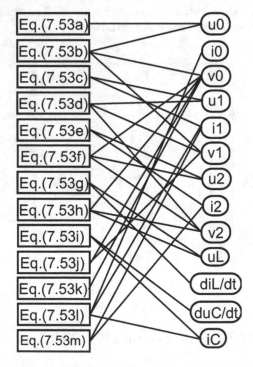

FIGURE 7.18. Structure digraph of electrical circuit with node potentials.

Figure 7.19 shows the partially colored digraph.

We again got stuck after two iterations. However, it seems that we made the problem worse rather than better. Just like last time, we again are left without an equation to compute i_C, but this time, we are also left with an equation, Eq.(7.53j), which has no unknowns left in it, although it has not

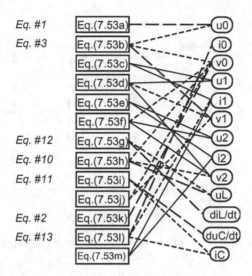

Eq. #1 — Eq.(7.53a)
Eq. #3 — Eq.(7.53b)
Eq.(7.53c)
Eq.(7.53d)
Eq.(7.53e)
Eq.(7.53f)
Eq. #12 — Eq.(7.53g)
Eq. #10 — Eq.(7.53h)
Eq. #11 — Eq.(7.53i)
Eq.(7.53j)
Eq. #2 — Eq.(7.53k)
Eq. #13 — Eq.(7.53l)
Eq.(7.53m)

u0, i0, v0, u1, i1, v1, u2, i2, v2, uL, diL/dt, duC/dt, iC

FIGURE 7.19. Partially colored structure digraph of electrical circuit with node potentials.

yet been made causal. We won't be able to use it for anything. Such an equation is called a *constraint equation*.

What has happened in this circuit? The capacitor has been placed in parallel with a voltage source. Consequently, the voltage across the capacitor cannot be chosen as an independent state variable. There is no freedom in choosing its initial condition.

7.7 Structural Singularity Elimination

Before we deal with the above circuit, let us choose a much simpler circuit that exhibits the same problems. Figure 7.20 shows the schematic of an electrical circuit with two capacitors in parallel.

Since the two capacitive voltages, u_1 and u_2, are the same, they don't qualify as independent state variables, as we cannot choose initial conditions for them independently. Hence we expect problems that are similar to those observed in the previous circuit.

The equations describing this circuit can be written as:

$$u_0 = f(t) \tag{7.54a}$$

$$u_R = R \cdot i_0 \tag{7.54b}$$

$$i_1 = C_1 \cdot \frac{du_1}{dt} \tag{7.54c}$$

$$i_2 = C_2 \cdot \frac{du_2}{dt} \tag{7.54d}$$

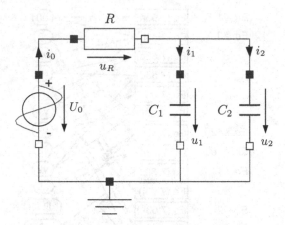

FIGURE 7.20. Schematic of electrical circuit with two capacitors in parallel.

$$u_0 = u_R + u_1 \qquad (7.54\text{e})$$

$$u_2 = u_1 \qquad (7.54\text{f})$$

$$i_0 = i_1 + i_2 \qquad (7.54\text{g})$$

If we choose u_1 and u_2 as state variables, then both u_1 and u_2 are considered known variables, and Eq.(7.54f) has no unknown left. Thus, it must be considered a *constraint equation*.

There are several different ways, how this problem can be solved [7.4]. We can turn the causality around on one of the capacitive equations, solving e.g. for the variable i_2, instead of du_2/dt. Consequently, the solver has to solve for du_2/dt instead of u_2, thus the *integrator* has been turned into a *differentiator*.

In the model equations, u_2 must be considered an unknown, whereas du_2/dt is considered a known variable. The equations can now easily be brought into causal form:

$$u_0 = f(t) \qquad (7.55\text{a})$$

$$i_2 = C_2 \cdot \frac{du_2}{dt} \qquad (7.55\text{b})$$

$$u_2 = u_1 \qquad (7.55\text{c})$$

$$u_R = u_0 - u_1 \qquad (7.55\text{d})$$

$$i_0 = \frac{1}{R} \cdot u_R \qquad (7.55\text{e})$$

$$i_1 = i_0 - i_2 \qquad (7.55\text{f})$$

$$\frac{du_1}{dt} = \frac{1}{C_1} \cdot i_1 \qquad (7.55\text{g})$$

with the block diagram as shown in Fig.7.21.

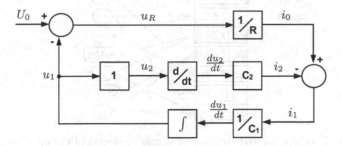

FIGURE 7.21. Block diagram of electrical circuit with parallel capacitors.

The solver generates u_1 out of du_1/dt by *numerical integration*, whereas it generates du_2/dt out of u_2 by *numerical differentiation*.

Numerical differentiation is a bad idea, at least if explicit formulae are being used. Using implicit formulae, numerical integration and differentiation are essentially the same, but as we already know, implicit formulae call for an iteration at every step.

Costas Pantelides [7.20] had a better idea. How about modifying the equation system such that the constraint equation disappears? In the above example, if:

$$u_2 = u_1 \tag{7.56}$$

at all times, then obviously, it must also be true that:

$$\frac{du_2}{dt} = \frac{du_1}{dt} \tag{7.57}$$

at all times. Thus, we can symbolically differentiate the constraint equation, and replace the constraint equation by its derivative. The new set of acausal equations takes the form:

$$u_0 = f(t) \tag{7.58a}$$
$$u_R = R \cdot i_0 \tag{7.58b}$$
$$i_1 = C_1 \cdot \frac{du_1}{dt} \tag{7.58c}$$
$$i_2 = C_2 \cdot \frac{du_2}{dt} \tag{7.58d}$$
$$u_0 = u_R + u_1 \tag{7.58e}$$
$$\frac{du_2}{dt} = \frac{du_1}{dt} \tag{7.58f}$$
$$i_0 = i_1 + i_2 \tag{7.58g}$$

with the partially colored structure digraph as shown in Fig.7.22.

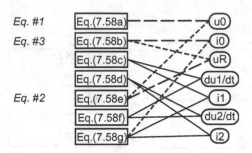

FIGURE 7.22. Partially colored structure digraph of electrical circuit with parallel capacitors after differentiation of the constraint equation.

The constraint equation has indeed disappeared. After partial causalization of the equations, we are now faced with an algebraic loop in four equations and four unknowns, a situation that we already know how to deal with.

Miraculously, we seem to have gotten rid of the constraint between the two capacitors. After the symbolic differentiation of the constraint equation, we seem to again have two integrators that we can integrate separately and independently.

Evidently, this cannot be true. The constraint on the capacitive voltages has not disappeared. It has only been hidden. It is true that we can now numerically integrate du_1/dt into u_1, and du_2/dt into u_2. However, we still must satisfy the original constraint equation when choosing the initial conditions for the two integrators.

The second integrator does not represent a true state variable. In fact, it is wasteful. We don't need two integrators, since the system has only one *degree of freedom*, i.e., one energy storage.

Let us thus modify the Pantelides algorithm once more. Instead of replacing the constraint equation by its derivative, we add the differentiated constraint equation as an additional equation to the set.

Hence we now have eight equations in seven unknowns. We have one equation too many, and consequently, we need to throw away one of them. We shall throw away one of the integrators, for example, the one that integrates du_2/dt into u_2.

We symbolize this by renaming the variable du_2/dt as du_2. du_2 is no longer a state derivative. It is simply an algebraic variable with a funny name. Hence both u_2 and du_2 are now unknowns, and we are thus faced with eight equations in eight unknowns. These are:

$$u_0 = f(t) \tag{7.59a}$$

$$u_R = R \cdot i_0 \tag{7.59b}$$

$$i_1 = C_1 \cdot \frac{du_1}{dt} \tag{7.59c}$$

$$i_2 = C_2 \cdot du_2 \tag{7.59d}$$

$$u_0 = u_R + u_1 \tag{7.59e}$$

$$u_2 = u_1 \tag{7.59f}$$

$$du_2 = \frac{du_1}{dt} \tag{7.59g}$$

$$i_0 = i_1 + i_2 \tag{7.59h}$$

with the partially colored structure digraph as shown in Fig.7.23.

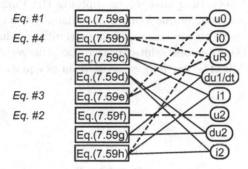

FIGURE 7.23. Partially colored structure digraph of electrical circuit with parallel capacitors after differentiation of the constraint equation.

Once again, we end up with an algebraic loop in four equations and four unknowns.

In the mathematical literature, *structurally singular* systems are called *higher–index problems*, or more precisely, structurally singular physical systems lead to mathematical descriptions that present themselves in the form of higher–index DAEs [7.1, 7.2, 7.19].

The *perturbation index* is a measure of the constraints among equations [7.10]. An *index–0 DAE* contains neither algebraic loops nor structural singularities. An *index–1 DAE* contains algebraic loops, but no structural singularities. A DAE with a perturbation index > 1, a so–called *higher–index DAE*, contains structural singularities[1].

The algorithm by Pantelides is a *symbolic index reduction* algorithm[2]. It reduces the perturbation index by one. Hence it may be necessary to

---

[1]A number of different definitions of structure indices are provided in the mathematical literature. A paper by Campbell and Gear [7.3] offers a good survey of this somewhat exotic issue. The different definitions all agree on the index of a *linear* DAE system, but sometimes disagree in the case of nonlinear DAE systems.

[2]The original paper by Pantelides did not concern itself with index reduction at all. It described an algorithm that could find, in a procedural fashion, a complete and consistent set of initial conditions for a DAE system. It was shown later in [7.4, 7.15] that the Pantelides algorithm can also be used as a symbolic index reduction algorithm.

apply the Pantelides algorithm more than once. For example, a mechanical system with constraints among positions or angles, such as a motor with a load, whereby the motor and the load are described separately by differential equations, leads to an index–3 DAE system. By applying the Pantelides algorithm once, the constraint gets reduced to a constraint between velocities or angular velocities, which are still state variables. By applying the Pantelides algorithm a second time, the constraint gets reduced to a constraint between accelerations or angular accelerations, which are no longer outputs of integrators, and therefore, are no longer state variables.

It is thus not surprising that, after applying the Pantelides algorithm, we ended up with an algebraic loop. This is usually the case.

Let us now return to the more complex circuit. We shall start with the version that contains a constraint equation, i.e., the version making use of the node potentials. The partially causalized set of equations can be written as follows:

$$u_0 = f(t) \tag{7.60a}$$
$$v_0 = 0 \tag{7.60b}$$
$$v_1 = u_0 - v_0 \tag{7.60c}$$
$$u_1 - R_1 \cdot i_1 = 0 \tag{7.60d}$$
$$u_1 + v_2 = v_1 \tag{7.60e}$$
$$u_2 - R_2 \cdot i_2 = 0 \tag{7.60f}$$
$$v_2 - u_2 = v_0 \tag{7.60g}$$
$$0 = u_C - v_1 + v_0 \tag{7.60h}$$
$$i_1 - i_2 = i_L \tag{7.60i}$$
$$u_L = v_2 - v_0 \tag{7.60j}$$
$$\frac{du_C}{dt} = \frac{1}{C} \cdot i_C \tag{7.60k}$$
$$\frac{di_L}{dt} = \frac{1}{L} \cdot u_L \tag{7.60l}$$
$$i_0 = i_1 + i_C \tag{7.60m}$$

The unknowns are written on the left side of the equal sign, thus equations with only one variable to the left of the equal sign are causal equations, those with more than one variable to the left of the equal sign are acausal equations, and those with zero variables to the left of the equal sign are constraint equations.

We differentiate the constraint equation, add it to the DAE system, and let go of an integrator associated with the constraint.

$$u_0 = f(t) \tag{7.61a}$$

$$v_0 = 0 \tag{7.61b}$$
$$v_1 = u_0 - v_0 \tag{7.61c}$$
$$u_1 - R_1 \cdot i_1 = 0 \tag{7.61d}$$
$$u_1 + v_2 = v_1 \tag{7.61e}$$
$$u_2 - R_2 \cdot i_2 = 0 \tag{7.61f}$$
$$v_2 - u_2 = v_0 \tag{7.61g}$$
$$0 = u_C - v_1 + v_0 \tag{7.61h}$$
$$0 = du_C - dv_1 + dv_0 \tag{7.61i}$$
$$i_1 - i_2 = i_L \tag{7.61j}$$
$$u_L = v_2 - v_0 \tag{7.61k}$$
$$du_C = \frac{1}{C} \cdot i_C \tag{7.61l}$$
$$\frac{di_L}{dt} = \frac{1}{L} \cdot u_L \tag{7.61m}$$
$$i_0 = i_1 + i_C \tag{7.61n}$$

In the process of differentiation, we introduced two new pseudo–derivatives[1], dv_1 and dv_0, for which we are lacking equations.

We now differentiate the causal equations that define v_1 and v_0, and add those to the set as well.

$$u_0 = f(t) \tag{7.62a}$$
$$v_0 = 0 \tag{7.62b}$$
$$dv_0 = 0 \tag{7.62c}$$
$$v_1 = u_0 - v_0 \tag{7.62d}$$
$$dv_1 = du_0 - dv_0 \tag{7.62e}$$
$$u_1 - R_1 \cdot i_1 = 0 \tag{7.62f}$$
$$u_1 + v_2 = v_1 \tag{7.62g}$$
$$u_2 - R_2 \cdot i_2 = 0 \tag{7.62h}$$
$$v_2 - u_2 = v_0 \tag{7.62i}$$
$$0 = u_C - v_1 + v_0 \tag{7.62j}$$
$$0 = du_C - dv_1 + dv_0 \tag{7.62k}$$
$$i_1 - i_2 = i_L \tag{7.62l}$$
$$u_L = v_2 - v_0 \tag{7.62m}$$
$$du_C = \frac{1}{C} \cdot i_C \tag{7.62n}$$

[1]The pseudo–derivatives are sometimes also called *dummy–derivatives* in the literature [7.15].

$$\frac{di_L}{dt} = \frac{1}{L} \cdot u_L \tag{7.62o}$$

$$i_0 = i_1 + i_C \tag{7.62p}$$

We again introduced one additional pseudo–derivative, du_0. Thus, the final set of equations can be written as:

$$u_0 = f(t) \tag{7.63a}$$

$$du_0 = \frac{df(t)}{dt} \tag{7.63b}$$

$$v_0 = 0 \tag{7.63c}$$

$$dv_0 = 0 \tag{7.63d}$$

$$v_1 = u_0 - v_0 \tag{7.63e}$$

$$dv_1 = du_0 - dv_0 \tag{7.63f}$$

$$u_1 - R_1 \cdot i_1 = 0 \tag{7.63g}$$

$$u_1 + v_2 = v_1 \tag{7.63h}$$

$$u_2 - R_2 \cdot i_2 = 0 \tag{7.63i}$$

$$v_2 - u_2 = v_0 \tag{7.63j}$$

$$0 = u_C - v_1 + v_0 \tag{7.63k}$$

$$0 = du_C - dv_1 + dv_0 \tag{7.63l}$$

$$i_1 - i_2 = i_L \tag{7.63m}$$

$$u_L = v_2 - v_0 \tag{7.63n}$$

$$du_C = \frac{1}{C} \cdot i_C \tag{7.63o}$$

$$\frac{di_L}{dt} = \frac{1}{L} \cdot u_L \tag{7.63p}$$

$$i_0 = i_1 + i_C \tag{7.63q}$$

If $f(t)$ is a known function of time, we can symbolically compute its derivative. On the other hand, if $f(t)$ stands for an input signal in a real–time simulation with hardware in the loop, we have a problem. In that case, we may need an additional sensor somewhere in the system that measures $df(t)/dt$, and add this signal as an additional real–time input to the simulation.

By now, we have 17 equations in 17 unknowns. The partially colored structure digraph is shown in Fig.7.24.

It worked. The Pantelides algorithm was able to reduce the DAE system to index–1. We ended up with an algebraic loop in five equations and five unknowns that can be tackled using any one among the techniques described in the previous sections of this chapter.

Let us now return to the original description of the circuit without node potentials. In that formulation, no constraint equation was visible. We only

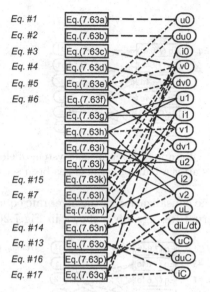

FIGURE 7.24. Partially colored structure digraph of electrical RLC circuit with node potentials after differentiation of the constraint equation.

knew that we had run into difficulties, because we recognized that we were left without an equation to compute the value of the variable i_c.

Let us write down the equations in their partially causalized form.

$$u_0 = f(t) \tag{7.64a}$$

$$u_1 - R_1 \cdot i_1 = 0 \tag{7.64b}$$

$$u_2 - R_2 \cdot i_2 = 0 \tag{7.64c}$$

$$u_1 + u_L = u_0 \tag{7.64d}$$

$$u_1 + u_2 = u_C \tag{7.64e}$$

$$u_L - u_2 = 0 \tag{7.64f}$$

$$i_1 - i_2 = i_L \tag{7.64g}$$

$$\frac{du_C}{dt} = \frac{1}{C} \cdot i_C \tag{7.64h}$$

$$\frac{di_L}{dt} = \frac{1}{L} \cdot u_L \tag{7.64i}$$

$$i_0 = i_1 + i_C \tag{7.64j}$$

We are left with six acausal equations in only five unknowns, since i_c doesn't show up anywhere in them. Thus, there still exists a constraint equation. However, it is hidden inside an algebraic system.

Let us draw the structure digraph of the algebraic system, and let us choose a residual equation and a tearing variable. The structure digraph is shown in Fig.7.25.

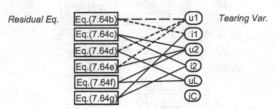

FIGURE 7.25. Structure digraph of algebraic subsystem of electrical RLC circuit without node potentials after a tearing variable has been selected.

We now complete the causalization of the algebraic equation system. The completely colored structure digraph is shown in Fig.7.26.

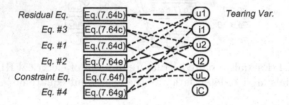

FIGURE 7.26. Completely colored structure digraph of algebraic subsystem of electrical RLC circuit without node potentials.

The constraint equation, Eq.(7.64f) has become clearly visible. We can now write down the equation system in its completely causalized form:

$$u_0 = f(t) \tag{7.65a}$$

$$u_L = u_0 - u_1 \tag{7.65b}$$

$$u_2 = u_C - u_1 \tag{7.65c}$$

$$i_2 = \frac{1}{R_2} \cdot u_2 \tag{7.65d}$$

$$i_1 = i_2 + i_L \tag{7.65e}$$

$$0 = u_L - u_2 \tag{7.65f}$$

$$u_1 = R_1 \cdot i_1 \tag{7.65g}$$

$$\frac{du_C}{dt} = \frac{1}{C} \cdot i_C \tag{7.65h}$$

$$\frac{di_L}{dt} = \frac{1}{L} \cdot u_L \tag{7.65i}$$

$$i_0 = i_1 + i_C \tag{7.65j}$$

We now apply the Pantelides algorithm. We start by differentiating the constraint equation, adding it to the equation system.

$$u_0 = f(t) \tag{7.66a}$$

$$u_L = u_0 - u_1 \tag{7.66b}$$

$$u_2 = u_C - u_1 \tag{7.66c}$$

$$i_2 = \frac{1}{R_2} \cdot u_2 \tag{7.66d}$$

$$i_1 = i_2 + i_L \tag{7.66e}$$

$$0 = u_L - u_2 \tag{7.66f}$$

$$0 = du_L - du_2 \tag{7.66g}$$

$$u_1 = R_1 \cdot i_1 \tag{7.66h}$$

$$\frac{du_C}{dt} = \frac{1}{C} \cdot i_C \tag{7.66i}$$

$$\frac{di_L}{dt} = \frac{1}{L} \cdot u_L \tag{7.66j}$$

$$i_0 = i_1 + i_C \tag{7.66k}$$

We introduced two new pseudo–derivatives, du_L and du_2. Hence we differentiate the equations defining u_L and u_2.

$$u_0 = f(t) \tag{7.67a}$$

$$u_L = u_0 - u_1 \tag{7.67b}$$

$$du_L = du_0 - du_1 \tag{7.67c}$$

$$u_2 = u_C - u_1 \tag{7.67d}$$

$$du_2 = du_C - du_1 \tag{7.67e}$$

$$i_2 = \frac{1}{R_2} \cdot u_2 \tag{7.67f}$$

$$i_1 = i_2 + i_L \tag{7.67g}$$

$$0 = u_L - u_2 \tag{7.67h}$$

$$0 = du_L - du_2 \tag{7.67i}$$

$$u_1 = R_1 \cdot i_1 \tag{7.67j}$$

$$\frac{du_C}{dt} = \frac{1}{C} \cdot i_C \tag{7.67k}$$

$$\frac{di_L}{dt} = \frac{1}{L} \cdot u_L \tag{7.67l}$$

$$i_0 = i_1 + i_C \tag{7.67m}$$

We introduced three new pseudo–derivatives, du_0, du_1, and du_C. We differentiate the equations defining u_0 and u_1, and we throw the integrator away that defines u_C.

$$u_0 = f(t) \tag{7.68a}$$

$$du_0 = \frac{df(t)}{dt} \tag{7.68b}$$

$$u_L = u_0 - u_1 \tag{7.68c}$$

$$du_L = du_0 - du_1 \tag{7.68d}$$

$$u_2 = u_C - u_1 \tag{7.68e}$$

$$du_2 = du_C - du_1 \tag{7.68f}$$

$$i_2 = \frac{1}{R_2} \cdot u_2 \tag{7.68g}$$

$$i_1 = i_2 + i_L \tag{7.68h}$$

$$0 = u_L - u_2 \tag{7.68i}$$

$$0 = du_L - du_2 \tag{7.68j}$$

$$u_1 = R_1 \cdot i_1 \tag{7.68k}$$

$$du_1 = R_1 \cdot di_1 \tag{7.68l}$$

$$du_C = \frac{1}{C} \cdot i_C \tag{7.68m}$$

$$\frac{di_L}{dt} = \frac{1}{L} \cdot u_L \tag{7.68n}$$

$$i_0 = i_1 + i_C \tag{7.68o}$$

We introduced another pseudo–derivative, di_1. Thus, we need to differentiate the equation defining i_1 as well.

$$u_0 = f(t) \tag{7.69a}$$

$$du_0 = \frac{df(t)}{dt} \tag{7.69b}$$

$$u_L = u_0 - u_1 \tag{7.69c}$$

$$du_L = du_0 - du_1 \tag{7.69d}$$

$$u_2 = u_C - u_1 \tag{7.69e}$$

$$du_2 = du_C - du_1 \tag{7.69f}$$

$$i_2 = \frac{1}{R_2} \cdot u_2 \tag{7.69g}$$

$$i_1 = i_2 + i_L \tag{7.69h}$$

$$di_1 = di_2 + \frac{di_L}{dt} \tag{7.69i}$$

$$0 = u_L - u_2 \tag{7.69j}$$

$$0 = du_L - du_2 \tag{7.69k}$$

$$u_1 = R_1 \cdot i_1 \tag{7.69l}$$

$$du_1 = R_1 \cdot di_1 \tag{7.69m}$$

$$du_C = \frac{1}{C} \cdot i_C \tag{7.69n}$$

$$\frac{di_L}{dt} = \frac{1}{L} \cdot u_L \tag{7.69o}$$

$$i_0 = i_1 + i_C \tag{7.69p}$$

We now introduced yet a new pseudo–derivative, di_2, and also a true derivative, di_L/dt.

As the constraint equation was hidden in one solid algebraic loop, we had to differentiate every single equation of that loop. We ended up with the following set of 17 equations in 17 unknowns:

$$u_0 = f(t) \tag{7.70a}$$

$$du_0 = \frac{df(t)}{dt} \tag{7.70b}$$

$$u_L = u_0 - u_1 \tag{7.70c}$$

$$du_L = du_0 - du_1 \tag{7.70d}$$

$$u_2 = u_C - u_1 \tag{7.70e}$$

$$du_2 = du_C - du_1 \tag{7.70f}$$

$$i_2 = \frac{1}{R_2} \cdot u_2 \tag{7.70g}$$

$$di_2 = \frac{1}{R_2} \cdot du_2 \tag{7.70h}$$

$$i_1 = i_2 + i_L \tag{7.70i}$$

$$di_1 = di_2 + \frac{di_L}{dt} \tag{7.70j}$$

$$0 = u_L - u_2 \tag{7.70k}$$

$$0 = du_L - du_2 \tag{7.70l}$$

$$u_1 = R_1 \cdot i_1 \tag{7.70m}$$

$$du_1 = R_1 \cdot di_1 \tag{7.70n}$$

$$du_C = \frac{1}{C} \cdot i_C \tag{7.70o}$$

$$\frac{di_L}{dt} = \frac{1}{L} \cdot u_L \tag{7.70p}$$

$$i_0 = i_1 + i_C \tag{7.70q}$$

Let us start from scratch with the causalization of this DAE system. Figure 7.27 shows the partially colored structure digraph of this set of equations.

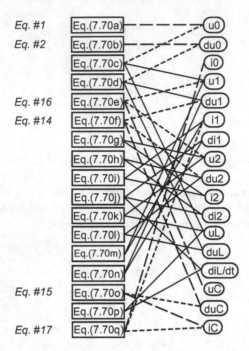

FIGURE 7.27. Partially colored structure digraph of electrical RLC circuit without node potentials.

It worked. The constraint equation has indeed disappeared. Instead we are now facing an algebraic loop in 11 equations and 11 unknowns.

Let us analyze this algebraic loop further, as this analysis will unveil yet another difficulty. The partially causalized equations can be written as follows:

$$u_0 = f(t) \tag{7.71a}$$

$$du_0 = \frac{df(t)}{dt} \tag{7.71b}$$

$$u_L + u_1 = u_0 \tag{7.71c}$$

$$du_L + du_1 = du_0 \tag{7.71d}$$

$$u_2 - R_2 \cdot i_2 = 0 \tag{7.71e}$$

$$du_2 - R_2 \cdot di_2 = 0 \tag{7.71f}$$

$$i_1 - i_2 = i_L \tag{7.71g}$$

$$di_1 - di_2 - \frac{di_L}{dt} = 0 \tag{7.71h}$$

$$u_L - u_2 = 0 \tag{7.71i}$$

$$du_L - du_2 = 0 \tag{7.71j}$$

$$u_1 - R_1 \cdot i_1 = 0 \tag{7.71k}$$

$$du_1 - R_1 \cdot di_1 = 0 \tag{7.71l}$$

$$u_L - l \cdot \frac{di_L}{dt} = 0 \tag{7.71m}$$

$$du_C = du_2 + du_1 \tag{7.71n}$$

$$i_C = C \cdot du_C \tag{7.71o}$$

$$u_C = u_2 + u_1 \tag{7.71p}$$

$$i_0 = i_1 + i_C \tag{7.71q}$$

Figure 7.28 shows the structure digraph of the algebraic subsystem after selecting a tearing variable and a residual equation.

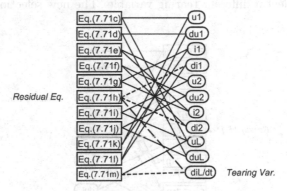

FIGURE 7.28. Structure digraph of algebraic subsystem of electrical RLC circuit after choosing a tearing variable and a residual equation.

We proceed with the usual graph–coloring algorithm. The partially colored structure digraph is shown in Fig.7.29.

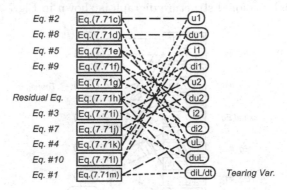

FIGURE 7.29. Partially colored structure digraph of algebraic subsystem of electrical RLC circuit.

We seem to again have ended up with a structural singularity. Equation (7.71g) is a constraint equation, whereas we are lacking an equation to compute the variable du_L.

Yet, this is a very different problem from the one discussed before. This constraint was caused by a poor selection of a tearing variable. Had we chosen a different tearing variable or a different residual equation, this problem would not have occurred. For this reason, we cannot simplify the heuristic procedure further. It is insufficient to look at the number of black (solid) lines attached to equations and the number of black (solid) lines attached to variables when selecting the residual equation and the tearing variable. For each proposed selection, we must pursue the consequences of that selection all the way to the end and be prepared to backtrack if we end up with a conflict.

Let us select a different tearing variable. The new selection is shown in Fig.7.30.

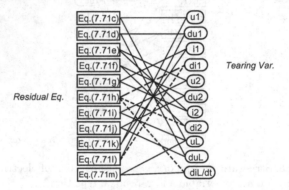

FIGURE 7.30. Structure digraph of algebraic subsystem of electrical RLC circuit after choosing a tearing variable and a residual equation.

The partially colored structure digraph is shown in Fig.7.31.

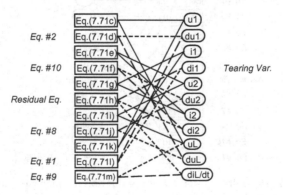

FIGURE 7.31. Partially colored structure digraph of algebraic subsystem of electrical RLC circuit.

We were able to causalize six of the eleven equations. We thus need to select a second residual equation and a second tearing variable, in order to complete the causalization of the algebraic equation system.

Dymola implements the Pantelides algorithm essentially in the form explained in this section. However as almost always, the devil is in the detail. For didactic reasons, we explained the algorithm by starting out with an individual constraint equation, which we differentiated and added to the set of equations. We then chose a pseudo–derivative, in order to ensure that we once again had the same number of unknowns as equations. We then checked, whether additional equations needed to be differentiated as well, since new pseudo–derivatives had been introduced in the process. Yet, this procedure already got us into trouble in one of the examples. For this reason, Dymola first determines *all* equations that need to be differentiated, and chooses the dummy derivative only in the very end.

Furthermore, a fixed choice of a pseudo–derivative may occasionally lead to a division by zero at run time. In fact, it can happen that no fixed choice of a pseudo–derivative avoids divisions by zero. In those cases, Dymola makes the choice of the state variables *dynamic*, switching from one selection to another during the course of the simulation run [7.14].

7.8 The Solvability Issue

The DAE literature talks about yet another issue, namely that of *solvability* [7.1]. Take for example the following DAE:

$$x - \dot{x}^2 = 0.0 \tag{7.72}$$

Converting Eq.(7.72) to ODE form, we obtain:

$$\dot{x} = \pm\sqrt{x} \tag{7.73}$$

Evidently, this ODE has only a real–valued solution as long as the initial value of x is positive. This constraint existed even in the DAE case. However, in the DAE formulation, the situation has become worse. The DAE formulation does not give us any hint, which of the two roots we should select. If we choose the positive root, $\dot{x}$ will also be positive, and x will keep growing. However, if we choose the negative root, $\dot{x}$ is negative, and x will decrease. Both solutions satisfy the DAE, and if the only information we have is the DAE, we can't tell which solution is for real. Even worse, it could happen that we should choose the positive root during some period of time, and the negative root during another. Thus, at any moment in time, we obtain a potential bifurcation in the solution depending on whether we choose the positive or the negative root.

To us, solvability is a non–issue. It is the typical worry of a mathematician who puts the mathematical formulation first, and then tries to interpret the ramifications of that formulation. Remember what we wrote earlier: mathematics is simply the language of physics. The reason why we are interested in differential equations and solving them is that we wish to gain a better understanding of physical phenomena in this universe. Consequently, the origin of our interest is always physics, not mathematics. Physics does not provide us with unsolvable riddles. Saying that a DAE is unsolvable is equivalent to saying that the phenomenon described by it is "defying causality" in the sense that the outcome of an experiment is non–deterministic, which in turn is almost equivalent to saying that the phenomenon is non–physical. True, chaos is for real [7.5]. We can observe chaotic phenomena in physics every day. However, chaotic phenomena are not described through unsolvable differential equations. Chaos only means that the solution in time is undecidable without infinite precision. However, the differential equation that produces a chaotic solution is perfectly deterministic [7.5]. Thus, philosophizing about the implications of solvability or non–solvability of DAEs is like discussing how many angels can dance on the tip of a needle. Or is it not?

Let us look at a simple pendulum as shown on Fig.7.32.

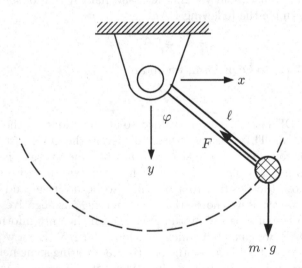

FIGURE 7.32. Mechanical pendulum.

The equations of motion for this pendulum can be described easily in DAE form:

$$m \cdot \frac{dv_x}{dt} = -\frac{F \cdot x}{\ell} \tag{7.74a}$$

$$m \cdot \frac{dv_y}{dt} = m \cdot g - \frac{F \cdot y}{\ell} \tag{7.74b}$$

$$\frac{dx}{dt} = v_x \tag{7.74c}$$

$$\frac{dy}{dt} = v_y \tag{7.74d}$$

$$x^2 + y^2 = \ell^2 \tag{7.74e}$$

These are five equations in the five unknowns dv_x/dt, dv_y/dt, dx/dt, dy/dt, and F. The four natural state variables: v_x, v_y, x, and y are assumed known.

We notice at once that Eq.(7.74e) is a constraint equation, since it doesn't contain any of the unknowns. We apply the Pantelides algorithm, and obtain the following set of six equations in six unknowns:

$$m \cdot \frac{dv_x}{dt} = -\frac{F \cdot x}{\ell} \tag{7.75a}$$

$$m \cdot \frac{dv_y}{dt} = m \cdot g - \frac{F \cdot y}{\ell} \tag{7.75b}$$

$$dx = v_x \tag{7.75c}$$

$$\frac{dy}{dt} = v_y \tag{7.75d}$$

$$x^2 + y^2 = \ell^2 \tag{7.75e}$$

$$2 \cdot x \cdot dx + 2 \cdot y \cdot \frac{dy}{dt} = 0 \tag{7.75f}$$

We decided to let go of the integrator for x, thus the six unknowns are dv_x/dt, dv_y/dt, dx, dy/dt, F, and x.

Eq.(7.75e) is no longer a constraint equation, as it can be solved for the new unknown x. Eq.(7.75f) can be solved for the unknown dx, but this leaves Eq.(7.75c) as a new constraint equation.

Evidently, the original problem was an index–3 problem, and the Pantelides algorithm needs to be applied a second time. We obtain the following set of nine equations in nine unknowns:

$$m \cdot dv_x = -\frac{F \cdot x}{\ell} \tag{7.76a}$$

$$m \cdot \frac{dv_y}{dt} = m \cdot g - \frac{F \cdot y}{\ell} \tag{7.76b}$$

$$dx = v_x \tag{7.76c}$$

$$d2x = dv_x \tag{7.76d}$$

$$\frac{dy}{dt} = v_y \tag{7.76e}$$

$$d2y = \frac{dv_y}{dt} \tag{7.76f}$$

$$x^2 + y^2 = \ell^2 \tag{7.76g}$$

$$x \cdot dx + y \cdot \frac{dy}{dt} = 0 \tag{7.76h}$$

$$dx^2 + x \cdot d2x + \left(\frac{dy}{dt}\right)^2 + y \cdot d2y = 0 \tag{7.76i}$$

In the differentiation of Eq.(7.76c), a new variable, $d2x$, was introduced. Thus, the equation defining dx, i.e., Eq.(7.76h), had to be differentiated as well. In that differentiation, again one more new variable, $d2y$, was introduced. Hence the equation defining dy/dt, i.e., Eq.(7.76e), had to be differentiated also. Finally, another integrator had to be eliminated, namely the one defining the variable v_x. The nine unknowns of this equation system are dv_x, x, F, dv_y/dt, dx, v_x, $d2x$, dy/dt, and $d2y$.

This set of equations represents an index–1 DAE problem that can be causalized using the tearing method. Figure 7.33 shows the partially causalized structure digraph of this DAE system.

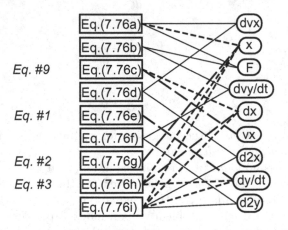

FIGURE 7.33. Partially causalized structure digraph of mechanical pendulum.

An algebraic loop in five equations and five unknowns remains. Figure 7.34 shows the completely causalized structure digraph after a residual equation and a tearing variable have been chosen. Since we have a choice, we decided to select a tearing variable that appears linearly in the residual equation.

We can read out the causal equations from the completely causalized structure digraph of Fig.7.34. They are:

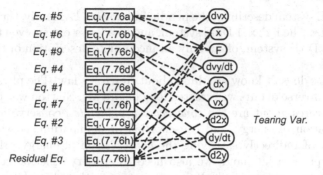

FIGURE 7.34. Completely causalized structure digraph of mechanical pendulum.

$$\frac{dy}{dt} = v_y \tag{7.77a}$$

$$x = \pm\sqrt{\ell^2 - y^2} \tag{7.77b}$$

$$dx = -\frac{y}{x} \cdot \frac{dy}{dt} \tag{7.77c}$$

$$dv_x = d2x \tag{7.77d}$$

$$F = -m \cdot \ell \cdot \frac{dv_x}{x} \tag{7.77e}$$

$$\frac{dv_y}{dt} = g - \frac{F \cdot y}{m \cdot \ell} \tag{7.77f}$$

$$d2y = \frac{dv_y}{dt} \tag{7.77g}$$

$$d2x = -\frac{dx^2 + \left(\frac{dy}{dt}\right)^2 + y \cdot d2y}{x} \tag{7.77h}$$

$$v_x = dx \tag{7.77i}$$

We are facing a new problem. We seem to have come across a *solvability* issue. At any point in time, there are two solutions to Eq.(7.77b), one of which is positive, whereas the other is negative. Yet, the physics behind the pendulum motion doesn't exhibit any ambiguity at all. The pendulum knows exactly, how to swing. It knows that we have to choose the positive root, whenever the pendulum is to the right of the joint, whereas we must choose the negative root otherwise.

Applying the Pantelides algorithm actually made the problem worse. The original index–3 DAE model at least knew that the position of the pendulum cannot jump, since x is the output of an integrator. The reduced index–1 DAE system no longer contains that information.

Evidently, the issue that we are facing here is not related to the physics of the pendulum, but only to the mathematical description thereof, i.e., to the DAE system describing the pendulum motion. Evidently, the simulation model, i.e., the index–1 DAE system, and to a lesser extent even the original index–3 DAE system, offers only an incomplete description of the physical reality.

Physics doesn't know anything about Newton's law. The physical reality of this universe of ours was created long before Mr. Newton was born. What physics cares about are the *conservation principles*: conservation of mass, conservation of energy, and conservation of momentum. Newton's law is one way of indirectly satisfying the conservation of energy principle. Yet for the problem at hand, we also need to conserve the momentum. The DAE system, as specified, does not capture, either directly or indirectly, the need for conserving the momentum.

For the example at hand, the problem can be solved easily by selecting a different set of state variables. Since the pendulum has one mechanical degree of freedom, we need two state variables. It turns out that φ and $\dot{\varphi}$ are a considerably smarter choice of a set of state variables than y and $\dot{y}$.

Unfortunately, the original model does not even contain φ as a variable. We need to add a description of the relationship between the variables currently captured and φ to the model. The easiest may be to replace the original constraint equation, Eq.(7.74e), by a set of two different equations:

$$m \cdot \frac{dv_x}{dt} = -\frac{F \cdot x}{\ell} \tag{7.78a}$$

$$m \cdot \frac{dv_y}{dt} = m \cdot g - \frac{F \cdot y}{\ell} \tag{7.78b}$$

$$\frac{dx}{dt} = v_x \tag{7.78c}$$

$$\frac{dy}{dt} = v_y \tag{7.78d}$$

$$x = \ell \cdot \sin(\varphi) \tag{7.78e}$$

$$y = \ell \cdot \cos(\varphi) \tag{7.78f}$$

These are six equations in the six unknowns dv_x/dt, dv_y/dt, dx/dt, dy/dt, F, and φ.

Since x and y are initially known variables, we can solve Eq.(7.78e) for φ, which then makes Eq.(7.78f) a constraint equation. Left to its own devices, the Pantelides algorithm will differentiate the constraint equation, while letting go of the integrator for y. In the process of differentiation, a new *algebraic* variable, $d\varphi$, is created, and therefore, Eq.(7.78e) needs to be differentiated as well:

$$m \cdot \frac{dv_x}{dt} = -\frac{F \cdot x}{\ell} \tag{7.79a}$$

$$m \cdot \frac{dv_y}{dt} = m \cdot g - \frac{F \cdot y}{\ell} \tag{7.79b}$$

$$\frac{dx}{dt} = v_x \tag{7.79c}$$

$$dy = v_y \tag{7.79d}$$

$$x = \ell \cdot \sin(\varphi) \tag{7.79e}$$

$$\frac{dx}{dt} = \ell \cdot \cos(\varphi) \cdot d\varphi \tag{7.79f}$$

$$y = \ell \cdot \cos(\varphi) \tag{7.79g}$$

$$dy = -\ell \cdot \sin(\varphi) \cdot d\varphi \tag{7.79h}$$

The Pantelides algorithm has no reason to select φ as a state variable on its own. It needs help. In Dymola [7.9], we can offer a *choice of preferred state variables* to the Pantelides algorithm. If we tell the algorithm that we wish to keep φ as a state variable, a true state derivative, $d\varphi/dt$, will be generated in the process of differentiation in place of the algebraic variable, $d\varphi$. The result of the operation will be:

$$m \cdot \frac{dv_x}{dt} = -\frac{F \cdot x}{\ell} \tag{7.80a}$$

$$m \cdot \frac{dv_y}{dt} = m \cdot g - \frac{F \cdot y}{\ell} \tag{7.80b}$$

$$\frac{dx}{dt} = v_x \tag{7.80c}$$

$$dy = v_y \tag{7.80d}$$

$$x = \ell \cdot \sin(\varphi) \tag{7.80e}$$

$$y = \ell \cdot \cos(\varphi) \tag{7.80f}$$

$$dy = -\ell \cdot \sin(\varphi) \cdot \frac{d\varphi}{dt} \tag{7.80g}$$

These are now seven equations in the seven unknowns dv_x/dt, dv_y/dt, dx/dt, dy, F, y, and $d\varphi/dt$. Since the integrator for y was eliminated, y is now an additional unknown. $d\varphi/dt$ was added as another unknown, but φ is no longer an unknown, since it is now the output of an integrator.

Since φ is now a known variable, Eq.(7.80e) has become a new constraint equation that needs to be differentiated. The result of this operation is:

$$m \cdot \frac{dv_x}{dt} = -\frac{F \cdot x}{\ell} \tag{7.81a}$$

$$m \cdot \frac{dv_y}{dt} = m \cdot g - \frac{F \cdot y}{\ell} \tag{7.81b}$$

$$dx = v_x \tag{7.81c}$$

$$dy = v_y \tag{7.81d}$$

$$x = \ell \cdot \sin(\varphi) \tag{7.81e}$$

$$dx = \ell \cdot \cos(\varphi) \cdot \frac{d\varphi}{dt} \tag{7.81f}$$

$$y = \ell \cdot \cos(\varphi) \tag{7.81g}$$

$$dy = -\ell \cdot \sin(\varphi) \cdot \frac{d\varphi}{dt} \tag{7.81h}$$

We now have eight equations in the eight unknowns dv_x/dt, dv_y/dt, dx, dy, F, x, y, and $d\varphi/dt$. A second integrator, the one defining variable x was thrown out in the process.

Since v_x and v_y are still known variables, Eqs.(7.81c–d) need to be solved for dx and dy, respectively. We can then solve Eq.(7.81f) for $d\varphi/dt$, and consequently, Eq.(7.81h) has become a new constraint equation that needs to be differentiated. Since we told the Pantelides algorithm that we wish to preserve $d\varphi/dt$ as a state variable, a true state derivative, $d^2\varphi/dt^2$ is generated in the process of differentiation. The result of the operation is:

$$m \cdot \frac{dv_x}{dt} = -\frac{F \cdot x}{\ell} \tag{7.82a}$$

$$m \cdot dv_y = m \cdot g - \frac{F \cdot y}{\ell} \tag{7.82b}$$

$$dx = v_x \tag{7.82c}$$

$$dy = v_y \tag{7.82d}$$

$$d2y = dv_y \tag{7.82e}$$

$$x = \ell \cdot \sin(\varphi) \tag{7.82f}$$

$$dx = \ell \cdot \cos(\varphi) \cdot \frac{d\varphi}{dt} \tag{7.82g}$$

$$y = \ell \cdot \cos(\varphi) \tag{7.82h}$$

$$dy = -\ell \cdot \sin(\varphi) \cdot \frac{d\varphi}{dt} \tag{7.82i}$$

$$d2y = -\ell \cdot \sin(\varphi) \cdot \frac{d^2\varphi}{dt^2} - \ell \cdot \cos(\varphi) \cdot \left(\frac{d\varphi}{dt}\right)^2 \tag{7.82j}$$

By now, we have 10 equations in the 10 unknowns dv_x/dt, dv_y, dx, dy, F, x, y, v_y, $d^2\varphi/dt^2$, and $d2y$. While differentiating the constraint equation, a new algebraic variable, $d2y$, was introduced. Hence the equation defining dy, Eq.(7.82d) had to be differentiated as well. This pointed to the integrator to be thrown out. It is the integrator defining v_y. Hence variable v_y has now also become an unknown. $d^2\varphi/dt^2$ was added as another unknown replacing the former unknown $d\varphi/dt$, which has now become a known variable, since it is the output of an integrator.

Since $d\varphi/dt$ is now a known variable, yet another constraint equation was introduced. It is Eq.(7.82g). This equation needs to be differentiated as well. The result of the operation is:

$$m \cdot dv_x = -\frac{F \cdot x}{\ell} \tag{7.83a}$$

$$m \cdot dv_y = m \cdot g - \frac{F \cdot y}{\ell} \tag{7.83b}$$

$$dx = v_x \tag{7.83c}$$

$$d2x = dv_x \tag{7.83d}$$

$$dy = v_y \tag{7.83e}$$

$$d2y = dv_y \tag{7.83f}$$

$$x = \ell \cdot \sin(\varphi) \tag{7.83g}$$

$$dx = \ell \cdot \cos(\varphi) \cdot \frac{d\varphi}{dt} \tag{7.83h}$$

$$d2x = \ell \cdot \cos(\varphi) \cdot \frac{d^2\varphi}{dt^2} - \ell \cdot \sin(\varphi) \cdot \left(\frac{d\varphi}{dt}\right)^2 \tag{7.83i}$$

$$y = \ell \cdot \cos(\varphi) \tag{7.83j}$$

$$dy = -\ell \cdot \sin(\varphi) \cdot \frac{d\varphi}{dt} \tag{7.83k}$$

$$d2y = -\ell \cdot \sin(\varphi) \cdot \frac{d^2\varphi}{dt^2} - \ell \cdot \cos(\varphi) \cdot \left(\frac{d\varphi}{dt}\right)^2 \tag{7.83l}$$

This is the final set of 12 equations in the 12 unknowns dv_x, dv_y, dx, dy, F, x, y, v_x, v_y, $d^2\varphi/dt^2$, $d2x$, and $d2y$. It constitutes an implicit index–1 DAE system.

Dymola [7.9] performs one more level of symbolic preprocessing. If it finds a *trivial equation* of the type $a = b$, it throws it out, keeps only one of the variables in the model, and replaces all occurrences of the other by the former. This operation results in:

$$m \cdot dv_x = -\frac{F \cdot x}{\ell} \tag{7.84a}$$

$$m \cdot dv_y = m \cdot g - \frac{F \cdot y}{\ell} \tag{7.84b}$$

$$x = \ell \cdot \sin(\varphi) \tag{7.84c}$$

$$v_x = \ell \cdot \cos(\varphi) \cdot \frac{d\varphi}{dt} \tag{7.84d}$$

$$dv_x = \ell \cdot \cos(\varphi) \cdot \frac{d^2\varphi}{dt^2} - \ell \cdot \sin(\varphi) \cdot \left(\frac{d\varphi}{dt}\right)^2 \tag{7.84e}$$

$$y = \ell \cdot \cos(\varphi) \tag{7.84f}$$

$$v_y = -\ell \cdot \sin(\varphi) \cdot \frac{d\varphi}{dt} \tag{7.84g}$$

$$dv_y = -\ell \cdot \sin(\varphi) \cdot \frac{d^2\varphi}{dt^2} - \ell \cdot \cos(\varphi) \cdot \left(\frac{d\varphi}{dt}\right)^2 \qquad (7.84h)$$

Hence we end up with a set of eight equations in the eight unknowns dv_x, dv_y, F, x, y, v_x, v_y, and $d^2\varphi/dt^2$.

Figure 7.35 shows the partially causalized structure diagram of this system.

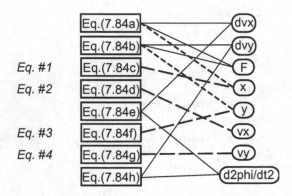

FIGURE 7.35. Partially causalized structure digraph of mechanical pendulum.

An algebraic loop in four equations and four unknowns remains. Figure 7.36 shows the completely causalized structure digraph after a suitable residual equation and tearing variable have been chosen.

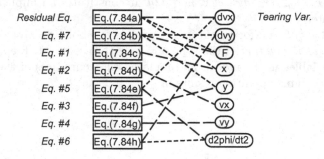

FIGURE 7.36. Completely causalized structure digraph of mechanical pendulum.

The causal equations can be read out of the structure digraph of Fig.7.36. They are:

$$x = \ell \cdot \sin(\varphi) \qquad (7.85a)$$

$$v_x = \ell \cdot \cos(\varphi) \cdot \frac{d\varphi}{dt} \qquad (7.85b)$$

$$y = \ell \cdot \cos(\varphi) \qquad (7.85c)$$

$$v_y = -\ell \cdot \sin(\varphi) \cdot \frac{d\varphi}{dt} \tag{7.85d}$$

$$\frac{d^2\varphi}{dt^2} = \frac{dv_x}{\ell \cdot \cos(\varphi)} + \frac{\sin(\varphi)}{\cos(\varphi)} \cdot \left(\frac{d\varphi}{dt}\right)^2 \tag{7.85e}$$

$$dv_y = -\ell \cdot \sin(\varphi) \cdot \frac{d^2\varphi}{dt^2} - \ell \cdot \cos(\varphi) \cdot \left(\frac{d\varphi}{dt}\right)^2 \tag{7.85f}$$

$$F = \frac{m \cdot g \cdot \ell}{y} - \frac{m \cdot \ell \cdot dv_y}{y} \tag{7.85g}$$

$$dv_x = -\frac{F \cdot x}{m \cdot \ell} \tag{7.85h}$$

With this choice of the set of state variables, all of the equations are linear in the variables they are being solved for. Consequently, there is no ambiguity, and the solvability problem has disappeared. This model can be simulated without difficulties for all values of φ and $\dot{\varphi}$. Clearly, the solvability issue was not related to the physics of the pendulum motion at all. It was purely a mathematical artifact caused by an unfortunate selection of state variables.

Of course, it would have been a yet better idea to formulate Newton's law directly in rotational coordinates. In that case, the resulting model would have been of index 1 or lower right from the beginning, and we would not have had to invoke the Pantelides algorithm at all.

Does this approach resolve all solvability issues in modeling mechanical systems? Unfortunately, this question must be answered in the negative. For multibody systems without closed kinematic loops, i.e., for tree–structured robots, it is always possible to avoid all solvability issues by choosing the relative positions and velocities of the joints as state variables. However, the same does no longer hold true for multibody systems with closed kinematic loops. The kinematic loops lead to large and highly nonlinear algebraic loops that must be solved by tearing. It is not always possible to choose the residual equations and tearing variables of these loops such that all loop equations are linear in the variables that they need to be solved for.

In fact, there exist fairly simple mechanical devices with closed kinematic loops, for which it can be shown that there does not exist a minimal set of state variables, in which all solvability issues can be avoided. One way how such problems have been dealt with in the past is by selecting redundant state variables together with some switching mechanisms that decide when to use which variables during the simulation.

This is precisely, what Dymola now does on its own. Whenever there is a potential problem with a fixed selection of state variables, Dymola postpones the decision until run time [7.14]. For the same reason, newer versions of Dymola will be perfectly capable of simulating the pendulum problem in its original formulation without any help from the user. Dymola recognizes the potential solvability issue, postpones the selection of states, and toggles between x and y at run time as needed.

We shall deal with switching models in Chapter 9 of this book. In Chapter 8, we shall look at these problems from yet another angle.

7.9 Summary

In this chapter, we have presented a number of interlinked algorithms that can be used to convert even higher–index DAE systems to ODE form.

The most central among these algorithms is the algorithm by Tarjan, an algorithm based on graph theory to partially sort a DAE system both horizontally and vertically. The algorithm also finds minimal subsets of algebraically coupled equation systems that need to be solved simultaneously. Although the algorithm is based on graph theory, it can be easily implemented algebraically using linked lists. The algorithm furthermore discovers constraint equations, i.e., can be used to detect higher–index problems.

If a higher–index problem has been detected, the algorithm by Pantelides can be employed to reduce the perturbation index, until all structural singularities have been resolved.

A heuristic procedure has been presented that allows to find suitable tearing variables for the algebraically coupled subsystems.

The algorithms presented in this chapter are similar to those that have been implemented in the model compiler of *Dymola* [7.8, 7.9], an object–oriented physical system modeling and simulation environment.

The algorithms are highly computationally efficient and well tested. Dymola is capable of converting DAE systems consisting of tens of thousands of equations to ODE form within seconds on a modern PC, while applying these algorithms.

7.10 References

[7.1] Kathryn E. Brenan, Stephen L. Campbell, and Linda R. Petzold. *Numerical Solution of Initial–Value Problems in Differential–Algebraic Equations.* North–Holland, New York, 1989. 256p.

[7.2] Pawel Bujakiewicz. *Maximum Weighted Matching for High Index Differential Algebraic Equations.* PhD thesis, Delft Institute of Technology, The Netherlands, 1995.

[7.3] Stephen L. Campbell and C. William Gear. The Index of General Nonlinear DAEs. *Numerische Mathematik,* 72:173–196, 1995.

[7.4] François E. Cellier and Hilding Elmqvist. Automated Formula Manipulation Supports Object–oriented Continuous System Modeling. *IEEE Control Systems,* 13(2):28–38, 1993.

[7.5] François E. Cellier. *Continuous System Modeling*. Springer Verlag, New York, 1991. 755p.

[7.6] Iain S. Duff, Albert M. Erisman, and John K. Reid. *Direct Methods for Sparse Matrices*. Oxford University Press, Oxford, United Kingdom, 1986. 341p.

[7.7] Hilding Elmqvist and Martin Otter. Methods for Tearing Systems of Equations in Object–oriented Modeling. In *Proceedings European Simulation Multiconference*, pages 326–332, Barcelona, Spain, 1994.

[7.8] Hilding Elmqvist. *A Structured Model Language for Large Continuous Systems*. PhD thesis, Dept. of Automatic Control, Lund Institute of Technology, Lund, Sweden, 1978.

[7.9] Hilding Elmqvist. *Dymola — Dynamic Modeling Language, User's Manual*. DynaSim AB, Research Park Ideon, Lund, Sweden, 2004.

[7.10] Ernst Hairer, Christian Lubich, and Michel Roche. *The Numerical Solution of Differential–Algebraic Systems by Runge-Kutta Methods*. Springer–Verlag, Berlin, Germany, 1989. 139p.

[7.11] Johann Joss. *Algorithmisches Differenzieren*. PhD thesis, Diss ETH 5757, Swiss Federal Institute of Technology, Zürich, Switzerland, 1976. 69p.

[7.12] Gabriel Kron. *Diakoptics: The Piecewise Solution of Large–Scale Systems*. Macdonald Publishing, London, United Kingdom, 1963. 166p.

[7.13] Richard S. H. Mah. *Chemical Process Structures and Information Flows*. Butterworth Publishing, London, United Kingdom, 1990. 500p.

[7.14] Sven Erik Mattsson, Hans Olsson, and Hilding Elmqvist. Dynamic Selection of States in Dymola. In *Proceedings Modelica Workshop*, pages 61–67, Lund, Sweden, 2000.

[7.15] Sven Erik Mattsson and Gustaf Söderlind. Index Reduction in Differential–Algebraic Equations Using Dummy Derivatives. *SIAM Journal on Scientific Computing*, 14(3):677–692, 1993.

[7.16] Martin Otter, Hilding Elmqvist, and François E. Cellier. Modeling of Multibody Systems with the Object–Oriented Modeling Language Dymola. *J. Nonlinear Dynamics*, 9(1):91–112, 1996.

[7.17] Martin Otter, Hilding Elmqvist, and François E. Cellier. 'Relaxing' – A Symbolic Sparse Matrix Method Exploiting the Model Structure in Generating Efficient Simulation Code. In *Proceedings Symposium on Modeling, Analysis, and Simulation, CESA'96, IMACS Multi-Conference on Computational Engineering in Systems Applications*, volume 1, pages 1–12, Lille, France, 1996.

[7.18] Martin Otter and Clemens Schlegel. Symbolic generation of efficient simulation codes for robots. In *Proceedings Second European Simulation Multi-Conference*, pages 119–122, Nice, France, 1988.

[7.19] Martin Otter. *Objektorientierte Modellierung mechatronischer Systeme am Beispiel geregelter Roboter*. PhD thesis, Dept. of Mech. Engr., Ruhr–University Bochum, Germany, 1994.

[7.20] Constantinos Pantelides.   The Consistent Initialization of of Differential–Algebraic Systems. *SIAM Journal of Scientific and Statistical Computing*, 9(2):213–231, 1988.

[7.21] Robert Tarjan. Depth–first search and linear graph algorithms. *SIAM Journal of Computation*, 1(2):146–160, 1972.

7.11   Homework Problems

[H7.1] Electrical Circuit, Horizontal and Vertical Sorting

Given the electrical circuit shown in Fig.H7.1a.

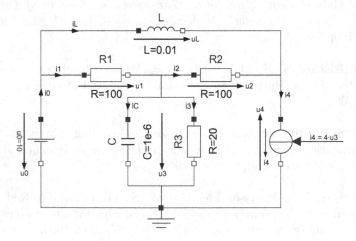

FIGURE H7.1a. Electrical circuit.

The circuit contains a constant voltage source, u_0, and a dependent current source, i_4, that depends on the voltage across the capacitor, C, and the resistor, R_3.

Write down the element equations for the seven circuit elements. Since the voltage u_3 is common to two circuit elements, these equations contain 13 rather than 14 unknowns. Add the voltage equations for the three meshes and the current equations for three of the four nodes. One current equation is redundant. Usually, the current equation for the ground node is therefore omitted.

In this way, you end up with 13 equations in the 13 unknowns. Draw the structure digraph of the DAE system, and apply the Tarjan algorithm to sort the equations both horizontally and vertically. Write down the causal equations, i.e., the resulting ODE system.

Simulate the ODE system across 50 μsec using RKF4/5 with zero initial conditions on both the capacitor and the inductor.

Plot the voltage u_3 and the current i_C on two separate subplots as functions of time.

[H7.2] Horizontal and Vertical Sorting, Newton Iteration

Given the following model in three nonlinear equations and three unknowns:

$$\mathcal{F}_1(x_1, x_3) = 0.0 \qquad (H7.2a)$$
$$\mathcal{F}_2(x_2) = 0.0 \qquad (H7.2b)$$
$$\mathcal{F}_3(x_1, x_2) = 0.0 \qquad (H7.2c)$$

Write down the structure incidence matrix, $\mathbf{S}$, of this nonlinear model.

Draw the structure digraph, and sort the equations both horizontally and vertically using the Tarjan algorithm.

Write down the causal equations and their structure incidence matrix, $\hat{\mathbf{S}}$, which should now be in lower–triangular form.

Find two permutation matrices, $\mathbf{P}$ and $\mathbf{Q}$, such that:

$$\hat{\mathbf{S}} = \mathbf{P} \cdot \mathbf{S} \cdot \mathbf{Q} \qquad (H7.2d)$$

A *permutation matrix* is a matrix, in which every row and column contains exactly one element with a value of 1, whereas all other elements have values of 0.

As the structure incidence matrix, $\hat{\mathbf{S}}$, is in lower–triangular form, we can set the simulation up by specifying three Newton iterations in one variable each, rather than one Newton iteration in three variables. This is much more economical.

Set up the Newton iterations by introducing symbolic functions denoting the Hessians.

[H7.3] Hydraulic System, Algebraic Differentiation

Given the hydraulic system shown in Fig.H7.3a.

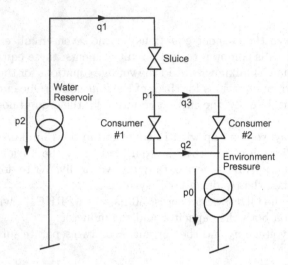

FIGURE H7.3a. Hydraulic system.

A water reservoir generates a pressure of p_2. A sluice reduces the pressure to p_1, which is the water pressure that the consumers see. p_0 is the pressure of the environment, i.e., the air pressure.

The sluice and the consumers can be represented by nonlinear turbulent resistance elements. The turbulent hydraulic resistance characteristic is shown in Fig.H7.3b.

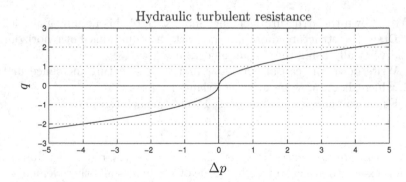

FIGURE H7.3b. Turbulent hydraulic resistance.

It shows the relationship between the pressure drop, Δp, and the flow rate, q. Mathematically, the relationship can be described by the formula:

$$q = k \cdot \text{sign}(\Delta p) \cdot \sqrt{|\Delta p|} \qquad (H7.3a)$$

or if the inverse computational causality is required:

$$\Delta p = \frac{1}{k} \cdot \text{sign}(q) \cdot q^2 \qquad (H7.3b)$$

Write down the nonlinear equations describing this system. You need six equations in the six unknowns p_0, p_1, p_2, q_1, q_2, and q_3.

Draw the structure digraph and isolate the nonlinear algebraic loop. You'll find an algebraic loop in four equations and four unknowns.

We shall first solve this equation system directly, i.e., without tearing. Set up a vector zero function, $\mathcal{F}(\mathbf{x})$ of length four, where the vector $\mathbf{x}$ stands for the four unknowns of the algebraic equation system. Find a symbolic expression for the Hessian, $\mathcal{H}(\mathbf{x})$, which is a matrix of dimensions 4×4. Write down the linear equation system that needs to be solved once per iteration step of the Newton iteration.

We now repeat the problem, this time with a tearing approach. Choose an appropriate tearing variable and residual equation, and causalize the equation system. Set up an appropriate scalar zero function in the tearing variable, and set up the Newton iteration such that it iterates over all equations, yet only uses the tearing variable as an iteration variable. Find a symbolic expression for the Hessian, which is now also a scalar. Use algebraic differentiation to compute the Hessian. Since the Newton iteration is scalar, you can come up with a closed–form symbolic expression for the next iteration of the tearing variable.

[H7.4] Linear System, Newton Iteration

Given the linear equation system:

$$\mathbf{A} \cdot \mathbf{x} = \mathbf{b} \tag{H7.4a}$$

$\mathbf{A}$ is assumed to be a nonsingular square matrix.

We wish to solve for the unknown vector $\mathbf{x}$ by Newton iteration. Set up the Newton iteration using symbolic expressions for the Jacobian and the Hessian. Prove that the Newton iteration indeed converges to the correct solution within a single step for arbitrary initial conditions.

[H7.5] Electrical Circuit, Tearing

Given the electrical circuit shown in Fig.H7.5a.

We wish to find a symbolic expression for the current i_3 as a function of the input voltage u_0 and the five resistance values.

Write down all the equations governing this circuit. Draw the structure digraph. You end up with an algebraic equation system in 10 equations and 10 unknowns. Use the heuristic procedure presented in this chapter to find appropriate tearing variables. You'll need two of them.

Use the substitution technique to come up with two symbolic expressions in the two tearing variables. These can be solved symbolically by matrix inversion. If one of the tearing variables was the current i_3, you are done. Otherwise, find a symbolic expression for i_3 in function of the two tearing variables, and substitute the previously found expressions once more into that new expression.

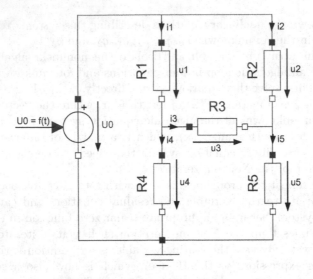

FIGURE H7.5a. Electrical resistance circuit.

[H7.6] Electrical Circuit, Relaxation

We wish to solve Problem [H7.5] once more, but this time using the relaxation algorithm.

Using the tearing structure found in Problem [H7.5], write the 10 equations in 10 unknowns in a matrix–vector form, such that you obtain only two non–zero elements above the diagonal of the matrix.

Apply symbolic Gaussian elimination without pivoting to this system to come up with a sequence of expressions to compute the tearing variables. At the end, use substitution to reduce this sequence of symbolic expressions to two expressions for the two tearing variables.

[H7.7] Electrical Circuit, Structural Singularity

Given the circuit shown in Fig.H7.7a containing three sinusoidal current sources.

Write down the complete set of equations describing this circuit. Draw the structure digraph and begin causalizing the equations. Determine a constraint equation.

Apply the Pantelides algorithm to reduce the perturbation index to 1. Then apply the tearing algorithm with substitution to bring the perturbation index down to 0.

Write down the structure incidence matrices of the index–1 DAE and the index–0 ODE systems, and show that they are in BLT form, and in LT form, respectively.

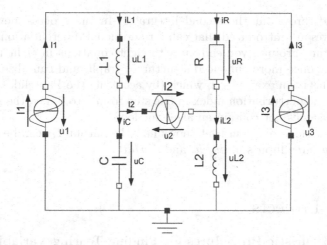

FIGURE H7.7a. Electrical structurally singular circuit.

[H7.8] Chemical Reactions, Pantelides Algorithm

The following set of DAEs:

$$\frac{dC}{dt} = K_1(C_0 - C) - R \tag{II7.8a}$$

$$\frac{dT}{dt} = K_1(T_0 - T) + K_2 R - K_3(T - T_C) \tag{H7.8b}$$

$$0 = R - K_3 \exp\left(\frac{-K_4}{T}\right) C \tag{H7.8c}$$

$$0 = C - u \tag{H7.8d}$$

describes a chemical isomerization reaction. C is the reactant concentration, T is the reactant temperature, and R is the reactant rate per unit volume. C_0 is the feed reactant concentration, and T_0 is the feed reactant temperature. u is the desired concentration, and T_C is the control temperature that we need to produce u. We want to turn the problem around (inverse model control) and determine the necessary control temperature T_C as a function of the desired concentration u. Thus, u will be an input to our model, and T_C is the output. The problem formulation was taken right out of [7.1].

Draw the structure digraph. You shall notice at once that one of the equations, Eq.(H7.8d), has no connections to it. Thus, it is a constraint equation that needs to be differentiated, while an integrator associated with the constraint equation needs to be thrown out.

We now have five equations in five unknowns. Draw the enhanced structure digraph, and start causalizing the equations. You shall notice that a second constraint equation appears. Hence the original DAE system had been an index–3 DAE system. Differentiate that constraint equation as well, and throw out the second integrator. In the process, new pseudo–derivatives are introduced that call for additional differentiations.

This time around, you end up with eight equations in eight unknowns. Draw the once more enhanced structure digraph, and causalize the equations. This is an example, in which (by accident) the Pantelides algorithm reduces the perturbation index in one step from 2 to 0, i.e., the final set of equations does not contain an algebraic loop.

Draw a block diagram that shows how the output T_C can be computed from the three inputs u, du/dt, and d^2u/dt^2.

7.12 Projects

[P7.1] Heuristic Procedures for Finding Tearing Variables

Study alternate strategies for finding small sets of tearing variables and residual equations. As the size of a DAE system generated by an object–oriented physical system modeling tool, such as Dymola [7.8, 7.9], can be very large, often containing thousands if not tens of thousands of equations, the computational efficiency of the heuristic procedure is very important.

[P7.2] Computation of Inverse Hessian

In Chapter 6, we have discussed approaches to numerically approximate the Hessian matrix. In this chapter, we have looked at an alternate approach making use of algebraic differentiation.

Study under what conditions it is more economical to approximate the Hessian numerically, and when a symbolic computation using algebraic differentiation should be used.

[P7.3] Solution of Linear Equation Systems

We have presented two different approaches to dealing with the solution of linear equation systems. On the one hand, we have presented a tearing approach, on the other, we have looked at a relaxation technique. Both techniques can be interpreted as symbolic sparse matrix algorithms. They have furthermore much in common. It was shown that the problem of finding small sets of tearing variables is identical to that of finding a small number of non–zero elements above the diagonal of the matrix in the relaxation approach.

Although we have shown by means of a few examples that the remaining linear systems in the tearing variables can be solved by substitution, this technique is not recommendable, as it invariably leads to an explosion in the size of the formulae. An alternate technique was also presented. It may make more sense to iterate over the entire set of equations, while using a Newton iteration on the tearing variables only.

In the case of linear systems, this requires an iteration rather than a closed–form solution, but the iteration may be acceptable as it converges in a single step if the Jacobian and Hessian are computed exactly, e.g. by means of algebraic differentiation.

The relaxation approach, on the other hand, leads to a closed–form solution of the linear equation system without requiring substitution. Hence this approach may be preferable at times.

Study under which conditions Newton iteration of a linear system is more economical, and when a relaxation approach may be cheaper.

7.13   Research

[R7.1] Pantelides and Small Equation Systems

In the modeling of multi–body systems (MBS), extensive research has focused on the generation of small sets of simulation equations. If the state variables are chosen carelessly in modeling a tree–structured robot, the number of simulation equations grows with the fourth power of the number of degrees of freedom (i.e., the number of articulations) of the robot. Yet, it is possible to choose the state variables such that the number of simulation equations grows only linearly in the number of articulations. Algorithms that behave in this fashion are called *order–n* algorithms in the literature [7.18, 7.19].

For this reason, the MBS library of Dymola [7.8, 7.9], which was developed by Martin Otter, does not make use of the Pantelides algorithm to resolve structural singularities. Instead, the model equations are formulated such that the structural singularities are resolved manually already at the time of the model formulation.

This places a heavy burden on the modeler. It would be better if the Pantelides algorithm could be made smart enough so that it would select the integrators to be thrown out such that an equation system is generated that is as small as possible.

Furthermore, the approach described by Otter only works in the case of tree–structured robots. If the MBS contains kinematic loops, the approach needs to be modified.

Study ways to automate the generation of efficient simulation code when using the Pantelides algorithm for index reduction.

[R7.2] Symbolic Model Compilation and Run–Time Errors

One of the biggest drawbacks of heavy symbolic preprocessing of the model equations in the generation of efficiently executable simulation code lies in the problem of tracing back run–time exceptions to original model equations.

When compiling a Dymola [7.8, 7.9] object–oriented model of a physical system into explicit ODE form, it happens frequently that the user receives an error message at the end of the model compilation of the type: "There are 3724 equations in 3725 unknowns." Of course, such a model cannot be simulated.

Unfortunately, it may be quite difficult to trace the error message back to the original model. Usually, a connection has been omitted somewhere. Yet, the simulation code no longer contains the information, where the error might be located.

When the equations are made causal, the compiler will tell the user, which is the variable, for which no equation was left over. However, that information may be quite arbitrary.

Similarly, when the simulation dies with a division by zero, it may no longer be easy for the user to recognize, which equation was responsible for the problem, as the error message will point at the simulation code, not at the original model equations. By that time, the equations may have changed their appearance so drastically that they have become unrecognizable.

Furthermore, many of the equations in the final model were not even explicitly present among the original model equations. They were automatically generated from the topological connections among submodels.

Study how the algorithms presented in this chapter can be enhanced so that they preserve as much information as possible about the original model equations for the purpose of presenting the user with error messages in terms that he or she can relate to.

8

Differential Algebraic Equation Solvers

Preview

In the previous chapter, we have discussed symbolic algorithms for converting implicit and even higher–index DAE systems to explicit ODE form. In this chapter, we shall look at these very same problems once more from a different angle. Rather than converting implicit DAEs to explicit ODE form, we shall try to solve the DAE systems directly. Solvers that are capable of dealing with implicit DAE descriptions directly have been coined *differential algebraic equation solvers* or DAE solvers. They are the focus point of this chapter.

8.1 Introduction

Let us look once more at the homework problem [H6.2]. In that problem, we simulated a parabolic PDE in one space dimension with a nonlinear boundary condition due to radiation. Because of the nonlinear boundary condition, we required one Newton iteration in a single unknown, T_{21}, per function evaluation. However, since the ODE problem after conversion of the PDE problem using the MOL approach is stiff, we also must employ an implicit integration algorithm, such as a BDF method. Consequently, we require a second Newton iteration over many variables once every integration step. Finally, if the simulation is to be error–controlled, we may need to reject some of the integration steps after the two Newton iterations converged, in order to repeat the step with a reduced step size. The three simulation loops are illustrated in Fig.8.1.

How accurately should we perform all these iterations? Clearly, if the relative error requested for the numerical integration is to be met by the outermost loop, then the internal loops must be computed at least as accurately. On the other hand, if e.g. the iteration of the integration algorithm is still far away from convergence, why should we perform the internal iteration within the individual function evaluation very accurately already?

Clearly, the different iterations and tolerances are closely interrelated. It seems awkward that we should have to keep track of different iterations and different error tolerances that are all part of one and the same process. Maybe, we should take a step back and reconsider all these issues in the

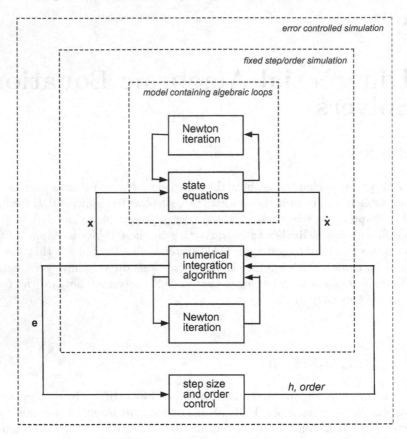

FIGURE 8.1. The three simulation loops.

new light of their seeming complexity to ascertain whether this complexity
is truly necessary.

We may start by asking ourselves, whether algebraic loops, or any other
numerical processes that require iterations, do really exist in this physical
world of ours. Isn't the physical world truly *causal*, i.e., isn't it true that
each event has one or several causes, and that a strictly sequential ordering
is possible between causes and effects? Don't iterations defy the principle
of strict causality?

Mutual causal dependencies do indeed exist in physics and are rather
common. The relationship between voltage and current in a resistor is non–
causal. It is not true that the potential difference at the two ends of the
resistor makes current flow, or that the current flowing through the resistor causes a voltage drop. These are simply two different facets of one and
the same physical phenomenon. Yet "causal loops" do not truly exist in
the physical world. If we place two resistors in series, this will create an
algebraic loop in our model. Yet, physics doesn't understand the concept

of a loop. The idea of a loop implies a sequence of execution, i.e., a causes b, which in turn causes c, which is responsible for a. Physics doesn't understand the concept of a "sequence of execution." Physics is by its very nature completely non–causal. All phenomena observed are byproducts of the big balance equations that we call the conservation principles: conservation of energy, conservation of mass, and conservation of momentum.

If "causal loops" show up in our models, they are artifactual. They are byproducts of the way in which we are dealing with the equations. Our way of thinking is strictly cause–effect oriented, and this is also how we have built our digital computers. We try to turn everything into cause–effect relationships. Sometimes, this is not possible. Causal loops, and the need for iteration, are our way of expressing this problem. Clearly, there does not exist a natural (physical) way of looking at causal loops.

Let us go back right to the foundations of continuous system simulation, and ask ourselves where the so–called state–space description of a physical system came from. It originated with the desire to separate the process of *modeling* (in a simple–minded way of looking at things, the process of generating a state–space model out of physical observations) from that of *simulation* (the process of translating the state–space model into trajectory behavior).

It all made sense. In the context of using explicit integration algorithms (all integration algorithms that were used in the early days were explicit algorithms), this separation comes quite naturally. The state–space model computes $\dot{\mathbf{x}}(t_k)$ out of $\mathbf{x}(t_k)$, and the integration algorithm in turn computes $\mathbf{x}(t_{k+1})$ out of $\mathbf{x}(t_k)$ and $\dot{\mathbf{x}}(t_k)$ — a meaningful and clean separation of duties.

By the time implicit integration algorithms were introduced, this separation was no longer as clean and crisp and beautiful. We suddenly had to deal with a causal loop, since the state–space model and the integration algorithm now operated on the same time instant, i.e., they had to co–operate to find *simultaneously* $\mathbf{x}(t_{k+1})$ and $\dot{\mathbf{x}}(t_{k+1})$. However, tradition had imprinted this separation so deeply into the brains of the simulation practitioners of that epoch that no–one bothered to raise the question whether this separation was still useful, or whether it might not even be detrimental to our task.

Let us check what happens if we let go of the constraint that state–space models have to be formulated such that they compute the state derivatives explicitly. Instead, we are going to use the implicit model:

$$\mathbf{f}(\mathbf{x}, \dot{\mathbf{x}}, \mathbf{u}, t) = 0.0 \tag{8.1}$$

Let us apply the BDF3 algorithm:

$$\mathbf{x}_{k+1} = \frac{6}{11} h \cdot \dot{\mathbf{x}}_{k+1} + \frac{18}{11} \mathbf{x}_k - \frac{9}{11} \mathbf{x}_{k-1} + \frac{2}{11} \mathbf{x}_{k-2} \tag{8.2}$$

to the model of Eq.(8.1). We can solve Eq.(8.2) for $\dot{\mathbf{x}}_{k+1}$:

$$\dot{\mathbf{x}}_{k+1} = \frac{1}{h} \left[\frac{11}{6} \cdot \mathbf{x}_{k+1} - 3\mathbf{x}_k + \frac{3}{2}\mathbf{x}_{k-1} - \frac{1}{3}\mathbf{x}_{k-2} \right] \tag{8.3}$$

and plug Eq.(8.3) into Eq.(8.1). We obtain a nonlinear vector equation $\mathbf{f}$ in the unknown parameter vector $\mathbf{x}_{k+1}$:

$$\mathcal{F}(\mathbf{x}_{k+1}) = 0.0 \tag{8.4}$$

which can be solved *directly* by Newton iteration.

In this new formulation, the distinction between iterating on the implicit integration step and iterating on nonlinear function evaluations has vanished. It becomes quite evident that these were not two separate processes, but only two different facets of one and the same process. It is this –very fruitful– idea, which had first been proposed by Bill Gear in 1971 in a frequently cited paper [8.13], that we shall pursue in this chapter in more detail.

8.2   Multi–step Formulae

We have learnt that implicit integration algorithms are most useful for dealing with stiff systems. Consequently, a DAE formulation in place of the former ODE formulation will be particularly fruitful in the context of the simulation of stiff systems, and it should contain a numerical formula that has been designed for dealing with stiff systems, such as a BDF algorithm.

Since we ultimately want to solve for $\mathbf{x}_{k+1}$, we eliminate $\dot{\mathbf{x}}_{k+1}$ from the implicit state–space model of Eq.(8.1), and this means that the integration formula will now have to be solved for $\mathbf{f}_{k+1}$ rather than for $\mathbf{x}_{k+1}$ as in the ODE case. Thus, we are now looking at *numerical differentiation formulae* rather than *numerical integration formulae*.

We have already seen what a DAE implementation of a BDF algorithm could look like. In order to assess the validity of this approach, we should ask ourselves what the stability and accuracy properties of such a BDF implementation are.

Let us start with a discussion of stability properties. The linear version of Eq.(8.1) can now be written as:

$$\mathbf{A} \cdot \mathbf{x} + \mathbf{B} \cdot \dot{\mathbf{x}} = 0.0 \tag{8.5}$$

Thus, our standard linear test problem has now two matrices, $\mathbf{A}$ and $\mathbf{B}$. Plugging Eq.(8.2) into Eq.(8.5), we obtain:

$$\mathbf{A} \cdot \mathbf{x}_{k+1} + \frac{11\mathbf{B}}{6h} \cdot \left(\mathbf{x}_{k+1} - \frac{18}{11}\mathbf{x}_k + \frac{9}{11}\mathbf{x}_{k-1} - \frac{2}{11}\mathbf{x}_{k-2} \right) = 0.0 \tag{8.6}$$

or:

$$\left(-\mathbf{B} - \frac{6\mathbf{A}h}{11}\right)\mathbf{x}_{k+1} = -\frac{18\mathbf{B}}{11}\mathbf{x}_k + \frac{9\mathbf{B}}{11}\mathbf{x}_{k-1} - \frac{2\mathbf{B}}{11}\mathbf{x}_{k-2} \qquad (8.7)$$

We already know that Newton iteration does not affect the stability domain of a method. Thus, we can solve Eq.(8.7) for $\mathbf{x}_{k+1}$ by use of matrix inversion without modifying the stability domain of the method. We find:

$$\mathbf{x}_{k+1} = \left(-\mathbf{B} - \frac{6\mathbf{A}h}{11}\right)^{-1} \cdot \left(-\frac{18\mathbf{B}}{11}\mathbf{x}_k + \frac{9\mathbf{B}}{11}\mathbf{x}_{k-1} - \frac{2\mathbf{B}}{11}\mathbf{x}_{k-2}\right) \qquad (8.8)$$

Let us first discuss the simplest case:

$$\mathbf{B} = -\mathbf{I}^{(n)} \qquad (8.9)$$

In this case, Eq.(8.5) degenerates to the explicit linear test problem:

$$\dot{\mathbf{x}} = \mathbf{A} \cdot \mathbf{x} \qquad (8.10)$$

and Eq.(8.8) becomes:

$$\mathbf{x}_{k+1} = \left(\mathbf{I}^{(n)} - \frac{6\mathbf{A}h}{11}\right)^{-1} \cdot \left(\frac{18}{11}\mathbf{x}_k - \frac{9}{11}\mathbf{x}_{k-1} + \frac{2}{11}\mathbf{x}_{k-2}\right) \qquad (8.11)$$

which is identical to the equation that had been used in Chapter 4 to determine the stability domain of BDF3. Consequently, at least in this simple situation, the stability domain is not at all affected by the DAE formulation.

Let us assume next that $\mathbf{B}$ is a non–singular matrix. In this case, Eq.(8.5) can be rewritten as:

$$\dot{\mathbf{x}} = -\mathbf{B}^{-1} \cdot \mathbf{A} \cdot \mathbf{x} \qquad (8.12)$$

Will the inversion of $\mathbf{B}$ have an effect on the stability domain? We can determine the stability domain of the method in the following way. We choose the eigenvalues of $-\mathbf{B}^{-1} \cdot \mathbf{A}$ along the unit circle of the complex plane, then apply the so found $\mathbf{A}$– and $\mathbf{B}$–matrices to the $\mathbf{F}$–matrix:

$$\mathbf{F} = \begin{pmatrix} \mathbf{O}^{(n)} & \mathbf{I}^{(n)} & \mathbf{O}^{(n)} \\ \mathbf{O}^{(n)} & \mathbf{O}^{(n)} & \mathbf{I}^{(n)} \\ \frac{2}{11}\left(\mathbf{B} + \frac{6}{11}\mathbf{A}h\right)^{-1}\mathbf{B} & -\frac{9}{11}\left(\mathbf{B} + \frac{6}{11}\mathbf{A}h\right)^{-1}\mathbf{B} & \frac{18}{11}\left(\mathbf{B} + \frac{6}{11}\mathbf{A}h\right)^{-1}\mathbf{B} \end{pmatrix} \qquad (8.13)$$

and determine h such that the dominant eigenvalues of $\mathbf{F}$ are on the unit circle. We arbitrarily chose several different non–singular $\mathbf{B}$–matrices of dimensions 2×2, and computed the corresponding $\mathbf{A}$–matrices using:

$$\mathbf{A} = -\mathbf{B} \cdot \begin{pmatrix} 0 & 1 \\ -1 & 2\cos(\alpha) \end{pmatrix} \qquad (8.14)$$

We then plugged these matrices into Eq.(8.13), and computed the stability domains. It turned out that, in every single case, the stability domain was exactly the same as in the ODE case. This is generally true. Non–singular **B**–matrices do not influence the numerical stability properties of the method in any way.

These are good news indeed. Notice that we just solved the *algebraic loop problem* once and for all — at least in the context of stiff system simulation. There is no longer any need to apply a Newton iteration to algebraic loops that form part of the state–space model, and then apply a separate Newton iteration around the first one for bringing the implicit integration scheme to convergence. The two iterations have turned out to be two different facets of one and the same process.

Let us now look at the case where **B** is singular. Let the rank of **B** be $r < n$. We can then perform a singular value decomposition on the matrix **B**, as indicated in Fig.8.2.

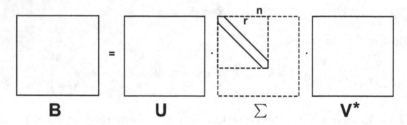

FIGURE 8.2. Singular value decomposition.

where **U** and **V** are two unitary matrices (each row vector is orthogonal to all other row vectors and of length 1.0, the same applies to all column vectors), and **Σ** is a diagonal matrix. Since both **U** and **V** have full rank, i.e.:

$$\text{rank}(\mathbf{U}) = \text{rank}(\mathbf{V}) = n \qquad (8.15)$$

the **Σ**–matrix has the same rank as **B**, thus:

$$\text{rank}(\mathbf{\Sigma}) = \text{rank}(\mathbf{B}) = r \qquad (8.16)$$

V* denotes the Hermitian transpose (the conjugate complex transpose) of **V**. Since:

$$\mathbf{B} = \mathbf{U} \cdot \mathbf{\Sigma} \cdot \mathbf{V}^* \qquad (8.17)$$

Eq.(8.5) becomes:

$$\mathbf{A} \cdot \mathbf{x} + \mathbf{U} \cdot \mathbf{\Sigma} \cdot \mathbf{V}^* \cdot \dot{\mathbf{x}} = 0.0 \qquad (8.18)$$

Since the inverse of a unitary matrix is its Hermitian transpose, we can rewrite Eq.(8.18) as:

$$\mathbf{U}^* \cdot \mathbf{A} \cdot \mathbf{x} + \mathbf{\Sigma} \cdot \mathbf{V}^* \cdot \dot{\mathbf{x}} = 0.0 \qquad (8.19)$$

Let us now perform a variable substitution:

$$\mathbf{z} = \mathbf{V}^* \cdot \mathbf{x} \qquad (8.20)$$

Plugging Eq.(8.20) into Eq.(8.19), we obtain:

$$\mathbf{U}^* \cdot \mathbf{A} \cdot \mathbf{V} \cdot \mathbf{z} + \mathbf{\Sigma} \cdot \dot{\mathbf{z}} = 0.0 \qquad (8.21)$$

or:

$$\tilde{\mathbf{A}} \cdot \mathbf{z} + \mathbf{\Sigma} \cdot \dot{\mathbf{z}} = 0.0 \qquad (8.22)$$

A graphical representation of Eq.(8.22) is shown in Fig.8.3.

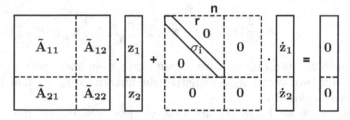

FIGURE 8.3. Linear test problem after variable substitution.

Eq.(8.22) can be decomposed into:

$$\tilde{\mathbf{A}}_{11} \cdot \mathbf{z}_1 + \tilde{\mathbf{A}}_{12} \cdot \mathbf{z}_2 + \mathbf{\Sigma}_{11} \cdot \dot{\mathbf{z}}_1 = 0.0 \qquad (8.23a)$$
$$\tilde{\mathbf{A}}_{21} \cdot \mathbf{z}_1 + \tilde{\mathbf{A}}_{22} \cdot \mathbf{z}_2 = 0.0 \qquad (8.23b)$$

If $\tilde{\mathbf{A}}_{22}$ is non–singular, we can solve Eq.(8.23b) for $\mathbf{z}_2$:

$$\mathbf{z}_2 = -\tilde{\mathbf{A}}_{22}^{-1} \cdot \tilde{\mathbf{A}}_{21} \cdot \mathbf{z}_1 \qquad (8.24)$$

and plugging Eq.(8.24) into Eq.(8.23a), we obtain:

$$\dot{\mathbf{z}}_1 = \mathbf{\Sigma}_{11}^{-1} \cdot \left(\tilde{\mathbf{A}}_{12} \cdot \tilde{\mathbf{A}}_{22}^{-1} \cdot \tilde{\mathbf{A}}_{21} - \tilde{\mathbf{A}}_{11} \right) \cdot \mathbf{z}_1 \qquad (8.25)$$

In the new state vector, $\mathbf{z}$, it becomes evident that only a subset of its variables, namely the vector $\mathbf{z}_1$ of length r is described by means of differential equations. The remaining variables appear only algebraically in Eq.(8.22). This means that the system does, in reality, not contain n different state variables or energy storages, but only r. Thus, in the terminology introduced in the previous chapter: a singular $\mathbf{B}$–matrix corresponds to a *structurally singular model*, i.e., a *higher–index model*.

In the new state vector, $\mathbf{z}_1$, the situation is now the same as before, when $\mathbf{B}$ was assumed non–singular. Hence the stability domain is indeed not affected at all by the choice of the $\mathbf{B}$–matrix.

We need not worry about *accuracy*. Since the numerical differentiation formula is n^{th}-order accurate, and since we have full control over what happens during the Newton iteration, the DAE solver using an nth-order accurate BDF method must obviously as a whole be n^{th}-order accurate.

Until this point, we have focused our interest on DAE formulations of BDF algorithms. What happens if we decide to use other types of implicit multi–step techniques, such as the Adams–Moulton family of methods? Let us look at the case of AM3:

$$\mathbf{x}(t_{k+1}) \approx \mathbf{x}(t_k) + \frac{h}{12}\left(5\mathbf{f_{k+1}} + 8\mathbf{f_k} - \mathbf{f_{k-1}}\right) \tag{8.26}$$

We can turn this formula around, and obtain:

$$\dot{\mathbf{x}}(t_{k+1}) = \mathbf{f_{k+1}} \approx \frac{12}{5h}\left(\mathbf{x}(t_{k+1}) - \mathbf{x}(t_k)\right) - \frac{8}{5}\mathbf{f_k} + \frac{1}{5}\mathbf{f_{k-1}} \tag{8.27}$$

Plugging Eq.(8.27) into Eq.(8.1), we obtain again a nonlinear vector function in the unknown vector $\mathbf{x_{k+1}}$, whereas the quantities $\mathbf{x_k}$, $\mathbf{f_k}$, and $\mathbf{f_{k-1}}$ can be treated as known.

The problems with Eq.(8.27) are that we have eliminated the state derivatives from our system of equations, thus, we don't really know what values to use for $\mathbf{f_k}$ and $\mathbf{f_{k-1}}$. One way to overcome this difficulty is to solve two Newton iterations in sequence:

$$\mathcal{F}_1\left(\mathbf{x}(t_{k+1})\right) = \mathbf{f}\left(\mathbf{x}(t_{k+1}), \frac{12}{5h}\mathbf{x}(t_{k+1}) - \frac{12}{5h}\mathbf{x}(t_k) - \frac{8}{5}\mathbf{w}(t_k)\right.$$
$$\left. + \frac{1}{5}\mathbf{w}(t_{k-1}), \mathbf{u}(t_{k+1}), t_{k+1}\right) = 0.0 \tag{8.28a}$$

$$\mathcal{F}_2\left(\mathbf{w}(t_{k+1})\right) = \mathbf{f}\left(\mathbf{x}(t_{k+1}), \mathbf{w}(t_{k+1}), \mathbf{u}(t_{k+1}), t_{k+1}\right) = 0.0 \tag{8.28b}$$

Equation (8.28a) determines $\mathbf{x}(t_{k+1})$. In this iteration, $\mathbf{u}(t_{k+1})$, $\mathbf{x}(t_k)$, $\mathbf{w}(t_k)$, and $\mathbf{w}(t_{k-1})$ are assumed known. Equation (8.28b) then evaluates $\mathbf{w}(t_{k+1})$. During that iteration, $\mathbf{x}(t_{k+1})$ can be assumed known as well. Clearly, $\mathbf{w}$ is just another name for $\dot{\mathbf{x}}$.

The BDF case was special, since in the BDF formulae, the state derivative vector shows up only once. In that case, the DAE formulation becomes particularly simple and efficient to implement, and half of the variables (the state derivatives) can be eliminated from the set of variables to be computed. An integration formula that shares this property is called a *one–leg method*. In general, linear one–leg methods can be written for the ODE case of Eq.(2.1) as:

$$\frac{1}{h}\sum_{j=0}^{n}\alpha_j\mathbf{x_{k-j+1}} = \mathbf{f}\left(\sum_{j=0}^{n}\beta_j\mathbf{x_{k-j+1}}, \sum_{j=0}^{n}\beta_j\mathbf{u_{k-j+1}}, \sum_{j=0}^{n}\beta_jt_{k-j+1}\right) \tag{8.29}$$

The left–hand side of Eq.(8.29) represents the state derivative vector evaluated at the intermediate time $\sum_{j=0}^{n} \beta_j t_{k-j+1}$. Equation (8.29) is turned
into a numerical integration formula by solving the expression on the left–
hand side for x_{k+1} moving all other terms to the right–hand side.

Equation (8.29) naturally extends to the following DAE formulation:

$$\mathcal{F}(x_{k+1}) = f\left(\sum_{j=0}^{n} \beta_j x_{k-j+1}, \frac{1}{h}\sum_{j=0}^{n} \alpha_j x_{k-j+1}, \sum_{j=0}^{n} \beta_j u_{k-j+1}, \sum_{j=0}^{n} \beta_j t_{k-j+1}\right)$$

$$= 0.0 \tag{8.30}$$

Let us now discuss the (linear) stability properties of AM3 in its DAE
formulation. To this end, we plug Eqs.(8.28a–b) into Eq.(8.5). We still
want to assume $\mathbf{B}$ to be non–singular. We obtain:

$$x(t_{k+1}) \approx \left(\mathbf{B} + \frac{5}{12}\mathbf{A}h\right)^{-1} \cdot \left(\mathbf{B}x(t_k) + \frac{2}{3}\mathbf{B}h w(t_k)\right.$$

$$\left. - \frac{1}{12}\mathbf{B}h w(t_{k-1})\right) \tag{8.31a}$$

$$w(t_{k+1}) \approx -\mathbf{B}^{-1}\mathbf{A}x(t_{k+1}) \tag{8.31b}$$

and by letting:

$$z_k = \begin{pmatrix} x_k \\ w_{k-1} \\ w_k \end{pmatrix} \tag{8.32}$$

we obtain the following $\mathbf{F}$–matrix:

$$\mathbf{F} = \begin{pmatrix} \mathbf{F}_{11} & \mathbf{F}_{12} & \mathbf{F}_{13} \\ \mathbf{O}^{(n)} & \mathbf{O}^{(n)} & \mathbf{I}^{(n)} \\ \mathbf{F}_{31} & \mathbf{F}_{32} & \mathbf{F}_{33} \end{pmatrix} \tag{8.33}$$

where:

$$\mathbf{F}_{11} = \left(\mathbf{B} + \frac{5}{12}\mathbf{A}h\right)^{-1}\mathbf{B} \tag{8.34a}$$

$$\mathbf{F}_{12} = -\frac{1}{12}\left(\mathbf{B} + \frac{5}{12}\mathbf{A}h\right)^{-1}\mathbf{B}h \tag{8.34b}$$

$$\mathbf{F}_{13} = \frac{2}{3}\left(\mathbf{B} + \frac{5}{12}\mathbf{A}h\right)^{-1}\mathbf{B}h \tag{8.34c}$$

$$\mathbf{F}_{31} = -\mathbf{B}^{-1}\mathbf{A}\left(\mathbf{B} + \frac{5}{12}\mathbf{A}h\right)^{-1}\mathbf{B} \tag{8.34d}$$

$$\mathbf{F}_{32} = \frac{1}{12}\mathbf{B}^{-1}\mathbf{A}\left(\mathbf{B} + \frac{5}{12}\mathbf{A}h\right)^{-1}\mathbf{B}h \tag{8.34e}$$

$$\mathbf{F}_{33} = -\frac{2}{3}\mathbf{B}^{-1}\mathbf{A}\left(\mathbf{B} + \frac{5}{12}\mathbf{A}h\right)^{-1}\mathbf{B}h \tag{8.34f}$$

We select $\mathbf{B}$ arbitrarily as a non–singular 2×2 matrix, and compute $\mathbf{A}$ in accordance with Eq.(4.37). We checked with different selections of $\mathbf{B}$. The results were always the same as shown in Fig.4.2. The (linear) stability behavior of the method was not influenced by the DAE formulation. This may at first look like a surprising result, since the $\mathbf{F}$–matrix of Eq.(4.41) used to compute the stability domain shown in Fig.4.2 is a 4×4 matrix, whereas the new $\mathbf{F}$–matrix of Eq.(8.33) is a 6×6 matrix. However, $\mathbf{w_k}$ is linear in $\mathbf{x_k}$, thus $\mathbf{F}$ is singular. All we did was to add two more spurious eigenvalues at the origin. These eigenvalues will never influence the stability behavior. Eigenvalues located at the origin of the discrete system correspond to eigenvalues of the continuous system located at $-\infty$. Such eigenvalues are completely harmless.

We have discussed how linear implicit multi–step methods can be converted from an ODE formulation to a DAE formulation by solving the formulae for $\dot{\mathbf{x}}_{k+1}$ instead of for $\mathbf{x}_{k+1}$. This leads to two separate Newton iterations in n variables each, where n is the order of the system to be simulated. In the case of the one–leg methods, such as the BDF formulae, the DAE formulation becomes particularly simple, since the state derivative vector can be eliminated from the model, and one of the two Newton iterations becomes unnecessary.

How about explicit multi–step formulae? Let us look at AB3:

$$\mathbf{x}(t_{k+1}) \approx \mathbf{x}(t_k) + \frac{h}{12} \left(23\mathbf{f}_k - 16\mathbf{f}_{k-1} + 5\mathbf{f}_{k-2}\right) \qquad (8.35)$$

Turning Eq.(8.35) around, we obtain:

$$\dot{\mathbf{x}}(t_{k+1}) \approx \frac{12}{23h} \left(\mathbf{x}(t_{k+2}) - \mathbf{x}(t_{k+1})\right) + \frac{16}{23}\mathbf{f}_k - \frac{5}{23}\mathbf{f}_{k-1} \qquad (8.36)$$

Thus, *explicit numerical integration* formulae turn in the conversion to DAE form into *overimplicit numerical differentiation* formulae. Using such a formula will turn the entire simulation run into one giant iteration loop, which is certainly not justifiable.

However, this brings up another idea. We could search for overimplicit numerical integration schemes, which, until now, were quite useless, and turn those around for use in a DAE solver.

The third–order accurate overimplicit Adams formula is:

$$\mathbf{x}(t_{k+1}) \approx \mathbf{x}(t_k) + \frac{h}{12} \left(-\mathbf{f}(t_{k+2}) + 8\mathbf{f}(t_{k+1}) + 5\mathbf{f}(t_k)\right) \qquad (8.37)$$

Solving for the newest state derivative, we obtain:

$$\dot{\mathbf{x}}(t_{k+1}) \approx \frac{12}{h} \left(\mathbf{x}(t_{k-1}) - \mathbf{x}(t_k)\right) + 8\mathbf{f}(t_k) + 5\mathbf{f}(t_{k-1}) \qquad (8.38)$$

which is a third–order accurate explicit numerical differentiation formula.

Similarly, we can find a third–order accurate overimplicit BDF formula:

$$\mathbf{x}(t_{k+1}) \approx \frac{57}{26}\mathbf{x}(t_k) - \frac{21}{13}\mathbf{x}(t_{k-1}) + \frac{11}{26}\mathbf{x}(t_{k-2}) + \frac{6h}{26}\mathbf{f}(t_{k+2}) \qquad (8.39)$$

which leads to the explicit third–order accurate numerical differentiation formula:

$$\dot{\mathbf{x}}(t_{k+1}) \approx \frac{1}{6h}\left(26\mathbf{x}(t_k) - 57\mathbf{x}(t_{k-1}) + 42\mathbf{x}(t_{k-2}) - 11\mathbf{x}(t_{k-3})\right) \qquad (8.40)$$

Unfortunately, both formulae are unstable in the vicinity of the origin when plugged into the DAE. Explicit numerical differentiation is not a recommended procedure because of its poor stability properties and should therefore be avoided. If numerical differentiation is a necessity in a simulation model (e.g., if an input variable needs to be differentiated), we strongly suggest the use of an implicit numerical differentiation formula together with a DAE formulation for the overall simulation problem.

We have learnt that implicit multi–step formulae lend themselves splendidly for use in DAE solvers. To this end, they simply need to be turned around and solved for the newest value of the state derivative vector instead of the newest value of the state vector. Both AMi and BDFi formulae can be used in DAE solvers. However, the BDF formulae are more attractive since they, being one–legged formulae, allow to eliminate the state derivative vector from the simulation program altogether.

We remember that BDFi formulae are inefficient for use in non–stiff ODEs due to their poor accuracy properties. This was documented in Fig.6.12. The problem certainly hasn't vanished by reformulating the model in a DAE format. Thus, we might suspect that the AMi formulae will still work better than the BDFi formulae also in non–stiff DAE simulation. However, whether this is true or not will depend on the relative cost to be paid for the second Newton iteration. Let us ponder this question.

We shall rerun the wave equation example of Eqs.(6.53a–e), this time in DAE format, using once a BDF3 and once an AM3 algorithm. We computed the global accuracy of the two algorithms using the same step sizes as in Table 6.4. The results are tabulated in Table 8.1.

Figure 8.4 shows these results graphically.

Although the entries in Table 8.1 look *exactly* like the corresponding entries in Table 6.4 (after all, these *are* the same methods applied to the same problem), the graph of Fig.8.4 looks a little different from that of Fig.6.12 due to the need for a second Newton iteration in the case of the DAE formulation of AM3.

Of course in the given example, the Newton iterations converge in a single step since the problem is linear and the Jacobian has been computed accurately. Whereas the "economy" of the BDF3 algorithm in terms of the number of function evaluations required does not change between the ODE and DAE formulations, the AM3 algorithm has become more expensive

h	BDF3	AM3
0.1	garbage	unstable
0.05	garbage	unstable
0.02	garbage	unstable
0.01	garbage	unstable
0.005	0.9469e-2	0.8783e-8
0.002	0.1742e-6	0.2149e-8
0.001	0.4363e-7	0.2120e-8

TABLE 8.1. Comparison of accuracy of integration algorithms.

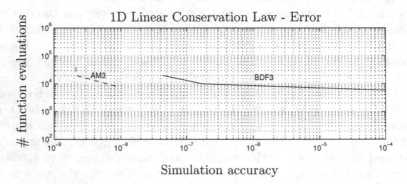

FIGURE 8.4. Cost–versus–accuracy plot for the 1D wave equation.

to compute in the DAE formulation due to the need for a second Newton iteration. The lesson to be learnt from this exercise: at least in the non–stiff case, it may well be worthwhile to convert the model first to ODE format, using the techniques described in the previous chapter of this book, before simulating it.

One problem we haven't discussed yet concerns the *initial conditions*. In the ODE case, it was sufficient for the user to specify initial values for the state vector $\mathbf{x}$ and for the input vector $\mathbf{u}$, and the state derivative vector could then be computed from the state–space model. In the DAE case, we don't have an explicit formula to compute the state derivative vector.

If we decide to use the BDF technique (or any other one–legged algorithm), we can eliminate the state derivative vector from the model altogether, and in this special case, we don't have a problem ... except during the startup period. Of course, we can use order control during the startup period, and then, we won't need the state derivative vector ever. Let us explain. The implicit differentiation formulae using the inverted BDF algorithms are:

$$\dot{\mathbf{x}}(t_{k+1}) \approx \frac{1}{h}\mathbf{x}(t_{k+1}) - \frac{1}{h}\mathbf{x}(t_k) \tag{8.41a}$$

$$\dot{\mathbf{x}}(t_{k+1}) \approx \frac{3}{2h}\mathbf{x}(t_{k+1}) - \frac{2}{h}\mathbf{x}(t_k) + \frac{1}{2h}\mathbf{x}(t_{k-1}) \tag{8.41b}$$

$$\dot{\mathbf{x}}(t_{k+1}) \approx \frac{11}{6h}\mathbf{x}(t_{k+1}) - \frac{3}{h}\mathbf{x}(t_k) + \frac{3}{2h}\mathbf{x}(t_{k-1}) - \frac{1}{3h}\mathbf{x}(t_{k-2}) \tag{8.41c}$$

where Eq.(8.41a) is first–order accurate, Eq.(8.41b) is second–order accurate, and Eq.(8.41c) is third–order accurate. If we wish to simulate the DAE model using a third–order accurate formula, we can use Eq.(8.41a) during the first step, then Eq.(8.41b) during the second, and from then on, we can use Eq.(8.41c). Plugging Eqs.(8.41a–c) into Eq.(8.1), we obtain:

$$\mathbf{f}\left(\mathbf{x}_1, \frac{1}{h}\mathbf{x}_1 - \frac{1}{h}\mathbf{x}_0, \mathbf{u}_1, t_1\right) = 0.0 \tag{8.42a}$$

$$\mathbf{f}\left(\mathbf{x}_2, \frac{3}{2h}\mathbf{x}_2 - \frac{2}{h}\mathbf{x}_1 + \frac{1}{2h}\mathbf{x}_0, \mathbf{u}_2, t_2\right) = 0.0 \tag{8.42b}$$

$$\mathbf{f}\left(\mathbf{x}_3, \frac{11}{6h}\mathbf{x}_3 - \frac{3}{h}\mathbf{x}_2 + \frac{3}{2h}\mathbf{x}_1 - \frac{1}{3h}\mathbf{x}_0, \mathbf{u}_3, t_3\right) = 0.0 \tag{8.42c}$$

$$\mathbf{f}\left(\mathbf{x}_4, \frac{11}{6h}\mathbf{x}_4 - \frac{3}{h}\mathbf{x}_3 + \frac{3}{2h}\mathbf{x}_2 - \frac{1}{3h}\mathbf{x}_1, \mathbf{u}_4, t_4\right) = 0.0 \tag{8.42d}$$

etc.

In each step, we perform one Newton iteration in the unknown state vector at the current time. In this way, the state derivative vector has been eliminated from the model once and for all.

Unfortunately using this approach, we are faced with the meanwhile well–known accuracy problems. We shall have to employ a very small step size initially in order to be able to meet our accuracy requirements. Moreover, the approach won't work in the case of the AMi algorithms. Those algorithms don't eliminate the state derivative vector (the $\mathbf{w}$–vector of Eqs.(8.28a–b)), and we need to find an estimate for $\mathbf{w}_0$ at time t_0 by means of Newton iteration. The problem here is that there is no guarantee that the Newton iteration will converge at all or will converge to the right solution if our initial guesses for the state derivative values are far off. Therefore, most DAE solvers on the market request that the user specify not only the initial values for the state vector, but also good initial guesses for the state derivative vector to be used as starting values for the first Newton iteration on Eq.(8.1) at time t_0.

How about using higher–order Runge–Kutta algorithms for startup? This may turn out to again be a smart idea, but we need to postpone the discussion of this approach until we have talked about the DAE format of the single–step algorithms.

Step–size control, order control, and the readout problem don't cause any difficulties beyond those that were already discussed in Chapter 4 of this text.

8.3   Single–step Formulae

In principle, DAE formulations of all single–step algorithms are straight-forward. For example, a DAE formulation of our standard explicit RK4 algorithm could be implemented in the following way:

$$\mathbf{f}\left(\mathbf{x_k}, \mathbf{k_1}, \mathbf{u}(t_k), t_k\right) = 0.0$$

$$\mathbf{f}\left(\mathbf{x_k} + \frac{h}{2}\mathbf{k_1}, \mathbf{k_2}, \mathbf{u}(t_k + \frac{h}{2}), t_k + \frac{h}{2}\right) = 0.0$$

$$\mathbf{f}\left(\mathbf{x_k} + \frac{h}{2}\mathbf{k_2}, \mathbf{k_3}, \mathbf{u}(t_k + \frac{h}{2}), t_k + \frac{h}{2}\right) = 0.0$$

$$\mathbf{f}\left(\mathbf{x_k} + h\mathbf{k_3}, \mathbf{k_4}, \mathbf{u}(t_k + h), t_k + h\right) = 0.0$$

$$\mathbf{x_{k+1}} = \mathbf{x_k} + \frac{h}{6}\left(\mathbf{k_1} + 2\mathbf{k_2} + 2\mathbf{k_3} + \mathbf{k_4}\right)$$

Thus, this is exactly the same formula that we were using in Chapter 3 of this text, except that each and every formerly explicit evaluation of the state derivative vector needs to be replaced by a Newton iteration.

In general, this is too costly. Let us take the example of BI3. BI3 contains one explicit RK3 step forward and approximately four RK3 steps backward due to the Newton iteration. Thus, BI3 contains altogether five RK3 steps, each requiring three function evaluations. Consequently, BI3 calls for 15 function evaluations per step. This was for the ODE formulation. However, in the DAE formulation, each of these function evaluations turns itself into a Newton iteration requiring approximately four function evaluations, thus, we are now looking at 60 function evaluations per step. If nothing else killed the efficiency of the BI algorithms, this certainly will.

None of the techniques discussed in Chapter 3 will lead to efficient DAE implementations. The techniques that are least affected by the DAE formulation are the *Richardson extrapolation techniques*. They won't become less efficient by the DAE formulation ... but they had been terribly inefficient already for the ODE case.

Is it hopeless then? Salvation comes from the *fully–implicit Runge–Kutta algorithms* [8.16]. Let us look at one type of these algorithms, namely the *Radau IIA algorithms*. They can be represented by the following Butcher tableaus:

$$
\begin{array}{c|cc}
1/3 & 5/12 & -1/12 \\
1 & 3/4 & 1/4 \\
\hline
x & 3/4 & 1/4
\end{array}
$$

$$
\begin{array}{c|ccc}
\frac{4-\sqrt{6}}{10} & \frac{88-7\sqrt{6}}{360} & \frac{296-169\sqrt{6}}{1800} & \frac{-2+3\sqrt{6}}{225} \\
\frac{4+\sqrt{6}}{10} & \frac{296+169\sqrt{6}}{1800} & \frac{88+7\sqrt{6}}{360} & \frac{-2-3\sqrt{6}}{225} \\
1 & \frac{16-\sqrt{6}}{36} & \frac{16+\sqrt{6}}{36} & \frac{1}{9} \\
\hline
x & \frac{16-\sqrt{6}}{36} & \frac{16+\sqrt{6}}{36} & \frac{1}{9}
\end{array}
$$

The method with the smaller Butcher tableau is a third–order accurate fully–implicit two–stage Runge–Kutta algorithm, whereas the method with the larger Butcher tableau is a fifth–order accurate fully–implicit three–stage Runge–Kutta algorithm.

The Butcher tableau of the third–order accurate method can be interpreted in the ODE case as:

$$\mathbf{k_1} = \mathbf{f}\left(\mathbf{x_k} + \frac{5h}{12}\mathbf{k_1} - \frac{h}{12}\mathbf{k_2}, \mathbf{u}(t_k + \frac{h}{3}), t_k + \frac{h}{3}\right) \tag{8.43a}$$

$$\mathbf{k_2} = \mathbf{f}\left(\mathbf{x_k} + \frac{3h}{4}\mathbf{k_1} + \frac{h}{4}\mathbf{k_2}, \mathbf{u}(t_k + h), t_k + h\right) \tag{8.43b}$$

$$\mathbf{x_{k+1}} = \mathbf{x_k} + \frac{h}{4}\left(3\mathbf{k_1} + \mathbf{k_2}\right) \tag{8.43c}$$

In the DAE formulation, the method can be written as:

$$\mathbf{f}\left(\mathbf{x_k} + \frac{5h}{12}\mathbf{k_1} - \frac{h}{12}\mathbf{k_2}, \mathbf{k_1}, \mathbf{u}(t_k + \frac{h}{3}), t_k + \frac{h}{3}\right) = 0.0 \tag{8.44a}$$

$$\mathbf{f}\left(\mathbf{x_k} + \frac{3h}{4}\mathbf{k_1} + \frac{h}{4}\mathbf{k_2}, \mathbf{k_2}, \mathbf{u}(t_k + h), t_k + h\right) = 0.0 \tag{8.44b}$$

$$\mathbf{x_{k+1}} = \mathbf{x_k} + \frac{h}{4}\left(3\mathbf{k_1} + \mathbf{k_2}\right) \tag{8.44c}$$

There is hardly any difference between the two formulations. In both cases, we are faced with a set of $2n$ coupled nonlinear equations in the $2n$ unknowns $\mathbf{k_1}$ and $\mathbf{k_2}$ that need to be solved simultaneously by Newton iteration. Just as in the case of the AMi algorithms, we need to provide the system with not only initial conditions for the state vector $\mathbf{x}(t_0)$, but also with a good estimate of the state derivative vector $\dot{\mathbf{x}}(t_0)$. We set initially:

$$\mathbf{k_1} = \mathbf{k_2} = \dot{\mathbf{x}}(t_0) \tag{8.45}$$

Let us now look at the accuracy and stability properties of the Radau IIA algorithms. To this end, we plug Eqs.(8.43a–c) into the linear test problem. We shall work with the ODE version, since it doesn't make any difference, which formulation we use. We obtain:

$$\mathbf{k_1} = \mathbf{A}\left(\mathbf{x_k} + \frac{5h}{12}\mathbf{k_1} - \frac{h}{12}\mathbf{k_2}\right) \tag{8.46a}$$

$$\mathbf{k_2} = \mathbf{A}\left(\mathbf{x_k} + \frac{3h}{4}\mathbf{k_1} + \frac{h}{4}\mathbf{k_2}\right) \tag{8.46b}$$

$$\mathbf{x_{k+1}} = \mathbf{x_k} + \frac{h}{4}\left(3\mathbf{k_1} + \mathbf{k_2}\right) \tag{8.46c}$$

or solved for the unknowns $\mathbf{k_1}$ and $\mathbf{k_2}$:

$$\mathbf{k_1} = \left[\mathbf{I}^{(n)} - \frac{2\mathbf{A}h}{3} + \frac{(\mathbf{A}h)^2}{6}\right]^{-1} \cdot \left(\mathbf{I}^{(n)} - \frac{\mathbf{A}h}{3}\right) \cdot \mathbf{A} \cdot \mathbf{x_k} \qquad (8.47a)$$

$$\mathbf{k_2} = \left[\mathbf{I}^{(n)} - \frac{2\mathbf{A}h}{3} + \frac{(\mathbf{A}h)^2}{6}\right]^{-1} \cdot \left(\mathbf{I}^{(n)} + \frac{\mathbf{A}h}{3}\right) \cdot \mathbf{A} \cdot \mathbf{x_k} \qquad (8.47b)$$

$$\mathbf{x_{k+1}} = \mathbf{x_k} + \frac{h}{4}\left(3\mathbf{k_1} + \mathbf{k_2}\right) \qquad (8.47c)$$

and therefore:

$$\mathbf{F} = \mathbf{I}^{(n)} + \left[\mathbf{I}^{(n)} - \frac{2\mathbf{A}h}{3} + \frac{(\mathbf{A}h)^2}{6}\right]^{-1} \cdot \left(\mathbf{I}^{(n)} - \frac{\mathbf{A}h}{6}\right) \cdot (\mathbf{A}h) \qquad (8.48)$$

Developing the denominator into a Taylor Series around $h = 0.0$, we find:

$$\mathbf{F} \approx \mathbf{I}^{(n)} + \mathbf{A}h + \frac{(\mathbf{A}h)^2}{2} + \frac{(\mathbf{A}h)^3}{6} + \frac{(\mathbf{A}h)^4}{36} \qquad (8.49)$$

Thus, the method is indeed third–order accurate (we proved this at least for linear systems), and the error coefficient is:

$$\varepsilon = \frac{1}{72}(\mathbf{A}h)^4 \qquad (8.50)$$

The fifth–order accurate Radau IIA method is characterized by the following $\mathbf{F}$–matrix:

$$\mathbf{F} = \mathbf{I}^{(n)} + \left[\mathbf{I}^{(n)} - \frac{3\mathbf{A}h}{5} + \frac{3(\mathbf{A}h)^2}{20} - \frac{(\mathbf{A}h)^3}{60}\right]^{-1} \left(\mathbf{I}^{(n)} - \frac{\mathbf{A}h}{10} + \frac{(\mathbf{A}h)^2}{60}\right)\mathbf{A}h$$
$$(8.51)$$

Developing the denominator into a Taylor Series around $h = 0.0$, we find:

$$\mathbf{F} \approx \mathbf{I}^{(n)} + \mathbf{A}h + \frac{(\mathbf{A}h)^2}{2} + \frac{(\mathbf{A}h)^3}{6} + \frac{(\mathbf{A}h)^4}{24} + \frac{(\mathbf{A}h)^5}{120} + \frac{11(\mathbf{A}h)^6}{7200} \qquad (8.52)$$

Thus, the method is indeed fifth–order accurate (at least for linear systems), and the error coefficient is:

$$\varepsilon = \frac{1}{7200}(\mathbf{A}h)^6 \qquad (8.53)$$

A frequently used fourth–order accurate fully–implicit Runge–Kutta algorithm is *Lobatto IIIC* with the Butcher tableau:

0	1/6	-1/3	1/6
1/2	1/6	5/12	-1/12
1	1/6	2/3	1/6
x	1/6	2/3	1/6

This method is characterized by the $\mathbf{F}$–matrix:

$$\mathbf{F} = \mathbf{I}^{(n)} + \left[\mathbf{I}^{(n)} - \frac{3\mathbf{A}h}{4} + \frac{(\mathbf{A}h)^2}{4} - \frac{(\mathbf{A}h)^3}{24}\right]^{-1} \left(\mathbf{I}^{(n)} - \frac{\mathbf{A}h}{4} + \frac{(\mathbf{A}h)^2}{24}\right)\mathbf{A}h$$

(8.54)

Developing the denominator into a Taylor Series around $h = 0.0$, we find:

$$\mathbf{F} \approx \mathbf{I}^{(n)} + \mathbf{A}h + \frac{(\mathbf{A}h)^2}{2} + \frac{(\mathbf{A}h)^3}{6} + \frac{(\mathbf{A}h)^4}{24} + \frac{(\mathbf{A}h)^5}{96}$$

(8.55)

Thus, the method is indeed fourth–order accurate (at least for linear systems), and the error coefficient is:

$$\varepsilon = \frac{1}{480}(\mathbf{A}h)^5$$

(8.56)

We plugged the three $\mathbf{F}$–matrices into our general–purpose stability domain plotting routine. The results are shown in Fig.8.5.

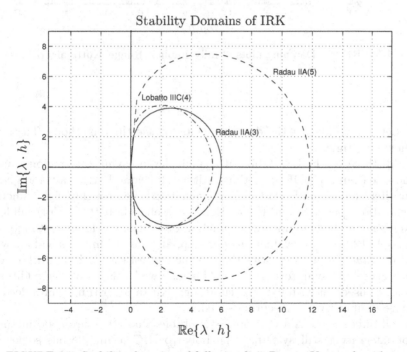

FIGURE 8.5. Stability domains of fully–implicit Runge–Kutta algorithms.

All three methods are A–stable, a desirable property that we hadn't been able to achieve with the higher–order BDF algorithms. The Radau techniques exhibit a somewhat larger unstable region in the right–half complex plane, which may be profitable at times.

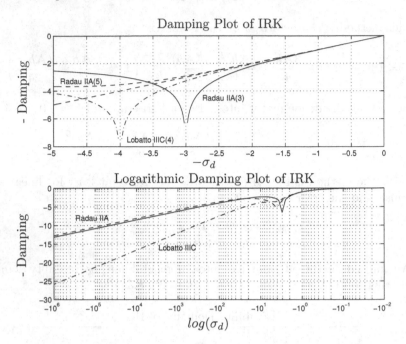

FIGURE 8.6. Damping plots of fully–implicit Runge–Kutta algorithms.

Let us also look at the damping plots for the three methods. These are shown on Fig.8.6.

All three methods exhibit satisfyingly large asymptotic regions, much more so than the BDF algorithms. Although Radau IIA(3) calls for a Newton iteration around two function evaluations, i.e., roughly eight function evaluations per step, and Radau IIA(5) as well as Lobatto IIIC(4) call for a Newton iteration around three function evaluations, adding up to approximately 12 function evaluations per step, all these techniques will allow us to use *much* larger step sizes than in the case of the BDF algorithms due to their large asymptotic regions. This may well balance off the additional cost. Consequently, fully–implicit Runge–Kutta algorithms can indeed be quite competitive in execution speed.

All three methods are obviously L–stable, thus, they are good methods for integrating stiff systems. The Lobatto IIIC technique has somewhat better damping characteristics than the Radau IIA algorithms for poles located far out in the left–half complex plane.

As the explicit RK algorithms are good starter algorithms for the Adams family of methods, and the BI algorithms are good starters for BDF techniques in ODE format, the fully–implicit Runge–Kutta algorithms can be used during startup of a BDF method in DAE format.

8.4 DASSL

DASSL is one of the most successful simulation codes on the market today. It implements the BDF formulae of orders one to five in their DAE format, as presented earlier in this chapter. DASSL is a variable–step, variable–order code that uses order buildup during the startup period. The code was written by Linda Petzold [8.4, 8.34].

The code has meanwhile been made the default simulator in Dymola [8.9, 8.10] in spite of its inefficiency when dealing with non–stiff problems due to the relatively large error coefficients of the BDF formulae, and in spite of its inefficiency when dealing with highly nonlinear problems that require frequent step–size adjustment, such as those that we shall look at in Chapter 9 of this book.

Why did the developers of Dymola choose a stiff–system solver as the default integration algorithm? Dymola was designed to be used in large–scale system modeling. Complex models are almost invariably stiff, as the complexity of the model usually arises from looking simultaneously at phenomena with different time constants. Furthermore, most engineering users don't know whether their models are stiff or not. Since a stiff–system solver is capable of dealing with non–stiff models as well (although not with optimal efficiency), whereas a non–stiff solver cannot deal with stiff models at all, it may be a good idea to use the vacuum cleaner approach, and offer, as the default simulation engine, a code that will be able to cope with most problems somewhat successfully. After all, computers have become fast in recent years, and therefore, optimal efficiency of the simulation engine may no longer be a prime requirement of a modeling and simulation environment.

Why did the developers of Dymola opt for DASSL as the default method rather than e.g. Radau IIA? From what we have learnt, we would expect Radau IIA to be much better suited than DASSL for dealing with highly nonlinear problems requiring frequent step–size adjustment. After all, most engineering models are highly nonlinear.

As we have mentioned earlier, the actual integration algorithm occupies maybe five percent of a production code. The other 95 percent of the code deal with step–size and order control, startup problems, readout problems, and other problems that we haven't looked at yet, such as discontinuity handling (the so–called root solving problem).

The reason is quite simple. There is no production code implementing the Radau IIA algorithms around that is as robust and well tested as the DASSL code. In fact, we haven't even talked yet about such issues as step–size control in implicit Runge–Kutta algorithms.

According to [8.4], DASSL is able to simulate problems of perturbation indices 0 and 1, whereas it may fail when confronted with higher–index problems. Dymola usually reduces the perturbation index of the model to zero, before simulating the model, i.e., although DASSL is capable of solving DAE problems directly, Dymola converts the model to explicit ODE form first, before handing it over to DASSL for simulation.

This decision again sacrifices efficiency for convenience. Multiple Newton iterations may be set up within each other, as Dymola may set up Newton iterations as part of the state–space model, and DASSL employs an overall Newton iteration as part of the simulation process. Yet, solving DAEs directly may be hard on the user, because in the DAE formulation, it is not always evident, how many initial conditions are needed, and where they must be specified. The conversion to explicit ODE form serves the purpose of ensuring that a complete and consistent set of initial conditions is available to properly initialize the simulation run.

Before bringing the discussion of DASSL to an end, let us discuss one more problem that DASSL users may face, a problem that is caused by exploiting the one–legged nature of the BDF formulae in setting up the DAE solver.

Let us look once more at an explicit linear state–space model:

$$\dot{x} = A \cdot x + B \cdot u \tag{8.57}$$

Let us use BDF3 in its ODE form to simulate this system:

$$x_{k+1} = \frac{6}{11}h \cdot \dot{x}_{k+1} + \frac{18}{11}x_k - \frac{9}{11}x_{k-1} + \frac{2}{11}x_{k-2} \tag{8.58}$$

We eliminate the state derivative vector from the Newton iteration by plugging Eq.(8.57) into Eq.(8.58):

$$x_{k+1}^{BDF3} = \frac{6}{11} \cdot A \cdot h \cdot x_{k+1} + \frac{6}{11} \cdot B \cdot h \cdot u_{k+1} + \frac{18}{11}x_k - \frac{9}{11}x_{k-1} + \frac{2}{11}x_{k-2} \tag{8.59}$$

We set the zero function for the Newton iteration up as follows:

$$\mathcal{F}(x_{k+1})^{ODE} = x_{k+1}^{true} - x_{k+1}^{BDF3} \tag{8.60}$$

i.e., we compute the difference between the true yet unknown value of x_{k+1}^{true} and the approximation of the value using the integration algorithm, x_{k+1}^{BDF3}. Thus, the Hessian can be computed as:

$$\mathcal{H}(x_{k+1})^{ODE} = \frac{\partial \mathcal{F}(x_{k+1})^{ODE}}{\partial x_{k+1}} = I^{(n)} - \frac{6}{11} \cdot A \cdot h \tag{8.61}$$

or more generally:

$$\mathcal{H}(x_{k+1})^{ODE} = I^{(n)} - \frac{6}{11} \cdot \mathcal{J} \cdot h \tag{8.62}$$

where $\mathcal{J}$ is the Jacobian of the system:

$$\mathcal{J}(x_{k+1}) = \frac{\partial f(x_{k+1})}{\partial x_{k+1}} \tag{8.63}$$

Let us now analyze the DAE formulation instead. We use the BDF3 formula in its derivative form:

$$\dot{x}_{k+1}^{\mathbf{BDF3}} = \frac{1}{h}\left[\frac{11}{6}\cdot x_{k+1} - 3x_k + \frac{3}{2}x_{k-1} - \frac{1}{3}x_{k-2}\right] \qquad (8.64)$$

We set the zero function up as follows:

$$\mathcal{F}(x_{k+1})^{\mathbf{DAE}} = \dot{x}_{k+1}^{\mathbf{st\_eq}} - \dot{x}_{k+1}^{\mathbf{BDF3}} \qquad (8.65)$$

Thus, we subtract the BDF approximation of the state derivative vector, $\dot{x}_{k+1}^{\mathbf{BDF3}}$ from the state derivative vector computed from the state equations, $\dot{x}_{k+1}^{\mathbf{st\_eq}}$, i.e., from Eq.(8.57). Both of these approximations are functions of x_{k+1}.

Hence the Hessian can now be computed as:

$$\mathcal{H}(x_{k+1})^{\mathbf{DAE}} = \frac{\partial \mathcal{F}(x_{k+1})^{\mathbf{DAE}}}{\partial x_{k+1}} = \mathbf{A} - \frac{11}{6h}\cdot\mathbf{I}^{(n)} \qquad (8.66)$$

or more generally:

$$\mathcal{H}(x_{k+1})^{\mathbf{DAE}} = \mathbf{J} - \frac{11}{6h}\cdot\mathbf{I}^{(n)} \qquad (8.67)$$

What happens when the step size, h, is made very small? In the ODE case, we find:

$$\lim_{h\to 0}\mathcal{H}(x_{k+1})^{\mathbf{ODE}} = \mathbf{I}^{(n)} \qquad (8.68)$$

Thus, the Hessian approaches the identity matrix as the step size approaches zero. In the DAE case, we find:

$$\lim_{h\to 0}\mathcal{H}(x_{k+1})^{\mathbf{DAE}} \to \infty \qquad (8.69)$$

As the step size approaches zero, the Hessian approaches infinity. For small step sizes, the Hessian is highly sensitive to a change in the step size. This forebodes trouble.

Although the ODE and DAE formulations of the BDF formulae are the same algorithms in theory, they may behave quite differently from a numerical point of view for small step sizes due to roundoff.

Let us now look at the most general case of a nonlinear implicit model of the type:

$$\mathcal{F}(x, \dot{x}, u, t) = 0 \qquad (8.70)$$

In accordance with Chapter 4 of this book, the different BDF algorithms can be written as:

$$\mathbf{x_{k+1}} = h \cdot \mathbf{f_{k+1}} + \mathbf{x_k} \tag{8.71a}$$

$$\mathbf{x_{k+1}} = \frac{2}{3} \cdot h \cdot \mathbf{f_{k+1}} + \frac{4}{3} \cdot \mathbf{x_k} - \frac{1}{3} \cdot \mathbf{x_{k-1}} \tag{8.71b}$$

$$\mathbf{x_{k+1}} = \frac{6}{11} \cdot h \cdot \mathbf{f_{k+1}} + \frac{18}{11} \cdot \mathbf{x_k} - \frac{9}{11} \cdot \mathbf{x_{k-1}} + \frac{2}{11} \cdot \mathbf{x_{k-2}} \tag{8.71c}$$

$$\text{etc.} \tag{8.71d}$$

Thus in general, we can write all of these equations in the form:

$$\mathbf{x_{k+1}} = \bar{h} \cdot \mathbf{f_{k+1}} + \text{old}(\mathbf{x}) \tag{8.72}$$

where $\bar{h}$ is proportional in the step size h, and $\text{old}(\mathbf{x})$ is a function of previous values of the state vector, which won't influence the Newton iteration at this time.

We can turn Eq.(8.72) around:

$$\mathbf{f_{k+1}} = \frac{\mathbf{x_{k+1}} - \text{old}(\mathbf{x})}{\bar{h}} \tag{8.73}$$

When DASSL is applied to the model of Eq.(8.70), it plugs Eq.(8.73) into Eq.(8.70):

$$\mathcal{F}\left(\mathbf{x_{k+1}}, \frac{\mathbf{x_{k+1}} - \text{old}(\mathbf{x})}{\bar{h}}, \mathbf{u_{k+1}}, t_{k+1}\right) = 0 \tag{8.74}$$

at time t_{k+1}, and iterates on $\mathbf{x_{k+1}}$. In setting up the Newton iteration, we don't actually need to perform the substitution, as we can see from Eq.(8.74) what contributions the state derivative vector produces in the computation of the Hessian:

$$\mathcal{H}(\mathbf{x_{k+1}}) = \mathcal{J}_\mathbf{x}(\mathbf{x_{k+1}}) + \frac{1}{\bar{h}} \cdot \mathcal{J}_\mathbf{\dot{x}}(\mathbf{x_{k+1}}) \tag{8.75}$$

where:

$$\mathcal{J}_\mathbf{x}(\mathbf{x_{k+1}}) = \left.\frac{\partial \mathcal{F}}{\partial \mathbf{x}}\right|_{x=x_{k+1}, \dot{x}=\dot{x}_{k+1}} \tag{8.76a}$$

$$\mathcal{J}_\mathbf{\dot{x}}(\mathbf{x_{k+1}}) = \left.\frac{\partial \mathcal{F}}{\partial \mathbf{\dot{x}}}\right|_{x=x_{k+1}, \dot{x}=\dot{x}_{k+1}} \tag{8.76b}$$

are partial Jacobians without substitution.

Hence we can set up the Newton iteration in the following way:

$$\left(\mathcal{J}_\mathbf{x} + \frac{1}{\bar{h}} \cdot \mathcal{J}_\mathbf{\dot{x}}\right) \cdot \delta^\ell = \mathcal{F}(\mathbf{x}^\ell, \mathbf{\dot{x}}^\ell, t) \tag{8.77a}$$

$$\mathbf{x}^{\ell+1} = \mathbf{x}^\ell - \delta^\ell \tag{8.77b}$$

$$\mathbf{\dot{x}}^{\ell+1} = \mathbf{\dot{x}}^\ell - \frac{1}{\bar{h}} \cdot \delta^\ell \tag{8.77c}$$

By multiplying Eq.(8.77a) by the step size h, we can write the linear system inside the Newton iteration as:

$$(\bar{h} \cdot \mathcal{J}_\mathbf{x} + \mathcal{J}_{\dot{\mathbf{x}}}) \cdot \delta^\ell = \bar{h} \cdot \mathcal{F}(\mathbf{x}^\ell, \dot{\mathbf{x}}^\ell, t) \qquad (8.78)$$

Which variables need to be included in the iteration vector, $\mathbf{x}$, of the Newton iteration? If the problem to be solved is an index–0 problem, the iteration vector is identical to the vector of independent state variables. However, if the problem to be solved is an index–1 problem, then at least the tearing variables of the algebraic loops need to be included in the iteration vector, $\mathbf{x}$, as well.

What happens if we let the normalized step size, $\bar{h}$, go to zero? The Hessian then degenerates to:

$$\lim_{\bar{h} \to 0} \mathcal{H} = \mathcal{J}_{\dot{\mathbf{x}}} \qquad (8.79)$$

which, in the case of an index–1 problem, unfortunately is a singular matrix. Thus, the smaller the step size, the more poorly conditioned the Newton iteration will become in the simulation of an index–1 problem.

Unfortunately, small step sizes will haunt us throughout Chapters 9 and 10 of this book, which is yet another reason, why the producers of Dymola chose to symbolically convert all DAEs to explicit ODE form prior to letting DASSL handle the simulation.

8.5 Inline Integration

You, the reader, may meanwhile have come to the conclusion that direct simulation of an index–1 DAE problem is a bad idea after all. Yet, the problems that we encountered are not directly related to the index–1 DAE problem, but rather to the way, in which DASSL was set up. When Linda Petzold developed the code, she still clung to the idea that the simulation engine must be separated from the model equations, in order to protect the hapless user. In 1983, when DASSL was developed, computers were still slow, memory was still expensive, and consequently, compilers were still limited in their capabilities.

It turns out that direct simulation of a stiff index–1 DAE problem may still be a good idea at times, but before we can attempt such a direct simulation, the final barrier between the simulation engine and the model equations must come down.

For the time being, let us restrict our discussion to the use of backward Euler, i.e., BDF1:

$$\mathbf{x}_{k+1}^{\mathbf{BE}} = \mathbf{x}_k + h \cdot \dot{\mathbf{x}}_{k+1} \qquad (8.80)$$

Let us look once more at the first of our three circuit problems. Its schematic is displayed in Fig.8.7.

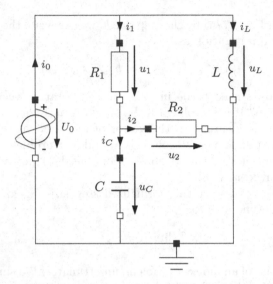

FIGURE 8.7. Schematic of electrical RLC circuit.

However this time around, we shall insert the integration equations directly into the model. The enhanced set of model equations can be written as follows:

$$u_0 = f(t) \tag{8.81a}$$
$$u_1 = R_1 \cdot i_1 \tag{8.81b}$$
$$u_2 = R_2 \cdot i_2 \tag{8.81c}$$
$$u_L = L \cdot di_L \tag{8.81d}$$
$$i_C = C \cdot du_C \tag{8.81e}$$
$$u_0 = u_1 + u_C \tag{8.81f}$$
$$u_L = u_1 + u_2 \tag{8.81g}$$
$$u_C = u_2 \tag{8.81h}$$
$$i_0 = i_1 + i_L \tag{8.81i}$$
$$i_1 = i_2 + i_C \tag{8.81j}$$
$$i_L = \text{pre}(i_L) + h \cdot di_L \tag{8.81k}$$
$$u_C = \text{pre}(u_C) + h \cdot du_C \tag{8.81l}$$

$\text{pre}(i_L)$ denotes the previous value of i_L. At time $t = 0$, we set $\text{pre}(i_L) = i_{L_0}$, i.e., we apply the initial conditions of the state variables to the vector of previous states, and evaluate the model equations for the first time at $t = h$.

By inserting ("inlining") the integrator equations into the model, we eliminated the *differential equations* altogether [8.8]. We are now faced with a set of *difference equations* that we need to solve once per step at times $t = h$, $t = 2h$, etc. di_L and du_C are no longer state derivatives. They have now turned into algebraic variables with funny names.

The structure digraph of the above difference equation (ΔE) model is shown in Fig.8.8.

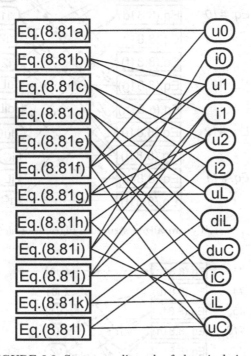

FIGURE 8.8. Structure digraph of electrical circuit.

Let us start to causalize the structure digraph. The results of our efforts are shown in Fig.8.9.

We were able to causalize five of our 12 equations, before encountering an algebraic loop. Whereas the original DAE problem had been an index–0 problem, i.e., a problem not leading to an algebraic loop, the converted ΔE problem contains an algebraic loop, which calls for a Newton iteration. This is the Newton iteration caused by the implicit integration algorithm.

Let us find a suitable tearing structure. We shall not use our usual heuristics. The reason is that we don't want the step size, h, to show up in the denominator of any equation. Thus, we shall use Eq.(8.81l) as our residual equation, which we solve for the tearing variable, u_C. It turns out that, with this choice, we are able to causalize all remaining equations. The results of the causalization are shown in Fig.8.10.

Using DASSL, we would have required two iteration variables, namely the two state variables, i_L and u_C. Using inline integration, we only require a single iteration variable, the tearing variable, u_C.

The fact that we were using backward Euler in the above analysis is

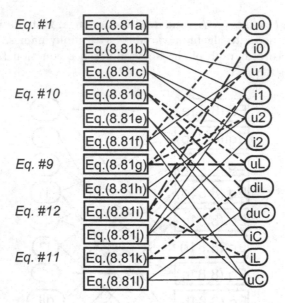

FIGURE 8.9. Partially causalized structure digraph of electrical circuit.

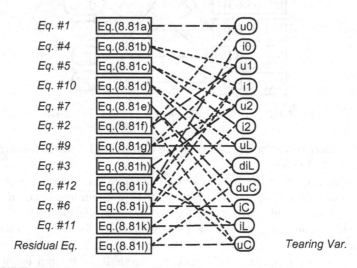

FIGURE 8.10. Completely causalized structure digraph of electrical circuit.

actually irrelevant. We could have used any BDF algorithm, or in fact, we even could have used a variable–step and variable–order BDF technique. All we would have had to do is to replace the step size h by the normalized step size $\bar{h}$, and pre($\mathbf{x}$) by old($\mathbf{x}$). Neither of these two substitutions modifies the structure digraph.

The causal set of ΔEs can be written as follows:

$$u_0 = f(t) \tag{8.82a}$$

$$u_1 = u_0 - u_C \tag{8.82b}$$

$$u_2 = u_C \tag{8.82c}$$

$$i_1 = \frac{1}{R_1} \cdot u_1 \tag{8.82d}$$

$$i_2 = \frac{1}{R_2} \cdot u_2 \tag{8.82e}$$

$$i_C = i_1 - i_2 \tag{8.82f}$$

$$du_C = \frac{1}{C} \cdot i_C \tag{8.82g}$$

$$u_C = \text{pre}(u_C) + h \cdot du_C \tag{8.82h}$$

$$u_L = u_1 + u_2 \tag{8.82i}$$

$$di_L = \frac{1}{L} \cdot u_L \tag{8.82j}$$

$$i_L = \text{pre}(i_L) + h \cdot di_L \tag{8.82k}$$

$$i_0 = i_1 + i_L \tag{8.82l}$$

Let us apply variable substitution to come up with a completely causal set of equations.

$$
\begin{aligned}
u_C &= \text{pre}(u_C) + h \cdot du_C \\
&= \text{pre}(u_C) + \frac{h}{C} \cdot i_C \\
&= \text{pre}(u_C) + \frac{h}{C} \cdot i_1 - \frac{h}{C} \cdot i_2 \\
&= \text{pre}(u_C) + \frac{h}{R_1 \cdot C} \cdot u_1 - \frac{h}{R_2 \cdot C} \cdot u_2 \\
&= \text{pre}(u_C) + \frac{h}{R_1 \cdot C} \cdot u_0 - \frac{h}{R_1 \cdot C} \cdot u_C - \frac{h}{R_2 \cdot C} \cdot u_C
\end{aligned}
$$

and therefore:

$$\left[1 + \frac{h}{R_1 \cdot C} + \frac{h}{R_2 \cdot C}\right] \cdot u_C = \text{pre}(u_C) + \frac{h}{R_1 \cdot C} \cdot u_0$$

or:

$$[R_1 \cdot R_2 \cdot C + h \cdot (R_1 + R_2)] \cdot u_C = R_1 \cdot R_2 \cdot C \cdot \text{pre}(u_C) + h \cdot R_2 \cdot u_0$$

which can be solved for u_C:

$$u_C = \frac{R_1 \cdot R_2 \cdot C}{R_1 \cdot R_2 \cdot C + h \cdot (R_1 + R_2)} \cdot \text{pre}(u_C) + \frac{h \cdot R_2}{R_1 \cdot R_2 \cdot C + h \cdot (R_1 + R_2)} \cdot u_0$$
$$(8.83)$$

If we let the step size go to zero, we find:

$$\lim_{h \to 0} u_C = \text{pre}(u_C) \qquad (8.84)$$

which is non–singular. Since the original DAE problem had been of index 0, this is not further surprising.

Let us now look at the second of our circuits. Its schematic is shown in Fig.8.11.

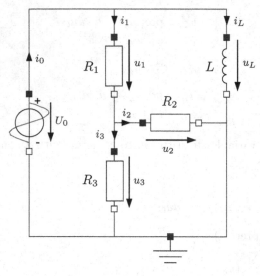

FIGURE 8.11. Schematic of modified electrical RLC circuit.

Remember this circuit represents an index–1 problem. Inlining the single integrator, we get the following set of acausal equations:

$$u_0 = f(t) \qquad (8.85a)$$
$$u_1 = R_1 \cdot i_1 \qquad (8.85b)$$
$$u_2 = R_2 \cdot i_2 \qquad (8.85c)$$
$$u_3 = R_3 \cdot i_3 \qquad (8.85d)$$
$$u_L = L \cdot di_L \qquad (8.85e)$$
$$u_0 = u_1 + u_3 \qquad (8.85f)$$
$$u_L = u_1 + u_2 \qquad (8.85g)$$
$$u_3 = u_2 \qquad (8.85h)$$

$$i_0 = i_1 + i_L \qquad (8.85\text{i})$$

$$i_1 = i_2 + i_3 \qquad (8.85\text{j})$$

$$i_L = \text{pre}(i_L) + h \cdot di_L \qquad (8.85\text{k})$$

Its structure digraph is shown in Fig.8.12.

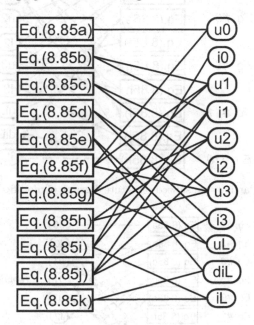

FIGURE 8.12. Structure digraph of modified electrical circuit.

We begin to causalize the structure digraph. The partially causalized structure digraph is shown in Fig.8.13.

We were able to causalize five of the 11 equations. Let us apply our heuristic procedure to select a first residual equation and a first tearing variable. The results of our efforts are shown in Fig.8.14.

A single tearing variable sufficed to causalize the entire equation system. DASSL would have required at least two iteration variables, the state variable, i_L, and the single tearing variable of the algebraic loop, i_3, of the index–1 DAE system. Inline integration is more economical. We get away with a single iteration variable, the tearing variable, i_1, of the ΔE system.

Let us write down the causal equations:

$$u_0 = f(t) \qquad (8.86\text{a})$$

$$u_1 = R_1 \cdot i_1 \qquad (8.86\text{b})$$

$$u_3 = u_0 - u_1 \qquad (8.86\text{c})$$

$$u_2 = u_3 \qquad (8.86\text{d})$$

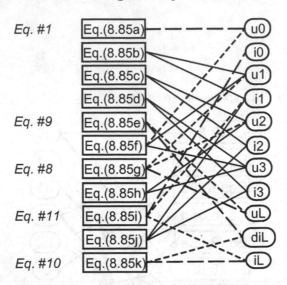

FIGURE 8.13. Partially causalized structure digraph of modified electrical circuit.

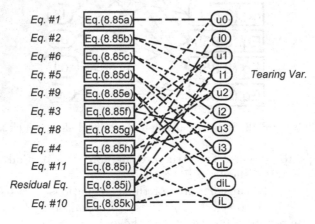

FIGURE 8.14. Completely causalized structure digraph of modified electrical circuit.

$$i_3 = \frac{1}{R_3} \cdot u_3 \qquad\qquad (8.86e)$$

$$i_2 = \frac{1}{R_2} \cdot u_2 \qquad\qquad (8.86f)$$

$$i_1 = i_2 + i_3 \qquad\qquad (8.86g)$$

$$u_L = u_1 + u_2 \qquad\qquad (8.86h)$$

$$di_L = \frac{1}{L} \cdot u_L \qquad\qquad (8.86i)$$

$$i_L = \mathrm{pre}(i_L) + h \cdot di_L \qquad\qquad (8.86j)$$

$$i_0 = i_1 + i_L \tag{8.86k}$$

Using the variable substitution technique, we can find a closed–form expression for the tearing variable:

$$i_1 = \frac{R_2 + R_3}{R_1 \cdot R_2 + R_1 \cdot R_3 + R_2 \cdot R_3} \cdot u_0 \tag{8.87}$$

The expression for i_1 is not even a function of the step size h, i.e., it is non–singular for any value of h.

Let us now analyze the third circuit. Its schematic is shown in Fig.8.15. Remember this is an index–2 problem.

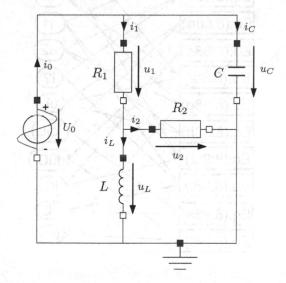

FIGURE 8.15. Schematic of once more modified electrical RLC circuit.

After inlining the integrator equations, the acausal equations present themselves in the following form:

$$u_0 = f(t) \tag{8.88a}$$
$$u_1 = R_1 \cdot i_1 \tag{8.88b}$$
$$u_2 = R_2 \cdot i_2 \tag{8.88c}$$
$$u_L = L \cdot di_L \tag{8.88d}$$
$$i_C = C \cdot du_C \tag{8.88e}$$
$$u_0 = u_1 + u_L \tag{8.88f}$$
$$u_C = u_1 + u_2 \tag{8.88g}$$
$$u_L = u_2 \tag{8.88h}$$
$$i_0 = i_1 + i_C \tag{8.88i}$$

$$i_1 = i_2 + i_L \tag{8.88j}$$

$$i_L = \mathrm{pre}(i_L) + h \cdot di_L \tag{8.88k}$$

$$u_C = \mathrm{pre}(u_C) + h \cdot du_C \tag{8.88l}$$

The structure digraph is shown in Fig.8.16.

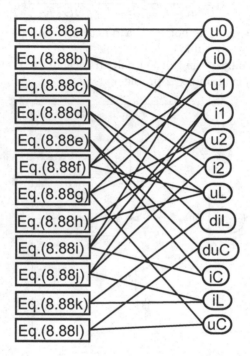

FIGURE 8.16. Structure digraph of once more modified electrical circuit.

We begin to causalize the structure digraph. The results of our efforts are shown in Fig.8.17.

We were able to causalize five of the 12 equations, before ending up with an algebraic loop. Evidently, since the computational causalities of all energy storage elements have been freed up after inlining the integrator equations, we don't obtain any constraint equation any longer.

Unfortunately, we already got ourselves into trouble, because Eq.(8.88l) needs to be solved for du_C:

$$du_C = \frac{u_C - \mathrm{pre}(u_C)}{h} \tag{8.89}$$

i.e., we ended up with the step size, h, in the denominator, which invariably will cause numerical difficulties, when we try to simulate the system using a small step size. We had no choice in the matter, as the derivative causality on the capacitor was dictated upon us.

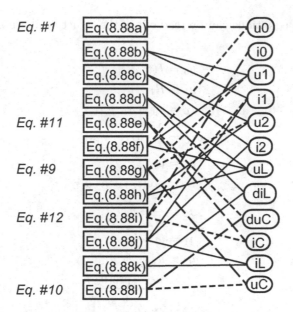

FIGURE 8.17. Partially causalized structure digraph of once more modified electrical circuit.

Let us nevertheless continue by applying our heuristic procedure for selecting a first residual equation and a first tearing variable. The results of our efforts are shown in Fig.8.18.

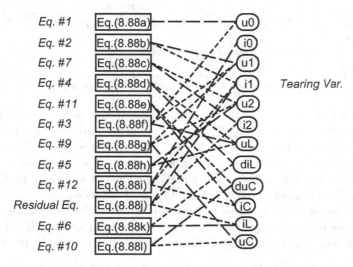

FIGURE 8.18. Completely causalized structure digraph of once more modified electrical circuit.

A single tearing variable suffices to causalize the entire equation system.

The causal equations can be read out of the structure digraph of Fig.8.18.

$$u_0 = f(t) \tag{8.90a}$$

$$u_1 = R_1 \cdot i_1 \tag{8.90b}$$

$$u_L = u_0 - u_1 \tag{8.90c}$$

$$di_L = \frac{1}{L} \cdot u_L \tag{8.90d}$$

$$u_2 = u_L \tag{8.90e}$$

$$i_L = \text{pre}(i_L) + h \cdot di_L \tag{8.90f}$$

$$i_2 = \frac{1}{R_2} \cdot u_2 \tag{8.90g}$$

$$i_1 = i_2 + i_L \tag{8.90h}$$

$$u_C = u_1 + u_2 \tag{8.90i}$$

$$du_C = \frac{u_C - \text{pre}(u_C)}{h} \tag{8.90j}$$

$$i_C = C \cdot du_C \tag{8.90k}$$

$$i_0 = i_1 + i_C \tag{8.90l}$$

Using the variable substitution technique, we can find a closed–form equation for the tearing variable, i_1.

$$i_1 = \frac{L + h \cdot R_2}{L \cdot (R_1 + R_2) + h \cdot R_2} \cdot u_0 + \frac{R_2 \cdot L}{L \cdot (R_1 + R_2) + h \cdot R_2} \cdot \text{pre}(i_L) \tag{8.91}$$

If we let the step size go to zero, we find:

$$\lim_{h \to 0} i_1 = \frac{1}{R_1 + R_2} \cdot u_0 + \frac{R_2}{R_1 + R_2} \cdot \text{pre}(i_L) \tag{8.92}$$

At least in the given example, inlining was able to solve also the higher–index problem directly. This discovery shall prove important in the context of the next chapter of this book. Yet, inlining the higher–index problem directly came at a price, as we ended up with the step size, h, in the denominator of one of the model equations. Thus, it is usually preferred to first apply the index reduction algorithm by Pantelides.

We have shown that inline integration can solve DAE problems directly and more economically than the standard version of DASSL[1] In all of these examples, we have used the backward Euler formula for inlining.

---

[1] The standard version of DASSL comes with a regular matrix solver and a band–matrix solver. In addition, DASSL offers an interface for supplying other matrix solvers externally. A sparse matrix solver can improve the efficiency of DASSL significantly fo large numbers of states [8.31].

However, this is not necessary. If we replace the true step size, h, by the normalized step size $\bar{h}$, and the previous value of the state vector, pre($\mathbf{x}$), by a combination of old state information, old($\mathbf{x}$), we can inline any and all of the BDF algorithms in exactly the same fashion.

If we wish to implement a step–size and/or order controlled algorithm, we can do so using the same techniques that were advocated in Chapter 4 of this book. Since both the new step size and the new order depend on previous state information only, the equations for step–size and order control do not need to be inlined. Only the integration formula itself must be inlined, which can be accomplished for all BDF algorithms in the manner demonstrated in this section.

8.6   Inlining Implicit Runge–Kutta Algorithms

How can the inlining technique be generalized to implicit Runge–Kutta algorithms as well? For each stage of the multi–stage algorithm, we need to replicate the entire set of equations once. Let us explain the technique by means of the first of the three circuit examples. We shall inline the third–order accurate Radau IIA algorithm. Since this is a two–stage algorithm, we need to write down the equations twice, once for each of the two stages, for the time instant, when that stage needs to be computed.

The first stage is computed at time $t = t_k + h/3$, whereas the second stage is computed at time $t = t_k + h = t_{k+1}$. The integrator formulae can thus be written as:

$$\mathbf{x}_{k+\frac{1}{3}} = \mathbf{x}_k + \frac{5h}{12} \cdot \dot{\mathbf{x}}_{k+\frac{1}{3}} - \frac{h}{12} \cdot \dot{\mathbf{x}}_{k+1} \qquad (8.93a)$$

$$\mathbf{x}_{k+1} = \mathbf{x}_k + \frac{3h}{4} \cdot \dot{\mathbf{x}}_{k+\frac{1}{3}} + \frac{h}{4} \cdot \dot{\mathbf{x}}_{k+1} \qquad (8.93b)$$

Hence the complete set of equations for the circuit example can be written as:

$$v_0 = f(t + \frac{h}{3}) \qquad (8.94a)$$

$$v_1 = R_1 \cdot j_1 \qquad (8.94b)$$

$$v_2 = R_2 \cdot j_2 \qquad (8.94c)$$

$$v_L = L \cdot dj_L \qquad (8.94d)$$

$$j_C = C \cdot dv_C \qquad (8.94e)$$

$$v_0 = v_1 + v_C \qquad (8.94f)$$

$$v_L = v_1 + v_2 \qquad (8.94g)$$

$$v_C = v_2 \qquad (8.94h)$$

$$j_0 = j_1 + j_L \tag{8.94i}$$

$$j_1 = j_2 + j_C \tag{8.94j}$$

$$u_0 = f(t + h) \tag{8.94k}$$

$$u_1 = R_1 \cdot i_1 \tag{8.94l}$$

$$u_2 = R_2 \cdot i_2 \tag{8.94m}$$

$$u_L = L \cdot di_L \tag{8.94n}$$

$$i_C = C \cdot du_C \tag{8.94o}$$

$$u_0 = u_1 + u_C \tag{8.94p}$$

$$u_L = u_1 + u_2 \tag{8.94q}$$

$$u_C = u_2 \tag{8.94r}$$

$$i_0 = i_1 + i_L \tag{8.94s}$$

$$i_1 = i_2 + i_C \tag{8.94t}$$

$$j_L = \text{pre}(i_L) + \frac{5h}{12} \cdot dj_L - \frac{h}{12} \cdot di_L \tag{8.94u}$$

$$v_C = \text{pre}(u_C) + \frac{5h}{12} \cdot dv_C - \frac{h}{12} \cdot du_C \tag{8.94v}$$

$$i_L = \text{pre}(i_L) + \frac{3h}{4} \cdot dj_L + \frac{h}{4} \cdot di_L \tag{8.94w}$$

$$u_C = \text{pre}(u_C) + \frac{3h}{4} \cdot dv_C + \frac{h}{4} \cdot du_C \tag{8.94x}$$

Thus, we end up with a difference equation (ΔE) system in 24 equations and 24 unknowns. Since the two stages are implicitly coupled to each other, they cannot be executed in sequence. They are simulated together leading to a model containing twice as many equations and unknowns [8.6, 8.38].

We are only interested in the variables of the second stage. At the end of the step, i_L and u_C need to be copied to the previous state vector, $\text{pre}(i_L)$ and $\text{pre}(u_C)$. Yet, we must compute the variables of the first stage simultaneously with those of the second stage due to the coupling between the two stages.

We shall refrain from drawing the structure digraph for this ΔE system. Let us summarize the results. 10 of the 24 equations can be causalized at once. The heuristic procedure chooses v_C as the first tearing variable, and Eq.(8.94v) as the first residual equation. With this choice, seven additional equations can be causalized. The procedure then chooses i_1 as the second tearing variable, and Eq.(8.94t) as the second residual equation. With this choice, the remaining seven equations can be causalized.

Hence we can simulate this problem using the third–order accurate Radau IIA algorithm with only two iteration variables in the Newton iteration.

8.7   Stiffly Stable Step–size Control of Radau IIA

A difficult problem with these types on numerical solvers concerns the control of the step size. To this end, it is necessary to find an estimate for the integration error, the order of approximation accuracy of which is one order higher than that of the solver itself. Typically, designers of such solvers will look for a second solver of the same or higher order of approximation accuracy to compare it against the solver to be used for the simulation.

While it is always possible to run two independent solvers in parallel for the purpose of step–size control, this approach is clearly undesirable, as it makes the solver highly inefficient. In explicit Runge–Kutta algorithms, it has become customary to search for an *embedding method*, i.e., a second solver that has most of the computations in common with the original solver, such that they share a large portion of the computational load between them. Unfortunately, this approach won't work in the case of fully implicit Runge–Kutta algorithms, since these algorithms are so compact and so highly optimized that there simply is not enough freedom left in these algorithms for embedding methods to co–exist with them.

One solution that comes to mind immediately is to use a Backward Difference Formula in parallel with the implicit Runge–Kutta technique. This solution can be implemented cheaply, because an appropriately accurate state derivative at time t_{k+1} can be obtained up front using the Runge–Kutta approximation, i.e., no Newton iteration is necessary. For example, the 3rd–order accurate Radau IIA algorithm could be accompanied by a 3rd–order accurate BDF solver implemented as:

$$x_{k+1}^{BDF} = \frac{18}{11}x_k - \frac{9}{11}x_{k-1} + \frac{2}{11}x_{k-2} + \frac{6}{11} \cdot h \cdot f_{k+1} \qquad (8.95)$$

where f_{k+1} is the function value evaluated from the state–space model at time t_{k+1}:

$$f_{k+1} = f(x_{k+1}^{IRK}, u_{k+1}, t_{k+1}) \qquad (8.96)$$

and x_{k+1}^{IRK} is the solution found by the Radau IIA algorithm. Unfortunately, such a solution inherits all the difficulties associated with step–size control in linear multi–step methods. Alternatively, the step size can only be modified once every n steps, where n is the order of the algorithm, which eliminates an important aspect of the elegance and efficiency of Runge–Kutta methods. For these reasons, we propose a different route.

Clearly, an embedding method cannot be found using only information that is being used by the Radau IIA algorithm. In each step, there are four pieces of information available: x_k, $x_{k+\frac{1}{3}}$, $\dot{x}_{k+\frac{1}{3}}$, and $\dot{x}_{k+1}$ to estimate x_{k+1}. Evidently, there is only one 3rd–order accurate polynomial going through these four pieces of information, and it is this polynomial that defines the Radau IIA algorithm. However, enough redundancy can be obtained to define an embedding algorithm if information from the two last steps is being used. In this case, the following eight pieces of information are available: x_{k-1}, $x_{k-\frac{2}{3}}$, x_k, $x_{k+\frac{1}{3}}$, $\dot{x}_{k-\frac{2}{3}}$, $\dot{x}_k$, $\dot{x}_{k+\frac{1}{3}}$, and $\dot{x}_{k+1}$. It was decided to look for 4th–order accurate polynomials that go through any five of these eight pieces of information. This technique defines 56 possible embedding methods. Out of these 56 methods, only six are stiffly stable. Two of those six techniques are not A–stable, i.e., have unstable regions within the left half complex $\lambda \cdot h$ plane. One method has a stability domain with a discontinuous derivative at the real axis, which is suspicious. The remaining three methods are:

$$x_{k+1}^1 = -\frac{25}{279}\,x_{k-1} + \frac{6}{31}\,x_{k-\frac{2}{3}} + \frac{250}{279}\,x_k + \frac{25h}{31}\,\dot{x}_{k+\frac{1}{3}} + \frac{65h}{279}\,\dot{x}_{k+1}$$

$$x_{k+1}^2 = -\frac{1}{36}\,x_{k-1} + \frac{16}{9}\,x_k - \frac{3}{4}\,x_{k+\frac{1}{3}} + h\,\dot{x}_{k+\frac{1}{3}} + \frac{2h}{9}\,\dot{x}_{k+1} \qquad (8.97)$$

$$x_{k+1}^3 = -\frac{2}{23}\,x_{k-\frac{2}{3}} + \frac{50}{23}\,x_k - \frac{25}{23}\,x_{k+\frac{1}{3}} + \frac{25h}{23}\,\dot{x}_{k+\frac{1}{3}} + \frac{5h}{23}\,\dot{x}_{k+1}$$

All of these three techniques have nice stability domains looping in the right half complex $\lambda \cdot h$ plane. Each of them is A–stable. It is possible to write these methods in the linear case as:

$$\mathbf{x}_{k+1} = \mathbf{F} \cdot \mathbf{x}_{k-1} \qquad (8.98)$$

The $\mathbf{F}$–matrices of the three methods can be expanded into Taylor series around $h = 0$. The three $\mathbf{F}$–matrices then take the form:

$$\mathbf{F}^1 \approx \mathbf{I}^{(n)} + 2\,\mathbf{A}h + 4\,\frac{(\mathbf{A}h)^2}{2} + \frac{2224}{279}\,\frac{(\mathbf{A}h)^3}{6} + \frac{877}{58}\,\frac{(\mathbf{A}h)^4}{24} \qquad (8.99a)$$

$$\mathbf{F}^2 \approx \mathbf{I}^{(n)} + 2\,\mathbf{A}h + 4\,\frac{(\mathbf{A}h)^2}{2} + \frac{73}{9}\,\frac{(\mathbf{A}h)^3}{6} + \frac{859}{54}\,\frac{(\mathbf{A}h)^4}{24} \qquad (8.99b)$$

$$\mathbf{F}^3 \approx \mathbf{I}^{(n)} + 2\,\mathbf{A}h + 4\,\frac{(\mathbf{A}h)^2}{2} + \frac{188}{23}\,\frac{(\mathbf{A}h)^3}{6} + \frac{374}{23}\,\frac{(\mathbf{A}h)^4}{24} \qquad (8.99c)$$

What would have been expected of a 4th–order accurate method is:

$$\mathbf{F} \approx \mathbf{I}^{(n)} + 2\,\mathbf{A}h + 4\,\frac{(\mathbf{A}h)^2}{2} + 8\,\frac{(\mathbf{A}h)^3}{6} + 16\,\frac{(\mathbf{A}h)^4}{24} \qquad (8.100)$$

since the expansion is over a double step. Unfortunately, neither of these three methods is even 3rd–order accurate. The problem is that although x_{k+1} is 3rd–order accurate, the first stage of the method, $x_{k+\frac{1}{3}}$ is only 2nd–order accurate. We evidently cannot expect the order of approximation accuracy of our 4th–order polynomials to be any higher than that of its supporting values, and indeed, all three of our 4th–order polynomials are only 2nd–order accurate.

Luckily, there are three such methods available. Hence it should be possible to blend them:

$$x_{k+1}^{blended} = \alpha \ x_{k+1}^1 + \beta \ x_{k+1}^2 + (1 - \alpha - \beta) \ x_{k+1}^3 \qquad (8.101)$$

such that the coefficients of the Taylor–series expansion of the blended method are correct up to the quartic term. Unfortunately, this doesn't work, because the three methods are not linearly independent of each other. There really are only two methods. The third one is a linear combination of the other two. However, it is possible to blend any two of these three methods with the solution found by Radau IIA:

$$x_{k+1}^{blended} = \alpha \cdot x_{k+1}^1 + \beta \cdot x_{k+1}^2 + (1 - \alpha - \beta) \cdot x_{k+1}^{Radau} \qquad (8.102)$$

These three techniques are indeed independent of each other. The resulting algorithm is:

$$x_{k+1}^{blended} = x_{k-1} - 2 \ x_{k-\frac{2}{3}} + 2 \ x_{k+\frac{1}{3}} - \frac{h}{2} \ \dot{x}_{k+\frac{1}{3}} + \frac{h}{2} \ \dot{x}_{k+1} \qquad (8.103)$$

This method is indeed 4th–order accurate. It has highly appealing coefficients. It has only one disadvantage. It is totally unstable everywhere.

It should be possible to find 4th–order accurate embedding methods spanned by the information collected from Radau IIA over two steps. Yet, for the purpose of step–size control, it is sufficient to find another 3rd–order accurate embedding method. To this end, it suffices to blend any two of the three algorithms found above:

$$x_{k+1}^{blended} = \vartheta \cdot x_{k+1}^1 + (1 - \vartheta) \cdot x_{k+1}^2 \qquad (8.104)$$

The resulting method is:

$$x_{k+1}^{blended} = -\frac{1}{13} \ x_{k-1} + \frac{2}{13} \ x_{k-\frac{2}{3}} + \frac{14}{13} \ x_k - \frac{2}{13} \ x_{k+\frac{1}{3}} + \frac{11h}{13} \ \dot{x}_{k+\frac{1}{3}} + \frac{3h}{13} \ \dot{x}_{k+1}$$
$$(8.105)$$

Also this method has beautifully simple rational coefficients. it is indeed 3rd–order accurate:

$$\mathbf{F} \approx \mathbf{I}^{(n)} + 2 \ \mathbf{A}h + 4 \ \frac{(\mathbf{A}h)^2}{2} + 8 \ \frac{(\mathbf{A}h)^3}{6} + \frac{149}{156} \cdot 16 \ \frac{(\mathbf{A}h)^4}{24} \qquad (8.106)$$

i.e., the error coefficient of the method is:

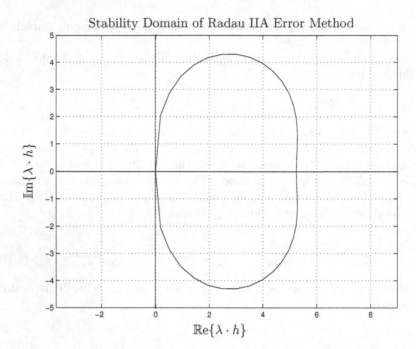

FIGURE 8.19. Stability domain of blended Radau IIA embedding method

$$\varepsilon = \frac{-7}{3744}(\mathbf{A}h)^4 \tag{8.107}$$

The stability domain of the blended method is given in Figure 8.19.

The blended method relies on h not changing its values between the two steps used in the approximation. It may be easiest to prevent the step size from changing two steps in a row. This seems a small price to pay. After the step size has remained constant for two consecutive steps, it is free to change in any way suitable. The code needed to perform step–size control can be merged with the model equations and the simulation equations, i.e., it can be inlined as well, but this is not truly necessary. The step–size control code can be kept in a separate routine called upon by the simulation engine whenever needed.

Which of the two approximations should be propagated to the next step? The error coefficient of the embedding method is considerably smaller than that of Radau IIA. Hence on a first glance, it seems reasonable to propagate the approximation of the embedding technique. However, there are two problems with this choice.

First, the embedding technique was designed assuming that the Radau IIA result would be propagated. If the embedding technique is being propagated, the $\mathbf{F}$–matrices change, and the blended method may no longer be 3rd–order accurate.

Second, Figure 8.20 shows the damping plot of the embedding method. Comparing it with the damping plot of Radau IIA, it can be seen that the embedding method is not L–stable, i.e., the damping does not approach infinity as the eigenvalues of the model move further and further to the left.

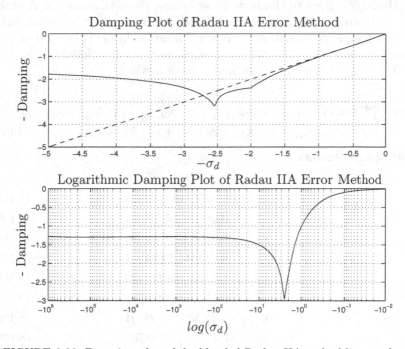

FIGURE 8.20. Damping plot of the blended Radau IIA embedding method

Thus, in spite of the smaller error coefficient, the embedding method should only be used for step–size control, not for propagation.

The fifth–order accurate Radau IIA method (Rad5) can be analyzed analogously. A single step of Rad5 stores six pieces of information: x_k, x_{1_k}, x_{2_k}, $\dot{x}_{1_k}$, $\dot{x}_{2_k}$, and $\dot{x}_{k+1}$, where x_{1_k} and x_{2_k} are the approximations of the two intermediate stages. There is only one 5th–order accurate polynomial going through these six pieces of information, and it is this polynomial that defines the Rad5 algorithm. Again, enough redundancy can be obtained to define an embedding algorithm if information from the two last steps is being used. In this case, the following 12 pieces of information are available: x_{k-1}, $x_{1_{k-1}}$, $x_{2_{k-1}}$, x_k, x_{1_k}, x_{2_k}, $\dot{x}_{1_{k-1}}$, $\dot{x}_{2_{k-1}}$, $\dot{x}_k$, $\dot{x}_{1_k}$, $\dot{x}_{2_k}$, and $\dot{x}_{k+1}$.

Searching for 6th–order polynomials going through seven of these twelve supporting values, there are 792 methods to be evaluated. Of those, 26 are A–stable methods that can be blended to form an alternate 5th–order accurate embedding method.

Although Rad5 as a whole is 5th–order accurate, its first two stages are only 3rd–order accurate. Thus, we should not expect any of these 6th–order polynomials to reach a higher order of approximation accuracy than three, and indeed, this is what we get. Hence we need to blend at least three of the methods to obtain a 5th–order accurate embedding method.

There exist 2600 combinations of blended methods from the 26 individual methods. We need to eliminate those among them that are not A–stable. We furthermore should choose a method with a small error coefficient and decent damping characteristics. It would be an additional benefit if we could come up with a method that has conveniently small rational coefficients.

A very good embedding method is the following:

$$
\begin{aligned}
\mathbf{x}_{k+1}^{\text{blended}} = {} & c_1 \cdot \mathbf{x}_{k-1} + c_2 \cdot \dot{\mathbf{x}}_{1_{k-1}} + c_3 \cdot \mathbf{x}_{2_{k-1}} + c_4 \cdot \dot{\mathbf{x}}_{2_{k-1}} + c_5 \cdot \mathbf{x}_k \\
& + c_6 \cdot \mathbf{x}_{1_k} + c_7 \cdot \dot{\mathbf{x}}_{1_k} + c_8 \cdot \mathbf{x}_{2_k} + c_9 \cdot \dot{\mathbf{x}}_{2_k} + c_{10} \cdot \dot{\mathbf{x}}_{k+1}
\end{aligned} \quad (8.108a)
$$

with the coefficients:

$$c_1 = -0.00517140382204 \tag{8.109a}$$
$$c_2 = -0.00094714677404 \tag{8.109b}$$
$$c_3 = -0.04060469717694 \tag{8.109c}$$
$$c_4 = -0.01364429384901 \tag{8.109d}$$
$$c_5 = +1.41786808325433 \tag{8.109e}$$
$$c_6 = -0.17475783086782 \tag{8.109f}$$
$$c_7 = +0.48299282769491 \tag{8.109g}$$
$$c_8 = -0.19733415138754 \tag{8.109h}$$
$$c_9 = +0.55942205973218 \tag{8.109i}$$
$$c_{10} = +0.10695524944855 \tag{8.109j}$$

We did program the computation of the coefficients also using MATLAB's symbolic toolbox, but the resulting expressions are quite awful, thus we decided to offer the numerical versions instead.

The blending method is indeed 5th–order accurate. It exhibits a nice convex A–stable stability domain, which is shown in Fig.8.21.

The damping plot exhibits a nice large asymptotic region and decent damping characteristics far out in the left–half complex $\lambda \cdot h$–plane. The method is not L–stable, but that is neither surprising nor truly necessary. The damping plot is presented in Fig.8.22.

8.8   Stiffly Stable Step–size Control of Lobatto IIIC

Let us now look at the Lobatto IIIC algorithm. Since the algorithm is less compact than the Radau IIA algorithms, it should be easier to find suitable

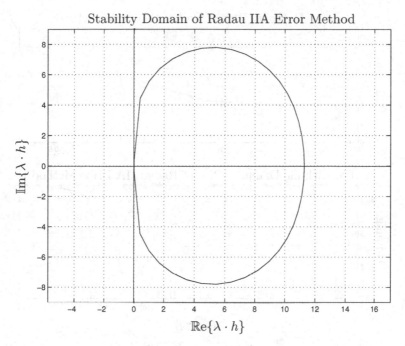

FIGURE 8.21. Stability domain of blended 5th–order Radau IIA embedding method

embedding methods. Yet, each algorithm is accompanied by its own set of difficulties.

First, we checked the order of approximation accuracy of the intermediate stages of the Lobatto IIIC algorithm. Unfortunately, they are only 2nd–order accurate. Hence we shall still need to blend three methods to raise the order of approximation accuracy of the embedding algorithm to four.

Secondly, although we again are working with 12 pieces of information across two steps, Lobatto IIIC has a peculiarity. It experiences a zero time advance between the third stage of one step and the first stage of the next. Although x_k and x_{1_k} represent the state vector at the same time instant, they are two different approximations. In particular, x_k is 4th–order accurate, whereas x_{1_k} is only 2nd–order accurate. The zero time advance reduces the flexibility in finding suitable error methods, as no individual error method can use both x_k and x_{1_k} simultaneously.

16 individual error methods were found that are all A–stable. Two of them are even L–stable. Of course, none of these error methods is of higher order of approximation accuracy than two.

We then proceeded to blend any three of these methods. The best among the 4th–order accurate blended methods is presented in the sequel. It can be written as:

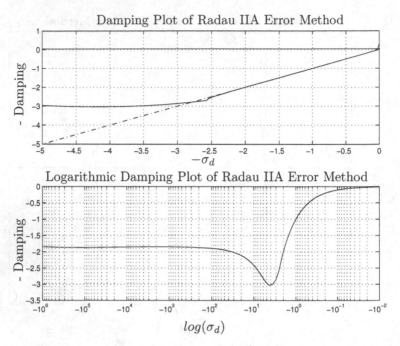

FIGURE 8.22. Damping plot of the blended 5th–order Radau IIA embedding method

$$
\begin{aligned}
\mathbf{x}_{k+1}^{blended} = {} & \frac{63}{4552} \cdot \mathbf{x}_{1_{k-1}} - \frac{91}{81936} \cdot \dot{\mathbf{x}}_{1_{k-1}} + \frac{1381}{81936} \cdot \dot{\mathbf{x}}_{2_{k-1}} + \frac{3101}{2276} \cdot \mathbf{x}_k \\
& - \frac{393}{4552} \cdot \mathbf{x}_{1_k} + \frac{775}{3414} \cdot \dot{\mathbf{x}}_{1_k} - \frac{165}{569} \cdot \mathbf{x}_{2_k} + \frac{62179}{81936} \cdot \dot{\mathbf{x}}_{2_k} \\
& + \frac{12881}{81936} \cdot \dot{\mathbf{x}}_{k+1}
\end{aligned}
\tag{8.110}
$$

The coefficients of the blending method were calculated using MATLAB's symbolic toolbox.

The embedding method offers a beautiful convex stability domain, as shown in Fig.8.23.

The damping characteristics of the embedding method are shown in Fig.8.24.

The method is characterized by a large asymptotic region and a decently large damping value far out in the left–half complex $\lambda \cdot h$ plane.

8.9   Inlining Partial Differential Equations

Let us return once more to the simulation of parabolic PDEs converted to sets of ODEs using the MOL approach. In Chapter 6 of this book, we

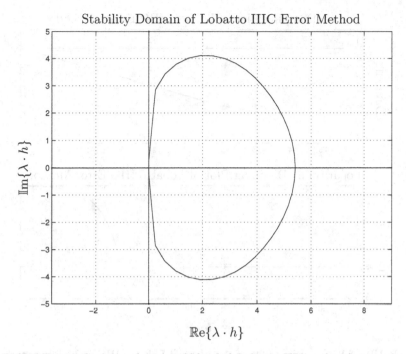

Stability Domain of Lobatto IIIC Error Method

FIGURE 8.23. Stability domain of blended Lobatto IIIC embedding method

simulated these types of problems using a stiff–system solver, such as a BDF algorithm. Whereas this approach worked quite well, the efficiency of the simulations was less than satisfactory. What killed our attempts at solving these problems efficiently was not the step size. The number of function evaluations was actually quite low, at least as long as we computed the Jacobian analytically. What made our simulations excruciatingly slow was the computation of the inverse Hessians.

Let us discuss once more the 1D heat diffusion problem discretized using 5th–order accurate central differences, as described in Eqs.(6.39a–h). We use 50 segments, $n = 50$.

Using the approach advertised in Chapter 6, we ended up with 50 ODEs, requiring a Hessian matrix of size 50×50 to be inverted. More precisely, a linear system of 50 equations in 50 unknowns had to be solved using Gaussian elimination during every iteration step.

Let us now apply inline integration to the problem. Let us start by inlining a variable step and variable order BDF algorithm. We can write the inlined ΔE system in matrix form as follows:

$$\dot{\mathbf{x}} = \mathbf{A} \cdot \mathbf{x} + \mathbf{b} \cdot u \qquad (8.111a)$$
$$\mathbf{x} = \text{old}(\mathbf{x}) + \bar{h} \cdot \dot{\mathbf{x}} \qquad (8.111b)$$

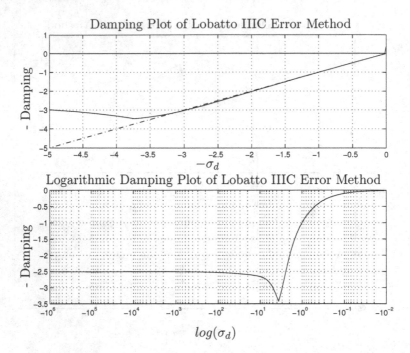

FIGURE 8.24. Damping plot of the blended Lobatto IIIC embedding method

where $\mathbf{A}$ is the matrix:

$$\mathbf{A} = \frac{n^2}{120\pi^2} \cdot \begin{pmatrix} -15 & -4 & 14 & -6 & 1 & \dots & 0 & 0 & 0 & 0 \\ 16 & -30 & 16 & -1 & 0 & \dots & 0 & 0 & 0 & 0 \\ -1 & 16 & -30 & 16 & -1 & \dots & 0 & 0 & 0 & 0 \\ 0 & -1 & 16 & -30 & 16 & \dots & 0 & 0 & 0 & 0 \\ \vdots & \vdots & \vdots & \vdots & \vdots & \ddots & \vdots & \vdots & \vdots & \vdots \\ 0 & 0 & 0 & 0 & 0 & \dots & 16 & -30 & 16 & -1 \\ 0 & 0 & 0 & 0 & 0 & \dots & -1 & 16 & -31 & 16 \\ 0 & 0 & 0 & 0 & 0 & \dots & 0 & -2 & 32 & -30 \end{pmatrix}$$

(8.112)

We can no longer hope to tear this equation system by hand. Thus, we encoded our heuristic procedure in a MATLAB routine, and applied that routine to the problem at hand. Since MATLAB works more naturally with matrices than with linked lists, we based our implementation on the structure incidence matrix instead of the structure digraph.

The structure incidence matrix for this problem is shown in Fig.8.25. We numbered the equations such that we started with the state variables, and concatenated them with the state derivatives.

Two trivial tearing structures come to mind immediately. We can either plug Eq.(8.111a) into Eq.(8.111b), and thereby eliminate the state derivatives from the set of iteration variables:

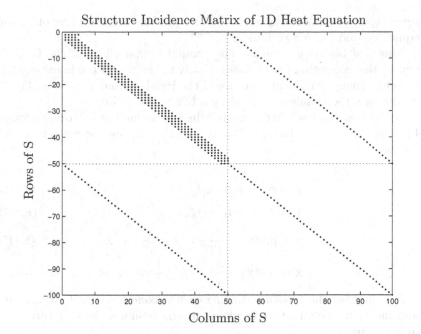

FIGURE 8.25. Structure incidence matrix of inlined 1D heat diffusion problem using a BDF algorithm.

$$\mathbf{x} = \text{old}(\mathbf{x}) + \bar{h} \cdot (\mathbf{A} \cdot \mathbf{x} + \mathbf{b} \cdot u) \qquad (8.113)$$

or alternatively, we can plug Eq.(8.111b) into Eq.(8.111a), and thereby eliminate the state variables from the set of iteration variables:

$$\dot{\mathbf{x}} = \mathbf{A} \cdot (\text{old}(\mathbf{x}) + \bar{h} \cdot \dot{\mathbf{x}}) + \mathbf{b} \cdot u \qquad (8.114)$$

In either case, we reduce the number of iteration variables back to 50, i.e., we end up with a Hessian of the same size as in Chapter 6.

Let us check, whether our heuristic procedure can do better. Applying the heuristic procedure as proposed, we obtain immediately a solution in 32 residual equations and 32 tearing variables. If we modify our heuristic procedure somewhat by searching for the number of additional equations to be causalized across all unknowns appearing in a candidate residual equation, rather than limiting the search to those variables with the largest number of black (solid) lines attached to them, we obtain a solution with 25 residual equations and 25 tearing variables.

Using this modified heuristic procedure, we have extended the search somewhat, thereby reducing the efficiency of the algorithm, but in return, we have obtained a more economical tearing structure.

We suspect that the solution with 25 iteration variables is indeed the optimal solution, but we are not sure of it. We did not attempt to solve the

np–complete exhaustive search across all possible combinations of residual equations and tearing variables.

This is a big improvement. The computational effort of the Gaussian elimination algorithm grows quadratically in the size of the linear equation system. Hence by reducing the size of the Hessian from 50×50 to 25×25, we increase the simulation speed by a full factor of four.

Let us now discuss what happens when we inline the 3rd–order accurate Radau IIA algorithm instead. Our ΔE system can now be written down as follows:

$$\dot{\mathbf{y}} = \mathbf{A} \cdot \mathbf{y} + \mathbf{b} \cdot u(t_{k+\frac{1}{3}}) \tag{8.115a}$$

$$\dot{\mathbf{x}} = \mathbf{A} \cdot \mathbf{x} + \mathbf{b} \cdot u(t_{k+1}) \tag{8.115b}$$

$$\mathbf{y} = \mathrm{pre}(\mathbf{x}) + \frac{5}{12} \cdot h \cdot \dot{\mathbf{y}} - \frac{1}{12} \cdot h \cdot \dot{\mathbf{x}} \tag{8.115c}$$

$$\mathbf{x} = \mathrm{pre}(\mathbf{x}) + \frac{3}{4} \cdot h \cdot \dot{\mathbf{y}} + \frac{1}{4} \cdot h \cdot \dot{\mathbf{x}} \tag{8.115d}$$

If we number the variables starting with $\mathbf{y}$, concatenating to it $\mathbf{x}$, then $\dot{\mathbf{y}}$, and finally $\dot{\mathbf{x}}$, the structure incidence matrix assumes the structure shown in Fig.8.26.

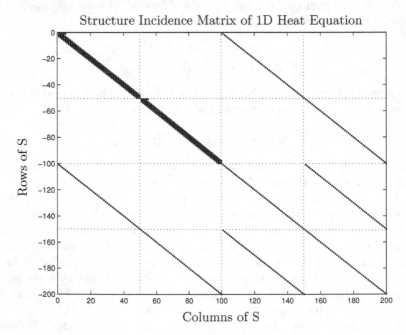

FIGURE 8.26. Structure incidence matrix of inlined 1D heat diffusion problem using Radau IIA.

Two trivial tearing structures come to mind. We can either plug the two sets of state equations into the two sets of integration equations, thereby eliminating all state derivatives from the set of iteration variables:

$$\mathbf{y} = \text{pre}(\mathbf{x}) + \frac{5}{12} \cdot h \cdot (\mathbf{A} \cdot \mathbf{y} + \mathbf{b} \cdot u(t_{k+\frac{1}{3}}))$$

$$- \frac{1}{12} \cdot h \cdot (\mathbf{A} \cdot \mathbf{x} + \mathbf{b} \cdot u(t_{k+1})) \tag{8.116a}$$

$$\mathbf{x} = \text{pre}(\mathbf{x}) + \frac{3}{4} \cdot h \cdot (\mathbf{A} \cdot \mathbf{y} + \mathbf{b} \cdot u(t_{k+\frac{1}{3}}))$$

$$+ \frac{1}{4} \cdot h \cdot (\mathbf{A} \cdot \mathbf{x} + \mathbf{b} \cdot u(t_{k+1})) \tag{8.116b}$$

or alternatively, we can plug the two sets of integration equations into the two sets of state equations, thereby eliminating all state variables from the set of iteration variables:

$$\dot{\mathbf{y}} = \mathbf{A} \cdot (\text{pre}(\mathbf{x}) + \frac{5}{12} \cdot h \cdot \dot{\mathbf{y}} - \frac{1}{12} \cdot h \cdot \dot{\mathbf{x}}) + \mathbf{b} \cdot u(t_{k+\frac{1}{3}}) \tag{8.117a}$$

$$\dot{\mathbf{x}} = \mathbf{A} \cdot (\text{pre}(\mathbf{x}) + \frac{3}{4} \cdot h \cdot \dot{\mathbf{y}} + \frac{1}{4} \cdot h \cdot \dot{\mathbf{x}}) + \mathbf{b} \cdot u(t_{k+1}) \tag{8.117b}$$

In either case, we end up with 100 iteration variables.

Let us see whether our heuristic procedure can do better. Unfortunately, the algorithm breaks down after having chosen about 60 tearing variables, and after having causalized about 120 equations. The heuristic procedure has maneuvered itself into a corner. Every further selection of a combination of residual equation and tearing variable leads to a structural singularity. Although the algorithm had been programmed to ignore selections that would lead to a structural singularity at once, it hadn't been programmed to backtrack beyond the last selection, i.e., throw earlier residual equations and tearing variables away to avoid future mishap.

This is why we wrote in Chapter 7 that the computational complexity of the heuristic procedure grows quadratically with the size of the equation system *for most applications*. It does so, if no backtracking is required.

I was curious how the tearing algorithm built into Dymola would fare when faced with this problem. I quickly programmed the equation system into Dymola Version 4.1d, and asked for a compilation. Whereas Dymola usually tears equation systems with tens of thousands of equations within a few seconds, it chewed on this problem for a very long time. I watched an entire movie (Animal Farm) on TV, while Dymola was thinking about the problem.

It turned out that the heuristic algorithm built into Version 4.1d of Dymola did not break down. Evidently, it is programmed to backtrack sufficiently to get itself out of a corner. Unfortunately after thinking hard, Dymola came up with one of the two trivial tearing structures.

The above paragraph had been written almost two years ago. Now, before sending the manuscript off to the printer, we decided to run the example once more through the current version of Dymola, which is Version 5.3d. This time around, Dymola came up with an answer after only six seconds of compilation time.

We had sent an earlier version of this chapter to Hilding Elmqvist. Whenever someone stumbles upon an example that the tearing algorithm does not handle well, the good folks up at Dynasim go into overdrive, trying to come up with an improved version of their tearing algorithm as fast as they can.

The answer, however, was still the same. Dymola chose one of the two trivial structures as the most suitable tearing structure for this system. We thus suspect that the trivial structures are indeed the optimal tearing structures in this case, but of course, we aren't sure. Going through an exhaustive search for finding the optimal tearing structure would be too painful to even consider.

Unfortunately, these are bad news. If indeed we pay for using Radau IIA instead of BDF3 with increasing the size of the Hessian by a factor of four, Radau IIA would have to be able to use step sizes that are on average at least 16 times larger than those used by BDF for the same accuracy. Otherwise, Radau IIA is not competitive for dealing with this problem. We doubt very much that Radau IIA will be able to do so.

PDE problems are notoriously difficult simulation problems. Although tearing is a very powerful symbolic sparse matrix technique, it cannot make an intrinsically difficult problem easy to solve.

8.10   Overdetermined DAEs

At this point, we shall resume the discussion of the mechanical pendulum problem that we began towards the end of Chapter 7. The mechanical pendulum schematic is presented once more in Fig.8.27.

We had already come up with a set of causal equations without solvability issues describing the motion of the mechanical pendulum. Let us use the variable substitution technique to come up with a closed–form formula for the tearing variable. Doing so, we find the following explicit ODE description of the pendulum problem.

$$x = \ell \cdot \sin(\varphi) \tag{8.118a}$$

$$v_x = \ell \cdot \cos(\varphi) \cdot \dot\varphi \tag{8.118b}$$

$$y = \ell \cdot \cos(\varphi) \tag{8.118c}$$

$$v_y = -\ell \cdot \sin(\varphi) \cdot \dot\varphi \tag{8.118d}$$

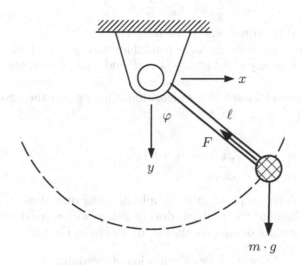

FIGURE 8.27. Mechanical pendulum.

$$dv_x = -\frac{x \cdot \ell \cdot \dot{\varphi}^2 + x \cdot \cos(\varphi) \cdot g}{x \cdot \sin(\varphi) + y \cdot \cos(\varphi)} \qquad (8.118e)$$

$$\ddot{\varphi} = \frac{dv_x}{\ell \cdot \cos(\varphi)} + \frac{\sin(\varphi)}{\cos(\varphi)} \cdot \dot{\varphi}^2 \qquad (8.118f)$$

$$dv_y = -\ell \cdot \sin(\varphi) \cdot \ddot{\varphi} - \ell \cdot \cos(\varphi) \cdot \dot{\varphi}^2 \qquad (8.118g)$$

$$F = \frac{m \cdot g \cdot \ell}{y} - \frac{m \cdot \ell \cdot dv_y}{y} \qquad (8.118h)$$

Eq.(8.118e) could have been simplified further, but this is unimportant for the discussion at hand. The formula, as presented above, is the one that Dymola will come up with, since it knows that $\sin^2 \varphi + \cos^2 \varphi = 1$, but it doesn't know to multiply both the numerator and the denominator with ℓ to eliminate the trigonometric functions from the expression.

We know that the pendulum, as described, is a conservative (Hamiltonian) system, since no friction was assumed anywhere. Hence the pendulum, once disturbed, should swing forever with the same frequency and amplitude. The total free energy, E_f:

$$E_f = E_p + E_k \qquad (8.119)$$

which is the sum of the potential energy, E_p, and the kinetic energy, E_k, should be constant. The potential energy can be modeled as:

$$E_p = m \cdot g \cdot (y_0 - y) \tag{8.120}$$

and the kinetic energy can be expressed using the formula:

$$E_k = \frac{1}{2} \cdot m \cdot v_x^2 + \frac{1}{2} \cdot m \cdot v_y^2 \tag{8.121}$$

Let us add these three equations to the model.

Let us simulate this problem for a pendulum with $g = 9.81 \ m/(sec^2)$, $m = 10 \ kg$, $\ell = 1 \ m$, $\varphi_0 = +45° = \pi/4 \ rad$, and $\dot{\varphi}_0 = 0 \ rad/sec$. Thus, $y_0 = \sqrt{2}/2 \ m$.

We shall inline the forward Euler algorithm, thus we add the two equations:

$$\dot{\varphi}_{k+1} = \dot{\varphi}_k + h \cdot \ddot{\varphi}_k \tag{8.122a}$$

$$\varphi_{k+1} = \varphi_k + h \cdot \dot{\varphi}_k \tag{8.122b}$$

We can now simulate the problem by simply iterating over these 13 equations. We shall simulate the problem during 10 sec with a fixed step size of $h = 0.01$. The results of this simulation are shown in Fig.8.28.

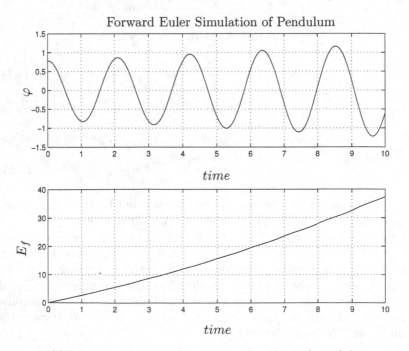

FIGURE 8.28. Inlined FE simulation of mechanical pendulum.

We just solved the world's energy crisis once and for all. Evidently, we are able to generate free energy out of thin air.

Let us see whether backward Euler fares any better. Instead of inlining Eqs.(8.122a–b), we inline the equations:

$$\dot\varphi = \mathrm{pre}(\dot\varphi) + h \cdot \ddot\varphi \qquad (8.123a)$$

$$\varphi = \mathrm{pre}(\varphi) + h \cdot \dot\varphi \qquad (8.123b)$$

Since the backward Euler algorithm is an implicit integration method, we expect to encounter another algebraic loop. The partially causalized structure digraph for this equation system is shown in Fig.8.29.

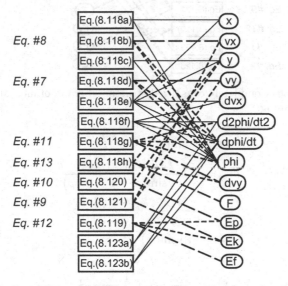

FIGURE 8.29. Partially causalized structure digraph of mechanical pendulum after BE inlining.

We indeed encountered an algebraic loop in six equations and six unknowns. Figure 8.30 shows the completely causalized equation system after a residual equation and a tearing variable have been chosen.

In mechanical systems, it is generally a good idea to select accelerations as tearing variables, and since the model equations had already been causalized before, i.e., each variable appears exactly once to the left side of the equal sign and does so in a linear fashion, it makes sense to use the equation that defines the angular acceleration $\ddot\varphi$ as the residual equation.

The causal equations can be read out of Fig.8.30. They are:

$$\dot\varphi = \mathrm{pre}(\dot\varphi) + h \cdot \ddot\varphi \qquad (8.124a)$$

$$\varphi = \mathrm{pre}(\varphi) + h \cdot \dot\varphi \qquad (8.124b)$$

$$y = \ell \cdot \cos(\varphi) \qquad (8.124c)$$

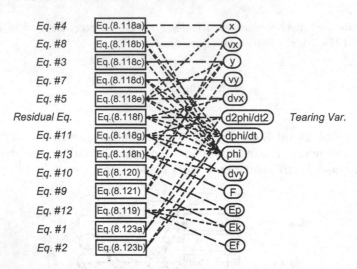

FIGURE 8.30. Completely causalized structure digraph of mechanical pendulum after BE inlining.

$$x = \ell \cdot \sin(\varphi) \tag{8.124d}$$

$$dv_x = -\frac{x \cdot \ell \cdot \dot{\varphi}^2 + x \cdot \cos(\varphi) \cdot g}{x \cdot \sin(\varphi) + y \cdot \cos(\varphi)} \tag{8.124e}$$

$$\ddot{\varphi} = \frac{dv_x}{\ell \cdot \cos(\varphi)} + \frac{\sin(\varphi)}{\cos(\varphi)} \cdot \dot{\varphi}^2 \tag{8.124f}$$

$$v_y = -\ell \cdot \sin(\varphi) \cdot \dot{\varphi} \tag{8.124g}$$

$$v_x = \ell \cdot \cos(\varphi) \cdot \dot{\varphi} \tag{8.124h}$$

$$E_k = \frac{1}{2} \cdot m \cdot v_x^2 + \frac{1}{2} \cdot m \cdot v_y^2 \tag{8.124i}$$

$$E_p = m \cdot g \cdot (y_0 - y) \tag{8.124j}$$

$$dv_y = -\ell \cdot \sin(\varphi) \cdot \ddot{\varphi} - \ell \cdot \cos(\varphi) \cdot \dot{\varphi}^2 \tag{8.124k}$$

$$E_f = E_p + E_k \tag{8.124l}$$

$$F = \frac{m \cdot g \cdot \ell}{y} - \frac{m \cdot \ell \cdot dv_y}{y} \tag{8.124m}$$

The first six of these equations, Eqs.(8.124a–f), constitute the algebraic loop. This time, we used Newton iteration in the single tearing variable, $\ddot{\varphi}$, to solve the algebraic loop. We computed the Hessian by means of algebraic differentiation.

The simulation results are shown in Fig.8.31.

It didn't work any better than before. This algorithm is losing energy, where it shouldn't. The result is easily explainable. This is a conservative system. The two eigenvalues of its Jacobian are located on the imaginary axis, at least on average. However, the numerical stability domain of the FE

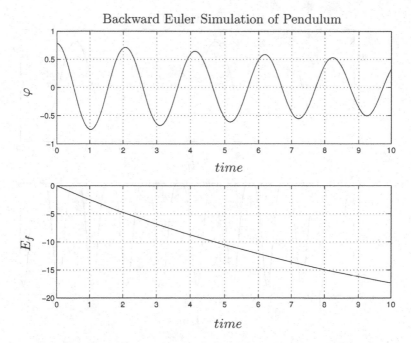

FIGURE 8.31. Inlined BE simulation of mechanical pendulum.

algorithm loops into the left–half complex plane. Thus, the two eigenvalues of the system are seen as mildly unstable, and the oscillation is growing. The algorithm adds energy to the system. On the other hand, the stability domain of the BE algorithm loops into the right–half complex plane. Consequently, the two marginally stable eigenvalues are seen as mildly damped, and the oscillation decays.

An F–stable algorithm, such as the BI technique, should be expected to work better. Let us implement BI2 as a cyclic method, toggling between a step of FE followed by a step of BE. The simulation results are presented in Fig.8.32.

The approach worked. Yet, it only worked, because we were able to analyze the problem and come up with a suitable solution. The code itself still has no inkling that it is supposed to conserve the free energy. It does so by accident rather than by design.

Let us try to change that. We shall force the backward Euler algorithm to preserve the free energy. To this end, we simply add the equation:

$$E_f = 0 \qquad (8.125)$$

to the set of equations.

This is a completely new situation. We haven't added any new variables to the set of equations. We only added another equation. Thus, we now have 14 equations in 13 unknowns. Clearly, this problem is constrained.

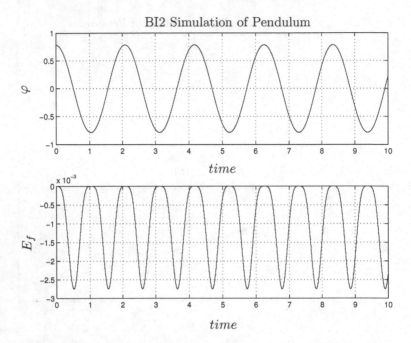

FIGURE 8.32. Inlined BI2 simulation of mechanical pendulum.

If we present this problem to the Pantelides algorithm, it will differentiate itself to death, or rather, until the compiler runs out of virtual memory. The Pantelides algorithm always adds exactly as many equations as variables, thus after each application of the algorithm, the number of equations is still one larger than the number of variables.

Inlining again saves our neck. We simply add the constraint equation to the iteration equations of the Newton iteration. Thus, the set of zero functions can now be written as:

$$\mathcal{F} = \begin{pmatrix} \ddot{\varphi}_{new} - \ddot{\varphi} \\ E_f \end{pmatrix} \qquad (8.126)$$

and therefore:

$$\mathcal{H} = \begin{pmatrix} \partial \ddot{\varphi}_{new}/\partial \ddot{\varphi} - 1 \\ \partial E_f/\partial \ddot{\varphi} \end{pmatrix} \qquad (8.127)$$

The Newton iteration can be written as:

$$\mathcal{H}^\ell \cdot \mathbf{dx}^\ell = \mathcal{F}^\ell \qquad (8.128a)$$

$$\mathbf{x}^{\ell+1} = \mathbf{x}^\ell - \mathbf{dx}^\ell \qquad (8.128b)$$

However, $\mathcal{H}$ is no longer a square matrix. It is now a rectangular matrix with 2 rows and 1 column. In general with n model equations and p constraints, the Hessian turns out to be a rectangular matrix with $n + p$ rows and n columns. Thus, Eq.(8.128a) is overdetermined. It cannot be satisfied exactly. The $\mathbf{dx}$–vector can only be determined in a least square sense. This can be accomplished by multiplying Eq.(8.128a) from the left with $\mathcal{H}^*$, i.e., with the Hermitian transpose of $\mathcal{H}$. If the rank of $\mathcal{H}$ is n, then $\mathcal{H}^* \cdot \mathcal{H}$ is a Hermitian matrix of full rank. Thus, we can compute $\mathbf{dx}$ as:

$$\mathbf{dx} = (\mathcal{H}^* \cdot \mathcal{H})^{-1} \cdot \mathcal{H}^* \cdot \mathcal{F} \qquad (8.129)$$

where $(\mathcal{H}^* \cdot \mathcal{H})^{-1} \cdot \mathcal{H}^*$ is called the *Penrose–Moore pseudoinverse* of $\mathcal{H}$. In MATLAB, this can be abbreviated as:

$$\mathbf{dx} = \mathcal{H} \backslash \mathcal{F} \qquad (8.130)$$

The results of the simulation are shown in Fig.8.33.

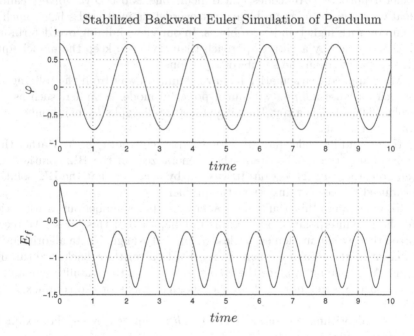

FIGURE 8.33. Inlined stabilized BE simulation of mechanical pendulum.

The oscillation has indeed been stabilized. Of course, the equation:

$$\mathcal{F} = 0 \qquad (8.131)$$

can no longer be solved precisely. The equation system does not contain enough freedom to do so. Yet, the error is minimized in a least square sense, and both the oscillation and the free energy are now stable by design. Initially, the approach still loses a bit of free energy, but the loss stops after the solution is stabilized. The solution using backinterpolation turns out to be better, but the solution using an overdetermined equation set is more robust.

There are DAE solvers on the market that can handle overdetermined DAEs, such as ODASSLRT (a "dialect" of DASSL) [8.12] and MEXX (a code using Richardson extrapolation) [8.20]. Overdetermined DAE solvers have become popular primarily among specialists of multibody dynamics, and the early codes tackling this problem indeed evolved in the engineering community. Most of these early codes were quite specialized. More recently, the problem was discovered by mainstream applied mathematicians [8.15], and it can therefore be expected that more general–purpose codes for the numerical solution of overdetermined DAE systems will soon become available.

Yet, the problem of merging overdetermined linear system solvers with general–purpose DAE codes is a difficult one. Most DAE solvers cannot deal with higher–index problems, yet overdetermined DAEs have much in common with higher–index problems. In our view, inlining overdetermined DAEs is generally a better approach than trying to keep the model equations separate from the simulator equations.

Most applied mathematicians have shunned away from the inlining approach, because inlining without a powerful model compiler, such as Dymola [8.9, 8.10], is a toy. Drawing structure digraphs by hand only works for toy problems.

Hairer and his colleagues thus went a different route [8.15]. Rather than constraining the DAE system, they generalized on the BI2 solution presented earlier in this section. It was not by accident that the BI2 solution produced the correct answer to the problem.

To understand this result, the reader needs to remember our introduction to the backinterpolation algorithms in Chapter 3 of this book. We recognized that we can implement a class of implicit Runge–Kutta algorithms by integrating the regular explicit Runge–Kutta algorithms backward through time. As we simply replace the step size h by $-h$, the stability domains of these methods get mirrored on the imaginary axis of the complex $\lambda \cdot h$–plane.

Some algorithms do not change, when h is replaced by $-h$. For example, the trapezoidal rule:

$$x_{k+1} = x_k + \frac{h}{2} \cdot (\dot{x}_k + \dot{x}_{k+1}) \tag{8.132}$$

turns into:

$$x_k = x_{k+1} - \frac{h}{2} \cdot (\dot{x}_{k+1} + \dot{x}_k) \tag{8.133}$$

i.e., the formula doesn't change. Such an ODE solver is called a *symmetric integration algorithm*. The stability domains of symmetric ODE solvers are symmetric to the imaginary axis. In particular, all of the F–stable integration algorithms introduced in Chapter 3 are symmetric ODE solvers.

Symmetric integration algorithms are not only symmetric to the imaginary axis w.r.t. their stability properties, but also w.r.t. their damping properties. Thus, symmetric integration algorithms are accompanied by symmetric order stars as well.

This symmetry can be exploited in the simulation of Hamiltonian systems. At least, if we carefully choose our step size to be in sync with the eigenfrequency of oscillation of the system, we can ensure that the damping errors committed during the integration over a full period cancel out, such that the solution at the end of one cycle coincides with that at the beginning of the cycle.

Yet, we still prefer the constrained solution proposed in this section, as it is considerably more robust. It works with any numerical integration scheme and enforces the physical constraint directly rather than indirectly.

8.11   Electronic Circuit Simulators

One important application area where DAEs are frequently used is electronic circuit modeling. Let us briefly relate the topics of Chapters 7 and 8 of this book to the discussions presented in Chapters 3 and 6 of the companion book on *Continuous System Modeling* [8.5].

We have seen that object–oriented modeling of electrical and electronic circuits invariably leads to implicit DAE descriptions. We have furthermore seen that the resulting sets of DAEs are often index–2 models.

You, the reader, may have come to the conclusion that whether or not a set of DAEs describing an electrical circuit presents itself as a higher–index model depends on the topology of the circuit. However, that conclusion is too simple. The perturbation index of a model is influenced by the abstraction mechanism chosen in its mathematical description. In the case of nonlinear systems, it even depends on the selection of state variables, as nonlinear transformations performed on a model can influence the perturbation index.

In an explicit ODE description of a system, the state variables are predetermined. They are simply the outputs of the integrators. However in an implicit DAE description, the answer is no longer as clear cut. First and second derivatives can show up multiple times anywhere within the implicit equations. How do we even know, how many degrees of freedom a DAE model really has? Here we have a true choice in deciding, which state variables to use, and the perturbation index of the model is dependent on that choice.

Inline integration blurs the situation even further. After inlining, all variables have become algebraic variables. The number of initial values needed to simulate an inlined model depends on the number of tearing variables. We need to specify one initial guess for each tearing variable, as well as initial conditions for all variables that appear in 'pre(.)' clauses.

In the companion book, we started out in Chapter 3 explaining the derivations of circuit equations in terms of either *mesh currents* or *cut-set potentials*. We chose a "tree" that defined either a minimal set of mesh currents or a minimal set of cutset potentials.

We now understand much better, what it is that we accomplished. We designed techniques to come up with small sets of *tearing variables*. The mesh currents assume the role of tearing variables, if we work with mesh equations, whereas the cutset potentials assume the same role, when we work with cutset (node) equations.

Branch currents and/or branch voltages are poor choices as state variables, because they frequently lead to higher–index DAE models. If we place two capacitors in parallel, we cannot choose the voltages across these two capacitors as independent state variables. Similarly, if we place two inductors in series, we cannot select the currents flowing through them as independent state variables. If we work with such selections, we invariable end up with index–2 models.

The problem disappears if we choose a subset of either mesh currents of cutset potentials as state variables. These are always independent of each other by design. The most difficult problem is to decide, which variables and how many to select as tearing variables.

Commercial electronic circuit simulators, such as Spice [8.26, 8.39] or Saber [8.24], work with the node potentials as their tearing variables [8.40]. Yet, rather than substituting the equations into each other, as we proposed in the companion book, they simply write the equations down as is, and iterate on them using either Newton iteration or at least fixed–point iteration.

Let us start with the simplest case: an electronic circuit without voltage sources and inductors. Spice [8.26, 8.39] uses all of the node potentials as tearing variables. In Spice, this is called the *nodeset*. Evidently, this is not a minimal set, but it makes the algorithm of finding the tearing variables trivial. In fact, there are even two different nodesets in use. The *reduced nodeset* contains all of the node potentials of user–defined nodes, whereas the *complete nodeset* also includes the internal nodes of the circuit expansions of the active devices (primarily BJT and MOS transistors).

We can formulate *Kirchhoff's Current Law (KCL)* as follows:

$$\mathbf{\Psi} \cdot \mathbf{i} = 0 \qquad (8.134)$$

where $\mathbf{i}$ is the vector of branch currents, and $\mathbf{\Psi}$ is the *reduced node incidence matrix*. $\mathbf{\Psi}$ has as many rows as there are nodes in the circuit, and it has as many columns as there are branches. It is called the *reduced* node incidence matrix, because it only considers those nodes and branches that are explicitly formulated in the model, excluding the expansions of the active devices. The element Ψ_{ij} has a value of $+1$, when the branch leaves the node, i.e., when the positive direction of the branch current is away from the node. If assumes a value of -1, if the branch arrives at the node, and it assumes a value of 0, if the branch is not connected to the node at all. Eq.(8.134) simply states that the sum of all currents into a node is zero.

The equation:

$$\mathbf{u} = \mathbf{\Psi}^{\mathbf{T}} \cdot \mathbf{v} \tag{8.135}$$

relates the node potentials $\mathbf{v}$ to the branch voltages $\mathbf{u}$.

Finally, the equation:

$$\mathbf{i} = \mathbf{g}(\mathbf{u}, \dot{\mathbf{u}}, \mathbf{v}, t) \tag{8.136}$$

captures the element law for each of the branches, relating the voltages and potentials to the currents.

Capacitors are implemented using the DAE formulation of a BDF formula, i.e.:

$$\dot{\mathbf{u}}_{\mathbf{C}} = \frac{\mathbf{u}_{\mathbf{C}} - \text{old}(\mathbf{u}_{\mathbf{C}})}{h} \tag{8.137}$$

Notice that Spice made use of this approach years before DASSL was written, but since the program was specialized to dealing with electronic circuits only, the mathematical community hardly paid any attention to it.

If all elements are either current sources, or resistors (including the nonlinear diodes), or capacitors, we are already in business. If we assume the node potentials, $\mathbf{v}$, as known, we can use Eq.(8.135) to determine all branch voltages, $\mathbf{u}$. We can then use the implicit numerical differentiators of Eq.(8.137) to compute the derivatives of the voltages for each of the capacitors. We can then use the elemental laws for each branch, Eq.(8.136), to compute the branch currents.

Hence we can set up the Newton iteration as follows:

$$\mathcal{F} = \mathbf{\Psi} \cdot \mathbf{i} = 0 \tag{8.138a}$$

$$\mathcal{H} = \mathbf{\Psi} \cdot \frac{\partial \mathbf{i}}{\partial \mathbf{v}} \tag{8.138b}$$

$$\mathbf{v}^{\ell+1} = \mathbf{v}^{\ell} - \mathcal{H} \backslash \mathcal{F} \tag{8.138c}$$

$\mathcal{H}$ is a square matrix with as many rows and columns as the circuit has nodes.

For the Newton iteration to converge properly, the circuit simulator will need a consistent set of initial values for all tearing variables. This is why Spice needs to compute an *OP–point*, i.e., a consistent set of initial conditions for the loop variables, before the "transient analysis" (simulation) can begin. If the initial OP–point does not converge, the program is in difficulties.

To overcome this problem, some Spice dialects offer automated *source ramping*. If all sources are initially set to zero and if all active devices are switched off, the initial nodeset is trivial: all node potentials are equal to zero. Then, the voltages are smoothly ramped up to their desired initial values in a pre–simulation run, and are then kept at their final (initial) values for some time to give the circuit a chance to settle into a steady state. The resulting node potentials are then used as the initial nodeset for the subsequent true transient analysis. Due to the special nature of circuit topologies, we have a simple and systematic way of determining a consistent set of initial conditions, a luxury that we do not have in all DAE simulations. Ramping had been described in Chapter 6 of the companion book.

How do we deal with inductors? Inductors can be implemented in similar ways as the capacitors. However, rather than using the BDF formula in its derivative form, we use it in its integral form:

$$\mathbf{i_L} = \text{old}(\mathbf{i_L}) + \bar{h} \cdot \frac{d\mathbf{i_L}}{dt} \tag{8.139}$$

Given the branch voltage, we compute the derivative of the current using the elemental law, then use the BDF formula in its integral form to find the current.

How do we deal with the ideal independent voltage sources? In the companion book, we proposed to move independent voltage sources into neighboring branches of the circuit. Commercial circuit simulators do it differently.

Let us assume an ideal voltage source is placed in branch i, which is located between node j and node k. The current through the voltage source is free to assume any value it needs to assume. It is not constrained by an elemental law.

Consequently, we can eliminate one row of the $\mathcal{F}$ vector, either the element number j, or the element number k, corresponding to either the jth or the kth row of the $\mathbf{\Psi}$ matrix. We add the equation specified by the eliminated row to the set of equations computing the currents, Eq.(8.136), solved for the unknown current through the voltage source.

Since the number of nodes has remained the same as before, we are now lacking one equation in $\mathcal{F}$ for the Newton iteration. We replace it by the new zero function:

$$v_j - v_k - u_i = 0 \tag{8.140}$$

We can now proceed as before.

This is a fairly generic description of how electronic circuit simulators may be implemented. The different simulators on the market vary in implementational details of how they make use of the above equations. In the circuit simulation literature, Eqs.(8.134–8.136) are generally called the *Sparse Tableau Equations* [8.14, 8.25, 8.26].

Many circuit simulators shun away from estimating the complete Hessian, and therefore, limit themselves to a fixed–point iteration only. In that case, we must iterate over all the loop variables, i.e., the tearing approach breaks down. In general, we are dealing with $n_n + 2 \, n_b$ equations in the same number of unknowns, where n_n denotes the number of nodes, and n_b

is the number of branches. It may therefore be advantageous to reduce the number of variables contained in the loop. To this end, Eq.(8.134) can be combined with Eq.(8.136) in the following way:

$$\mathbf{\Psi} \cdot \mathbf{g}(\mathbf{u}, \dot{\mathbf{u}}, \mathbf{v}, t) = 0 \qquad (8.141)$$

thereby eliminating the currents altogether from the iteration loop. Again, there exist different variations of this general scheme, usually referred to in the literature under the name *Modified Nodal Analysis (MNA)* [8.18, 8.25, 8.26].

All of the classical circuit simulators have in common that they limit the symbolic preprocessing to the interpretation of the network topology. The entire analysis is done numerically, using the equations pretty much as they come. Contrary to a general–purpose DAE solver, such as DASSL, the integration of the storage variables is performed in a decentralized manner, i.e., for each storage element separately.

The approach only works, because the structure of all equations is predetermined. For this reason, circuit simulators cannot be combined with other tools to form e.g. mechatronic system simulators. Even the thermal analysis offered by the traditional circuit simulators is fairly limited. They all allow a user to simulate a circuit at different temperatures (the circuit parameter values, such as R and C, can be specified as functions of temperature), but it is impossible to simulate how the flow of electrical current through the circuit heats up the circuit, and then simulate the effects of the change in temperature on the circuit's electrical performance simultaneously. This cannot be done, because the structure of the equations would change in such a way that it would violate the assumptions on which the modified nodal analysis are based.

A mixed symbolic and numerical approach, as pursued e.g. in Dymola, is therefore considerably more flexible and powerful. To preserve this generality, the developers of Dymola made it a point to make sure that Dymola knows absolutely nothing about physics. All it understands are mathematical algorithms. The entire physical knowledge is encoded in the models themselves, not in the underlying algorithms that are built (hardwired) into Dymola.

The price that we pay for this generality is small. Electronic circuit simulations performed by Dymola are as fast and accurate as their Spice or Saber counterparts. Furthermore, the maintenance of the electronic model library is considerably easier in Dymola than in either Spice or Saber, because of the object–oriented nature of the Dymola modeling environment. Furthermore, Dymola enables the user to simulate electronic circuits that are parts of larger mechatronic systems in a mechatronic system simulation. They also allow the electrical and thermal interactions of integrated circuits to be explored in full, e.g. in the design of packages [8.35, 8.36].

8.12   Multibody System Dynamics Simulators

Whereas Chapters 3 and 6 of the companion book describe fairly accurately the state–of–the–art of electronic circuit modeling, Chapters 4 and 7 *don't* describe state–of–the–art multibody system (MBS) dynamics modeling. This topic is simply too advanced to be presented in full in a general–purpose modeling class. The companion book limited a detailed discussion to one–dimensional devices. Unfortunately, this view does not extend smoothly to two– or even three–dimensional devices, such as robots or vehicles.

The problem is the following: Asking a user to come up with an ODE model describing the dynamics of a complicated MBS is not a practical proposition. DAE models, on the other hand, are fairly easy to derive. To this end, one simply describes the dynamics of each body separately, and adds the interactions between bodies as constraint equations. However, this usually leads to index 3 models with nasty nonlinear constraints. Relying on the Pantelides algorithm to blindly reduce the index down to index 1 will lead to an explosion in the complexity of the resulting equations, unless the user is very cautious about how he or she chooses the variables in the model. Furthermore, it often leads to models with solvability issues.

Selection of an appropriate coordinate system is absolutely essential. In the case of tree–structured robots, special selections of generalized coordinates both for the description of the *direct MBS dynamics* (motor torques are inputs, and positions and velocities of the end–effector are outputs), as well as for the description of *inverse MBS dynamics* (desired end–effector positions and velocities are inputs, and necessary motor torques to achieve those are outputs) have been found that don't lead to algebraic loops at all. Using these generalized coordinates, the number of equations will grow linearly in the number of bodies described in the model. Algorithms implementing this methodology are called *order–n* algorithms or order–f algorithms, depending on the particular reference consulted [8.2, 8.11, 8.21, 8.32].

MBS topologies with closed kinematic loops are more problematic, and the final word on how to efficiently model such systems hasn't been spoken yet. However, let us at least explain briefly how such systems are currently being modeled. Any tree–structured MBS can be brought into the form:

$$\mathbf{M}(\mathbf{q},t) \cdot \ddot{\mathbf{q}} = \mathbf{h}(\mathbf{q},\dot{\mathbf{q}},t) + \mathbf{f}(\mathbf{q},t) \qquad (8.142)$$

where $\mathbf{q}$ are the generalized positions (including angular positions) of the tree-structured MBS, $\mathbf{M}$ is the so–called mass matrix, $\mathbf{h}$ models the effects of body–fixed coordinate systems (Coriolis and centripetal forces) as well as friction phenomena, and $\mathbf{f}$ are the generalized forces (including torques) acting on the joints.

If an MBS has kinematic constraints (closed kinematic loops), we can first cut these constraints open, thereby transforming the kinematically constrained MBS into a tree–structured MBS. For this so modified MBS, we can derive Eq.(8.142), e.g. using the algorithm described in [8.21]. We then add the constraints as additional constraint equations back into the overall DAE description, thereby transforming the carefully formulated index 1 model back into an index 3 model. Luckily, there often aren't too many of these additional constraint equations, and the Pantelides algorithm may work quite decently. The resulting index 1 model can then either be solved directly using a DAE solver (possibly using a symbolically generated Hessian matrix), or we can try to reduce the model further to index 0 by solving the algebraic loops. Luckily, the special structure of mechanical manipulators suggests immediately a set of tearing variables, namely the generalized accelerations, $\ddot{q}$.

Meanwhile, an MBS library has been designed for Dymola [8.30] that enables even non–specialists of MBS dynamics to formulate efficient sets of DAEs for multibody systems in an effective object–oriented manner. The library contains models for most of the components that a user might need, such as different types of joints (revolute joints, prismatic joints, screws, etc.), rigid bodies and their connections, different types of force elements, and so on. A top–down description of the topology of an arbitrarily connected three–dimensional (or two–dimensional) tree–structured robot in an object–oriented fashion is made a fairly simple undertaking using the MBS library. The generated code compares favorably with other commercial MBS systems such as Adams [8.17, 8.27], or SD–Fast [8.19] in terms of run–time efficiency.

However, contrary to the more specialized tools, Dymola lends itself elegantly to modeling and simulation of general mechatronic systems, i.e., the drive chains, motors, and controllers of these robots can be described together with the MBS dynamics in a unified framework [8.3, 8.29, 8.30, 8.33].

Dymola does a fairly good job of coming up on its own with suitable tearing structures even in the case of closed kinematic loops. However, Dymola's multibody systems (MBS) library [8.30] still supports Dymola in this task by making sure that the (fully automated) tearing algorithm starts out with a suitable set of equations. Let us explain.

In the MBS library, *translational variables* are being carried along in the inertial frame, whereas *rotational variables* are described in a body–centric coordinate system. This by itself already helps with generating efficient simulation code. Yet, the decision requires that the library perform coordinate transformations from one body to the next in the rotational variables.

The coordinate transformation inside a joint model can be written as:

$$\mathbf{x}_2 = \mathbf{R} \cdot \mathbf{x}_1 \tag{8.143}$$

where the vectors $\mathbf{x}_1$ and $\mathbf{x}_2$ contain generalized coordinates of the bodies to the left and to the right, and the matrix $\mathbf{R}$ is a rotation matrix.

This process was demonstrated in the research section of Chapter 4 of the companion book for a six–degree–of–freedom Stanford robotic arm using *Denavit–Hartenberg (DH) coordinates* [8.7].

Depending on where the inertial frame is, we need to use either Eq.(8.143) or the inverse equation:

$$\mathbf{x_1} = \mathbf{R}^{-1} \cdot \mathbf{x_2} \qquad (8.144)$$

However, since $\mathbf{R}$ is an *orthogonal matrix*, the inverse of $\mathbf{R}$ can also be written as a transpose:

$$\mathbf{x_1} = \mathbf{R}^T \cdot \mathbf{x_2} \qquad (8.145)$$

which is more economical.

Unfortunately, Dymola, although offering a matrix manipulation language similar to that of MATLAB, doesn't understand the concept of orthogonal matrices. The reason is that Dymola, in order to provide full flexibility for causalizing equations, expands all matrix expressions into scalar expressions upon compilation. In the scalar version, the orthogonality of the $\mathbf{R}$–matrix is no longer easily visible. Thus, Dymola will do it the hard way and solve a linear equation system, whenever it needs the transformation equations in reversed causality.

In order to support Dymola in producing efficient simulation code, the MBS library keeps track of where the root (inertial frame) is, and provides the coordinate transformation to Dymola in a form similar to:

```
if rooted(frame_a) then
    x2 = R * x1
else
    x1 = transpose(R) * x2
end if;
```

where $frame\_a$ denotes the connector of the body to the left. In this way, Dymola starts out with the best suited equation set when looking for tearing variables using its built–in tearing algorithm.

8.13   Chemical Process Dynamics Simulators

Chemical processes are another prime candidate for DAE formulations. Here, the problems are again quite different. Chemical processes are modeled through highly nonlinear equations describing the (i) reaction rate dynamics, (ii) mass flow dynamics, (iii) thermal dynamics, and (iv) energy balance.

In Chapters 8 and 9 of the companion book, the basic equations describing chemical processes were introduced. However, these models mostly served the purpose of furthering the understanding of what is going on physically within a chemical reaction system. In reality, chemical reaction processes are invariably distributed parameter systems that should be described by PDEs. Also, there is no such thing as a homogeneous medium. Consequently, we are dealing with accuracy problems. If a chemical engineer can determine, in a simulation run, what is going on in the real process with an accuracy of 1%, he or she is very lucky.

For these reasons, it isn't warranted for practical simulations to deal with the exact equations. Why use a very complicated model if it is inaccurate anyway? Moreover, the energy balance equations have a much faster time constant than e.g. the mass balance equations. Therefore, chemical reaction processes are usually approximated by implicit ODEs describing average reaction rates, implicit ODEs describing mass continuity, implicit ODEs describing average temperatures, and algebraic constraint equations for energy balances and the equation of state [8.22, 8.37].

The result is a set of higher–index DAEs with nasty nonlinearities. Index reduction can usually be accomplished more easily here than in the case of mechanical systems, since the nonlinearities are usually polynomial rather than trigonometric. Consequently, the Pantelides algorithm will work fine. In fact, it is for chemical process engineers that Constantinos Pantelides developed his algorithm in the first place.

Solving the resulting algebraic loops is a different matter. Various tearing strategies have been described for such purposes [8.23], but they are very specialized and too complicated for our taste. Whenever we are confronted in physics with an equation or an algorithm that is very complicated, we should get suspicious that, probably, we are looking at the problem from a wrong angle. A typical example are the equations describing celestial dynamics when adopting a geostationary world view. We believe strongly that all physical laws governing this universe are basically simple. Complexity is introduced into this universe of ours by having many different –and simple– equations interact with each other. Equations can become more messy when we are forced to average or aggregate (as is the case in chemical process engineering), but the DAEs themselves are still fairly harmless. It is the conversion to explicit ODE form that makes them become truly messy ... and this has to do with the previously made simplifications (aggregations). Thus, we should probably abstain from trying to convert these equations to explicit ODE form.

So, if we stay with DAEs, where does tearing fit in? Isn't tearing a concept related to the transition from DAEs to ODEs? In chemical process engineering, tearing has mostly been used to simplify the process of fixed–point iteration of the resulting set of algebraic equations after inlining the integration method into the implicit state–space model.

A first attempt at object–oriented modeling of chemical process dynamics has recently been reported [8.28]. Bernt Nilsson didn't use Dymola for that purpose, but its twin brother, called Omola [8.1].

8.14   Summary

In this chapter, mixtures of symbolic and numerical tools for the treatment of differential and algebraic systems of equations were introduced. DAE formulations of dynamic models are very natural to applications in many areas of science and engineering. However, a direct approach to numerically dealing with the DAEs as they present themselves initially may not be the wisest thing to do. Automated symbolic preprocessing of the DAE models into a form that is better suited for the subsequent numerical integration is an exciting new development in modeling and simulation research.

A symbolic manipulation tool, Dymola, was introduced that has been specifically designed for such purpose. Dymola is the most advanced tool for that purpose currently on the market. While more generic symbolic formula manipulation programs, such as Mathematica or Reduce, could be used alternatively, they are not efficient for the task at hand. Dymola has already proven its utility in very large MBS models, for example.

The chapter introduced the most common numerical algorithms for dealing with the resulting set of index 1 DAEs, namely the BDF methods and fully–implicit Runge–Kutta algorithms, and explained how they work. It turns out that the solution of (potentially large) sets of algebraically coupled equations are at the heart of dealing with DAE systems. It was shown how the problem of numerical DAE solution is reduced to one of Newton iteration, and symbolic generation of the Hessian matrix was proposed as an additional tool for improvement of efficiency in numerical DAE solution.

By inlining the symbolic equations describing the integration algorithm into the model, the derivative operator disappears from the model altogether, and Dymola generates directly a set of difference equations that can be solved by simply looping over the model.

The chapter ended with three application areas: electronic circuits, multibody system dynamics, and chemical process engineering. These application areas demonstrate vividly the convenience and importance of a mixture of symbolic and numerical tools that can deal with DAE formulations.

Interestingly enough, many topics became simpler rather than more complicated when looking at them from a DAE rather than an ODE perspective. Yet, the research area of DAEs is much younger than that of ODEs. This can be explained easily. When scientists and engineers became interested in numerically simulating dynamic phenomena, no computers were available as yet. Armies of "applied mathematicians" (at that time not a highly respected tag for a mathematician) were employed and placed in a room for weeks in a row. Pipelining algorithms were designed such that the first mathematician could calculate the first step at time zero, then pass his or her result on to the next mathematician who would then solve step two at time zero, while the first mathematician would start on step one for time h, etc. Implicit algorithms don't lend themselves that easily to pipelining, and so, the focus was entirely on explicit algorithms, where

ODE formulations are most natural.

Later on when computers became available, engineers and scientists had become so used to ODE formulations that it took a while before they reconsidered the issue. The most exciting part of the story is that here is a research area where not too much has happened yet. It has become very difficult to hit a mark in numerical ODE solutions. Too many excellent mathematicians have already ploughed the field in hope of finding a leftover grain that might grow and bloom and flourish. This is not so in numerical DAE solution. This is therefore an excellent field for young applied mathematicians (a proud and respected lot by now) to do research in.

8.15   References

[8.1] Mats Andersson. Omola — An Object–Oriented Modelling Language. Technical Report TFRT–7417, Dept. of Automatic Control, Lund Institute of Technology, Lund, Sweden, 1989.

[8.2] Helmut Brandl, Rainer Johanni, and Martin Otter. A Very Efficient Algorithm for the Simulation of Robots and Similar Multibody–Systems Without Inversion of the Mass Matrix. In P. Kopacek, Inge Troch, and K. Desoyer, editors, *Theory of Robots*, pages 95–100. Pergamon Press, Oxford, United Kingdom, 1986.

[8.3] Dag Brück, Hilding Elmqvist, Hans Olsson, and Sven-Erik Mattsson. Dymola for Multi–Engineering Modeling and Simulation. In *Proceedings 2nd Intl. Modelica Conference*, pages 55:1–8, Munich, Germany, 2002.

[8.4] Kathryn E. Brenan, Stephen L. Campbell, and Linda R. Petzold. *Numerical Solution of Initial–Value Problems in Differential–Algebraic Equations*. North–Holland, New York, 1989. 256p.

[8.5] François E. Cellier. *Continuous System Modeling*. Springer Verlag, New York, 1991. 755p.

[8.6] François E. Cellier. Inlining Step–size Controlled Fully Implicit Runge–Kutta Algorithms for the Semi–analytical and Semi–numerical Solution of Stiff ODEs and DAEs. In *Proceedings Vth Conference on Computer Simulation*, pages 259–262, Mexico City, Mexico, 2000.

[8.7] Jacques Denavit and Richard S. Hartenberg. A Kinematic Notation for Lower–Pair Mechanisms Based on Matrices. *ASME Journal of Applied Mechanics*, 22(2):215–221, 1955.

[8.8] Hilding Elmqvist, Martin Otter, and François E. Cellier. Inline Integration: A New Mixed Symbolic/Numeric Approach for Solving Differential–Algebraic Equation Systems. In *Proceedings European Simulation Multiconference*, pages xxiii–xxxiv, Prague, Czech Republic, 1995.

[8.9] Hilding Elmqvist. *A Structured Model Language for Large Continuous Systems*. PhD thesis, Dept. of Automatic Control, Lund Institute of Technology, Lund, Sweden, 1978.

[8.10] Hilding Elmqvist. *Dymola — Dynamic Modeling Language, User's Manual, Version 5.3*. DynaSim AB, Research Park Ideon, Lund, Sweden., 2004.

[8.11] Roy Featherstone. The Calculation of Robot Dynamics Using Articulated–Body Inertias. *Internat. Journal of Robotics Research*, 2:13–30, 1983.

[8.12] Claus Führer and Ben J. Leimkuhler. Numerical Solution of Differential–Algebraic Equations for Constrained Mechanical Motion. *Numerische Mathematik*, 59:55–69, 1991.

[8.13] C. William Gear. The Simulataneous Numerical Solution of Differential–Algebraic Equations. *IEEE Trans. Circuit Theory*, CT–18(1):89–95, 1971.

[8.14] Gary D. Hachtel, Robert K. Brayton, and Fred G. Gustavson. The Sparse Tableau Approach to Network Analysis and Design. *IEEE Trans. Circuit Theory*, CT–18(1):101–118, 1971.

[8.15] Ernst Hairer, Christian Lubich, and Gerhard Wanner. *Geometric Numerical Integration: Structure–Preserving Algorithms for Ordinary Differential Equations*. Springer Verlag, Berlin, 2002. 515p.

[8.16] Ernst Hairer and Gerhard Wanner. *Solving Ordinary Differential Equations II: Stiff and Differential–Algebraic Problems*, volume 14 of *Series in Computational Mathematics*. Springer–Verlag, Berlin, Germany, 2nd edition, 1996. 632p.

[8.17] Russell C. Hibbeler. *Engineering Mechanics: Dynamics*. Prentice Hall, Upper Saddle River, New Jersey, 9th edition, 2001. 688p.

[8.18] Chung-Wen Ho, Albert E. Ruehli, and Pierce A. Brennan. The Modified Nodal Approach to Network Analysis. In *Proceedings IEEE Intl. Symposium on Circuits and Systems*, pages 505–509, San Francisco, California, 1974.

[8.19] Michael G. Hollars, Rosenthal Dan E., and Michael A. Sherman. SD/Fast: User's Manual. Technical report, Symbolic Dynamics, Inc., Mountain View, California, 2001.

[8.20] Christian Lubich. Extrapolation Integrators for Constrained Multibody Systems. *Impact on Computer Science and Engineering*, 3:213–234, 1991.

[8.21] Johnson Y. S. Luh, Michael W. Walker, and Richard P. Paul. On–Line Computational Scheme for Mechanical Manipulators. *Trans. ASME, Journal of Dynamic Systems Measurement and Control*, 102:69–76, 1980.

[8.22] William L. Luyben. *Process Modeling, Simulation, and Control for Chemical Engineers*. McGraw–Hill, New York, 1973.

[8.23] Richard S. H. Mah. *Chemical Process Structures and Information Flows*. Butterworth Publishing, London, United Kingdom, 1990.

[8.24] H. Alan Mantooth and Martin Vlach. Beyond Spice With Saber and MAST. In *Proceedings IEEE Intl. Symposium on Circuits and Systems*, pages 77–80, San Diego, California, 1993.

[8.25] William J. McCalla. *Fundamentals of Computer–Aided Circuit Simulation*. Kluwer Academic Publishers, Dordrecht, The Netherlands, 1988. 175p.

[8.26] Laurence W. Nagel. SPICE2: A Computer Program to Simulate Semiconductor Circuits. Technical Report ERL–M 520, Electronic Research Laboratory, University of California Berkeley, Berkeley, California, 1975.

[8.27] Dan Negrut and Harris Brett. ADAMS: Theory in a Nutshell. Technical report, Dept. of Mechanical Engineering, University of Michigan, Ann Arbor, Michigan, 2001.

[8.28] Bernt Nilsson. *Structured Modelling of Chemical Processes — An Object–Oriented Approach*. PhD thesis, Lund Institute of Technology, Lund, Sweden, 1989.

[8.29] Martin Otter, Hilding Elmqvist, and François E. Cellier. Modeling of Multibody Systems with the Object–Oriented Modeling Language Dymola. *J. Nonlinear Dynamics*, 9(1):91–112, 1996.

[8.30] Martin Otter, Hilding Elmqvist, and Sven Erik Mattsson. The New Modelica Multibody Library. In *Proceedings 3rd International Modelica Conference*, pages 311–330, Linköping, Sweden, 2003.

[8.31] Martin Otter, Sven Erik Mattsson, Hans Olsson, and Hilding Elmqvist. Simulator for Large Scale, Multi–physics Systems. Technical Report Deliverable D27, Report for Task 2.7, German Aerospace Center, Oberpfaffenhofen, Germany, 2002.

[8.32] Martin Otter and Clemens Schlegel. Symbolic generation of efficient simulation codes for robots. In *Proceedings Second European Simulation Multi–Conference*, pages 119–122, Nice, France, 1988.

[8.33] Martin Otter. *Objektorientierte Modellierung mechatronischer Systeme am Beispiel geregelter Roboter*. PhD thesis, Dept. of Mech. Engr., Ruhr–University Bochum, Germany, 1994.

[8.34] Linda R. Petzold. A Description of DASSL: A Differential/Algebraic Equation Solver. In R.S. Stepleman, editor, *Scientific Computing*, pages 65–68. North–Holland, Amsterdam, The Netherlands, 1983.

[8.35] Michael C. Schweisguth and François E. Cellier. A bond graph model of the bipolar junction transistor. In *Proceedings SCS Intl. Conference on Bond Graph Modeling and Simulation*, pages 344–349, San Francisco, California, 1999.

[8.36] Michael C. Schweisguth. Semiconductor Modeling with Bondgraphs. Master's thesis, Dept. of Electrical & Computer Engineering, University of Arizona, Tucson, Arizona, 1997.

[8.37] George Stephanopoulos. *Chemical Process Control: An Introduction to Theory and Practice*. Prentice–Hall, Englewood Cliffs, N.J., 1984. 696p.

[8.38] Vicha Treeaporn. Efficient Simulation of Physical System Models Using Inlined Implicit Runge–Kutta Algorithms. Master's thesis, Dept. of Electrical & Computer Engineering, University of Arizona, Tucson, Arizona, 2005.

[8.39] Paul W. Tuinenga. *Spice: A Guide to Circuit Simulation and Analysis Using PSpice*. Prentice Hall, Englewood Cliffs, N.J., 3rd edition, 1988. 288p.

[8.40] Jiri Vlach and Kishore Singhal. *Computer Methods for Circuit Analysis and Design*. Van Nostrand Reinhold, New York, 2nd edition, 1994. 712p.

8.16   Bibliography

[B8.1] Braden A. Brooks and François E. Cellier. Modeling of a Distillation Column Using Bond Graphs. In *Proceedings SCS International*

Conference on Bond Graph Modeling, pages 315–320, San Diego, California, 1993. SCS Publishing.

[B8.2] Roy Featherstone. *Robot Dynamics Algorithms*. Kluwer, Boston, Mass, 1997. 228p.

[B8.3] Steve Gallun. *Solution Procedures for Nonideal Equilibrium Stage Processes at Steady and Unsteady State Described by Algebraic or Differential–Algebraic Equations*. PhD thesis, Texas A&M University, 1979.

[B8.4] Ernst Hairer, Christian Lubich, and Michel Roche. *The Numerical Solution of Differential–Algebraic Systems by Runge–Kutta Methods*. Springer–Verlag, Berlin, Germany, 1989. 139p.

[B8.5] Daryl Hild and François E. Cellier. Object–Oriented Electronic Circuit Modeling Using Dymola. In *Proceedings OOS'94, SCS Object Oriented Simulation Conference*, pages 68–75, Tempe, Arizona, 1994.

[B8.6] Charles D. Holland and Athanasios I. Liapis. *Computer Methods for Solving Dynamic Separation Problems*. McGraw–Hill, New York, 1983. 475p.

[B8.7] Asghar Husain. *Chemical Process Simulation*. John Wiley & Sons, New York, 1986. 376p.

[B8.8] William L. Luyben. *Practical Distillation Control*. Van Nostrand Reinhold, New York, 1992. 533p.

[B8.9] Parviz E. Nikravesh. *Computer–Aided Analysis of Mechanical Systems*. Prentice–Hall, Englewood Cliffs, N.J., 1988. 370p.

[B8.10] Richard P. Paul. *Robot Manipulators: Mathematics, Programming, and Control — The Computer Control of Robot Manipulators*. MIT Press, Cambridge, Mass., 1981. 279p.

[B8.11] Mark W. Spong and Mathukumalli Vidyasagar. *Robot Dynamics and Control*. John Wiley & Sons, New York, 1989. 336p.

[B8.12] Michael W. Walker and David E. Orin. Efficient Dynamic Computer Simulation of Robotic Mechanisms. *Journal of Dynamic Systems, Measurement and Control*, 104:205–211, 1982.

8.17 Homework Problems

[H8.1] Inlining BDF3

Given the electrical circuit shown in Fig.H8.1a.

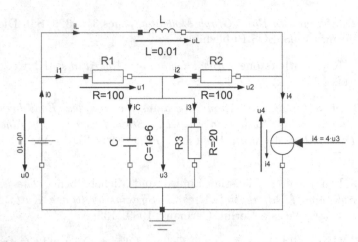

FIGURE H8.1a. Electrical circuit.

The circuit contains a constant voltage source, u_0, and a nonlinear (driven) current source, i_4, that depends on the voltage across the capacitor, C, and the resistor, R_3.

Write down the element equations for the seven circuit elements. Since the voltage u_3 is common to two circuit elements, these equations contain 13 rather than 14 unknowns. Add the voltage equations for the three meshes and the current equations for three of the four nodes. One current equation is redundant. Usually, the current equation for the ground node is therefore omitted. In this way, you end up with 13 equations in the 13 unknowns.

We wish to inline a fixed–step BDF3 algorithm, using order buildup during the startup phase. Draw the structure digraph of the inlined equation system, which now consists of 15 equations in 15 unknowns, and causalize it using the tearing method.

Simulate the ΔE system across 50 μsec using the inlined BDF3 algorithm with zero initial conditions on both the capacitor and the inductor. Choose a step size of $h = 0.5$ μsec. Use algebraic differentiation for the computation of the Hessian.

Plot the voltage u_3 and the current i_C on two separate subplots as functions of time.

[H8.2] Inlining Radau IIA

We wish to repeat Hw.[H8.1], this time inlining the 3rd–order accurate Radau IIA algorithm. Draw the structure digraph of the inlined equation system, which now consists of 30 equations in 30 unknowns, and causalize it using the tearing method.

Simulate the ΔE system across 50 μsec using the inlined Radau IIA algorithm with zero initial conditions on both the capacitor and the inductor. Choose a step size of $h = 0.5$ μsec. Use algebraic differentiation for the computation of the Hessian.

Plot the voltage u_3 and the current i_C on two separate subplots as functions of time.

[H8.3] Step–size Control for Radau IIA

We wish to augment the solution to Hw.[H8.2] by adding a step–size control algorithm.

Use Eq.(8.105) as the embedding method for the purpose of error estimation, and use Fehlberg's step–size control algorithm, Eq.(3.89), for the computation of the next step size. Of course, the formula needs to be slightly modified, since it assumes the error estimate to be 5th–order accurate, whereas in our algorithm, it is only 4th–order accurate. Remember that the step size can never be modified two steps in a row.

Simulate the ΔE system across 50 μsec using the step–size controlled inlined Radau IIA algorithm with zero initial conditions on both the capacitor and the inductor. Count the number of Newton iterations. Multiply that number with the number of statements inside the loop. This should provide you with a decent estimate of the computational efficiency of the method.

Plot the voltage u_3 and the current i_C on two separate subplots as functions of time.

[H8.4] Inlining Lobatto IIIC

We wish to repeat Hw.[H8.2], this time inlining the 4th–order accurate Lobatto IIIC algorithm. Draw the structure digraph of the inlined equation system, which now consists of 45 equations in 45 unknowns, and causalize it using the tearing method.

Simulate the ΔE system across 50 μsec using the inlined Lobatto IIIC algorithm with zero initial conditions on both the capacitor and the inductor. Choose a step size of $h = 0.5$ μsec. Use algebraic differentiation for the computation of the Hessian.

Plot the voltage u_3 and the current i_C on two separate subplots as functions of time.

[H8.5] Step–size Control for Lobatto IIIC

We wish to augment the solution to Hw.[H8.4] by adding a step–size control algorithm.

Use Eq.(8.110) as the embedding method for the purpose of error estimation, and use Fehlberg's step–size control algorithm, Eq.(3.89), for the computation of the next step size. Remember that the step size can never be modified two steps in a row.

Simulate the ΔE system across 50 μsec using the step–size controlled inlined Lobatto IIIC algorithm with zero initial conditions on both the capacitor and the inductor. Count the number of Newton iterations. Multiply that number with the number of statements inside the loop. This should provide you with a decent estimate of the computational efficiency of the method. MATLAB used to offer a better means of estimating the efficiency of a code by counting the number of floating point operations, using the built–in function *flops*. Unfortunately, this feature has been disabled in version 6 of MATLAB.

If you also solved Hw.[H8.3], you can compare the computational efficiency of the two algorithms for solving the given circuit problem against each other.

Plot the voltage u_3 and the current i_C on two separate subplots as functions of time.

[H8.6] Algebraic Differentiation

We wish to reproduce Fig.8.31 of this chapter. On purpose, we haven't shown you the details of how it has been derived. In particular, we didn't provide the symbolic equations for the computation of the Hessian by means of algebraic differentiation.

[H8.7] Stabilized BE Simulation of Overdetermined DAE System

We wish to reproduce Fig.8.33 of this chapter. On purpose, we haven't shown you the details of how it has been derived. In particular, we didn't provide you with a formula for when to end the Newton iteration. Since the linear system is now only solved in a least square sense, you can no longer test for $\|\mathcal{F}\|$ having decreased to a small value. The way we did it was to compute the norm of $\mathcal{F}$ and save that value between iterations. We then tested, whether the norm of $\mathcal{F}$ has no longer decreased significantly from one iteration to the next:

> **while** abs($\|\mathcal{F}^\ell\| - \|\mathcal{F}^{\ell-1}\|$) $< 1.0e - 6$,
>   perform iteration
> **end**,

8.18   Projects

[P8.1] Inlining DIRK

There exists yet another interesting class of implicit stiffly stable Runge–Kutta algorithms that we haven't discussed in this chapter. These are called *diagonally implicit Runge–Kutta algorithms*, and are usually abbreviated as DIRK algorithms. One of the more fashionable among the DIRK algorithms is HW–SDIRK(3)4 [8.16] with the Butcher tableau:

$$
\begin{array}{c|cccccc}
\frac{1}{4} & \frac{1}{4} & 0 & 0 & 0 & 0 \\[2mm]
\frac{3}{4} & \frac{1}{2} & \frac{1}{4} & 0 & 0 & 0 \\[2mm]
\frac{11}{20} & \frac{17}{50} & \frac{-1}{25} & \frac{1}{4} & 0 & 0 \\[2mm]
\frac{1}{2} & \frac{371}{1360} & \frac{-137}{2720} & \frac{15}{544} & \frac{1}{4} & 0 \\[2mm]
1 & \frac{25}{24} & \frac{-49}{48} & \frac{125}{16} & \frac{-85}{12} & \frac{1}{4} \\[2mm]
\hline
x & \frac{59}{48} & \frac{-17}{96} & \frac{225}{32} & \frac{-85}{12} & 0 \\[2mm]
\hat{x} & \frac{25}{24} & \frac{-49}{48} & \frac{125}{16} & \frac{-85}{12} & \frac{1}{4}
\end{array}
$$

HW–SDIRK(3)4 is a five–stage algorithm. DIRKs are much less compact than their IRK cousins, and therefore, allow proper embedding algorithms to exist within them. x represents a 3rd–order accurate method, whereas $\hat{x}$ represents a 4th–order accurate method.

DIRK methods are attractive alternatives to the IRK methods discussed in this chapter, since they can be implemented with one Newton iteration per stage, rather than with one Newton iteration across all stages.

Remember the dilemma that we were facing when we tried to inline parabolic PDEs. Inlining a BDF algorithm, we had to perform a Newton iteration in 25 tearing variables, whereas inlining the 3rd–order accurate Radau IIA algorithm, we had to perform a Newton iteration in 100 tearing variables. Thus, Radau IIA would need to be able to use step sizes that are at least 16 times as large as those used by BDF3 in order to be competitive.

Inlining HW–SDIRK(3)4, we would expect to require five Newton iterations, each in 25 tearing variables. Thus, we would need to use only five times as large step sizes as those employed by BDF3, in order to be competitive.

Find the **F**–matrices of the two embedded methods, and perform Taylor–series expansions to verify that the two methods are indeed 3rd–order and 4th–order accurate, respectively. Compute the error coefficient of the error–controlled method.

Plot the stability domains as well as the damping plots of the two individual methods. Decide, which of the two estimates should be propagated to the next step.

Show how HW–SDIRK(3)4 can be inlined by means of the problem discussed in Hw.[H8.1].

Simulate the circuit using the step–size controlled inlined HW–SDIRK(3)4 algorithm.

8.19   Research

[R8.1] Inlining Parabolic PDEs

Develop suitable heuristic procedures for finding small sets of tearing variables for inlining parabolic PDEs in multiple space dimensions.

As we have discussed in Chapter 6 of this book, the simulation of parabolic PDEs converted to sets of ODEs by the MOL approach often requires internal Newton iterations due to either nonlinear boundary conditions or irregular domain boundaries. Hence inlining them might be quite attractive.

The numerical PDE literature is full of descriptions of sparse matrix algorithms for improving the efficiency of the simulation of such problems. Tearing can also be viewed as a sparse matrix technique, although it is applied in a symbolic form.

Compare the computational efficiency of the ΔE simulation after inlining with that of alternative ODE simulations without inlining.

9

Simulation of Discontinuous Systems

Preview

In this chapter, we shall discuss how discontinuous models can be handled by the simulation software, and in particular by the numerical integration algorithm. Discontinuous models are extremely common in many areas of engineering, e.g. to describe dry friction phenomena or impact between bodies in mechanical engineering, or to describe switching circuits in electronics. In the first part of this chapter, we shall be dealing with the numerical aspects of integrating across discontinuities. Two types of discontinuities are introduced, time events and state events, that require different treatment by the simulation software. In the second part of this chapter, we shall discuss the modeling aspects of how discontinuities can be conveniently described by the user in an object–oriented manner, and what the compiler needs to do to translate these object–oriented descriptions down into event descriptions.

9.1 Introduction

As we have seen, all numerical integration algorithms used in today's simulation programs are based, either explicitly or implicitly, on Taylor–Series expansions. Simulation trajectories are always approximated by polynomials or rational functions in the step size h around the current time t_k.

This causes problems when dealing with discontinuous models, since polynomials never exhibit discontinuities at all, and also rational functions only exhibit occasional poles, but no discontinuities. Thus, if an integration algorithm tries to integrate across a discontinuity, it will invariably be in trouble.

Since the step size is finite, the integration algorithm doesn't recognize a discontinuity as such. It simply notices that the trajectory suddenly and unexpectedly changes its behavior by showing symptoms of a very steep gradient. Thus, the integration algorithm experiences the discontinuity as the sudden appearance of a new eigenvalue far out to the left in the complex plane. If the algorithm is step–size controlled, it will react to this observation by reducing the step–size in order to shrink the eigenvalue into the asymptotic region of the $(\lambda \cdot h)$–plane. Unfortunately, this new eigenvalue has the nasty habit of being evasive. Although the step size is made smaller and smaller, the eigenvalue doesn't allow itself to be captured. The integration algorithm thus experiences the discontinuity as a *singular point of infinite stiffness*.

The algorithm finally gives up, as its step size is either reduced to the smallest tolerable value, or because the step–size control is getting fooled. We shall see why this can easily happen. As a consequence, the discontinuity is passed through with a very small step size ... and the spooky phenomenon vanishes as fast as it appeared. The integration algorithm notices that the funny eigenvalue has disappeared again, and consequently will enhance the step size in the steps to come, until the appropriate optimal step size has been regained. It is in this fashion that the step–size control within the numerical integration algorithm is able to handle discontinuities ... and often, it does so with quite decent success.

Figure 9.1 illustrates how step–size control handles discontinuities.

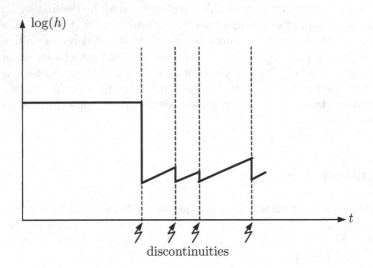

FIGURE 9.1. Discontinuity handling by step–size control.

Figure 9.1 shows the logarithm of the step size, h, plotted across simulated time, t. As the integration algorithm approaches a discontinuity, the step size is reduced until the algorithm judges the solution to be correct. After the discontinuity has passed, the step size is cautiously increased again until the next discontinuity is encountered. This is quite inefficient, but often produces decent results.

It is this lucky by–product of the step–size control mechanism that allowed the simulation software producers to get by for many years without spending too much of a thought on the problem of discontinuity handling. Unfortunately, things can go awfully awry as was demonstrated in [9.5].

9.2   Basic Difficulties

In the seventies, one of the authors was a Ph.D. student at ETH Zürich in Switzerland. He was working on a dissertation on exactly the topic of this chapter [9.5]. One day, a colleague of his, who had difficulties with his simulation program, came to see him. He had worked on his program for weeks and weeks, and it simply didn't want to run properly. He was another Ph.D. student, working on the design of a velocity controller for electrically driven locomotive engines [9.25]. When analyzing his friend's problem, he soon realized that his program exhibited difficulties that were closely related to the way the numerical integration algorithm handled the discontinuities in his model. Let us explain.

In Switzerland, electric train engines are operated by AC current with a frequency of $16\frac{2}{3}$ Hz. The amplitude of the voltage available to the engine is constant, thus velocity control cannot be achieved by simply modifying the voltage. An Ohmic voltage divider is out of the question, since we want to propel the engine, not heat it up. Variable transformers, on the other hand, are too large and bulky.

Previously, train engines in Switzerland had been equipped with a thyristor circuit controlling the firing angle of the thyristor. Figure 9.2 shows the circuit diagram of the thyristor circuit.

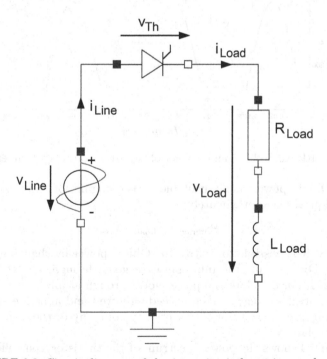

FIGURE 9.2. Circuit diagram of thyristor circuit for train speed control.

The partly Ohmic partly inductive load represents the engine. This model is simplified, but shall do to explain the difficulties with this approach. The thyristor is a switch element. It can be "fired" (i.e., closed) by applying a low voltage impulse to the thyristor gate. The thyristor then stays on until the current through the thyristor passes through zero. At zero current, the thyristor automatically opens again.

Figure 9.3 shows the current, i_{Load}, flowing through the load and the voltage, v_{Load} across the load, assuming that the thyristor is repetitively fired by an impulse applied once every period after a given firing angle α. In the example, we chose $\alpha = 30°$.

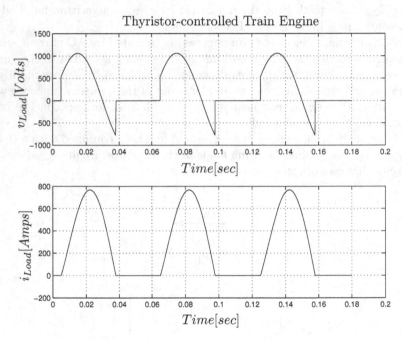

FIGURE 9.3. Voltage and current of thyristor–controlled train engine.

The Ohmic power made available to the engine for conversion to mechanical power is approximately:

$$P_{Ohmic} = v_{Load} \cdot i_{Load} \tag{9.1}$$

Evidently, it is possible to control the Ohmic power by changing the firing angle α. For $\alpha = 0°$, the full sine wave goes through, i.e., the power is maximized. For $\alpha \geq 180°$, no power goes through at all.

This control strategy worked exceedingly well and almost everyone was very happy ... except for the electricity company of the Canton of Uri. Let us explain.

Figure 9.4 shows the power spectrum of the thyristor–controlled voltage signal.

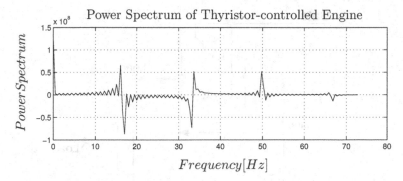

FIGURE 9.4. Power spectrum of thyristor–controlled voltage signal.

This was computed by simulating the above circuit across 1.5 seconds of simulated time using 1200 communication points. A *fast Fourier transform (FFT)* of P_{Ohmic} was then computed. Figure 9.4 shows the real part of the low frequency end of that spectrum plotted across frequency.

Roughly 17% of the power is DC power, 30% are at the base frequency, another 30% are at the 2nd harmonic, roughly 15% are at the 3rd harmonic, and 4% are at the 4th harmonic.

The 3rd harmonic thus carries a substantial percentage of the overall power of the signal. Unfortunately, the 3rd harmonic happens to be located at 50 Hz, i.e., precisely at the frequency, with which the electric power company delivers electric power to the households in Switzerland. It so happened that whenever one of these trains (usually equipped with two engines) drove up the St.Gotthard mountain, the electric counters in households located near the rails were reset to zero.

Next, the train engineers tried *burst control*. Figure 9.5 shows the circuit diagram of a burst–controlled engine.

Figure 9.6 shows the voltage across and current through the train engine when using burst control. The high–voltage circuitry is very similar to the one used in the previous approach. This time, we use two thyristors with a common gate control logic. However, the gate control of the thyristor now works differently. Rather than letting through a certain percentage of every period, the burst–controlled thyristor fires constantly during a certain number of periods, and then stops firing for the remainder of the burst.

It was decided to use bursts of eight periods. Consequently, the burst frequency is one eighth of the line frequency, i.e., $2\frac{1}{12}$ Hz. Out of these eight periods, a certain number of periods is being let through, and the remainder is stopped. In Fig.9.6, five out of every eight periods are let through. Evidently, engines using this speed control strategy cannot operate at an arbitrary percentage of the full power, but only at $\frac{1}{8}^{\text{th}}$, or $\frac{2}{8}^{\text{th}}$, or $\frac{3}{8}^{\text{th}}$, etc. of the full power.

The advantages of this simple solution were twofold. On the one hand, it solved the problem of resetting the electric counters, since the power

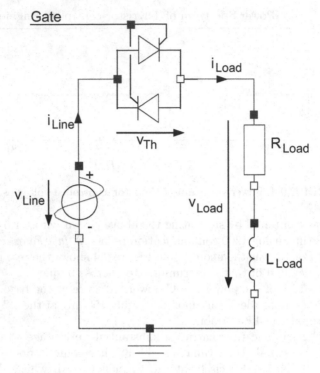

FIGURE 9.5. Circuit diagram of thyristor circuit for burst control.

spectrum no longer contains a significant amount of power at 50 Hz, and secondly, it was very cheap, since the (expensive) high–voltage circuitry needed very little modification. Only the (comparatively inexpensive) low–voltage circuitry needed to be replaced.

These circuits were installed in the trains that served the northern shore of Lake Zürich, on the line Zürich–Meilen–Rapperswil, and were used there for a number of years. When the train pulled out of the station, it operated during one burst (about 0.5 seconds) at $\frac{1}{8}^{\text{th}}$ of full power, then during the next burst at $\frac{2}{8}^{\text{th}}$, etc. These trains weren't able to accelerate smoothly. The speed changed abruptly, which the customers felt noticeably in their stomachs. It just wasn't very comfortable.

Thus, our colleague had been asked to come up with something better. He designed the circuitry shown in Fig.9.7.

This time, the engine is represented by something that drains current out of the net, i.e., as a current source. The representation is not accurate, but it is good enough for the task at hand. Also, the line frequency has been normalized to $\omega = 2\pi f = 1$ sec^{-1}, so that the same circuit would also work for other countries with different line frequencies. The impedance values have been adjusted accordingly.

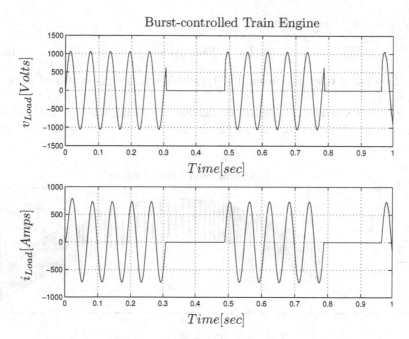

FIGURE 9.6. Voltage and current of burst–controlled train engine.

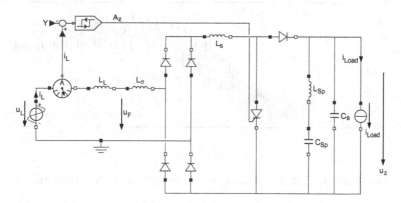

FIGURE 9.7. Circuit diagram of SCR circuit for train speed control.

The gate control logic is also shown on Fig.9.7. The line current, i_L, is controlled in such a way that it always remains in the vicinity of:

$$Y(t) = \frac{15 \cdot 10^6}{u_L} \sin \omega t \qquad (9.2)$$

For $A_z = 0.0$, the line current, i_L, grows rapidly until it crosses $(Y + B_T)$ in the positive direction. At that moment, A_z assumes a value of $A_z = 1.0$, and i_L decays quickly again until it reaches $(Y - B_T)$, where A_z takes a value of $A_Z = 0.0$ as before.

$$B_T = 200.0 \text{ Amps} \tag{9.3}$$

is the allowed tolerance around $Y(t)$, within which i_L is supposed to operate.

Figure 9.8 shows two signals of this circuit during the first half–period, namely the filter voltage u_F within the control loop, and the load voltage, u_z.

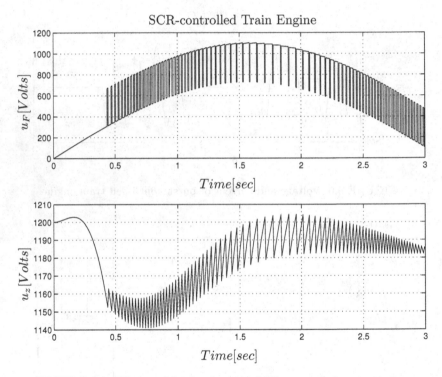

FIGURE 9.8. Filter and load voltage in SCR–controlled train engine.

If a numerical integration algorithm could fall into depression, this might be as good a reason for it to do so as any. What nightmarish curves to integrate over (!) The filter voltage, u_F, after an initial transitory phase, essentially follows a sine wave. It toggles back and forth between the sine wave itself and the same curve with a constant DC value of about 300 Volts superposed. The load voltage, u_z, is regulated to stay essentially at a constant value, in the given example somewhere around 1184 Volts. The power spectrum of the load is mostly DC, except for a small percentage located at frequencies much higher than 50 Hz.

However, these are not the results that our colleague had found, when he came to discuss his simulation results. Figure 9.9 presents the results that he had obtained.

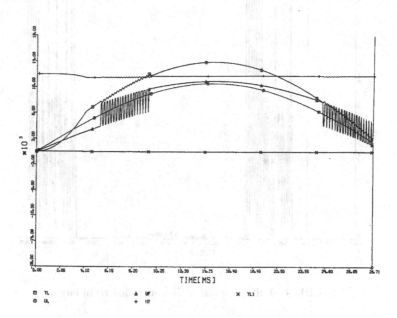

FIGURE 9.9. Filter voltage in SCR–controlled train engine.

This is an old plot that we scanned in from [9.5]. We were unable to reproduce precisely the results that our colleague had obtained, since they had been produced by an old simulation software, CSMP–III, that we don't have around any longer. It was a software specifically designed for use on IBM mainframes, machines that have been moth–balled long ago.

The graph shows the filter voltage, u_F, plotted over time together with some other signals. The reader notices that the curve looks similar to the newly obtained one, except during the time interval from about 8 msec to 24 msec, when the filter voltage on the old plot didn't exhibit the high frequency oscillation.

The simulation took forever to run. For this reason, we recommended to our friend to also plot the step size, h, used as a function of simulated time. It is shown in Figure 9.10.

The step size varies a lot over time, as the step–size control algorithm is being used to catch the discontinuities. However, it is quite evident from the plot that the simulation uses consistently a very small step size during the period from 8 msec to 24 msec, i.e., the period, during which the simulation results are incorrect. The simulation exhibits *creeping behavior.*

Somehow, the gate control had gotten stuck. The numerical integration algorithm was aware of that fact and tried to fix it by using very small step sizes, but was unable to do so. Thus, using the step–size control mechanism

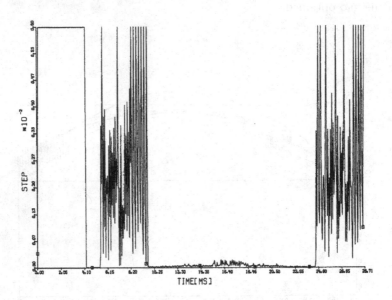

FIGURE 9.10. Step size in SCR–controlled train engine.

for handling discontinuities evidently is not only *inefficient*, it can also be *dangerous*. Notice that the simulation program did not produce any error message at all.

How can these results be explained? The step–size control mechanism of any step–size–controlled integration algorithm is based on an error estimate. This error estimate, for an nth–order algorithm, is something like:

$$\varepsilon = c \cdot h^{n+1} \tag{9.4}$$

Consequently, as we reduce the step size more and more, the error estimate *will* become smaller and smaller, irrespective of whether the integration makes any sense or not. Practically speaking, as we reduce the step size, the higher–order terms in the Taylor–Series expansion become less and less important until, finally, every integration algorithm behaves like Euler. Explicit algorithms will behave like forward Euler, whereas implicit algorithms may behave either like forward Euler or like backward Euler.

If we try to integrate across a discontinuity, the two formulae that are compared to each other for the purpose of step–size control, will eventually both behave like Euler, and at that time, they will agree on their "solution" ... not necessarily the *correct* solution, mind you, but at least a solution they both came up with. If two numerical codes agree on a solution to a problem, that may indeed indicate that the solution is correct ... but it may just as well simply mean that the two codes employ the same (possibly flawed) algorithm. Therefore, if two different numerical codes miraculously agree on a solution to almost machine resolution, we are usually much more suspicious of foul play than if their agreement is less spectacular.

Clearly, in the case of Fig.9.9, this is what happened. During some periods of time, the two algorithms agreed happily on the same –evidently quite wrong– solution. What happened was the following. The program used a step–size controlled explicit single–step algorithm, some variant of a fourth–order Runge–Kutta method, more precisely, it used the Runge–Kutta–Simpson method described in H3.16, a rather dubious method, as we now understand.

When the solution approached the threshold, the solution managed to switch several times back and forth within a single integration step. If the number of switchings happened to be *odd*, the step ended with the other model, and integration proceeded as desired. On the other hand, if the number of switchings was *even*, the step ended in the same switch position it had started out with, and the algorithm went through the same switching immediately again during the next step. This explains why the solution was creeping along the switching boundary, unable to leave it.

Abusing the step–size control for locating discontinuities is always quite inefficient. The reason is that the algorithm doesn't know, and cannot know, that a discontinuity is taking place. It must therefore assume the worst, namely that the system is highly nonlinear with rapidly changing eigenvalues of its Jacobian matrix. Consequently, the algorithm has to be cautious in increasing its step size again after the discontinuity has been cleared in order to avoid potential numerical instability problems that may be caused by a hyperactive step–size adjustment strategy. This is documented in Fig.9.1, where the step size remains constantly at too small a value since the next discontinuity is always encountered before the step size could regain its optimal value.

Abusing the step–size control for locating discontinuities can sometimes lead to incorrect results that may be difficult to identify as such, i.e., incorrect results may be produced and go unnoticed. The above application is a good example of that.

We evidently need something better.

9.3 Time Events

In many cases, we do know some time in advance when a discontinuity will take place. For example, in the case of the original thyristor gate control logic, we know that the thyristor will close exactly α° after the start of each period. It is just a question of providing this information to the integration algorithm. Discontinuities will from now on be called *discrete events*, and if we know when such an event will take place, we can *schedule* it to happen by entering the *event time* and the *event type* into a *calendar of forthcoming events*.

The event calendar is a linearly–linked list of events arranged in the order of increasing times of occurrence, thus the first event in the event calendar is always the *next event*. In the case of multiple simultaneous events, additional tie–breaking rules can be specified to decide which event comes first. The sequence may matter. For example, if a car arrives at a traffic light that simultaneously switches to red, it may make a big difference whether the simulation program decides that the car arrived first, or whether it decides that the light changed first. Therefore, tie–breaking rules should be implemented, and should be considered carefully.

The next event time is considered by the step–size control of the integration algorithm exactly like a communication or readout point. If the integration algorithm usually will adjust the step size in the vicinity of a readout point in order to hit the point accurately (mostly done in the case of single–step algorithms), then so should it treat the next event time. If the next event time falls in between the current time and the time when the next step should ordinarily end, the step size is reduced in order not to miss the communication point. If the next event can be reached by increasing the next step by not more than 10%, then this is justifiable in order to prevent a very short step thereafter. On the other hand, if the integration algorithm interpolates in order to visit the next communication point (mostly done in multi–step integration by use of the Nordsieck vector), then it should do the same in order to accommodate the next event time.

Notice that no discontinuity takes places while the event is being located. The discontinuity is not directly coded into the model, only the condition of its occurrence is. Thus, the trajectories seen by the integration algorithm are perfectly continuous, and the integration algorithm therefore has nothing to worry about.

Once the next event time has been located, the continuous simulation comes to a halt, and a discrete event section of the simulation program is visited that implements the consequences of the event taking place, i.e., sets the state variables to their new values, changes the current values of input functions, etc. It is this section that implements the discontinuities. A simulation program may contain many different discrete event sections, one for every event type.

The end result of event handling can be considered a new set of initial conditions, from which a completely new integration can start. Thus, a simulation run across a discontinuous model can be interpreted as a sequence of distinct strictly continuous simulation runs, separated by discrete events.

The recipe is so trivial that one would assume that all serious continuous system simulation languages (CSSLs) would meanwhile have adopted it ... or faced the destiny of natural attrition. However, due to the soi–disant "event handling" capabilities of the step–size control algorithms themselves, many simulation software designers never bothered to look into the issue ... and so far they got away with it. Well, hopefully this book will finally change all of this.

Let us consider once more the thyristor–controlled train engine model. The gate needs to be closed after $\alpha°$. Thus, the time of the first time event that closes the gate takes place at:

$$t_{\text{period}} = \frac{1}{2\pi f} \tag{9.5a}$$

$$t_{\text{event}} = \frac{\alpha}{360} \cdot t_{\text{period}} \tag{9.5b}$$

Since we know from the beginning of the simulation, when this event is going to take place, the event can be scheduled in the initialization portion of the simulation program.

Thus, the *initialization section* of the simulation program could contain the statements (in pseudo–code):

> *Gate* = *open*
> **schedule** *CloseGate* **at** *t_event*

The *event description section* of the simulation program would then close the gate, and schedule the next gate closing event one period later:

> *Gate* = *closed*
> **schedule** *CloseGate* **at** t + *t_period*

The variable *Gate* can be referred to from within the continuous–time simulation model. This is not dangerous, since discrete states behave exactly like parameters or constants as far as the integration algorithm is concerned. They never change their values while the integration is proceeding. They only change their values in between segments of numerical integration, i.e., at event times.

We haven't talked yet about the gate opening event. We cannot handle the gate opening event in the same fashion as the gate closing event, because we don't know beforehand, *when* the gate will open. We only know, *under what condition* this will be the case, namely when the current that flows through the thyristor becomes negative.

The gate opening event will be discussed in due course.

9.4   Simulation of Sampled–data Systems

A typical application of time events is the simulation of sampled–data control systems. A continuous–time plant is being controlled by one or several discrete–time controllers that may operate on the same or on different frequencies (multi–rate sampling).

A typical application is shown in Fig.9.11.

A robot arm is to be controlled by one or several computers. The innermost control loop serves the purposes of stabilization, linearization, and

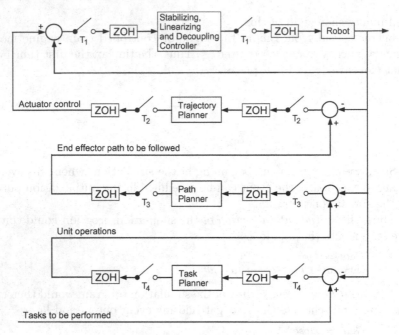

FIGURE 9.11. Robot control.

decoupling. The purpose of this controller is to make the larger control issues easier to tackle. The signal needs to be sampled in short time intervals, T_1, in order to keep the control loop stable. This first controller is then added to the plant, i.e., the next higher–level controller considers the innermost control loop part of the plant to be controlled. Its purpose is to translate a desired path into control signals for the actuators of the motors that drive the individual joints of the robot arm. This controller solves the dynamic control problem. It can operate at a slightly slower sampling rate, T_2, than the stabilizing controller. The next higher–level controller solves the static control problem. It translates descriptions of individual unit operations into desired end–effector positions expressed as functions of time. It again can operate at a somewhat reduced sampling rate, T_3. Finally, the task planner decomposes complex tasks into series of unit operations that it then submits to the path planner for execution. The task planner can operate at a considerably slower sampling rate, T_4. Thus:

$$T_1 \leq T_2 \leq T_3 \ll T_4 \tag{9.6}$$

Figure 9.11 is somewhat stylized. There are multiple signals to be fed back, and after decoupling, there may be multiple control loops, one for each joint.

The simulation program will contain a single dynamic block describing the motion of the robot arm itself together with its motors and drive trains in between events. The program also contains four separate discrete blocks, one for each controller, that are executed at different, yet previously known, points in time. All four discrete controllers are probably scheduled to be executed for the first time during initialization of the simulation. Thus, we are confronted with four simultaneous events, and it will be important that the task planner is executed first, then the path planner, then the trajectory planner, and finally the stabilizer, since the inner control loops need the set points from the outer control loops to function properly. Each controller will, as part of its event description, schedule the next execution of itself to occur T_i time units into the future. At a later time, it is probably better to resolve ties by assigning a higher priority to the inner control loops, since they are more time–critical.

At any point in time, there are thus scheduled four different time events to take place at different time instants in the future. These are maintained by the so–called *event queue*, which is usually implemented as a linear linked list with pointers back and forth, in which future events are placed in ascending order of execution time using additional rules for tie breaking.

9.5   State Events

Frequently, the time of occurrence of a discontinuity is not known in advance. For example in the thyristor circuit, it is not known in advance when the thyristor will open again. All we know is that it will open when the current passes through zero. Thus, we know the *event condition*, rather than the *event time*, specified in terms of a function of continuously varying simulation variables.

Event conditions are usually specified implicitly, i.e., in the form of *zero–crossing functions*. A state event occurs when a variable associated with it crosses through zero. Multiple zero–crossing functions may be associated with a single event type.

The zero–crossing functions must be tested continuously during simulation. Thus, they are part of the continuous system simulation environment. To this end, many of the numerical ODE solvers currently on the market offer so–called *root solvers*. Variables to be tested for zero crossing are placed in a vector. These variables are monitored constantly during simulation, and if one of them passes through zero, an iteration is started to determine the zero–crossing time with a pre–specified precision.

Since we don't know when event conditions become true, we cannot reduce the step size to hit them accurately. Instead, we need some sort of iteration (or interpolation) mechanism to locate the event time. Thus, when an event condition is alerted during the execution of an integration step, it influences the step–size control mechanism of the integration algorithm by forcing the continuous simulation to iterate (or interpolate) to the earliest zero–crossing within the current integration step.

9.5.1 Multiple Zero Crossings

Figure 9.12 illustrates the iteration of event conditions, assuming that multiple zero crossings have taken place within a single integration step.

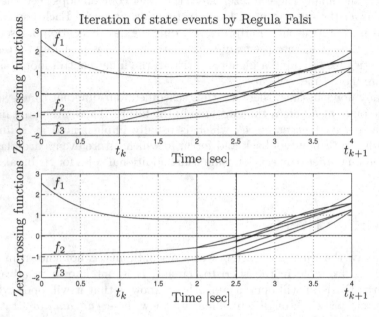

FIGURE 9.12. Iteration of multiple event conditions using Regula Falsi.

Figure 9.12 shows three different zero–crossing functions, f_1, f_2, and f_3. At time $t_k = 1.0$, f_1 is positive, whereas f_2 and f_3 are negative. We perform an integration step of length $h = 3.0$. At time $t_{k+1} = 4.0$, f_1 is still positive, whereas both f_2 and f_3 are now also positive, i.e., two zero crossings have taken place within this integration step.

We connect the end points of each zero–crossing function, determine, where these straight lines cross through zero, and choose the smallest of these time instants as the next time point. Mathematically:

$$t_{\text{next}} = \min_{\forall i} \left[\frac{f_i(t_{k+1}) \cdot t_k - f_i(t_k) \cdot t_{k+1}}{f_i(t_{k+1}) - f_i(t_k)} \right] \qquad (9.7)$$

where i stretches over all functions with a zero crossing within the interval. Thus, we repeat the last time step with a step size of $h = t_{\text{next}} - t_k$.

If no zero crossing has taken place during the reduced step from t_k to t_{next}, we accept t_{next} as t_k and repeat the algorithm using the remainder of the interval.

If more than one zero crossing has taken place during the reduced interval, we reduce t_{k+1} to t_{next}, and apply the same algorithm once more to the so reduced interval.

If exactly one zero crossing has taken place during the reduced interval, we have simplified the problem to that of finding the zero crossing of a single zero–crossing function, for which a number of algorithms can be used that shall be presented in due course.

The algorithm converges always, as the interval is reduced during each iteration step. Unfortunately, it is not possible to estimate the number of iteration steps needed until convergence has been reached using this method. Convergence can indeed be quite slow.

Another algorithm that is sometimes used instead is the *Golden Section* method. The Golden Section method has the advantage that, in each iteration step, the interval is reduced by a fixed ratio. Thus, the interval will soon become quite small. This is how it works.

Already the ancient Greeks had discovered that there exists a special rectangle with the property that if one cuts off a square, the remaining rectangle has the same proportions as the original one. This is shown in Fig.9.13.

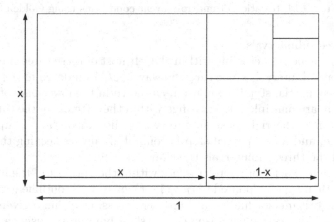

FIGURE 9.13. The Golden Section.

Thus:

$$\frac{x}{1} = \frac{1 - x}{x} \tag{9.8}$$

which leads to $x = 0.618$.

This idea can be applied to the problem of isolating individual zero–crossing functions. The method is shown in Fig.9.14.

Once the iteration algorithm has been triggered by multiple zero crossings within a single step, the interval is subdivided by calculating two partial steps, one of length $(1 - x) \cdot h$, the other of length $x \cdot h$. Both of these partial steps start at time t_k. In this way, the interval is subdivided

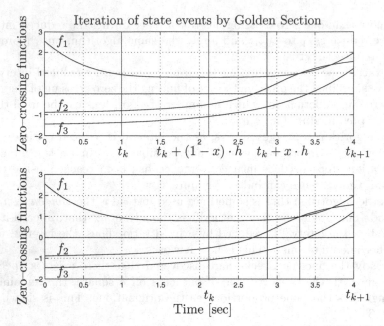

FIGURE 9.14. Iteration of multiple event conditions using Golden Section.

into three subintervals.

If there is no zero crossing within the leftmost of these three subintervals, then that subinterval can be thrown away, i.e. t_k is updated to $t_k+(1-x)\cdot h$, and a new partial step is computed, as shown in the lower part of Fig.9.14.

If there are multiple zero crossings within the leftmost of the three subintervals, then the rightmost subinterval is discarded, t_{k+1} is updated to $t_k + x \cdot h$, and a new partial step is computed, always keeping the proportions of the three subintervals the same.

If there is exactly one zero crossing within the leftmost of the four subintervals, then t_{k+1} is updated to $t_k+(1-x)\cdot h$, and we continue with any one of the algorithms for finding a single zero crossing within a given interval.

The Golden Section algorithm can be slightly improved using a *Fibonacci Series* instead, but this is hardly ever worth it. The Fibonacci Series shrinks the interval slightly faster than the Golden Section technique, but it can be shown that the Fibonacci Series is always less than one iteration step ahead of Golden Section, and it is only better at all, if we decide up front how many iteration steps we are going to perform altogether.

9.5.2 Single Zero Crossings, Single–step Algorithms

Of course, any of the techniques presented so far for isolating individual zero–crossing functions can also be used to find the zero crossings themselves. Yet, this may be inefficient, as all of these techniques offer only *linear convergence speed.*

All simulation variables in a state–space model can ultimately be expressed in terms of state variables and inputs only. This also applies to the zero–crossing functions. Thus, we could use, in the determination of the zero crossings, not only the values of the zero–crossing functions themselves at different points in time, but also the values of their derivatives.

A first algorithm that exploits this possibility is the well-known *Newton iteration* algorithm that we have used so often already in this book, albeit for different purposes. Figure 9.15 documents the approach.

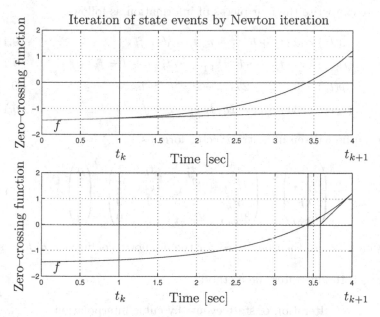

FIGURE 9.15. Iteration on single zero–crossing functions using Newton iteration.

Once a zero–crossing function has been isolated, we can use either t_k or t_{k+1} as the starting point of a Newton iteration.

The good news about Newton iteration is that the algorithm exhibits a *quadratic convergence speed.* Thus, Newton iteration converges much more rapidly than either Regula Falsi or Golden Section, if the algorithm converges at all.

Unfortunately, and contrary to the previously introduced algorithms, the Newton iteration algorithm does not always converge. In the given example, if we start at t_{k+1}, the algorithm converges quickly, whereas if we start at t_k, already the next step takes the algorithm far outside the interval $[t_k, t_{k+1}]$.

Furthermore, it may not be easy to determine upfront, whether or not the algorithm will converge on a given example. For these reasons, Newton iteration may not be the method of choice to be used as a root solver.

A better approach may be to use the derivative values at both ends of the interval $[t_k, t_{k+1}]$ simultaneously. Since we have access to four pieces of information: f_k, df_k/dt, f_{k+1}, and df_{k+1}/dt, we can lay a third-order polynomial through these four pieces of information and solve for its roots.

The interpolation polynomial can thus be written as:

$$p(t) = a \cdot t^3 + b \cdot t^2 + c \cdot t + d \qquad (9.9)$$

with the derivative:

$$\dot{p}(t) = 3a \cdot t^2 + 2b \cdot t + c \qquad (9.10)$$

Thus, we can write the four pieces of information as follows:

$$p(t_k) = a \cdot t_k^3 + b \cdot t_k^2 + c \cdot t_k + d = f_k \qquad (9.11a)$$

$$p(t_{k+1}) = a \cdot t_{k+1}^3 + b \cdot t_{k+1}^2 + c \cdot t_{k+1} + d = f_{k+1} \qquad (9.11b)$$

$$\dot{p}(t_k) = 3a \cdot t_k^2 + 2b \cdot t_k + c = \dot{f}_k = h_k \qquad (9.11c)$$

$$\dot{p}(t_{k+1}) = 3a \cdot t_{k+1}^2 + 2b \cdot t_{k+1} + c = \dot{f}_{k+1} = h_{k+1} \qquad (9.11d)$$

which can be written in matrix/vector form as:

$$
\begin{pmatrix} f_k \\ f_{k+1} \\ h_k \\ h_{k+1} \end{pmatrix} =
\begin{pmatrix}
t_k^3 & t_k^2 & t_k & 1 \\
t_{k+1}^3 & t_{k+1}^2 & t_{k+1} & 1 \\
3t_k^2 & 2t_k & 1 & 0 \\
3t_{k+1}^2 & 2t_{k+1} & 1 & 0
\end{pmatrix} \cdot
\begin{pmatrix} a \\ b \\ c \\ d \end{pmatrix} \qquad (9.12)
$$

Equation 9.12 can then be solved for the unknown coefficients a, b, c, and d.

Figure 9.16 illustrates the method.

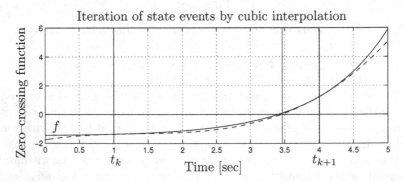

FIGURE 9.16. Iteration on single zero–crossing functions using cubic interpolation.

The method converges even faster than Newton iteration, as it exhibits *cubic convergence speed*. Furthermore, it is guaranteed to converge, just like Regula Falsi and Golden Section.

The cubic polynomial must have at least one real solution within the interval $[t_k, t_{k+1}]$. Possibly there are three real solutions within the interval, in which case any one of them could be used as the next evaluation time, t_{next}.

Yet, we may be able to improve on that method even a little further. One drawback of the proposed technique is that we need to solve for the roots of a cubic polynomial to determine a real root that lies inside the interval $[t_k, t_{k+1}]$.

Instead of fitting a cubic polynomial as proposed above, we could also fit an *inverse cubic polynomial* of the type:

$$t(p) = a_1 \cdot p^3 + b_1 \cdot p^2 + c_1 \cdot p + d_1 \tag{9.13}$$

which can simply be evaluated for $p = 0$. Thus, the next evaluation time can be computed as:

$$t_{\text{next}} = t(p = 0) = d_1 \tag{9.14}$$

The following four pieces of information are at our disposal:

$$t_k = t(f_k) \tag{9.15a}$$

$$t_{k+1} = t(f_{k+1}) \tag{9.15b}$$

$$u_k = \frac{dt(f_k)}{df} = \frac{1}{h_k} \tag{9.15c}$$

$$u_{k+1} = \frac{dt(f_{k+1})}{df} = \frac{1}{h_{k+1}} \tag{9.15d}$$

We know that:

$$t_k = a_1 \cdot f_k^3 + b_1 \cdot f_k^2 + c_1 \cdot f_k + d_1 \tag{9.16a}$$

$$t_{k+1} = a_1 \cdot f_{k+1}^3 + b_1 \cdot f_{k+1}^2 + c_1 \cdot f_{k+1} + d_1 \tag{9.16b}$$

$$u_k = 3a_1 \cdot f_k^2 + 2b_1 \cdot f_k + c_1 \tag{9.16c}$$

$$u_{k+1} = 3a_1 \cdot f_{k+1}^2 + 2b_1 \cdot f_{k+1} + c_1 \tag{9.16d}$$

or in matrix form:

$$\begin{pmatrix} t_k \\ t_{k+1} \\ u_k \\ u_{k+1} \end{pmatrix} = \begin{pmatrix} f_k^3 & f_k^2 & f_k & 1 \\ f_{k+1}^3 & f_{k+1}^2 & f_{k+1} & 1 \\ 3f_k^2 & 2f_k & 1 & 0 \\ 3f_{k+1}^2 & 2f_{k+1} & 1 & 0 \end{pmatrix} \cdot \begin{pmatrix} a_1 \\ b_1 \\ c_1 \\ d_1 \end{pmatrix} \tag{9.17}$$

which could be solved directly for the four unknowns by means of Gaussian elimination.

Yet, we can do even better. We shall use *inverse Hermite interpolation*. The scheme is called inverse interpolation, since we fit the inverse function with the polynomial. The polynomials that we shall use to span our base are Hermite polynomials.

We shall introduce a new variable ϕ of the type:

$$\phi = coef_1 \cdot f + coef_2 \tag{9.18}$$

such that:

t	f	ϕ
t_k	f_k	0.0
t_{k+1}	f_{k+1}	1.0
t_{next}	0.0	$\hat{\phi}$

TABLE 9.1. Variable transformation.

We find that:

$$coef_1 = \frac{1}{f_{k+1} - f_k} \tag{9.19a}$$

$$coef_2 = -\frac{f_k}{f_{k+1} - f_k} = \hat{\phi} \tag{9.19b}$$

We now construct four auxiliary polynomials in ϕ:

$$p_i(\phi) = \alpha_i \cdot \phi^3 + \beta_i \cdot \phi^2 + \gamma_i \cdot \phi + \delta_i \tag{9.20a}$$

$$\frac{dp_i(\phi)}{d\phi} = 3\alpha_i \cdot \phi^2 + 2\beta_i \cdot \phi + \gamma_i \tag{9.20b}$$

such that:

$$p_1(0) = 1 \quad ; \quad p_1(1) = 0 \quad ; \quad \frac{dp_1(0)}{d\phi} = 0 \quad ; \quad \frac{dp_1(1)}{d\phi} = 0 \tag{9.21a}$$

$$p_2(0) = 0 \quad ; \quad p_2(1) = 1 \quad ; \quad \frac{dp_2(0)}{d\phi} = 0 \quad ; \quad \frac{dp_2(1)}{d\phi} = 0 \tag{9.21b}$$

$$p_3(0) = 0 \quad ; \quad p_3(1) = 0 \quad ; \quad \frac{dp_3(0)}{d\phi} = 1 \quad ; \quad \frac{dp_3(1)}{d\phi} = 0 \tag{9.21c}$$

$$p_4(0) = 0 \quad ; \quad p_4(1) = 0 \quad ; \quad \frac{dp_4(0)}{d\phi} = 0 \quad ; \quad \frac{dp_4(1)}{d\phi} = 1 \tag{9.21d}$$

It is easy to verify that these polynomials are:

$$p_1(\phi) = 2\phi^3 - 3\phi^2 + 1 \tag{9.22a}$$

$$p_2(\phi) = -2\phi^3 + 3\phi^2 \tag{9.22b}$$

$$p_3(\phi) = \phi^3 - 2\phi^2 + \phi \tag{9.22c}$$

$$p_4(\phi) = \phi^3 - \phi^2 \tag{9.22d}$$

The inverse Hermite interpolation polynomial:

$$p(\phi) = a_2 \cdot \phi^3 + b_2 \cdot \phi^2 + c_2 \cdot \phi + d_2 \tag{9.23}$$

now expressed as a function of ϕ rather than of f, can be written in these auxiliary polynomials as:

$$p(\phi) = t_k \cdot p_1(\phi) + t_{k+1} \cdot p_2(\phi) + s_k \cdot p_3(\phi) + s_{k+1} \cdot p_4(\phi) \qquad (9.24)$$

where:

$$s_k = \frac{dt_k}{d\phi} = \frac{1}{d\phi_k/dt} = \frac{1}{coef_1 \cdot (df_k/dt)} = \frac{f_{k+1} - f_k}{h_k} \qquad (9.25a)$$

$$s_{k+1} = \frac{f_{k+1} - f_k}{h_{k+1}} \qquad (9.25b)$$

In order to obtain the desired zero–crossing time, t_{next}, we simply evaluate Eq.(9.24) at $\phi = \hat{\phi}$.

Figure 9.17 illustrates the method.

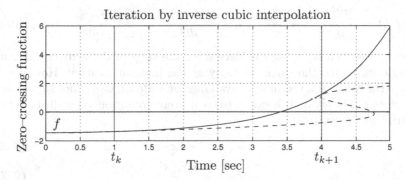

FIGURE 9.17. Iteration on single zero–crossing functions using inverse cubic interpolation.

Inverse Hermite interpolation is certainly more elegant than direct cubic interpolation. Unfortunately, the simplification in the computation came at a dire price, as we lost our guaranteed convergence. We can no longer guarantee that the solution is to be found within the interval $[t_k, t_{k+1}]$, and in the given example, this indeed is not the case.

Notice that all of these techniques were used only to determine the next time instant, t_{next}, for evaluating the zero–crossing function. The actual computation of the zero–crossing function is done by means of numerical integration, i.e., using the same higher–order numerical integration scheme used throughout the simulation. Thus, no approximation accuracy is lost in the process.

9.5.3 Single Zero Crossings, Multi–step Algorithms

In the case of multi–step algorithms, we may be able to do even better [9.3]. At the end of the step that puts the event conditions on alert, i.e., at time t_{k+1}, we have the Nordsieck vector available. Thus, we can write:

$$\mathcal{F}(\hat{h}) = \mathcal{F}_i(t_{\text{next}}) = \mathcal{F}_i(t_{k+1}) + \hat{h}\frac{d\mathcal{F}_i(t_{k+1})}{dt} + \frac{\hat{h}^2}{2}\frac{d^2\mathcal{F}_i(t_{k+1})}{dt^2}$$

$$+ \frac{\hat{h}^3}{6}\frac{d^3\mathcal{F}_i(t_{k+1})}{dt^3} + \cdots = 0.0 \qquad (9.26)$$

This is a function in the unknown $\hat{h}$ that can be solved by Newton iteration. We set:

$$\hat{h}^0 = 0.5 \cdot (t_k - t_{k+1}) \qquad (9.27)$$

and iterate:

$$\hat{h}^{\ell+1} = \hat{h}^\ell - \frac{\mathcal{F}(\hat{h}^\ell)}{\mathcal{H}(\hat{h}^\ell)} \qquad (9.28)$$

where:

$$\mathcal{H}(\hat{h}) = \frac{d\mathcal{F}(\hat{h})}{d\hat{h}} = \frac{d\mathcal{F}_i(t_{k+1})}{dt} + \hat{h}\frac{d^2\mathcal{F}_i(t_{k+1})}{dt^2} + \frac{\hat{h}^2}{2}\frac{d^3\mathcal{F}_i(t_{k+1})}{dt^3} + \cdots \quad (9.29)$$

Using this technique, we can determine the time of the zero–crossing in a single step with the same accuracy as the integration itself. However, we have the Nordsieck vector only available for *state variables*, not for *algebraic variables*. Therefore, it is useful to treat event conditions as additional state variables, by writing:

$$x_{n+i} = \mathcal{F}_i(\mathbf{x}) \qquad (9.30a)$$

$$\dot{x}_{n+i} = \frac{d\mathcal{F}_i(\mathbf{x})}{dt} \qquad (9.30b)$$

For the benefit of improved accuracy, it is probably a good idea to keep both equations in the model rather than integrating Eq.(9.30b) into Eq.(9.30a). However, the variables will be treated like additional state variables, and will be maintained by the integration algorithm in its data base of old values. In this way, it is possible to compute the Nordsieck vector for event conditions whenever needed.

We shall need to compute Eq.(9.30b) anyway, since otherwise, we cannot conveniently apply an iteration procedure other than Regula Falsi or Golden Section.

9.5.4 Non–essential State Events

Sometimes, it may be a good idea to even add Eq.(9.30b) as a *non–essential event condition* to the set of event conditions. Figure 9.18 illustrates the reason for this suggestion.

f_1 is an essential event condition, whereas $f_2 = \dot{f}_1$ is a non–essential event condition. A non–essential event condition is an event condition that doesn't have an event action associated with it.

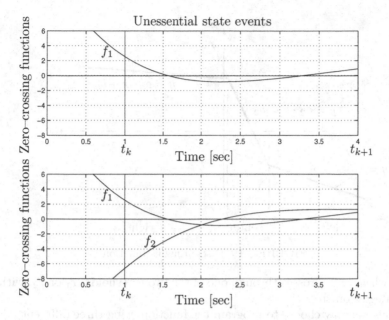

FIGURE 9.18. Non–essential event conditions.

Had we only formulated f_1 as a zero–crossing function, the event at time $t = 1.55$ would have been missed, because the essential zero–crossing function, f_1, crosses through zero twice within the single integration step from time $t_k = 1.0$ to time $t_{k+1} = 4.0$.

Adding the non–essential zero–crossing function, f_2, to the set of zero–crossing functions solves the dilemma, because f_2 exhibits a zero crossing, whenever f_1 goes through an extremum.

During the iteration of the non–essential event condition, f_2, the algorithm will discover that also f_1 crosses through zero, and will iterate on that zero crossing first, as it happens earlier.

9.6   Consistent Initial Conditions

Figure 9.19 shows a piecewise linear function with three segments. In the "left" region, $y = a_1 \cdot x + b_1$, in the "center" region, $y = a_2 \cdot x + b_2$, and in the "right" region, $y = a_3 \cdot x + b_3$.

Traditionally, we would describe such a function using an if–statement:

> **if** $x < x_1$ **then** $y = a_1 \cdot x + b_1$
>   **else if** $x < x_2$ **then** $y = a_2 \cdot x + b_2$
>   **else** $y = a_3 \cdot x + b_3$;

However, we know meanwhile that, if the variable y is used in a state–space model, this will force the step–size control mechanism to reduce the step

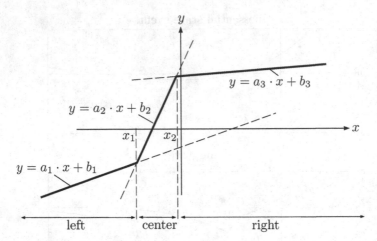

FIGURE 9.19. Discontinuous function.

size, whenever x crosses through one of the two thresholds, x_1 or x_2, within an integration step.

Thus, we may choose to program the function using three different models, one for each region, with appropriate zero–crossing functions describing the conditions for switching from one region to the next.

In pseudo–code, we might write:

```
case region
    left :    y = a₁ · x + b₁;
              schedule Center when x − x₁ == 0;
    center :  y = a₂ · x + b₂;
              schedule Left when x − x₁ == 0;
              schedule Right when x − x₂ == 0;
    right :   y = a₃ · x + b₃;
              schedule Center when x − x₂ == 0;
end;
```

together with the three discrete event descriptions:

```
event Left
    region := left;
end Left;

event Center
    region := center;
end Center;

event Right
    region := right;
end Right;
```

The *schedule*–statements are used in this pseudo–code to describe zero–crossing functions. The variable *region* is not a continuously changing variable. From the point of view of the continuous simulation, it assumes the role of a parameter. Its value can only change within a discrete event description.

This should work except for one little detail. The variable *region* is a *discrete state variable* that needs to be initialized. Somewhere in the initial region of the simulation program, we would need a statement such as:

```
if x < x₁ then region := left;
    else if x < x₂ then region := center;
    else region := right;
```

Will this code always work? Unfortunately, the answer to that question is no. One problem that we haven't considered yet is that x may reach one of the thresholds without actually crossing through it.

Let us assume that:

$$x(t) = \frac{x_2 - x_1}{2} \cdot \sin(t) + \frac{x_1 + x_2}{2} \tag{9.31}$$

In this case, x will always remain in the center region. It will only just reach the two thresholds, x_1 and x_2, every once in a while.

The event description, as programmed above, would make the model switch regions, each time a threshold is reached. One of the more difficult problems associated with the simulation of discontinuous functions is to know, in which region the model operates after the event has been processed, i.e., to find a consistent set of initial conditions after event handling.

The problem is by no means an academic one. Consider the case of a set of bowling balls resting on a guide rail. They are all in contact with each other. A new ball arrives with velocity v that hits the first of these balls. We all know what will happen: the new ball will come to rest at once, and the last of the previously resting balls will move away with the same speed v. Yet, convincing a simulation program that this is what must happen is anything but trivial.

One way to deal with this problem is to define a narrow band around each of the zero–crossing functions. The event is detected when the function crosses through zero, at which time the event is being processed. Yet, before starting with the next continuous–time simulation segment, trial steps are taken to determine whether or not the zero-crossing functions will leave the bands placed around the zero crossing as expected. It happens frequently that one event immediately triggers other events that change the condition on the original event again.

An example of this problem might be a robot arm with sticking friction in each of its articulations [9.11]. Once the force in an articulation overcomes sticking friction, the articulation starts to move. Yet, this immediately changes the forces in neighboring articulations. As a consequence, another neighboring articulation may come out of sticking friction also, which changes the forces in the articulations once again, with the possible effect that the original articulation returns back to its sticking region.

Modeling this situation correctly is anything but trivial. Let us attempt this task. Figure 9.20 shows a typical friction model with sticking friction, dry (Coulomb) friction, and viscous friction.

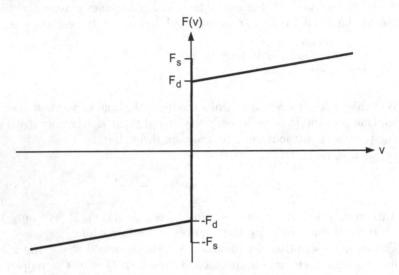

FIGURE 9.20. Friction characteristic.

There are three different regions (modes) of this nonlinear model: a *backward* mode, a *sticking* mode, and a *forward* mode. While the velocity of the articulation is zero, the articulation operates in its sticking mode. It will remain in this mode, until the sum of forces applied to this articulation becomes either larger than the positive sticking friction force, F_s, or smaller than the negative sticking friction force, $-F_s$. When this happens, the articulation comes out of sticking friction, and changes its operational mode to moving either *forward* or *backward*, in which the friction force is computed as the sum of a dry (Coulomb) friction component, $\pm F_d$, and a linear (viscous) friction component. Once the model operates in one of its two moving modes, it will remain in that mode, until the velocity of the articulation crosses through zero, at which time the model will return to its sticking mode.

Yet, this model is still too simple, as it does not account for the possibility that the result of coming out of sticking friction might be to return to sticking friction immediately again, after having freed up another articulation in the process.

A more complete model is shown in Figure 9.21, which exhibits a *state transition diagram* of the friction characteristic.

The new model possesses six different modes. Beside from the three modes used in the earlier model, we also have a *Start Forward* mode, and a *Start Backward* mode. If the sum of all forces applied to the articulation

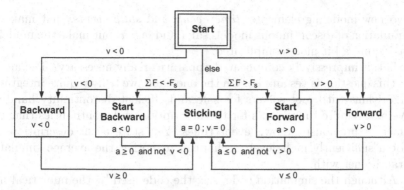

FIGURE 9.21. State transition diagram of friction characteristic.

is larger than zero, the articulation leaves the sticking mode. However, it doesn't proceed immediately to the *Forward* mode yet. Instead, it enters a transitory mode, called the *Start Forward* mode. As the sum of all forces is larger than zero, also the acceleration, in accordance with *Newton's law*, is now positive. However, the velocity is initially still zero. It will only become positive through integration, as time proceeds forward.

However, before integration starts again, the new forces are propagated to the neighboring bodies, possibly taking some other articulations also out of their sticking mode. As the condition for mode switching is programmed in the form of *state conditions*, rather than *time events*, integration needs to start, in order for this propagation to take place.

As integration starts, multiple zero–crossing functions may be triggered during the first new integration step. One would be to take the original articulation from the *Start Forward* mode to the *Forward* mode. Another may be to take a neighboring articulation from the *Sticking* mode to the *Start Forward* mode.

The model must make sure that the latter event takes precedence over the former. This is accomplished by recognizing the velocity as being positive only, after the velocity has become larger than some fudge factor $v > \varepsilon$, which implements the narrow band around the zero crossings that we wrote about earlier.

The *Start* mode implements the initialization of the discrete state variable. The discrete state variable starts out in its *Start* mode, from where it proceeds immediately to one of the other modes, depending on the initial velocity.

9.7 Object–oriented Descriptions of Discontinuities

What a mess have we created here! In order to protect the integration algorithm from having to integrate across discontinuities, we introduced two new modeling elements: *time events* and *state events*, that make the simulation of discontinuous models safer and faster, but make the modeling task quite a bit more complicated.

Is this impressively complicated apparatus really necessary? The answer to this question is yes and no. On the one hand, we truly require integration algorithms with root solvers for safe and efficient discontinuity handling. We also require time events for the description of discontinuities that take place at previously known event times. Yet, state event descriptions are not a sufficiently high–level mechanism to bother the average simulation practitioner with.

Although the simulation code, i.e., the code used by the numerical integration algorithm, may have to be complex and messy, this doesn't mean that the modeler has to manually enter it in this fashion.

Returning once more to the example of Fig. 9.19. What is wrong with a description of the type:

$$y = \text{if } x < x_1 \text{ then } a_1 \cdot x + b_1$$
$$\text{else if } x < x_2 \text{ then } a_2 \cdot x + b_2$$
$$\text{else } a_3 \cdot x + b_3;$$

to describe what this function does? It expresses perfectly well and in an unambiguous fashion, what the model is supposed to do. Can't we build a *model compiler* that takes such a description, and translates it down to the level of state events at compile time?

This is the approach that was taken in the design of the Dymola modeling environment [9.11, 9.12], and indeed, the syntax of the program segment shown above is that of Dymola.

Already in the previous two chapters, we encountered the need for symbolic preprocessing of model equations, in order to obtain numerically suitable simulation code. Although we applied these symbolic graph coloring algorithms in a manual fashion, by manually causalizing the structure digraph, this can obviously only be done for toy problems, such as the simple electrical circuits used to introduce the algorithms.

In a realistically complex model, such as a six–degree–of–freedom robot arm, leading to possibly 10,000 equations initially, it must be possible to apply all of these algorithms in a completely automated fashion. This is what the Dymola model compiler does [9.4]. The algorithms implemented in Dymola [9.12] are essentially those that were introduced in the previous two chapters.

Yet, Dymola is capable of performing considerably more complex model compilations, as it decomposes object–oriented descriptions of discontinuous models into suitable event descriptions at compile time.

Up to this point, we were able to either describe our algorithms in MATLAB, or apply them manually, as we did with some of the algorithms in the previous two chapters. Now, we don't have that luxury any longer, as even simple functions, such as the friction model introduced earlier, quickly become too involved to conveniently describe them as a collection of event descriptions.

Thus, we shall need to introduce some of the low–level modeling constructs of Dymola [9.12] at this time to be able to describe the necessary discontinuity handling algorithms in a suitably compact fashion.

9.7.1 The Computational Causality of if–Statements

We have seen in the previous two chapters that the computational causality of statements should not be predetermined, but must be allowed to vary depending on the embedding of the objects containing these statements within their environment.

The equal sign of an equation is not to be interpreted as an *assignment* in the usual sense of sequential programming languages, but rather as an *equality* in the algebraic sense.

Hence in a Dymola program, it doesn't matter whether Ohm's law is formulated as:

$$u = R * i$$

or:

$$i = u/R$$

or finally:

$$0 = u - R * i$$

Dymola will treat each of these statements in exactly the same fashion. It will turn equations around symbolically as needed.

It may now have become clear, why the Dymola syntax for the *if*–statement of the nonlinear characteristic of Fig. 9.19 is:

$$y = \textbf{if } x < x_1 \textbf{ then } a_1 \cdot x + b_1$$
$$\textbf{else if } x < x_2 \textbf{ then } a_2 \cdot x + b_2$$
$$\textbf{else } a_3 \cdot x + b_3;$$

rather than:

$$\textbf{if } x < x_1 \textbf{ then } y = a_1 \cdot x + b_1$$
$$\textbf{else if } x < x_2 \textbf{ then } y = a_2 \cdot x + b_2$$
$$\textbf{else } y = a_3 \cdot x + b_3;$$

Dymola needs to ensure that each branch of the *if*–statement computes the same variable, as otherwise, the *vertical sorting* algorithm would invariably fail.

Do *if*–statements have a fixed computational causality, or is it possible to turn them around in the same way as we turn around algebraic equations?

To answer this question, let us translate the above *if*–statements to an event description that looks a bit different from the one used before. To this end, we shall introduce three additional integer variables, m_l, m_c, and m_r, whose values are linked to the linguistic discrete state variable, *region*, in the following way:

region	m_l	m_c	m_r
left	1	0	0
center	0	1	0
right	0	0	1

Using these new variables, the event description of the nonlinear characteristic can be rewritten as follows:

$y = m_l \cdot (a_1 \cdot x + b_1) + m_c \cdot (a_2 \cdot x + b_2) + m_r \cdot (a_3 \cdot x + b_3);$
case *region*
  left :   **schedule** *Center* **when** $x - x_1 == 0;$
  center : **schedule** *Left* **when** $x - x_1 == 0;$
          schedule *Right* **when** $x - x_2 == 0;$
  right :   **schedule** *Center* **when** $x - x_2 == 0;$
end;

together with the three discrete event descriptions:

event *Left*
  region := *left*;
  $m_l = 1;$ $m_c = 0;$ $m_r = 0;$
end *Left*;

event *Center*
  region := *center*;
  $m_l = 0;$ $m_c = 1;$ $m_r = 0;$
end *Center*;

event *Right*
  region := *right*;
  $m_l = 0;$ $m_c = 0;$ $m_r = 1;$
end *Right*;

In this way, the former *if*–statement has been converted to the algebraic statement:

$$y = m_l \cdot (a_1 \cdot x + b_1) + m_c \cdot (a_2 \cdot x + b_2) + m_r \cdot (a_3 \cdot x + b_3) \qquad (9.32)$$

which can be turned around in the usual way:

$$x = \frac{y - m_l \cdot b_1 - m_c \cdot b_2 - m_r \cdot b_3}{m_l \cdot a_1 - m_c \cdot a_2 - m_r \cdot a_3} \tag{9.33}$$

as long as none of the three slopes is flat, i.e., as long as none of the parameters a_1, a_2, or a_3 is equal to zero.

9.7.2   Multi–valued Functions

The *if*–statements that we have introduced so far don't allow the description of multi–valued functions, such as the dry hysteresis function of Fig. 9.22.

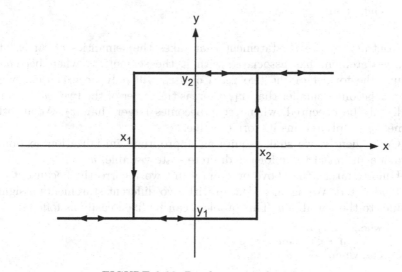

FIGURE 9.22. Dry hysteresis function.

A possible event description for the dry hysteresis function could look as follows:

```
y = ylast;
case region
    up :      schedule Down when x − x1 == 0;
    down :    schedule Up when x − x2 == 0;
end;
```

together with the two discrete event descriptions:

```
event Up
    region := up;
    ylast := y2;
end Left;
```

```
event Down
    region := down;
    y_last := y_1;
end Center;
```

Dymola offers a *when*–statement that allows to encode such an event description in a compact form. We could try to encode the dry hysteresis function as follows:

```
when x < x_1
    y = y_1;
end when;

when x > x_2
    y = y_2;
end when;
```

Contrary to the *if*–statement that takes the semantics of "if is," the *when*–statement has associated with it the semantics "when becomes." Thus, the former of the two *when* clauses will only be executed, whenever x becomes smaller than x_1, whereas the latter of the two *when* clauses will only be executed, whenever x becomes larger than x_2. At all other times, y simply retains its former value.

Consequently, we shall require an appropriate initialization section to provide an initial value for the discrete state variable, y.

Unfortunately, the above program won't work correctly, because it cannot be sorted. We again ended up with two different statements assigning values to the variable y. This problem can be fixed easily as follows:

```
when x < x_1 or x > x_2
    y = if x < 0 then y_1 else y_2;
end when;
```

Here, y assumes a new value if and only if either x becomes smaller than x_1 or if x becomes larger than x_2. The new value of y will be y_1, if x is at that time smaller than 0, else y assumes a value of y_2.

9.8   The Switch Equation

Let us now try to describe the electrical switch of Fig. 9.23.

When the switch is *open*, the current flowing through it is zero. When it is *closed*, the voltage across it is zero.

An elegant way to describe the switch properties in Dymola using a single statement would be:

```
0 = if switch == open then i else u;
```

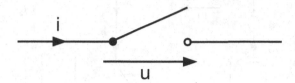

FIGURE 9.23. Electrical switch.

Let us convert the *if*–statement to an equivalent algebraic statement. To this end, we introduce an integer variable, m_o, with the following values:

switch	m_o
open	1
closed	0

Using the new variable m_o, we can rewrite the switch equation as follows:

$$0 = m_o \cdot i + (1 - m_o) \cdot u \qquad (9.34)$$

The algebraic switch equation can be made causal in two different ways:

$$i = \frac{m_o - 1}{m_o} \cdot u \qquad (9.35a)$$

$$u = \frac{m_o}{m_o - 1} \cdot i \qquad (9.35b)$$

Unfortunately, neither of these two equations will work correctly in both switch positions. Equation (9.35a) will lead to a division by zero, whenever the switch closes, whereas Eq.(9.35b) will lead to a division by zero, whenever the switch opens.

The switch equation confronts us with a new problem. The correct computational causality of the switch equation depends on the numerical value of a parameter. In the given example, it depends on the numerical value of m_o.

In previous chapters, we have learnt that the computational causality of all equations is fixed, except for those that show up inside an *algebraic loop*.

Hence we may postulate that:

Switch equations must always be placed inside algebraic loops.

Let us illustrate this concept by means of a simple circuit example, as shown in Fig. 9.24.

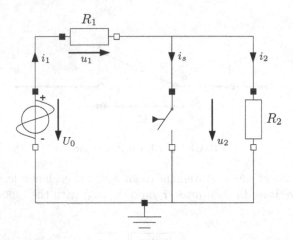

FIGURE 9.24. Electrical circuit containing a switch.

The circuit operates correctly in both switch positions. If the switch is open, the resistor across the voltage source assumes a value of $R_1 + R_2$, otherwise it assumes a value of R_1 only.

We can read out the equations from this circuit:

$$U_0 = f(t) \tag{9.36a}$$
$$u_1 = R_1 \cdot i_1 \tag{9.36b}$$
$$u_2 = R_2 \cdot i_2 \tag{9.36c}$$
$$U_0 = u_1 + u_2 \tag{9.36d}$$
$$i_1 = i_s + i_2 \tag{9.36e}$$
$$0 = m_o \cdot i_s + (1 - m_o) \cdot u_2 \tag{9.36f}$$

The structure digraph of this equation system is shown on Fig. 9.25.

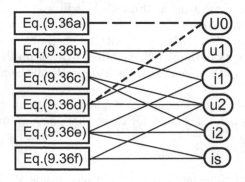

FIGURE 9.25. Partially causalized structure digraph of switching circuit.

The equation system indeed contains an algebraic loop in five equations and five unknowns, and as expected, the switch equation shows up inside the algebraic loop.

This time around, we didn't use our normal heuristics for choosing a suitable tearing structure. We want our switch equation to serve as the residual equation, solving it for whichever variable works better. In the given example, we chose u_2 as the tearing variable, since this allowed us to causalize all remaining equations. The resulting set of causal equations is:

$$U_0 = f(t) \tag{9.37a}$$

$$i_2 = \frac{1}{R_2} \cdot u_2 \tag{9.37b}$$

$$u_1 = U_0 - u_2 \tag{9.37c}$$

$$i_1 = \frac{1}{R_1} \cdot u_1 \tag{9.37d}$$

$$i_s = i_1 - i_2 \tag{9.37e}$$

$$u_2 = \frac{m_o}{m_o - 1} \cdot i_s \tag{9.37f}$$

where Eq.(9.37f) is the residual equation, and u_2 serves as the tearing variable.

Using the variable substitution method, we find the following replacement equation for the residual equation:

$$u_2 = \frac{m_o \cdot R_2}{m_o \cdot (R_1 + R_2) + (m_o - 1) \cdot R_1 \cdot R_2} \cdot U_0 \tag{9.38}$$

Equation(9.38) is indeed the correct equation in both switch positions, since, when the switch is *closed*, i.e., $m_o = 0$:

$$u_2 = 0 \tag{9.39}$$

and when the switch is *open*, i.e., $m_o = 1$:

$$u_2 = \frac{R_2}{R_1 + R_2} \cdot U_0 \tag{9.40}$$

No division by zero is obtained in either of the two switch positions.

9.9 Ideal Diodes and Parameterized Curve Descriptions

Ideal diodes are ideal electrical switches complemented by an internal logic for determining the switch position. An ideal diode closes its switch, when the voltage across the diode from the anode to the cathode becomes positive, and it opens its switch again, when the current through the diode passes through zero, if at that time the voltage across the diode is negative.

An ideal diode can be modeled in Dymola as follows:

$0 = m_o \cdot i_d + (1 - m_o) \cdot u_d;$
$m_o = $ **if** $u_d <= 0$ **and not** $i_d > 0$ **then** 1 **else** 0;

A yet more compact way to describe this model would be:

$0 = $ **if** *OpenSwitch* **then** i_d **else** $u_d;$
OpenSwitch $= u_d <= 0$ **and not** $i_d > 0;$

OpenSwitch is here a Boolean variable, the value of which is computed in the above Boolean expression. If *OpenSwitch* is *true*, the switch is considered *open*.

The latest example exhibits a third way for encoding state–event descriptions, beside from the previously introduced *if*–statements and *when*–statements. Any Boolean function of real–valued variables is automatically converted to a state–event description by Dymola's model compiler.

In reality, this is even the *only* way to produce state–event descriptions, as Dymola extracts the conditions of *if*– and *when*–statements into separate Boolean statements prior to expanding them.

How are Boolean functions of real–valued variables converted to zero–crossing functions? In the simplest cases, such as:

$$B_1 = x > x_2 \tag{9.41}$$

i.e., cases in which the Boolean expression is formed by a single relational operator, the conversion is trivial, as B_1 is almost in the correct form already. The corresponding zero–crossing function can be written as:

$$f_1 = x - x_2 \tag{9.42}$$

The case:

when $x < x_1$ **or** $x > x_2$
$\quad y = $ **if** $x < 0$ **then** y_1 **else** $y_2;$
end when;

is a bit more difficult to handle. The condition of the *when*–statement gets extracted into the Boolean function:

$$B_2 = x < x_1 \lor x > x_2 \tag{9.43}$$

which gets then converted to the following zero–crossing function:

$$f_2 = \text{ if } B_2 \text{ then } 1 \text{ else } -1 \tag{9.44}$$

Whenever B_2 switches from *true* to *false* or vice–versa, f_2 crosses through zero.

Unfortunately, f_2 is anything but a smooth function. In fact, the gradient of f_2 is zero everywhere except at the zero crossing itself, where it is infinite. Thus, no higher–order method, such as *cubic interpolation* or *inverse Hermite interpolation* can be used on such a zero–crossing function. Only first–order methods, such as the *Regula Falsi* or the *Golden Section* method can be used, and of those, even only the Golden Section method can be used efficiently.

A better solution would have been to generate two separate zero–crossing functions:

$$f_{2a} = x - x_1 \qquad (9.45a)$$
$$f_{2b} = x - x_2 \qquad (9.45b)$$

that are both being associated with the same event action:

$$y = \textbf{if } x < 0 \textbf{ then } y_1 \textbf{ else } y_2 \qquad (9.46)$$

The Dymola user can enforce that the model is being translated in this fashion by employing a slightly different model syntax:

```
when { x < x₁ , x > x₂ }
    y = if x < 0 then y₁ else y₂;
end when;
```

Using this syntax, each of the set of conditions of the *when*–statement is converted independently to a separate zero–crossing function. All of these zero–crossing functions are associated with the same event action.

Unfortunately, even with the enhanced syntax, the problem:

```
0 = if OpenSwitch then i_d else u_d;
OpenSwitch = u_d <= 0 and not i_d > 0;
```

cannot be converted to a set of smooth zero–crossing functions. The proposed technique works only in the case of a set of simple Boolean expressions that are connected by *or*–conditions. Another approach must thus be taken.

To this end, we shall apply a *parameterized curve description*, as advocated in [9.22]. Figure 9.26 displays the diode characteristic in the $i_d(u_d)$ plane.

The curve is *parameterized* by adding an additional variable, s, to the model, defined such that $s = u_d$ whenever the diode is blocking, and $s = i_d$, whenever the diode is conducting. This allows us to program a smooth zero–crossing function in terms of the newly introduced variable s:

```
u_d = if OpenSwitch then s else 0;
i_d = if OpenSwitch then 0 else s;
```

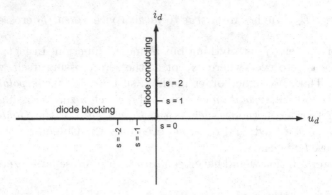

FIGURE 9.26. Diode characteristic.

$OpenSwitch = s < 0;$

This is how the ideal diode has been modeled in Dymola's standard electrical library.

An algebraic version of that model can be written as:

$u_d = m_o \cdot s;$
$i_d = (1 - m_o) \cdot s;$
$m_o = \textbf{if } s < 0 \textbf{ then } 1 \textbf{ else } 0;$

which is the version that we shall work with here, as these equations are easier to analyze.

Let us illustrate the use of the ideal diode model by means of the simple half–way rectifier circuit of Fig.9.27.

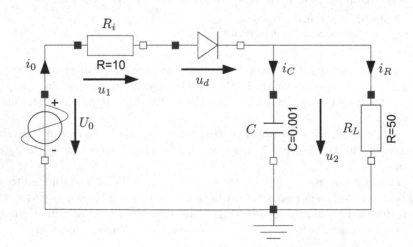

FIGURE 9.27. Half–way rectifier circuit.

We can read the equations from that circuit:

$$u_0 = f(t) \tag{9.47a}$$

$$u_1 = R_i \cdot i_0 \tag{9.47b}$$

$$u_2 = R_L \cdot i_R \tag{9.47c}$$

$$i_C = C \cdot \frac{du_2}{dt} \tag{9.47d}$$

$$u_0 = u_1 + u_d + u_2 \tag{9.47e}$$

$$i_0 = i_C + i_R \tag{9.47f}$$

$$u_d = m_o \cdot s \tag{9.47g}$$

$$i_0 = (1 - m_o) \cdot s \tag{9.47h}$$

The partially causalized structure digraph of this equation system is shown in Fig. 9.28. We ended up with an algebraic loop in four equations and four unknowns. The switch equations are contained within the algebraic loop.

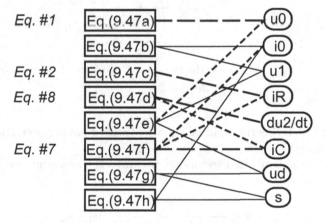

FIGURE 9.28. Partially causalized structure digraph.

We now must choose a suitable tearing structure. Once again, we won't use our normal heuristics. Instead, we want to make sure that the variable s is being selected as the tearing variable. We choose one of the two switch equations, e.g. Eq.(9.47h), as the corresponding residual equation.

The completely causalized structure digraph of this equation system is shown in Fig. 9.29.

Thus, the horizontally and vertically sorted equations can be written as:

$$u_0 = f(t) \tag{9.48a}$$

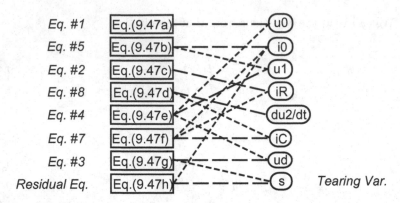

FIGURE 9.29. Completely causalized structure digraph.

$$i_R = \frac{1}{R_L} \cdot u_2 \qquad\qquad (9.48b)$$

$$u_d = m_o \cdot s \qquad\qquad (9.48c)$$

$$u_1 = u_0 - u_d - u_2 \qquad\qquad (9.48d)$$

$$i_0 = \frac{1}{R_i} \cdot u_1 \qquad\qquad (9.48e)$$

$$s = \frac{1}{1 - m_o} \cdot i_0 \qquad\qquad (9.48f)$$

$$i_C = i_0 - i_R \qquad\qquad (9.48g)$$

$$\frac{du_2}{dt} = \frac{1}{C} \cdot i_C \qquad\qquad (9.48h)$$

where Eq.(9.48f) is the residual equation, with s having been chosen as the tearing variable.

Using the substitution technique, we find a replacement equation for the residual equation:

$$s = \frac{1}{m_o + (1 - m_o) \cdot R_i} \cdot (u_0 - u_2) \qquad\qquad (9.49)$$

which is correct in both switch positions.

The following equation system results:

$$u_0 = f(t) \qquad\qquad (9.50a)$$

$$i_R = \frac{1}{R_L} \cdot u_2 \qquad\qquad (9.50b)$$

$$s = \frac{1}{m_o + (1 - m_o) \cdot R_i} \cdot (u_0 - u_2) \qquad\qquad (9.50c)$$

$$u_d = m_o \cdot s \qquad\qquad (9.50d)$$

$$u_1 = u_0 - u_d - u_2 \tag{9.50e}$$

$$i_0 = \frac{1}{R_i} \cdot u_1 \tag{9.50f}$$

$$i_C = i_0 - i_R \tag{9.50g}$$

$$\frac{du_2}{dt} = \frac{1}{C} \cdot i_C \tag{9.50h}$$

which can be simulated without any difficulties using any numerical integration algorithm with a root solver.

There is only a single zero–crossing function:

$$f = s \tag{9.51}$$

with the associated event action:

event *Toggle*
  $m_o := 1 - m_o;$
end *Toggle*;

The correct initial value of the discrete state variable, m_o, is assigned to that variable in an appropriate initialization section of the simulation program.

The voltage across the capacitor is shown in Fig. 9.30 as a function of time.

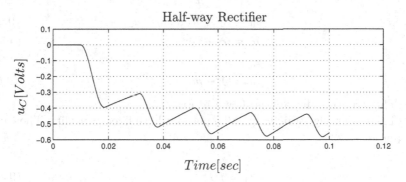

FIGURE 9.30. Voltage across capacitor of half–way rectifier circuit.

The default algorithm used in Dymola is DASSLRT, an implementation of the well–known DASSL algorithm supplemented with a root solver.

9.10   Variable Structure Models

Let us repeat the previous analysis for the slightly different circuit of Fig. 9.31.

The following set of equations characterizes this circuit:

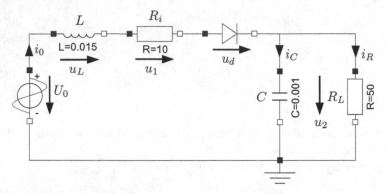

FIGURE 9.31. Half–way rectifier circuit with line inductance.

$$u_0 = f(t) \tag{9.52a}$$

$$u_1 = R_i \cdot i_0 \tag{9.52b}$$

$$u_2 = R_L \cdot i_R \tag{9.52c}$$

$$i_C = C \cdot \frac{du_2}{dt} \tag{9.52d}$$

$$u_L = L \cdot \frac{di_0}{dt} \tag{9.52e}$$

$$u_0 = u_L + u_1 + u_d + u_2 \tag{9.52f}$$

$$i_0 = i_C + i_R \tag{9.52g}$$

$$u_d = m_o \cdot s \tag{9.52h}$$

$$i_0 = (1 - m_o) \cdot s \tag{9.52i}$$

The structure digraph is shown in Fig. 9.32.

There is no algebraic loop. All equations have fixed causality. The causal equations are:

$$u_0 = f(t) \tag{9.53a}$$

$$u_1 = R_i \cdot i_0 \tag{9.53b}$$

$$i_R = \frac{1}{R_L} \cdot u_2 \tag{9.53c}$$

$$s = \frac{1}{1 - m_o} \cdot i_0 \tag{9.53d}$$

$$i_C = i_0 - i_R \tag{9.53e}$$

$$u_d = m_o \cdot s \tag{9.53f}$$

$$u_L = u_0 - u_1 - u_d - u_2 \tag{9.53g}$$

$$\frac{du_2}{dt} = \frac{1}{C} \cdot i_C \tag{9.53h}$$

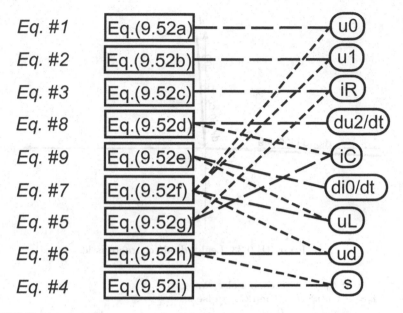

FIGURE 9.32. Causalized structure digraph of half–way rectifier circuit with line inductance model.

$$\frac{di_0}{dt} = \frac{1}{L} \cdot u_L \tag{9.53i}$$

These equations unfortunately cannot be simulated, since Eq.(9.53d) leads to a division by zero, as soon as the switch opens.

What happened? The current through the inductor is a state variable. Thus, the inductor computes the current i_0, which means that the causality of the diode is fixed. The diode has no choice but to compute the voltage u_d.

If we replace the diode by a manual switch, we see at once what happens. If we try to open the switch, while current is flowing through it, we'll draw an arc, because the current through the inductance cannot go instantly to zero. The arc can be modeled as a nonlinear resistor, the value of which increases, as the gap widens. This resistance drives the current to zero. Yet, this effect was not included in the model equations.

With a diode, this cannot happen, as the diode always opens at the moment, when the current passes through zero. Yet, our model doesn't know this. Since the logic for when the diode switch opens or closes is not contained in the continuous model equations, but forms part of the event description, the continuous model equations are identical in the case of the diode and the manual switch.

Dymola tackles this problem by offering in its standard electrical library a *leaky diode* model, as shown in Fig. 9.33.

The leaky diode can be modeled using the equations:

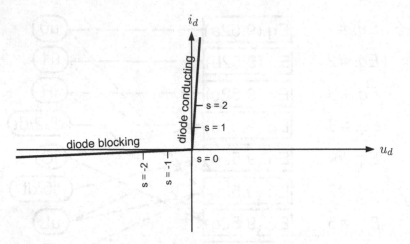

FIGURE 9.33. Leaky diode characteristic.

$u_d = $ **if** $OpenSwitch$ **then** s **else** $R_0 \cdot s$;
$i_d = $ **if** $OpenSwitch$ **then** $G_0 \cdot s$ **else** s;
$OpenSwitch = s < 0$;

or formulated algebraically:

$u_d = [m_o + (1 - m_o) \cdot R_0] \cdot s$;
$i_d = [m_o \cdot G_0 + (1 - m_o)] \cdot s$;
$m_o = $ **if** $s < 0$ **then** 1 **else** 0;

R_0 is the resistance of the wires connected to the switch, when the switch is closed, and G_0 is the conductance of the air in the gap, while the switch is open.

The leaky diode model doesn't change the causalities of the equation system, i.e., the structure digraph of the model using the leaky diode is exactly the same as that using the ideal diode. However, the leaky diode avoids the division by zero.

The causal equations of the model using the leaky diode are:

$$u_0 = f(t) \tag{9.54a}$$

$$u_1 = R_i \cdot i_0 \tag{9.54b}$$

$$i_R = \frac{1}{R_L} \cdot u_2 \tag{9.54c}$$

$$s = \frac{1}{m_o \cdot G_0 + (1 - m_o)} \cdot i_0 \tag{9.54d}$$

$$i_C = i_0 - i_R \tag{9.54e}$$

$$u_d = [m_o + (1 - m_o) \cdot R_0] \cdot s \tag{9.54f}$$

$$u_L = u_0 - u_1 - u_d - u_2 \qquad (9.54\text{g})$$

$$\frac{du_2}{dt} = \frac{1}{C} \cdot i_C \qquad (9.54\text{h})$$

$$\frac{di_0}{dt} = \frac{1}{L} \cdot u_L \qquad (9.54\text{i})$$

This model is valid in both switch positions, i.e., it can be simulated. Unfortunately, whenever the original model containing an ideal diode exhibits a division by zero, the new model containing a leaky diode becomes very stiff. The degree of stiffness is directly related to the values of the two leakage parameters, R_0 and G_0. The smaller the leakage parameters are chosen, the stiffer the model will become. Hence we would prefer to use the ideal model, if we could.

What is so special about this model? When the switch is closed, i.e., while the diode is conducting, the model exhibits second–order dynamics. However, once the switch opens, i.e., while the diode blocks the current, we are faced with only first–order dynamics. The inductor does not contribute to the dynamics in that case.

We call a model that exhibits different structural properties, such as a varying number of differential equations depending on the position of some switches a *variable structure model*.

Variable structure systems are very common, e.g. in mechanical engineering. All systems involving clutches are by their very nature variable structure systems. In electrical engineering, most switching power converters are variable structure systems.

The way, the equations of our system were formulated, Eqs.(9.52a–i), it doesn't look like these equations contain a *structural singularity* though. There is no constraint equation to be found. The singularity looks to be *parametric* in nature, thus the Pantelides algorithm [9.23] cannot be applied to solve it.

9.11 Mixed–mode Integration

One way to tackle this problem, while preserving the use of an ideal diode model, is to relax the causality on the inductor, by inlining the integrator that is associated with it. This approach was first proposed in [9.18].

An approach to simulation by applying different integration algorithms to different integrators contained in the model is called simulation by mixed–mode integration [9.24].

The system equations now take the form:

$$u_0 = f(t) \qquad (9.55\text{a})$$

$$u_1 = R_i \cdot i_0 \qquad (9.55\text{b})$$

$$u_2 = R_L \cdot i_R \tag{9.55c}$$

$$i_C = C \cdot \frac{du_2}{dt} \tag{9.55d}$$

$$i_0 = \text{pre}(i_o) + \frac{h}{L} \cdot u_L \tag{9.55e}$$

$$u_0 = u_L + u_1 + u_d + u_2 \tag{9.55f}$$

$$i_0 = i_C + i_R \tag{9.55g}$$

$$u_d = m_o \cdot s \tag{9.55h}$$

$$i_0 = (1 - m_o) \cdot s \tag{9.55i}$$

The partially causalized structure digraph of this model is given in Fig. 9.34.

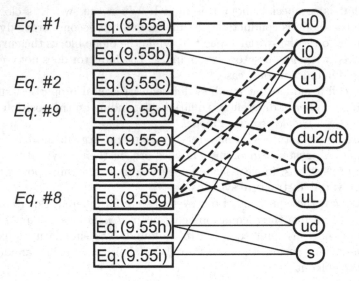

FIGURE 9.34. Partially causalized structure digraph of half–way rectifier circuit with inlined inductor.

Only four of the nine equations could be causalized directly. There now appeared an algebraic loop, which includes the switch equations.

We need to choose s as a tearing variable, because otherwise, the equation computing s will invariably contain either the term m_o or the term $(1 - m_o)$ alone in the denominator, which consequently leads to a division by zero in one of the two switch positions.

We can choose either Eq.(9.55h) or Eq.(9.55i) as the associated residual equation. If we choose Eq.(9.55h) as the residual equation, we can causalize all of the remaining equations. Unfortunately, Eq.(9.55e) will in that case be solved for u_L, which we don't like, since it leaves h alone in the denominator.

Thus, we chose Eq.(9.55i) as the associated residual equation. The further causalized structure digraph is shown in Fig. 9.35.

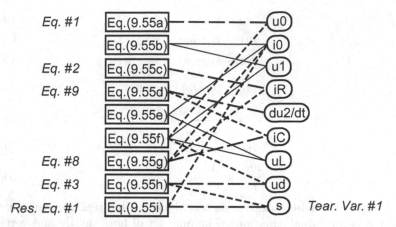

FIGURE 9.35. Partially causalized structure digraph of half–way rectifier circuit with inlined inductor.

There remains a second algebraic loop in three equations and three unknowns. This time, we choose Eq.(9.55e) as the new residual equation, and i_0 as the tearing variable. In this way, we can force the causality on the inlined integrator equation as well. The completely causalized structure digraph is shown in Fig. 9.36.

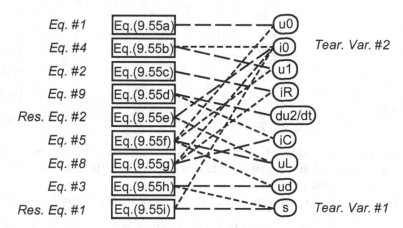

FIGURE 9.36. Completely causalized structure digraph of half–way rectifier circuit with inlined inductor.

The causalized equations can be read out of the structure digraph:

$$u_0 = f(t) \tag{9.56a}$$

$$i_R = \frac{1}{R_L} \cdot u_2 \tag{9.56b}$$

$$u_d = m_o \cdot s \qquad (9.56\text{c})$$

$$u_1 = R_i \cdot i_0 \qquad (9.56\text{d})$$

$$u_L = u_0 - u_1 - u_d - u_2 \qquad (9.56\text{e})$$

$$i_0 = \text{pre}(i_o) + \frac{h}{L} \cdot u_L \qquad (9.56\text{f})$$

$$s = \frac{1}{1 - m_o} \cdot i_0 \qquad (9.56\text{g})$$

$$i_C = i_0 - i_R \qquad (9.56\text{h})$$

$$\frac{du_2}{dt} = \frac{1}{C} \cdot i_C \qquad (9.56\text{i})$$

Using the variable substitution technique, we find replacement equations for the two residual equations. The final set of horizontally and vertically sorted equations presents itself as follows:

$$u_0 = f(t) \qquad (9.57\text{a})$$

$$i_R = \frac{1}{R_L} \cdot u_2 \qquad (9.57\text{b})$$

$$s = \frac{L \cdot \text{pre}(i_o) + h \cdot (u_0 - u_2)}{h \cdot m_o + (L + h \cdot R_i) \cdot (1 - m_o)} \qquad (9.57\text{c})$$

$$u_d = m_o \cdot s \qquad (9.57\text{d})$$

$$i_0 = (1 - m_o) \cdot s \qquad (9.57\text{e})$$

$$u_1 = R_i \cdot i_0 \qquad (9.57\text{f})$$

$$u_L = u_0 - u_1 - u_d - u_2 \qquad (9.57\text{g})$$

$$i_C = i_0 - i_R \qquad (9.57\text{h})$$

$$\frac{du_2}{dt} = \frac{1}{C} \cdot i_C \qquad (9.57\text{i})$$

Let us analyze this set of equations a bit further. The only potentially dangerous equation is Eq.(9.57c). Let us discuss, how this equation behaves in the two switch positions.

If the switch is closed, $m_o = 0$, Eq.(9.57c) degenerates to:

$$s = \frac{L \cdot \text{pre}(i_o) + h \cdot (u_0 - u_2)}{L + h \cdot R_i} \qquad (9.58)$$

which is completely harmless for all values of the step size, h.

If the switch is open, $m_o = 1$, Eq.(9.57c) degenerates to:

$$s = \frac{L}{h} \cdot \text{pre}(i_o) + u_0 - u_2 \qquad (9.59)$$

This equation is correct for all values of the step size, h, if switching occurs at a moment, when the current, i_0, goes through zero, as this will always be the case for a diode. However, if switching occurs for any other value of i_0, only one step will be incorrect, since during that first step, the current i_0 will be reduced to zero due to Eq.(9.57e). Thus, already one step later, the solution is again accurate. There is no stiffness problem using this approach.

9.12   State Transition Diagrams

Let us now return to the discussion of friction phenomena, an important application of discontinuous models in mechanical engineering.

Before a possible general model for the friction element can be presented, the friction phenomenon needs to be carefully analyzed. According to Fig.9.20, the friction force is a known applied force if the velocity v is different from zero. In that situation, the computational causality of the friction model is such that the velocity is an input to the model, whereas the friction force is its output.

When the velocity becomes zero, the two bodies, between which the friction force is acting, become stuck. In this situation, the model changes its structure: A new equation, $v = 0.0$, and a new unknown force, F_c, are added to the model. The constraint force F_c is determined such that the new constraint on the velocity, $v = 0.0$, is always met.

This is a new situation as compared to the electrical switch, because the electrical switch toggles between two different equations one of which is always active. Thus, the number of equations remains the same. In contrast, the friction element adds one equation and one variable to the model, when v becomes 0, and removes them again, when abs(F_c) becomes larger than the threshold value F_s.

Simulation environments do not usually allow to add/remove variables during integration. Therefore, a dummy equation is added, which becomes active, when the constraint equation, $v = 0.0$, is removed. The dummy equation is used to provide a unique –but arbitrary– value for F_c during sliding motion. For example, $F_c = 0.0$ is as good a value as any.

The friction force F can thus be defined through the following equations:

$$F = \text{if } v > 0 \text{ then } c_f \cdot v + F_d \text{ else}$$
$$\text{if } v < 0 \text{ then } c_f \cdot v - F_d \text{ else } F_c$$
$$0 = \text{if } Sticking \text{ then } v \text{ else } F_c$$

The third equation is our meanwhile well–known *switch equation*.

The model is so simple, it looks like magic ... and it also works like magic, i.e., it doesn't. In a Newtonian world, it is not sufficient to describe how the rabbit is being pulled out of the magician's hat. We also need to describe how it got into the hat in the first place, at which time, unfortunately, the magic is gone.

There are two problems with the above model. First, we haven't come up yet with an equation for the discrete state variable *Sticking*. Evidently, it won't do to say that:

$$Sticking = v == 0 \qquad (9.60)$$

as this would simply state that whenever v equals zero, then v equals zero, which is undoubtedly a true statement, but unfortunately, it isn't a very useful one.

The second problem with this model is that the *then*–branch of the switch equation is constrained, since the velocity v is a state variable. Thus, the causality of the switch equation is fixed, which invariably leads to a division by zero in one of the two switch positions.

Let us tackle the latter problem first. We already know one possible solution to this problem. We could relax the causality on the velocity, v, by inlining the integrator that integrates the acceleration, a, into the velocity, v. Yet, there is a better way.

While in the sticking position, the velocity, v, remains constantly zero. Thus, also the acceleration, a, must remain constantly zero. We can thus replace the former switch equation in the velocity, v, by a modified switch equation in the acceleration, a, as follows:

F = **if** $v > 0$ **then** $c_f \cdot v + F_d$ **else**
　　 if $v < 0$ **then** $c_f \cdot v - F_d$ **else** F_c
0 = **if** *Sticking* **then** a **else** F_c

This looks like a generalization of the Pantelides algorithm [9.23]. We seem to have partially differentiated the switch equation. Unfortunately, this technique rarely works. The Pantelides algorithm can only be generalized to the case of *conditional index changes* modeled by means of switch equations, if either both branches of the *if*–statement formulating the switch equation are constrained, or if the unconstrained branch is unimportant.

In the case of the friction model, the *then*–branch of the switch equation is constrained, as it is a function of state variables only, whereas the *else*–branch is unimportant. While the model is not sticking, we don't care what value the variable F_c assumes. Thus, there is no need to differentiate the *else*–branch of the switch equation simultaneously with the *then*–branch.

There is still a small problem with this formulation though. Since the friction model enters its *Sticking* region when the velocity passes through zero, the velocity may numerically not be exactly equal to zero, after the model entered its "*Sticking*" region. Therefore, the position, x, may slowly drift away.

This problem can be easily fixed by adding:

F = **if** $v > 0$ **then** $c_f \cdot v + F_d$ **else**
　　 if $v < 0$ **then** $c_f \cdot v - F_d$ **else** F_c
0 = **if** *Sticking* **then** a **else** F_c
when *Sticking* **then**
　　 reinit$(v, 0)$;
end when;

to the model. Thus, when the friction model enters its *Sticking* region, the velocity, v, is explicitly re–initialized to 0.

Let us now tackle the other problem. We haven't defined yet, how the discrete state variable, *Sticking*, is computed by the model. To this end, we need to define, how the switching between the sliding and the sticking phases takes place.

It is advantageous to split the friction force law into the following five different regions:

region:		region conditions:	
Forward	:	$v > 0$	and $F = c_f \cdot v + F_d$
StartForward	:	$v = 0$ and $a > 0$	and $F = +F_d$
Sticking	:	$v = 0$ and $a = 0$	and $F \in [-F_s, +F_s]$
StartBackward	:	$v = 0$ and $a < 0$	and $F = -F_d$
Backward	:	$v < 0$	and $F = c_f \cdot v - F_d$

Regions *Forward* and *Backward* describe the sliding phase and are defined by a non–zero velocity. Region *Sticking* denotes the sticking phase and is defined by identically vanishing velocity and acceleration. Regions *StartForward* and *StartBackward* define the transition from sticking to sliding. These regions are characterized by a zero velocity. The difference to the sticking phase is that the acceleration is no longer fixed to zero. The above five regions cannot be encoded directly, because the equality relation "=" appears in the definition. It is not meaningful to test computed real–valued variables for being equal to zero.

Hence an indirect approach will be used. The switching between the five regions is described by a *deterministic finite state machine (DFSM)* [9.1]. The *state transition diagram* of the DFSM was shown earlier in this chapter. It is repeated here.

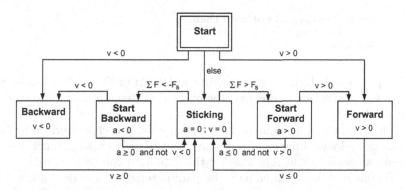

FIGURE 9.37. State transition diagram of friction characteristic.

The DFSM has six states, corresponding to the five regions of the model and a *Start* state. Starting from any one state of the DFSM and using one of the mutually exclusive conditions, a new state of the DFSM is reached in an unambiguous fashion. None of the switching conditions contains the equality relation.

A valid Dymola code can be easily derived from a DFSM by defining a boolean variable (a discrete state variable) for every state of the DFSM and by encoding the state transitions leading into or out of each state as boolean expressions determining the next value of that state.

$$F = \textbf{if } Forward \qquad \textbf{then } c_f \cdot v + F_d \textbf{ else}$$
$$\quad \textbf{if } Backward \qquad \textbf{then } c_f \cdot v - F_d \textbf{ else}$$
$$\quad \textbf{if } StartForward \quad \textbf{then } +F_d \qquad \textbf{else}$$
$$\quad \textbf{if } StartBackward \textbf{ then } -F_d \qquad \textbf{else } F_c;$$

$$0 = \textbf{ if } Sticking \textbf{ or } Start \textbf{ then } a \textbf{ else } F_c;$$

$$Forward \qquad = \text{pre}(Start) \qquad\qquad \textbf{and } v > 0 \textbf{ or}$$
$$\qquad\qquad\qquad \text{pre}(StartForward) \quad \textbf{and } v > 0 \textbf{ or}$$
$$\qquad\qquad\qquad \text{pre}(Forward) \qquad\qquad \textbf{and not } v <= 0;$$

$$Backward \qquad = \text{pre}(Start) \qquad\qquad \textbf{and } v < 0 \textbf{ or}$$
$$\qquad\qquad\qquad \text{pre}(StartBackward) \quad \textbf{and } v < 0 \textbf{ or}$$
$$\qquad\qquad\qquad \text{pre}(Backward) \qquad\quad \textbf{and not } v >= 0;$$

$$StartForward \quad = \text{pre}(Sticking) \qquad\quad \textbf{and } F_c > +F_s \textbf{ or}$$
$$\qquad\qquad\qquad \text{pre}(StartForward) \qquad \textbf{and not}$$
$$\qquad\qquad\qquad (v > 0 \textbf{ or } a <= 0 \textbf{ and not } v > 0);$$

$$StartBackward = \text{pre}(Sticking) \qquad\qquad \textbf{and } F_c < -F_s \textbf{ or}$$
$$\qquad\qquad\qquad \text{pre}(StartBackward) \quad \textbf{and not}$$
$$\qquad\qquad\qquad (v < 0 \textbf{ or } a >= 0 \textbf{ and not } v < 0);$$

$$Sticking \qquad = \textbf{not } (Start \textbf{ or}$$
$$\qquad\qquad\qquad Forward \textbf{ or } StartForward \textbf{ or}$$
$$\qquad\qquad\qquad Backward \textbf{ or } StartBackward);$$

when *Sticking* **and not** *Start* **then**
    reinit$(v, 0)$;
end when;

Comparing this Dymola model with the DFSM of Fig.9.37, it can be seen that the translation of one into the other is systematic and quite straight-forward.

This model can be simulated. Unfortunately, it is characterized by fairly complicated switching conditions that lead to zero–crossing functions that aren't smooth. Let us see, whether this situation can be rectified.

To this end, we shall employ the parameterized curve description technique once again. Figure 9.38 shows a slightly simplified friction characteristic that has been parameterized in similar ways as with the diode characteristic introduced earlier in this chapter.

The curve parameter is defined as follows:

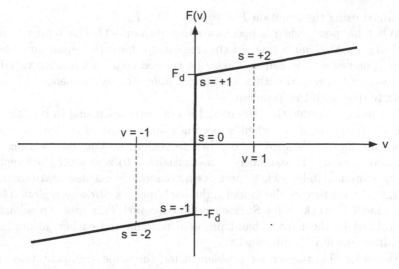

FIGURE 9.38. Simplified friction characteristic with curve parameterization.

region	s
forward	$v + 1$
sticking	F/F_d
backward	$v - 1$

Curve parameters can be defined in any way that is most suitable. They don't have to be equidistantly spaced, and they can even adopt different units in different regions, as the example demonstrates. Using the new variable, s, we can define the simplified friction model as follows [9.22]:

$$
\begin{aligned}
&Forward \ \ = s > +1; \\
&Backward = s < -1; \\
&v = \textbf{if } Forward \ \ \textbf{then } s - 1 \qquad\qquad \textbf{else} \\
&\quad\ \ \textbf{if } Backward \textbf{ then } s + 1 \qquad\quad\ \textbf{else } 0; \\
&F = \textbf{if } Forward \ \ \textbf{then } c_f \cdot (s - 1) + F_d \textbf{ else} \\
&\quad\ \ \textbf{if } Backward\textbf{ then } c_f \cdot (s + 1) - F_d \textbf{ else } F_d \cdot s;
\end{aligned}
$$

This model is correct in the sense that it describes unambiguously our intentions of what the model is supposed to accomplish. Thus, we might expect that a decent model compiler would be capable of translating the model down to an event description that can be properly simulated.

Unfortunately, the Dymola model compiler, as it is currently implemented, is unable to do so. There are two problems with this model. Let us explain.

While the model operates in its *Forward* region, the velocity, v, is a state variable, thus can be assumed known. Hence the curve parameter, s, can be computed from the equation $s = v + 1$, and the friction force can be obtained using the equation $F = c_f \cdot (s - 1) + F_d$.

What happens, when s becomes smaller than $+1$? The model is now entering its *Sticking* region. In this region, we have the equation; $v = 0$. Thus, the velocity, v, can no longer be treated as a known state variable, and we are confronted with a *conditional index change*. Somehow, we shall have to deal with this problem.

Let us now assume that the model is currently operating in its *Sticking* region. What happens, when s becomes larger than $+1$? The model is now entering its *Forward* region. In this region, we compute s using the equation $s = v + 1$, and since v was initialized to zero after the event, s returns immediately back to one. As a consequence, a new state event is triggered that throws the model right back into its *Sticking* region. Thus, the model is stuck in its *Sticking* region forever! This problem seems to be related to the narrow band problem encountered earlier, although it manifests itself a bit differently.

We can tackle the former problem using the same argumentation that had been used already in the previous model: If the velocity, v, is constantly equal to zero over a period of time, then also the acceleration, a, must be constantly equal to zero during that time period.

Thus, we can describe the *Sticking* region and its immediate surroundings by looking at the acceleration, rather than the velocity. This concept is illustrated in Fig. 9.39.

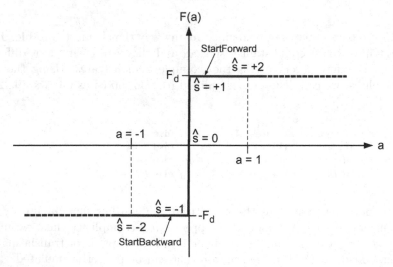

FIGURE 9.39. Sticking region of simplified friction characteristic with curve parameterization.

Since this is a different friction curve from the one shown before, the model uses a different parameter for its curve parameterization, $\hat{s}$. The model can be described using the same techniques introduced earlier:

$StartForward\ \ \ = \hat{s} > +1;$
$StartBackward = \hat{s} < -1;$
$a = $ **if** $StartForward$ **then** $\hat{s} - 1$ **else**
  if $StartBackward$ **then** $\hat{s} + 1$ **else** $0;$
$F = $ **if** $StartForward$ **then** $+F_d$ **else**
  if $StartBackward$ **then** $-F_d$ **else** $F_d \cdot \hat{s};$

We shall use a DFSM to describe the switching between the three main regions of the model, as illustrated in Fig. 9.40.

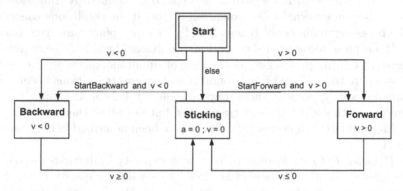

FIGURE 9.40. Deterministic finite state machine modeling the switching events of the simplified friction characteristic.

This is a much simplified version of the DFSM of Fig. 9.37 used by the earlier model. The new DFSM has only four instead of six discrete states (regions). The *StartForward* and *StartBackward* modes of operation are no longer considered separate regions. Instead, they are contained within the *Sticking* region model. They only represent different aspects of the *Sticking* region.

We shall not offer an encoding of the DFSM at this point, but instead, we shall leave this problem for one of the exercises at the end of this chapter.

Unfortunately, the simplified DFSM still contains two mixed switching conditions, describing the conditions under which the model leaves the *Sticking* region. These switching conditions prevent Dymola from generating smooth zero–crossing functions in those cases. Yet, the problem is not too damaging numerically, because these switchings occur always as an almost immediate result of a previous switching to one of the two transitory modes, *StartForward* and *StartBackward*, for which smooth zero–crossing functions in the curve parameter, $\hat{s}$, had been defined.

9.13   Petri Nets

We shall now demonstrate that it is always possible to decompose complex (combined) event conditions into sets of simple event conditions that consist of a single relational operator only. Thus, all zero–crossing functions can be made smooth. To this end, we shall introduce a new model description tool: the *Petri net*.

Petri nets [9.20] consist of two modeling elements: *places* and *transitions*. Places are holders of *tokens*. Each place maintains a discrete state variable that counts the number of tokens currently held by the place. Transitions connect places. When a transition *fires*, it takes some tokens out of places connected at its inputs, and places some new tokens at places connected at its outputs in accordance with some logic to be defined. A transition may fire, when an external firing condition is true, if the conditions concerning the necessary numbers of tokens held by its input places are true as well.

If one place feeds several transitions, additional logic may be required to determine firing preferences in the case of simultaneous events, i.e., in the case where the external firing conditions of several transitions become true simultaneously, because there may be enough tokens in the input place to fire one or the other of these transitions, but not all of them.

Many different dialects of Petri nets have been described in the literature [9.21].

Bounded Petri nets are Petri nets with capacity limitations imposed on its places. *Normal Petri nets* are Petri nets with a capacity limit of one imposed on each place. In a normal Petri net, the discrete state counting the number of tokens contained in a place can thus be represented as a Boolean state. If the state has a value of *true*, there is a token located at the place. If the state has a value of *false*, there is no token at the place.

Priority Petri nets resolve the ambiguity associated with multiple transitions being able to fire simultaneously by associating a prioritization scheme to these transitions.

Normal priority Petri nets (NPPNs) are normal Petri nets employing prioritization schemes in all of their transitions.

A NPPN place with two inputs and two outputs has been depicted in Fig. 9.41.

The place passes state information, s_i, to all neighboring transitions, and in turn receives firing information, f_i, back from these transitions.

The NPPN place could be governed by the following equations:

$$s_1 = \mathrm{pre}(p_1) \tag{9.61a}$$

$$s_2 = \mathrm{pre}(p_1) \textbf{ or } f_1 \tag{9.61b}$$

$$s_3 = \mathrm{pre}(p_1) \tag{9.61c}$$

$$s_4 = \mathrm{pre}(p_1) \textbf{ and not } f_3 \tag{9.61d}$$

$$p_1 = [\mathrm{pre}(p_1) \textbf{ and not } (f_3 \textbf{ or } f_4)] \textbf{ or } f_1 \textbf{ or } f_2 \tag{9.61e}$$

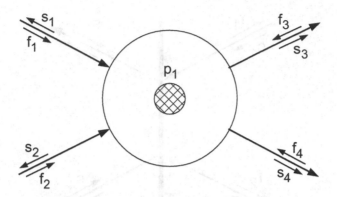

FIGURE 9.41. NPPN place with two inputs and two outputs.

The logic of these equations goes as follows. The place first provides the first input transition, t_1, with its state information. Transition t_1 needs to know this information, because, due to the single–token capacity limitation, it cannot fire, unless the place, p_1, is currently unoccupied. The place receives the firing information, f_1, back from transition t_1. If t_1 fires, it means that it is going to place a new token at p_1.

The place then provides the appropriate state information, s_2, to the second input transition, t_2. Transition t_2 is assigned a lower priority than transition t_1. Transition t_2 is not allowed to fire if either there is already a token at place p_1, or if the other input transition, t_1, decided to fire, because if both transitions were to fire simultaneously, they both would try to place a token at p_1, which would violate the imposed capacity limit of one.

The place then provides its state information to the first output transition, t_3. Transition t_3 is allowed to fire if a token is currently at p_1. If it fires, it will take the token away from place p_1.

The place then provides the appropriate state information, s_4, to the second output transition, t_4. Transition t_4 is assigned a lower priority than transition t_3. Transition t_4 is not allowed to fire, unless there is currently a token at place p_1 and transition t_3 has not decided to fire, because if both transitions were to fire simultaneously, they both would fight over who gets to remove the token from p_1.

Finally, the place must update its own state information. If there was a token at p_1 before, and neither of the two output transitions, t_3 or t_4, has taken it away, or, if one of the two input transitions, t_1 or t_2, has placed a new token at p_1, there will be a token at that place during the next cycle.

Let us now look at a transition with two input places and two output places. It has been depicted in Fig. 9.42.

The logic governing the transitions could be the following. The transition is allowed to fire along all of its connections, when the external firing

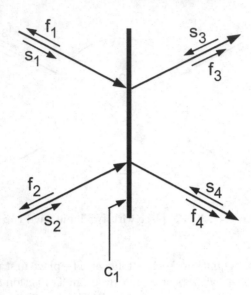

FIGURE 9.42. NPPN transition with two inputs and two outputs.

condition, c_1, is true, and if each of the input places holds a token (or more precisely, if the state information arriving from all of the input places is *true*), and if none of the output places holds a token (or more precisely, if the state information of none of the output places is *true*).

This logic can be described by the following set of equations:

$$fire = c_1 \text{ and } s_1 \text{ and } s_2 \text{ and not } (s_3 \text{ or } s_4) \tag{9.62a}$$
$$f_1 = fire \tag{9.62b}$$
$$f_2 = fire \tag{9.62c}$$
$$f_3 = fire \tag{9.62d}$$
$$f_4 = fire \tag{9.62e}$$

DFSMs can be modeled as normal priority Petri nets with the additional constraints that there is only one token in the system that is initially located at the *Start* place. Furthermore, DFSMs map to NPPNs, in which each transition is associated with exactly one input place and one output place.

Let us model the DFSM of Fig. 9.37 as a Petri net. The corresponding NPPN representation is depicted in Fig. 9.43.

We immediately recognize what the external firing conditions, c_i, represent. These are the conditions that are associated with state transitions in the DFSM. Hence those are the edge–triggered Boolean variables associated with the zero–crossing functions.

What have we gained by this representation? In the past, we had many different discrete event blocks representing the actions to be taken, when one or the other of the zero–crossing functions triggered an event. This is

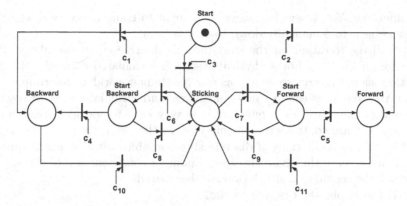

FIGURE 9.43. Petri net representation of friction characteristic.

no longer the case. All of the discrete equations governing both *places* and *transitions* are valid at every event, since they were formulated as functions of the current location of the tokens, i.e., they were functions of the discrete state that the system is currently operating in.

Thus, every discontinuous model, as complex as it may be, can be described by exactly three sets of equations. There are the implicitly defined algebraic and differential equations describing the continuous subsystem. There is the set of zero–crossing functions that are all evaluated in parallel, while the continuous subsystem is being simulated. If a state event is being triggered by one of them, an iteration (or interpolation) takes place to locate the event time as accurately as necessary. At that moment, the third set of simultaneous equations is being executed. These are the (possibly implicitly defined) algebraic and difference equations describing the discrete subsystem.

The discrete equations are executed iteratively, until no discrete state changes occur any longer. When this happens, we have found our new initial state, from which we can start the continuous simulation afresh.

A simulation model that has been compiled into this form, can be simulated in an organized and systematic fashion based on a synchronous data flow [9.22].

It may not be convenient for the end user of the modeling and simulation environment to describe his or her model in this fashion. Different application domains make use of different modeling formalisms that users are familiar with. It is the job of the *model compiler* to dissect the model description that the user supplies, and translate it down to sets of simultaneous equations that can be simulated without numerical difficulties.

As this book concerns itself with the set of algorithms underlying a powerful modeling and simulation environment, such as Dymola [9.12], we had to show step by step, how model equations need to be preconditioned, until they are finally in a form such that they can be simulated without

difficulties. Yet, it was no longer convenient to translate every model that
we came across manually down to such a form.

Will the iteration on the simultaneous discrete equations always converge? If the model of a physical system is formulated correctly, the iteration should always converge, as our Newtonian world is deterministic in
nature. Yet, it is easy to make mistakes, and formulate a set of discrete
equations that will not converge. It is very easy to specify logical conditions that are contradicting themselves. In the Petri–net implementation,
this leads to oscillations of discrete state variables with infinite frequency,
i.e., it prevents the algorithm from finding a consistent initial state, from
which the continuous simulation can be started.

For example, the discrete "equation":

$$p_1 = \textbf{not } \text{pre}(p_1) \tag{9.63}$$

should not be contained in the set of discrete equations, as this will lead
to an oscillation between the two states *true* and *false* that will never end.
If we mean to toggle between two discrete states as a response to a state–
event being triggered (a fairly common situation), we need to model this
using two separate places with transitions back and forth that get fired by
zero–crossing functions.

How can complex zero–crossing functions be reduced to simple ones? *or*–
conditions can be mapped to a set of parallel transitions located between
the same two places. They can thus be easily implemented. *and*–conditions
are harder to implement, as they would require transitions to be placed in
series with each other. Unfortunately, this cannot be done without introducing a new place between them. Thus, *and*–conditions invariably call for
an increase in the number of discrete states.

We shall demonstrate this concept by means of the DFSM of Fig. 9.40.
We recognize that we wouldn't need the *and*–conditions on the zero–crossing
functions in this example, if we were to have available separate discrete
states called *StartForward* and *StartBackward*. Thus, we shall decompose
the state *Slipping* again into three separate discrete states. Luckily, we
know the conditions for switching between them.

We don't need to draw the modified Petri net, as is looks *exactly* like the
one of Fig. 9.43. Only the interpretation of the zero–crossing functions is
now different. They are:

$$c_1 = v < 0 \tag{9.64a}$$
$$c_2 = v > 0 \tag{9.64b}$$
$$c_3 = v == 0 \tag{9.64c}$$
$$c_4 = v < 0 \tag{9.64d}$$
$$c_5 = v > 0 \tag{9.64e}$$
$$c_6 = \hat{s} < -1 \tag{9.64f}$$

$$c_7 = \hat{s} > 1 \tag{9.64g}$$
$$c_8 = \hat{s} >= -1 \tag{9.64h}$$
$$c_9 = \hat{s} <= 1 \tag{9.64i}$$
$$c_10 = v >= 0 \tag{9.64j}$$
$$c_11 = v <= 0 \tag{9.64k}$$

As expected, all of the zero–crossing functions are now simple functions consisting of a single relational operation only.

9.14   Summary

In this chapter, we have dealt with heavily discontinuous models. We have shown that integration algorithms should be spared from having to deal with discontinuous models directly. Two types of event descriptions were introduced, the time events and the state events, that enable the simulation software to treat discontinuous models in a safe and efficient manner, while protecting the integration algorithms from them. Special root finding algorithms were discussed that are particularly well suited to locate state events.

Event descriptions are quite general, and can be used to deal with most types of discontinuities adequately from a numerical point of view. Exceptions may be the propagation of discontinuous functions through conservation equations. If a step enters an ideal wave equation, a discontinuity will occur that travels through space with time. Consequently, the event times will be infinitely dense, which, from a practical point of view, doesn't make any sense. Adequate handling of discontinuities in hyperbolic PDEs is a very difficult task, and no good answer has been found to date for tackling this challenging problem. The best answer currently available is to apply a variable transformation that will ensure that the waves travel at least along the axes of the coordinate system rather than in an arbitrary direction, which boils down to using the *method of characteristics*. However, even this approach doesn't *solve* the problem. It only alleviates it somewhat.

It was also shown that event descriptions are awkward when dealing with complex engineering models. They are low–level constructs that should not be viewed as modeling elements, but only as intermediate descriptions that are automatically being generated by the model compiler on the way of transforming the model, as specified by the user, into a simulation program that can be executed safely and efficiently using numerical integration software.

Higher–level constructs were introduced in the form of object–oriented *if*–expressions and *when*–clauses, and several fairly advanced applications of these tools have been demonstrated.

It was finally shown that, although the description mechanisms using these constructs are general and convenient, currently available modeling software (i.e., Dymola) is still unable to translate all possible (and physically meaningful) models described using these constructs down into properly executable simulation code. Variable structure models may be contaminated by conditional index changes that require special handling, such as inlining those integrators that are responsible for the partial constraint on a switch equation. Sometimes it is also possible to apply a generalized version of the Pantelides algorithm instead. Whereas it should be possible to at least automate the former approach using the inlining technique, this has not yet been attempted in the current version of the Dymola model compiler.

9.15   References

[9.1] Alfred V. Aho, Ravi Sethi, and Jeffrey D. Ullman. *Compilers: Principles, Techniques, and Tools*. Addison-Wesley, Reading, Massachusetts, 1986. 500p.

[9.2] Ilia Nikolaevich Bronshtein and Konstantin Adolfovich Semendiaev. *A Guide–Book to Mathematics*. H. Deutsch Publishing, Frankfurt am Main, Germany, 1971.

[9.3] Michael B. Carver. Efficient Handling of Discontinuities and Time Delays in Ordinary Differential Equation Simulations. In Mohammed H. Hamza, editor, *Proceedings Simulation'77*, pages 153–158, Montreux, Switzerland, 1977. Acta Press.

[9.4] François E. Cellier and Hilding Elmqvist. Automated Formula Manipulation Supports Object–oriented Continuous System Modeling. *IEEE Control Systems*, 13(2):28–38, 1993.

[9.5] François E. Cellier. *Combined Continuous/Discrete System Simulation by Use of Digital Computers: Techniques and Tools*. PhD thesis, Swiss Federal Institute of Technology, Zürich, Switzerland, 1979.

[9.6] John R. Dormand and Peter J. Prince. Runge–Kutta Triples. *J. of Computational and Applied Mathematics*, 12A(9):1007–1017, 1986.

[9.7] John R. Dormand and Peter J. Prince. Runge–Kutta–Nyström Triples. *J. of Computational and Applied Mathematics*, 13(12):937–949, 1987.

[9.8] Steven L. Dvorak, Richard W. Ziolkowski, and Donald G. Dudley. Ultra–Wideband Electromagnetic Pulse Propagation in a Homogeneous Cold Plasma. *Radio Science*, 32(1):239–250, 1997.

[9.9] Edda Eich-Söllner. *Projizierende Mehrschrittverfahren zur numerischen Lösung von Bewegungsgleichungen technischer Mehrkörpersysteme mit Zwangsbedingungen und Unstetigkeiten.* PhD thesis, Universität Augsburg, Augsburg, Germany, 1991.

[9.10] Edda Eich-Söllner. Convergence Results for a Coordinate Projection Method Applied to Mechanical Systems With Algebraic Constraints. *SIAM J. of Numerical Analysis*, 30(5):1467–1482, 1993.

[9.11] Hilding Elmqvist, François E. Cellier, and Martin Otter. Object–Oriented Modeling of Hybrid Systems. In *Proceedings ESS'93, European Simulation Symposium*, pages xxxi–xli, Delft, The Netherlands, 1993.

[9.12] Hilding Elmqvist. *Dymola — Dynamic Modeling Language, User's Manual, Version 5.3.* DynaSim AB, Research Park Ideon, Lund, Sweden., 2004.

[9.13] Gerald Grabner and Andrés Kecskeméthy. Reliable Multibody Collision Detection Using Runge–Kutta Integration Polynomials. In *Proceedings International Conference on Andvances in Computational Multibody Dynamics*, Lisbon, Portugal, 2003.

[9.14] Nicola Guglielmi and Ernst Hairer. Implementing Radau–IIA Methods for Stiff Delay Differential Equations. *Computing*, 67:1–12, 2001.

[9.15] Mary Kathleen Horn. *Developments in High Order Runge–Kutta–Nyström Formulas.* PhD thesis, University of Texas at Austin, 1977.

[9.16] Mary Kathleen Horn. Fourth– and Fifth–Order Scaled Runge–Kutta Algorithms for Treating Dense Output. *SIAM J. of Numerical Analysis*, 20:558–568, 1983.

[9.17] Katsushi Ito and Makiko Nisio. On Stationary Solutions of a Stochastic Differential Equation. *J. of Mathematics of Kyoto University*, 4(1):1–75, 1964.

[9.18] Matthias Krebs. Modeling of Conditional Index Changes. Master's thesis, Dept. of Electrical & Computer Engineering, University of Arizona, Tucson, Ariz., 1997.

[9.19] Shengtai Li and Linda R. Petzold. Moving Mesh Methods with Upwinding Schemes for Time–Dependent PDEs. *J. of Computational Physics*, 131:368–377, 1997.

[9.20] Pieter J. Mosterman, Martin Otter, and Hilding Elmqvist. Modeling Petri Nets As Local Constraint Equations For Hybrid Systems Using Modelica. In *Proceedings SCSC'98, Summer Computer Simulation Conference*, pages 314–319, Reno, Nevada, 1998.

[9.21] Tadao Murata. Petri Nets: Properties, Analysis and Applications. *Proceedings of the IEEE*, 77(4):541–580, 1989.

[9.22] Martin Otter, Hilding Elmqvist, and Sven Erik Mattsson. Hybrid Modeling in Modelica Based On Synchronous Data Flow Principle. In *Proceedings IEEE, International Symposium on Computer Aided Control System Design*, pages 151–157, Kohala Coast, Hawaii, 1999.

[9.23] Constantinos Pantelides.   The Consistent Initialization of of Differential–Algebraic Systems. *SIAM Journal of Scientific and Statistical Computing*, 9(2):213–231, 1988.

[9.24] Anton Schiela and Hans Olsson. Mixed–mode Integration for Real–time Simulation. In *Proceedings Modelica'2000 Workshop*, pages 69–75, Lund, Sweden, 2000.

[9.25] Hans Schlunegger.   *Untersuchung eines netzrückwirkungsarmen, zwangskommutierten Triebfahrzeugstromrichters zur Einspeisung eines Gleichstromzwischenkreises aus dem Einphasennetz*.   PhD thesis, Swiss Federal Institute of Technology, Zürich, Switzerland, 1977.

[9.26] Lawrence F. Shampine, Ian Gladwell, and Richard W. Brankin. Reliable Solutions of Special Event Location Problems for ODEs. *ACM Transactions on Mathematical Software*, 17(1):11–25, 1991.

9.16   Bibliography

[B9.1] Brian Armstrong-Hélouvry.   *Control of Machines With Friction*. Kluwer Academic Publishers, Boston, Mass., 1991.

[B9.2] Carlos A. Canudas de Wit, Hans Olsson, Karl Johan Åström, and Pablo Lischinsky. A New Model for Control of Systems With Friction. In *Proceedings International Conference on Control Theory and Its Applications*, pages 225–229, Kibbutz Maab Hachamisha, Israel, 1993.

[B9.3] René David and Hassane Alla. *Petri Nets and Grafcet*. Prentice–Hall, Upper Saddle River, N.J., 1992.

[B9.4] Martin Otter, Hilding Elmqvist, and François E. Cellier. Modeling of Multibody Systems With the Object–Oriented Modeling Language Dymola. *Journal of Nonlinear Dynamics*, 9(1):91–112, 1996.

[B9.5] Martin Otter. *Objektorientierte Modellierung mechatronischer Systeme am Beispiel geregelter Roboter*. PhD thesis, Dept. of Mech. Engr., Ruhr–University Bochum, Germany, 1994.

[B9.6] Friedrich Pfeiffer and Christoph Glocker. *Multibody Dynamics With Unilateral Contacts*. John Wiley & Sons, New York, N.Y., 1996. 318p.

[B9.7] Muhammad H. Rashid. *Spice for Power Electronics and Electric Power*. Prentice–Hall, Englewood Cliffs, N.J., 1994.

[B9.8] Jiri Vlach and Kishore Singhal. *Computer Methods for Circuit Analysis and Design*. Van Nostrand Reinhold, New York, second edition, 1994.

9.17 Homework Problems

[H9.1] Runge–Kutta–Fehlberg with Root Solver

Implement in MATLAB the RKF4/5 algorithm introduced in Chapter 3 of this book together with the optimistic step–size control algorithm of Eq.(3.89).

Add a root solver (RKF4/5RT) to the method that is based on an implementation of the *Regula Falsi* algorithm.

[H9.2] Runge–Kutta–Fehlberg with Root Solver

Repeat Hw.[H9.1]. This time around, we wish to add a root solver based on an implementation of the *Golden Section* algorithm to the method.

[H9.3] Runge–Kutta–Fehlberg with Root Solver

Repeat Hw.[H9.1]. This time around, we wish to add a root solver based on an implementation of *direct cubic interpolation* to the method.

[H9.4] Direct Hermite Interpolation

We wish to improve the solution to Hw.[H9.3]. Rather than solving for the coefficients of the cubic interpolation polynomial directly using matrix inversion, we want to define a set of spanning polynomials, similar to the way introduced earlier in the chapter in the implementation of the *inverse Hermite interpolation* algorithm.

[H9.5] The Mechanical Loose Element

The functioning of a mechanical loose element is illustrated graphically in Fig.H9.5a.

The output, y, lags behind the input, x, by no more than the distance, d. If the direction of x changes, y remains constant, until it again lags behind by d, now in the opposite direction.

Model the loose element using *if*– and *when*–statements such that the model equations can be sorted appropriately.

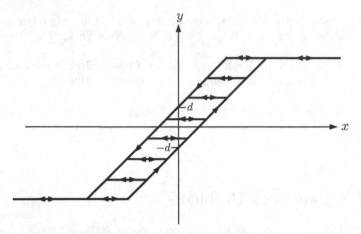

FIGURE H9.5a. Mechanical loose element.

[H9.6] Quantization With Hysteresis

The hysteretic quantization function is illustrated graphically in Fig.H9.6a.

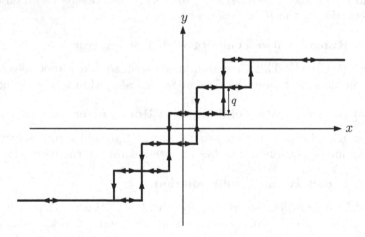

FIGURE H9.6a. Hysteretic quantization function.

The output, y, stays always in the vicinity of the input, x. The distance between them is never greater than half of the quantization distance, $q/2$. Yet, whereas x can change continuously over time, y is a discrete state variable.

Model the hysteretic quantization element using *if*- and *when*-statements such that the model equations can be sorted appropriately.

[H9.7] Thyristor

We wish to model the thyristor described earlier in the chapter by means of *if*-statements. The thyristor element is depicted in Fig. H9.7a.

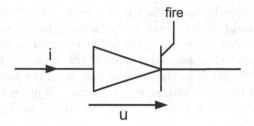

FIGURE H9.7a. Thyrisor.

The thyristor *is a* diode with a modified firing logic. The diode can only close when the external Boolean variable *fire* has a value of *true*. The opening logic is the same as for the regular diode.

Since the thyristor *is a* diode, we can use the same parameterized curve description that we used for the regular diode. Only the switching condition is modified.

Convert all *if*–statements of the thyristor model to their algebraic equivalents. Write down all of the equations governing the thyristor–controlled rectifier circuit of Fig. H9.7b.

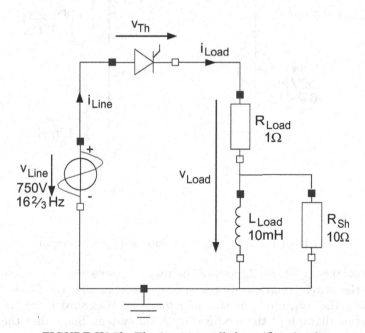

FIGURE H9.7b. Thyrisor–controlled rectifier circuit.

Draw the structure digraph of the resulting equation system, and show that the switch equations indeed appear inside an algebraic loop.

Choose a suitable tearing structure, and solve the equations both horizontally and vertically using the variable substitution technique.

Using any one of the integration algorithms of Hw.[H9.1–4], simulate the model in MATLAB across 0.2 seconds. The external control variable of the thyristor, *fire*, is to be assigned a value of *true* from the angle of 30° until the angle of 45°, and from the angle of 210° until the angle of 225° during each period of the line voltage, v_{Line}. During all other times, it is set to *false*. Plot the load voltage, v_{Load}, as well as the load current, i_{Load}, as functions of time.

[H9.8] Thyristor

We wish to repeat the simulation of Hw.[H9.7] for the modified thyristor–controlled rectifier circuit of Fig. H9.8a.

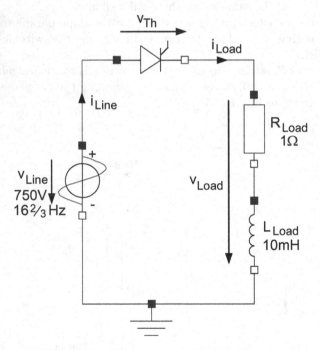

FIGURE H9.8a. Thyrisor–controlled rectifier circuit.

Draw the structure digraph of the resulting equation system, and show that the switch equations do not appear inside an algebraic loop.

Inline the integrator for the inductor using backward Euler. Draw the structure digraph of the modified equation system. Show that the switch equations now indeed appear inside an algebraic loop.

Choose a suitable tearing structure, and solve the equations both horizontally and vertically using the variable substitution technique.

Simulate the model in MATLAB across 0.2 seconds. The external control variable of the thyristor, *fire*, is to be assigned a value of *true* from the angle of 30° until the angle of 45°, and from the angle of 210° until the angle of 225° during each period of the line voltage, v_{Line}. During all other times, it is set to *false*. Since there is no integrator left in the model, you cannot use RKF4/5RT any longer. Instead, you need to program the iteration on the zero–crossing function directly into the simulation program. Plot the load voltage, v_{Load}, as well as the load current, i_{Load}, as functions of time.

[H9.9] Zener Diode

No diode can hold current against an arbitrarily strong electrical field. Thus, if the negative voltage across the diode becomes too large, we are confronted with *avalanche breakdown*. The diode suddenly starts conducting again.

A Zener diode makes use of the avalanche breakdown phenomenon, by constructing a diode such that avalanche breakdown occurs early and at a well defined voltage.

Zener diodes are not used like regular diodes, but rather as reverse diodes. Thus, the voltage, in a Zener diode, is defined positive from the cathode to the anode, rather than from the anode to the cathode.

Figure H9.9a shows the Zener diode element together with its voltage and current conventions.

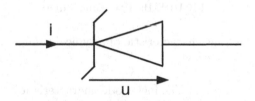

FIGURE H9.9a. Zener diode.

The current/voltage characteristic of the ideal Zener diode is shown in Fig. H9.9b.

The voltage u_B is the *breakdown voltage* of the device.

Zener diodes are commonly placed in parallel with delicate equipment, such as electro–motors. Their purpose is to protect the equipment from potential damage caused by high voltage.

Use the parameterized curve description technique to derive a model of the ideal Zener diode.

[H9.10] Tunnel Diode

A tunnel diode is a regular diode with a tunnelling effect in the conducting area of the device. The tunnel diode element is shown in Fig. H9.10a together with its voltage and current conventions.

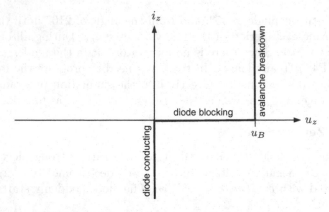

FIGURE H9.9b. Ideal Zener diode characteristic.

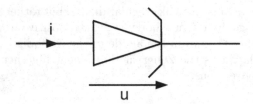

FIGURE H9.10a. Tunnel diode.

The current/voltage characteristic of a typical tunnel diode are shown in Fig. H9.10b.

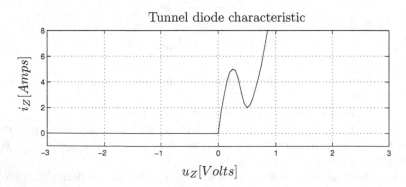

FIGURE H9.10b. Typical tunnel diode characteristic.

When the voltage across the tunnel diode becomes positive, the tunnel diode, just like a regular diode, starts conducting. Yet, the current doesn't grow as rapidly as in the case of a regular diode. With increasing voltage, the current first starts growing, then it decays once more (the tunnelling effect), before it starts growing rapidly like with a regular diode.

Tunnel diodes are sometimes used for constructing nonlinear oscillator circuits.

We wish to idealize the tunnel diode. To this end, we shall describe it by the idealized characteristic of Fig. H9.10c.

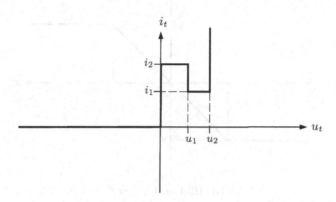

FIGURE H9.10c. Ideal tunnel diode characteristic.

Derive a model of the ideal tunnel diode using the parameterized curve description technique.

[H9.11] Friction

Translate the DFSM of Fig. 9.40 into a set of Boolean expressions governing the four states and their transitions.

Integrate this model with the model of the simplified friction characteristic of Fig. 9.39 developed in the chapter, and convince yourself by manual simulation that the integrated model represents the simplified friction characteristic correctly under all operating conditions.

[H9.12] Dry Hysteresis

Given the dry hysteresis function of Fig. 9.22. Let us assume that $x_1 = y_1 = -1$, and $x_2 = y_2 = +1$. We wish to drive that model using the input:

$$x(t) = 2 \cdot \cos(t) \quad\quad\quad (H9.12a)$$

Derive a Petri net description of the dry hysteresis function. Develop generic synchronous data flow models for the different types of places and transitions encountered in the model.

Extract all of the equations of the discrete model as well as the zero–crossing functions. Implement the model in MATLAB using a suitable algorithm for state–event detection.

Simulate the model in MATLAB across 10 seconds of simulated time, and plot y as a function of x.

[H9.13] Limiter Function

Given the limiter function of Fig. H9.13a.

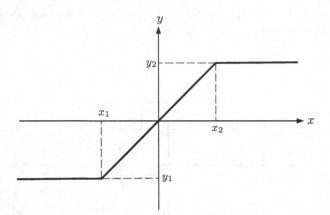

FIGURE H9.13a. Limiter function.

Let us assume that $x_1 = y_1 = -1$, and $x_2 = y_2 = +1$. We wish to drive that model using the input:

$$x(t) = 2 \cdot \cos(t) \qquad \text{(H9.13a)}$$

Derive a Petri net description of the limiter function. Develop generic synchronous data flow models for the different types of places and transitions encountered in the model.

Extract all of the equations of the discrete model as well as the zero-crossing functions. Implement the model in MATLAB using a suitable algorithm for state–event detection.

Simulate the model in MATLAB across 10 seconds of simulated time, and plot y as a function of x.

9.18 Projects

[P9.1] State Event Localization

In this chapter, we talked little about the use of linear multi–step methods in the simulation of discontinuous models. The reason is that the overhead associated with restarting such a method after an event has occurred is too large to make these methods attractive for the simulation of models containing frequent discontinuities.

Yet, multi–step techniques have an advantage over single–step algorithms due to the availability of the Nordsieck vector. The Nordsieck vector makes it possible to find the zero crossing of a zero–crossing function expressed as a state variable through *interpolation* instead of *iteration*. This can be done using an interpolation polynomial of the same order of approximation accuracy as the integration method itself. Therefore, zero crossings found in this way are almost as accurate as those found by iteration. They may still be a little less accurate, because the iteration technique involves a reduction of the integration step size in the vicinity of the event, whereas the interpolation method does not.

We had to use iteration in the case of the Runge–Kutta algorithms, because the solution is only available to us with full approximation accuracy at the end of the interval, not at any point in between.

The problem of finding interpolation algorithms for Runge–Kutta methods was first tackled by Horn [9.15, 9.16]. The most commonly used codes today offering implementations of explicit Runge–Kutta algorithms with *dense output* interpolation algorithms are codes based on DOPRI4/5 [9.6, 9.7].

The DOPRI4/5 algorithm is characterized by the Butcher tableau:

0	0	0	0	0	0	0	0
$\frac{1}{5}$	$\frac{1}{5}$	0	0	0	0	0	0
$\frac{3}{10}$	$\frac{3}{40}$	$\frac{9}{40}$	0	0	0	0	0
$\frac{4}{5}$	$\frac{44}{45}$	$\frac{-56}{15}$	$\frac{32}{9}$	0	0	0	0
$\frac{8}{9}$	$\frac{19372}{6561}$	$\frac{-25360}{2187}$	$\frac{64448}{6561}$	$\frac{-212}{729}$	0	0	0
1	$\frac{9017}{3168}$	$\frac{-355}{33}$	$\frac{46732}{5247}$	$\frac{49}{176}$	$\frac{-5103}{18656}$	0	0
1	$\frac{35}{384}$	0	$\frac{500}{1113}$	$\frac{125}{192}$	$\frac{-2187}{6784}$	$\frac{11}{84}$	0
x_1	$\frac{5179}{57600}$	0	$\frac{7571}{16695}$	$\frac{393}{640}$	$\frac{-92097}{339200}$	$\frac{187}{2100}$	$\frac{1}{40}$
x_2	$\frac{35}{384}$	0	$\frac{500}{1113}$	$\frac{125}{192}$	$\frac{-2187}{6784}$	$\frac{11}{84}$	0

where:

$$f_1(q) = 1 + q + \frac{1}{2}q^2 + \frac{1}{6}q^3 + \frac{1}{24}q^4 + \frac{1097}{120000}q^5 + \frac{161}{120000}q^6 + \frac{1}{24000}q^7$$

$$f_2(q) = 1 + q + \frac{1}{2}q^2 + \frac{1}{6}q^3 + \frac{1}{24}q^4 + \frac{1}{120}q^5 + \frac{1}{600}q^6$$

In DOPRI4/5, usually the 5th–order accurate algorithm is propagated, whereas the 4th–order accurate algorithm is used for step–size control purposes.

Dormand and Prince determined that a third algorithm can be added without adding an additional stage:

$$x_3(\sigma) = x_n + \sigma \cdot h \cdot \sum_{i=1}^{7} \hat{b}_i(\sigma) \cdot f_i \qquad \text{(P9.1b)}$$

where:

$$\sigma \in [0, 1]$$

The third approximation polynomial, $x_3(\sigma)$, is parameterized in an additional parameter σ. It offers a 5th–order accurate smooth interpolation polynomial valid anywhere between t_n and t_{n+1}, where σ denotes the percentage of the step taken, i.e.

$$x(\sigma) = x(t_\sigma) = x(t_n + \sigma \cdot h) \qquad \text{(P9.1c)}$$

Thus:

$$x_3(\sigma = 0) = x_n \qquad \text{(P9.1d)}$$
$$x_3(\sigma = 1) = x_2 = x_{n+1} \qquad \text{(P9.1e)}$$

The coefficients $\hat{b}_i$ are cubic polynomials in σ [9.13]:

$$\hat{b}_1 = -\frac{435\sigma^3 - 1184\sigma^2 + 1098\sigma - 384}{384} \qquad \text{(P9.1f)}$$

$$\hat{b}_2 = 0 \qquad \text{(P9.1g)}$$

$$\hat{b}_3 = \frac{500\sigma(6\sigma^2 - 14\sigma + 9)}{1113} \qquad \text{(P9.1h)}$$

$$\hat{b}_4 = -\frac{125\sigma(9\sigma^2 - 16\sigma + 6)}{192} \qquad \text{(P9.1i)}$$

$$\hat{b}_5 = \frac{729\sigma(35\sigma^2 - 64\sigma + 26)}{6784} \qquad \text{(P9.1j)}$$

$$\hat{b}_6 = -\frac{11\sigma(3\sigma - 2)(5\sigma - 6)}{84} \qquad \text{(P9.1k)}$$

$$\hat{b}_7 = \frac{\sigma(\sigma - 1)(5\sigma - 3)}{2} \qquad \text{(P9.1l)}$$

Dense output interpolation was originally designed as a means to facilitating the display of smoother output curves. Yet, the technique is very useful for the localization of zero–crossing functions in discontinuous models as well. Other applications concern the simulation of *delay–differential equations*, and also aspects of *real–time simulation*, as we shall demonstrate in the next chapter of this book.

Assuming that the derivatives of all zero–crossing functions have been added to the model as additional state equations, the zero–crossing functions themselves are state variables, for which dense interpolation is available.

Assuming further that $x_n \cdot x_{n+1} < 0$ for any of the zero–crossing states, we can find the corresponding next event time t_{next} by computing the value $\hat{\sigma}$, for which $x_3(\hat{\sigma}) = 0$. Then, $t_{\text{next}} = t_{\hat{\sigma}} = t_n + \hat{\sigma} \cdot h$.

Develop effective algorithms for determining $\hat{\sigma}$, and compare the computational efficiency of the interpolation technique with that of the earlier introduced iteration techniques.

[P9.2] State Event Detection

We have demonstrated in this chapter that state events may be missed, if the corresponding zero–crossing functions exhibit two zero crossings that are only separated by a short distance in time.

One approach to dealing with this problem, as demonstrated in this chapter, is through adding *unimportant state events* to the set of events to be iterated upon by appending the derivative of the original zero–crossing function as an additional zero–crossing function to the set.

Yet, this is not the only way of tackling this problem. Another approach has been described in the literature that might be worth considering as an alternative.

Given an nth–order polynomial:

$$p_0(t) = t^n + a_{n-1} \cdot t^{n-1} + a_{n-2} \cdot t^{n-2} + \cdots + a_1 \cdot t + a_0 \qquad \text{(P9.2a)}$$

We can define the following series of polynomials:

$$p_1(t) = \frac{d}{dt} p_0(t) \qquad \text{(P9.2b)}$$

and:

$$p_2(t) = -\text{rem}\left(\frac{p_0(t)}{p_1(t)}\right) \qquad \text{(P9.2c)}$$

$$\vdots \qquad \text{(P9.2d)}$$

$$p_m(t) = -\text{rem}\left(\frac{p_{m-2}(t)}{p_{m-1}(t)}\right) \qquad \text{(P9.2e)}$$

where the *rem*–operator denotes the remainder of the polynomial division. Such a series of polynomials is called a *Sturm sequence* [9.2].

If we wish to determine, how many zero crossings the polynomial $p_0(t)$ has in the time interval $[t_a, t_b]$, we can evaluate the polynomials of the Sturm sequence for $t = t_a$ and for $t = t_b$. We count the number of sign changes in the values of the Sturm sequence separately at both ends. The

difference between the number of sign changes at both ends equals the
number of zero crossings of the polynomial $p_0(t)$ in the interval $[t_a, t_b]$.

If the zero–crossing function has been defined as a state variable, and if
we simulate the model using a Runge–Kutta triple, we have an nth–order
interpolation polynomial available, as was shown in Pr.[P9.1].

Thus, we can define the Storm sequence of that interpolation polynomial
and determine accurately, how many zero crossings occur within the time
interval $[t_n, t_{n+1}]$ [9.26].

Study, how the Sturm sequence can be implemented most effectively.

Compare algorithms for detection of *short–living state events* that are
based on augmented sets of zero–crossing functions with methods based on
the Sturm sequence for their computational efficiency and reliability.

[P9.3] Delay–Differential Equations

Delay–differential equations are frequently encountered in geological engineering applications and also in chemical process engineering models. In
these types of applications, it happens frequently that one process generates some material that is then transported to another process, where it is
being used as an input. Other applications of delay–differential equations
include the remote control of equipment in space, where the communication
delays have to be taken into account.

In all of these cases, we encounter delay–differential equations of the
form:

$$\dot{x}_2(t) = f(x_1(t - \Delta)) \qquad (P9.3a)$$

The problem here is that the time instant $t - \Delta$ may not be an output
point, or even the end of an integration step.

In many applications, a small error in the delay, Δ does not matter.
However when it does matter, i.e., if there is a feedback loop back from x_2
to x_1, then we have a problem.

If the model is simulated using a linear multi–step algorithm, it no longer
suffices to store the state variables at each output point. We need to store
the entire Nordsieck vector at the end of each integration step for at least
Δ time units, so that we can appropriately interpolate to evaluate x_1 at
time $t - \Delta$.

If the model is simulated using a single–step algorithm, it may again be
preferable to use one of the Runge–Kutta triples. However in that case,
we would need to store the solution of every stage of the algorithm at
the end of each integration step for at least Δ time units, so that we can
appropriately interpolate to evaluate x_1 at time $t - \Delta$.

Although this technique doesn't create any principle difficulties, it causes
significant computational overhead. The issue thus is how solvers for delay–
differential equations can be implemented in a computationally efficient
way. *Circular shift registers* are one approach that comes to mind, but this
may not be the only one, or even the best approach to dealing with this

problem.

Study computationally efficient ways of data storage and retrieval for the numerical simulation of delay–differential equations, and modify existing codes to implement those.

9.19   Research

[R9.1] Stiff Discontinuous Models

If a discontinuous model is stiff, we must use an implicit integration algorithm to simulate it. Although we could use a code, such as DASSLRT, this may be quite inefficient, because linear multi–step methods are hardly ever suitable for dealing with heavily discontinuous models due to the overhead and inaccuracy associated with the start–up algorithm needed after each event.

Thus, it is important to extend the idea of an interpolation polynomial to obtain dense output from the explicit Runge–Kutta algorithms to implicit ones, such as the Radau–IIA, or Lobatto–IIIC algorithms introduced earlier in this book.

The problem has been recognized, and a number of research groups are currently working on this issue. First results have recently been published [9.14].

Yet, the problem is still essentially unsolved. The reason is that the interpolated result needs to be propagated to the next step. Thus, it is insufficient to prove that the interpolated result is n^{th}–order accurate. We ought to prove in addition that it is also numerically stable.

[R9.2] Discontinuous Hyperbolic PDEs

Whereas we have discussed in this chapter the problems associated with the detection and localization of state events, we always made the assumption that the event times are somewhat spaced out, i.e., within a finite time interval, the number of events must remain finite.

Unfortunately, this assumption does not always hold true. If we apply a discontinuity to a hyperbolic PDE, such as the wave equation, the discontinuity travels through the medium with time, i.e., at any point in time, the discontinuity can be located somewhere in the medium. Hence the event times are no longer spread out.

Some researchers have applied moving grid methods to these types of problems [9.19]. Others have applied frequency–domain techniques [9.8]. Yet, whereas these techniques may be suitable to track steep wave fronts, neither of these techniques is geared to dealing with true discontinuities.

How do we know that inaccuracies in estimating, where the discontinuity is located at any point in time will not propagate through the solution and accumulate as time passes? Do we have any handle on the numerical stability problems associated with these types of situations?

There must exist better ways to calculate with arbitrary accuracy, where the discontinuity is located when, and tackle the problem by subdividing the domain into "left" and "right" regions in space, and "before" and "after" domains in time, and extrapolate (interpolate) to the location of the discontinuity from all sides.

[R9.3] Sliding Motion

Sliding motion is a second type of problem that can lead to events with infinite frequency of occurrence.

In this chapter, we have encountered creeping behavior of a simulation code implementing a discontinuous model twice.

The first time was in the context of the train engine model. However, the creeping behavior only occurred because we had implemented the discontinuity handling incorrectly. Once we solved that problem, the creeping behavior went away.

The second time, we ran into a similar problem was in the context of one of our friction models, where we found that coming out of sticking friction caused the model to be thrown back into sticking friction immediately again. This happened, in spite of the fact that the model is formally correct.

Here, we were able to tackle the problem by introducing two additional discrete states, *StartForward* and *StartBackward*. Once these states had been introduced and the state transition logic had been undated appropriately, the problem went away.

Is this the worst that can happen? Unfortunately, the answer to this question is negative. Let us explain this assertion by means of an example.

Figure R9.3a shows a flying vehicle on a slow collision course with a sloped wall.

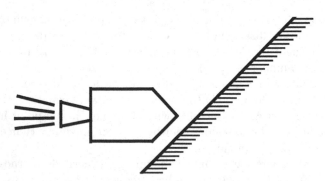

FIGURE R9.3a. Sliding motion.

Once the vehicle arrives at the wall, it either gets stuck there, or if the thrust is sufficiently large to overcome sticking friction, it will glide up the slope, as it has no choice in the matter.

Unfortunately, it is a rather difficult problem to convince the simulation code that this is what must happen. If Newton's law is being formulated separately for the horizontal and vertical motions, the vehicle cannot move forward at all, as the wall is in the way. It can only move upward. However as it moves upward, it no longer remains in contact with the wall. Thus, the vehicle starts moving forward again. However by doing so, it bumps immediately back into the wall. The model ends up with state events occurring at infinite frequency. Of course, the problem will go away, if we modify the coordinate system to coincide with the slope of the ramp.

Although the example looks somewhat academic, the problem itself is quite realistic, and these types of problems indeed occur frequently in mechanical systems with closed kinematic loops, such as the simulation of a car moving on a road. As all four wheels are in contact with the ground, we are faced with multiple closed kinematic loops. If the car drives around a bend, two of the wheels need to move a little faster than the other two. Any numerical discrepancy between the simulated motion and the physical constraint will invariably lead to the type of behavior explained above.

These types of problems have been studied in recent years [9.9, 9.10]. Yet, no fully automated algorithms have been designed that can detect these problems and modify the problem formulation automatically and on the fly in such a way as to remove the events occurring with infinite frequency.

[R9.4] Simulation of Noisy Models

A third type of problems that will lead, in the theoretical limit, to a series of events occurring with infinite frequency is the simulation of models with noise.

Most continuous–system simulation software offers at least uniform and Gaussian distributed random number generators that enable the modeler to superpose noise to some input signals of his or her model. The noise signal may e.g. be used to describe the headwind facing a helicopter in flight, or it may be used to describe the unevenness of a road along which a vehicle is driving. In the case of the helicopter, the purpose of including the headwind may be to test the robustness of the control algorithm. In the case of the road vehicle, it may be to simulate the behavior of the shock absorbers.

Unfortunately, the random number generator is a rather dubious modeling element, as it changes its behavior as a function of the integration step size used.

Uncorrelated white noise ought to have a frequency spectrum that is totally flat at all frequencies. Yet, plotting the frequency spectrum of a random number generator used in a simulation model, we notice that the spectrum eventually decays as $1/f$. The bandwidth of the random number generator is band–limited by the sampling rate. The smaller we choose the step size, the larger the bandwidth of the random number generator will become.

Although some highly theoretical investigations have looked at the analytical solutions of stochastic differential equations [9.17], this is not useful for our purpose.

The problem that we are confronted with is that we cannot use event handling mechanisms to deal with random signals. Yet, if we ignore them, they will invariably get entangled with the step–size control of the variable–step integration algorithms.

Very little research has been done to date that looks at this problem from a practical perspective.

10

Real–time Simulation

Preview

In this chapter, we shall discuss the special requirements of real–time simulation, i.e., of simulation runs that keep abreast of the passing of real time, and that can accommodate driving functions (input signals) that are generated outside the computer and that are read in by means of analog to digital (A/D) converters.

Until now, computing speed has always been a soft constraint — slow simulation meant expensive simulation, but now, it becomes a very hard constraint. Simulation becomes a *race against time*. If we cannot complete the computations associated with one integration step *before* the real–time clock has advanced by h time units, where h is the current step size of the integration algorithm, the simulation is out of sync, and we just lost the race.

Until now, we always tried to make simulation more comfortable for the user. For example, we introduced step–size controlled algorithms so that the user wouldn't have to worry any more about whether or not the numerical integration meets his or her accuracy requirements. The algorithm would do so on its own. In the context of real–time simulation, we may not be able to afford all this comfort any longer. We may have to throw many of the more advanced features of simulation over board in the interest of saving time, but of course, this means that we have to understand even better ourselves how simulation works in reality.

10.1 Introduction

Several very important applications of simulation require real-time performance.

A *flight simulator* for training purposes is useless if it cannot produce a reflection of the performance of the real aircraft or helicopter or space craft in real time. The trainee uses the simulator because learning often is synonymous with making mistakes ... and mistakes may be too costly when working with the real system.

Model Reference Adaptive Controllers (MRACs) make use of a model of an idealized plant, the reference plant, trying to make the real plant behave as similar as possible to the reference plant [10.34]. However, this requires that the reference plant model be simulated in real time in parallel with the real plant, both being driven simultaneously by the same input signals.

A *watchdog monitor* [10.3, 10.2, 10.47] of a nuclear power station reasons about the sanity of the plant. It has some knowledge of how the plant is supposed to operate, and looks out for significant discrepancies between expected and observed plant behavior. To this end, the watchdog monitor maintains a model of the power plant that it runs in parallel with the real plant, comparing its outputs to the measurement data extracted from the real plant. The watchdog monitor thus contains a real–time simulation of a model of the correctly working power plant. Once it discovers a significant aberration in real plant behavior, it kicks off a *fault discriminator* program that, again in real time, tries to narrow down the source of the fault, i.e., seeks to determine, which of the subsystems of the real plant is malfunctioning. It maintains real-time simulations of abstractions of models of all subsystems that permit it to localize errors to a particular subsystem. Once this has been accomplished, a *fault isolation* program is kicked off that invokes a real–time simulation of a more refined model of the faulty subsystem including models of faulty behavior with the aim of identifying the kind of error that is most likely to have occurred within the faulty subsystem [10.11, 10.12, 10.45].

Conceptually, the implementation of real–time simulation software is straightforward. It contains only four new components:

1. The *real–time clock* is responsible for the synchronization of real time and simulated time. The real–time clock is programmed to send a trigger impulse once every h time units of real time, where h is the current step size of the integration algorithm, and the simulation program is equipped with a busy waiting mechanism that is launched as soon as all computations associated with the current step have been completed, and that checks for arrival of the next trigger signal. The new step will not begin until the trigger signal has been received.

2. The *analog to digital (A/D) converters* are read at the beginning of each integration step to update the values of all external driving functions. This corresponds effectively to a sample and hold (S/H) mechanism. The inputs are updated once at the beginning of every integration step and are then kept constant during the entire step.

3. The *digital to analog (D/A) converters* are set at the end of each integration step, i.e., the newest output information is put out through the D/A converters for inspection by the user, or for driving real hardware (for so–called *hardware–in–the–loop* simulations.

4. *External events* are time events that are generated outside the simulation. External events are used for asynchronous communication with the simulation program, e.g. for the modification of parameter values, or for handling asynchronous readout requests, or for communication between several asynchronously running computer programs

either on the same or different computers. External events are usually postponed to the end of the current step and replace a portion of the busy waiting period.

Figure 10.1 illustrates the different tasks that take place during the execution of an integration step.

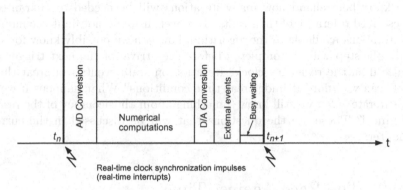

FIGURE 10.1. Task scheduling within integration step.

Once the message from the real–time clock has arrived indicating that the real time has advanced to time t_k, the simulation program first reads all the A/D converters to update the values of all input functions to the current time. It then performs the actual numerical computations associated with the step, calling upon the numerical integration routine and the routine that implements the state–space model. Once this is accomplished, the results are written out to the D/A converters. At this time, the "regular" business associated with the current step are over. The algorithm now consults the "mailbox" in which external events that may have arrived in the meantime are stored, and handles those. Once this has been accomplished, the algorithm has nothing more left to do and enters a "busy waiting" loop in which it repetitively checks the mailbox for arrival of the next message from the real–time clock.

The interprocessor and intertask communication mechanisms can actually be implemented in many different ways. In some cases, it may be desirable to use the waiting time of the processor for background tasks, rather than waste it in a busy waiting loop. In that case, it is not sufficient for the real–time clock to send a message to the simulation program. Instead, it must use the interrupt mechanism of the processor on which the simulation is running to interrupt whatever other task the processor is currently working on.

The difficulties of real–time simulation are *not* of a conceptual nature. They have to do with keeping track of real time. How can we guarantee that all that needs to be accomplished during the integration step can be completed prior to the arrival of the next trigger impulse?

In the previous chapters of this book, we introduced more and more bells and whistles that would help us in being able to guarantee the *correctness* of the simulation results obtained, but all these additional tools were accompanied by some run–time overhead, and in many cases, the amount of time needed to bring these algorithms to completion was not fixed. For example, if we decide to use an implicit integration algorithm, how can we know beforehand how many iterations will be needed to guarantee a prescribed tolerance of the results? However, if we do not limit the number of iterations available to the algorithm, how can we possibly know for sure that the step will be completed before the arrival of the next trigger impulse from the real–time clock? Iteration on state events is a great thing. Yet, can we afford it under real–time conditions? What happens if we do not iterate? Can we still know something about the accuracy of the results obtained? These are the questions that will be discussed in the current chapter.

10.2   The Race Against Time

There are two questions that we can ask ourselves in the context of racing against real time: (i) How can we guarantee that all computations necessary to end the current integration step in time are indeed completed before the next trigger impulse from the real–time clock arrives? (ii) What happens if we don't meet the schedule? Let me first address the second question since it is somewhat easier to deal with.

There are basically four things that we can do if we don't meet the schedule. We can:

1. increase the step size, h, in order to make more time for the tasks that need to be accomplished,

2. make the function evaluation more efficient, i.e., optimize the program that represents our state–space model,

3. improve the speed of the integration algorithm, e.g. by reducing the number of function evaluations necessary during one step, and finally

4. buy ourselves a faster computer.

The last solution may sound like a last resort, but in these times of cheap hardware and expensive software and manpower, it may actually often be the wisest thing to do.

The first solution is interesting. Until now, the step size was always bounded from the top due to accuracy and stability constraints. Now suddenly, the step size is also bounded from the bottom. We cannot reduce the step size to a value smaller than the total real time needed to perform all the computations associated with simulating the model across one step plus the real time needed for dealing with the administration of the simulation during that step. If it happens that the lower bound is larger than the upper, then we are in real trouble.

The second solution is one that has, over the years, been most actively pursued by Granino Korn, who wrote a large number of articles on the issue of how to obtain "cheap" (in the sense of fast) approximations for all kind of functions. He also treated this topic in several of his books [10.26, 10.27].

Many engineering models, such as models used in flight simulators or models of thermal power plants are full of two– and three–dimensional tables representing static characteristics that have been deduced by measurements and for which no explicit formulae are known. The need to interpolate in large three–dimensional tables is a nightmare for designers of real–time simulation software, since these interpolations can be very time consuming, and since the time needed to find the right entries in the table between which to interpolate is not even constant, but depends on the numerical values of the current arguments. Recent advances in neural network technology make it now possible to design feedforward neural networks trained e.g. through accelerated backpropagation algorithms that approximate two– and three–dimensional static functions with arbitrary precision. The training of these networks is slow, but this can be done off–line. Once trained, neural networks are very efficient at run time, providing for very fast multidimensional function evaluation capabilities. Also in this arena, it was Granino Korn who did pioneering work in combining fast neural network technology with high–speed simulation capabilities [10.28].

Finally, the most prominent researchers who dealt (and are still dealing) with the third solution are Jon Smith [10.43] and Bob Howe [10.22, 10.32, 10.33]. Since this approach deals with the numerical integration algorithms themselves, it is most relevant to this textbook, and therefore, we shall talk more about this approach in the current chapter.

Yet, before studying the way of improving the speed of the algorithms, we shall analyze the different methods in order to focus only on those that show suitable features for real–time simulation.

10.3 Suitable Numerical Integration Methods

In real–time simulation, it is not sufficient to obtain a good approximation of the values of the state variables. These approximations are in fact useless, if they arrive too late. We need to make sure that all of the computations associated with a single integration step are completed within the allowed time slot.

To this end, the total number of calculations performed by a single integration step must be bounded, and all of the iterative processes should be cut after a fixed number of iterations. It is evident that this will affect the accuracy of the algorithms, but it is better to obtain a solution with some remaining error, than not be able to obtain it at all within the allowed time [10.18].

Taking into account these considerations, the following analysis tries to examine the different features of the methods introduced in previous chapters of this book in order to discuss their pros and cons in the context of real–time simulation. The analysis is primarily based on Schiela's diploma thesis [10.40].

- *Multi–step methods*. Multi–step methods use information from the previous steps to compute a high–order approximation of the next step. This idea is based on the assumption that the differential equation is smooth, since the approximation uses a polynomial function. Unfortunately, many real–time simulations receive input signals from the real world that are not very smooth. Therefore, multi–step methods may give inaccurate results in such cases.

  On the other hand, multi–step methods reduce the number of function evaluations per step, which is a crucial factor in the real–time context. For this reason alone, and in spite of the fact that some accuracy may be sacrificed in this way, explicit linear multi–step methods, such as Adams–Bashforth, and among them especially those of low order of approximation accuracy, are widely used in real–time simulation [10.23].

- *Explicit single–step methods*. These methods are compatible with the requirements mentioned earlier. Their computational effort is relatively low and constant. The number of calculations per step can be easily estimated. Furthermore, the methods can deal fairly well with discontinuous input signals, since they do not use information from the past. Thus, for non–stiff ordinary differential equations, explicit single–step methods may constitute the best choice.

  However as we already know, these methods have problems with stiff systems. For mildly stiff problems, one remedy is to use integration step sizes that are a fraction of the sample interval, but then the efficiency decays with increasing stiffness.

  A different strategy for stiff systems is to modify the model so that the stiffness decreases. In some cases, the fast dynamics do not significantly influence the overall solution, and under such circumstances, the fast modes can be removed from the model. However, this is not generally the case for stiff systems, and it is always a questionable tactic to change the model in order to get it simulated.

- *Implicit single–step methods*. As we know, implicit methods require solving a system of nonlinear equations at each step, which implies the use of iterative methods, such as Newton iteration. Therefore, the computational effort for each step cannot be estimated reliably, as it depends on the (theoretically unbounded) number of iterations. Hence implicit methods are not suitable for the purpose of real–time simulation. Nevertheless, algorithms based on implicit methods can be used in real–time simulation, provided that the number of iterations is kept bounded to a fixed value.

However, we must also take into account that, by limiting the number of iterations, we modify the stability domain of these methods.

- *High–order methods.* In most real–time applications, the sampling intervals are small compared to the time scales of interest, and the required accuracy is usually rather low. One important reason for using small sampling intervals is to be able to accommodate real–time input. Input signals must be sampled frequently, since they cannot be reliably interpolated. Taking into account that reducing the amount of calculations at each step is a crucial factor, high–order algorithms will not be suitable for real–time simulation, except under very particular circumstances. It is therefore rare to find real–time simulations that make use of integration methods of orders of approximation accuracy greater than two or three.

- *Variable step methods.* In real–time simulations, we do not have the luxury to be able to change the step size, as it is synchronized with the sampling rate and severely restricted by the real–time specifications of the problem. The best thing that we can do for "controlling" the integration error is to estimate it and log these estimates, so that the quality of the results obtained can at least be judged *a–posteriori*.

  The numerical integration error can be estimated on–line, by comparing the actual simulation with another real–time simulation using a bigger step size. This idea was proposed by Bob Howe [10.23], making use of interpolation techniques to obtain the values of the control run at the sampling times of the actual simulation.

After analyzing all of these features, only *low–order explicit methods* seem well suited for real–time simulation.

In absence of stiffness, discontinuities, or badly nonlinear implicit equations, those methods work properly.

It can happen that the system dynamics are fast compared with the computer clock frequency. In such a case, there is little that we can do except try to optimize the way, in which the calculations are made, or buy ourselves a faster computer.

Leaving high–bandwidth applications aside, real–time simulation of non–stiff smooth systems does not call for any special treatment from a simulation methodology point of view.

Unfortunately, many systems in engineering applications are in fact stiff and, as we already know, explicit methods show poor performance in their integration.

We are unable to solve this problem without using implicit principles but, for the reasons explained above, we must avoid iterative solutions.

These considerations lead us to *semi–implicit* or *linearly implicit* methods and –in a further step– to *multi–rate integration*.

10.4   Linearly Implicit Methods

Linearly implicit or semi–implicit methods exploit the fact that implicit methods applied to linear systems do not require a theoretically unbounded number of iterations. Indeed, the resulting implicit equations can be solved by means of matrix inversion.

A widely used linearly–implicit method is given by the semi–implicit Euler formula [10.40, 10.38]:

$$\mathbf{x_{k+1}} = \mathbf{x_k} + h \cdot [\mathbf{f}(\mathbf{x_k}, t_k) + \mathcal{J}_{\mathbf{x_k}, t_k} \cdot (\mathbf{x_{k+1}} - \mathbf{x_k})] \tag{10.1}$$

where

$$\mathcal{J}_{\mathbf{x_k}, t_k} = \left. \frac{\partial \mathbf{f}}{\partial \mathbf{x}} \right|_{\mathbf{x_k}, t_k} \tag{10.2}$$

is the Jacobian matrix evaluated at $(\mathbf{x_k}, t_k)$.

Notice that

$$\mathcal{J}_{\mathbf{x_k}, t_k} \cdot (\mathbf{x_{k+1}} - \mathbf{x_k}) \approx \mathbf{f}(\mathbf{x_{k+1}}, t_{k+1}) - \mathbf{f}(\mathbf{x_k}, t_k) \tag{10.3}$$

and therefore:

$$\mathbf{f}(\mathbf{x_k}, t_k) + \mathcal{J}_{\mathbf{x_k}, t_k} \cdot (\mathbf{x_{k+1}} - \mathbf{x_k}) \approx \mathbf{f}(\mathbf{x_{k+1}}, t_{k+1}) \tag{10.4}$$

Thus, the linearly implicit Euler approximates the implicit Euler method. Moreover, in the linear case:

$$\dot{\mathbf{x}} = \mathbf{A} \cdot \mathbf{x} \tag{10.5}$$

we have:

$$\mathbf{x_{k+1}} = \mathbf{x_k} + h \cdot [\mathbf{A} \cdot \mathbf{x_k} + \mathcal{J}_{\mathbf{x_k}, t_k} \cdot (\mathbf{x_{k+1}} - \mathbf{x_k})] = \mathbf{x_k} + h \cdot \mathbf{A} \cdot \mathbf{x_{k+1}} \tag{10.6}$$

which exactly coincides with Backward Euler. This implies that the stability domain of the linearly implicit Euler method also coincides with the stability domain of Backward Euler.

Equation (10.1) can be rewritten as:

$$(I - h \cdot \mathcal{J}_{\mathbf{x_k}, t_k}) \cdot \mathbf{x_{k+1}} = (I - h \cdot \mathcal{J}_{\mathbf{x_k}, t_k}) \cdot \mathbf{x_k} + h \cdot \mathbf{f}(\mathbf{x_k}, t_k) \tag{10.7}$$

which shows that $\mathbf{x_{k+1}}$ can be obtained by solving a linear system of equations.

The value of $\mathbf{x_{k+1}}$ can also be obtained as:

$$\mathbf{x_{k+1}} = \mathbf{x_k} + h \cdot (I - h \cdot \mathcal{J}_{\mathbf{x_k},t_k})^{-1} \cdot \mathbf{f}(\mathbf{x_k}, t_k) \qquad (10.8)$$

The formula given by Eq.(10.8) is similar to Forward Euler, but differs in the presence of the term $(I - h \cdot \mathcal{J}_{\mathbf{x_k},t_k})^{-1}$.

¿From a computational point of view, that term implies that the algorithm has to calculate the Jacobian at each step and then either solve a linear equation system or invert a matrix.

Despite the fact that those calculations may turn out to be quite expensive, the computational effort is predictable, which makes the method well suited for real–time simulation.

Taking into account that the stability domain coincides with that of Backward Euler, this method results appropriate for the simulation of stiff and differential algebraic problems. Low–order linearly implicit methods may indeed often be the best choice for real–time simulation. However, they share one drawback with implicit methods: if the size of the problem is large, then the solution of the resulting linear equation system is computationally expensive.

Due to this fact, many different techniques were proposed that optimize how and how often the Jacobian is being evaluated and the linear equation system is being solved [10.19].

We shall not discuss those techniques here for two reasons. First, many of these techniques are designed to make the numerical integration faster *on average*. We are not interested in such approaches in the context of real–time simulation, because we must ensure that the algorithm converges *always* within the allotted time. Second, the statement that the stability domain of the linearly implicit Euler algorithm is the same as that of Backward Euler is only true if the *exact Jacobian* is being used in every step. For example, if we were to approximate the Jacobian by the zero matrix, the method would have the stability domain of Forward Euler.

Many of the implicit algorithms make use of a so–called *modified Newton iteration*. In one variant of that approach, the underlying Hessian matrix is being approximated by a diagonal matrix to make its inversion cheap and painless. This can be done. The price to be paid for this luxury is that the Newton iteration will converge more slowly, i.e., we have to spend more iteration steps, while each individual iteration step is now cheaper. Whether or not this pays off, depends on the application at hand.

Some authors proposed to apply this technique in the case of semi–implicit algorithms as well by approximating the Jacobian through its diagonal elements. We do not recommend this approach. In most cases, this will be the kiss of death, as the stability domain of the method using a diagonal approximation of the Jacobian will most likely loop in the left–half $\lambda \cdot h$ plane, i.e., the method will no longer be stiffly stable.

Other authors proposed to carefully look at the structure of the Jacobian, and at least zero out some of the smallest non–vanishing elements in it. This technique is called *sparsing* [10.40, 10.38], as it makes the Jacobian more sparse, thereby enabling a cheaper linear system solution using either numerical or symbolic sparse matrix techniques. It was shown that sparsing can indeed reduce the computational effort needed to complete the calculations of a single integration step by a significant amount. Yet, the technique must be applied cautiously, as it doesn't take much for the stiffly stable nature of the algorithms to be lost in the process. It thus may generally pay off to work with an accurate computation of the exact Jacobian in each step.

Having said that, we are of course free in how we compute the exact Jacobian, and which technique we use for solving the resulting linear equation system. The Jacobian can either be computed symbolically, leaving it up to the model compiler to find the appropriate expressions by symbolic (algebraic) differentiation, or it can be approximated numerically. Furthermore, Jacobian matrices are usually sparse, because not every state derivative depends on every state variable. Thus, we can use either numerical sparse matrix algorithms in solving the resulting linear system, or we can use one of the two symbolic sparse sparse matrix techniques introduced earlier in this book, i.e., tearing [10.17] or relaxation [10.35].

The improvements achieved by those techniques allow some large stiff systems to be simulated in real time using semi–implicit algorithms. However, there are still larger and more complicated systems, in which these ideas are not enough to win the race against time.

Fortunately, stiffness in large systems is often connected to the presence of some identifiable slow and fast sub–models. In those cases, we can use that information to our advantage by splitting the system and applying different step sizes or even different integration algorithms to the different parts. These ideas lead to the concepts of multi–rate and mixed–mode integration.

Finally, we should mention that several higher–order semi–implicit versions of both multi–step [10.6] and Runge–Kutta [10.1, 10.46] methods have been reported in the literature. We shall not explore these algorithms further, since their principles are similar to those of the linearly implicit Euler method. However, we shall derive one of these methods in a homework problem.

Least suitable among all of the linearly implicit stiffly stable algorithms for the task at hand are those algorithms that are F–stable, in particular the *trapezoidal rule* and its one–legged twin, the *implicit midpoint rule*. The reason for this assertion is the following. Since we cannot use a variable–step algorithm, we are bound to end up with a numerical error in each integration step that is caused by the fixed step size and that is essentially uncontrollable. It is thus recommended to use an algorithm with some additional *artificial damping* to prevent error accumulation [10.18].

10.5 Multi–rate Integration

There are many cases, in which the stiffness is due to the presence of a sub–system with very fast dynamics compared to the rest of the system. Typical examples of this can be found in multi–domain physical systems, since the components of different physical domains usually involve distinct time constants.

For example, if we wish to study the thermal properties of an integrated circuit package, we shall recognize that the electrical time constants of the device are faster in comparison with the thermal time constants by several orders of magnitude. Yet, we cannot ignore the fast time constants, since they are the cause of the heating. In some cases, such as switching power converters, the heating of the device grows with the frequency of switching, i.e., while no switching takes place, the thermal effects are minimal [10.41].

Let us introduce the idea with the following example. Figure 10.2 shows a lumped model of a transmission line fed by a Van–der–Pol oscillator (this example is a variant of an example offered in [10.36]).

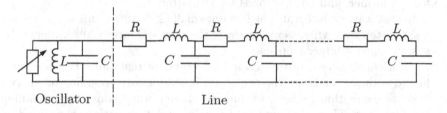

FIGURE 10.2. Van–der–Pol oscillator and transmission line.

We shall assume that the nonlinear resistor of the oscillator circuit satisfies the law:

$$i_R = k \cdot u_R^3 - u_R \qquad (10.9)$$

Then the system can be described by the following set of state equations:

$$\frac{di_L}{dt} = \frac{1}{L} u_C \qquad (10.10a)$$

$$\frac{du_C}{dt} = \frac{1}{C}(u_C - k \cdot u_C^3 - i_L - i_1) \qquad (10.10b)$$

$$\frac{di_1}{dt} = \frac{1}{L} u_C - \frac{R}{L} i_1 - \frac{1}{L} u_1 \qquad (10.10c)$$

$$\frac{du_1}{dt} = \frac{1}{C} i_1 - \frac{1}{C} i_2 \qquad (10.10d)$$

$$\frac{di_2}{dt} = \frac{1}{L} u_1 - \frac{R}{L} i_2 - \frac{1}{L} u_2 \qquad (10.10e)$$

$$\frac{du_2}{dt} = \frac{1}{C} i_2 - \frac{1}{C} i_3 \qquad (10.10f)$$

$$\vdots$$

$$\frac{di_n}{dt} = \frac{1}{L}u_{n-1} - \frac{R}{L}i_n - \frac{1}{L}u_n \qquad (10.10\text{g})$$

$$\frac{du_n}{dt} = \frac{1}{C}i_n \qquad (10.10\text{h})$$

Here, u_C and i_L are the voltage and current of the capacitor and inductance in the oscillator. Similarly, u_j and i_j are the voltage and current of the capacitors and inductances at the jth stage of the transmission line.

Let us assume that the transmission line has 5 stages (i.e., $n = 5$), and the parameters are $L = 10$ mH, $C = 1$ mF, $R = 10\Omega$ and $k = 0.04$.

If we wish to simulate the system using the Forward Euler method, we need to use a step size no greater than $h = 10^{-4}$ seconds. Otherwise, the oscillator output (u_C) is computed with an error that is totally unacceptable.

However, using the input signal generated by the oscillator, the transmission line alone can be simulated with a step size that is 10 times bigger.

Thus, we decided to split the system into two subsystems, the oscillator circuit and the transmission line, using two different step sizes: 10^{-4} seconds for the former, and 10^{-3} seconds for the latter.

In that way, we integrate the fast but small (2nd–order) sub–system using a small step size, whereas we integrate the slow and large (10th–order) sub–system using a larger step size.

As a consequence, during each millisecond of real time, the computer has to evaluate ten times the two scalar functions corresponding to the two first state equations, whereas it only needs to evaluate once the remaining ten functions. Thus, the number of floating–point operations is reduced by about a factor of four compared with a regular simulation using a single step size throughout.

The simulation results are shown in Figs.10.3–10.4.

We can generalize this procedure to systems of the form:

$$\dot{\mathbf{x}}_{\mathbf{f}}(t) = \mathbf{f_f}(\mathbf{x_f}, \mathbf{x_s}, t) \qquad (10.11\text{a})$$

$$\dot{\mathbf{x}}_{\mathbf{s}}(t) = \mathbf{f_s}(\mathbf{x_f}, \mathbf{x_s}, t) \qquad (10.11\text{b})$$

where the sub–indexes, f and s, stand for "fast" and "slow," respectively.

Then, the use of the multi–rate version of Forward Euler with inlining results in a set of difference equations of the form:

$$\mathbf{x_f}(t_i + (j+1) \cdot h) = \mathbf{x_f}(t_i + j \cdot h) + h \cdot \mathbf{f_f}(\mathbf{x_f}(t_i + j \cdot h),$$
$$\mathbf{x_s}(t_i + j \cdot h), t_i + j \cdot h) \qquad (10.12\text{a})$$

$$\mathbf{x_s}(t_i + k \cdot h) = \mathbf{x_s}(t_i) + h \cdot \mathbf{f_s}(\mathbf{x_f}(t_i), \mathbf{x_s}(t_i), t_i) \qquad (10.12\text{b})$$

where k is the (integer) ratio of the two step sizes, $j = 0 \ldots k - 1$, and $h = t_{i+1} - t_i$ is the step–size of the slow sub–system.

Equations (10.12a–b) do not specify, how $\mathbf{x_s}(t_i + j \cdot h)$ is being calculated, since the variables of the slow sub–system are not evaluated at the intermediate time instants.

Multirate Simulation (Oscillator)

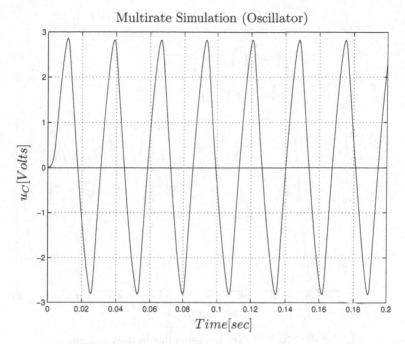

FIGURE 10.3. Van–der–Pol oscillator voltage.

In our example, we chose $\mathbf{x_s}(t_i + j \cdot h) = \mathbf{x_s}(t_i)$, i.e., we used the last calculated value. A more accurate solution might involve using some form of extrapolation technique.

This last problem is known as the *interfacing* problem [10.30]. It is related to the way, in which the fast and slow sub–systems are interconnected with each other.

In our case, we used the Forward Euler method. Similar approaches have been reported in the literature based on the 2nd–order explicit Adams–Bashforth technique [10.24], including also some improvements for parallel implementation.

In spite of the improvement achieved in this case using multi–rate integration, we must not forget that the example we analyzed was not very demanding, since the speed of the fast sub–system is not much higher than that of the slow sub–system. We already know that explicit algorithms won't work in more strongly stiff systems.

In those cases, as previously discussed, semi–implicit methods may be a better choice in the real–time context. However, we know that in large systems, those methods have a drawback, as they need to invert a potentially very large matrix.

A solution that combines both ideas, multi–rate and semi–implicit integration, consists in splitting the system into a fast and a slow part, while applying a semi–implicit method to the fast sub–system, whereas the

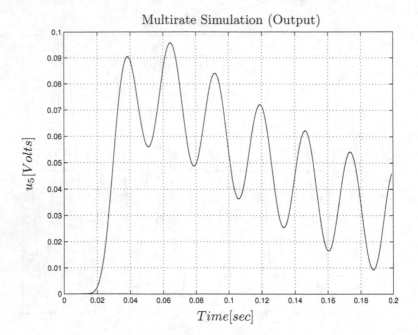

FIGURE 10.4. Transmission line output voltage.

slow sub–system is being simulated using an explicit integration algorithm. These types of schemes are referred to in the literature as *mixed–mode* integration algorithms.

We shall discuss mixed–mode integration in due course, but let us first pursue another avenue.

10.6 Inline Integration

Figure 10.5 shows the same circuit as Fig. 10.2 with the inclusion of an additional RC load at the end of the transmission line.

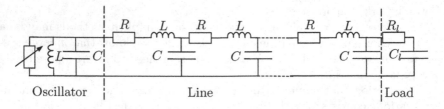

FIGURE 10.5. Van–der–Pol oscillator and transmission line.

The state equations are similar to the previous case, but now we have:

$$\frac{di_L}{dt} = \frac{1}{L}u_C \tag{10.13a}$$

$$\frac{du_C}{dt} = \frac{1}{C}(u_C - k \cdot u_C^3 - i_L - i_1) \tag{10.13b}$$

$$\frac{di_1}{dt} = \frac{1}{L}u_C - \frac{R}{L}i_1 - \frac{1}{L}u_1 \tag{10.13c}$$

$$\frac{du_1}{dt} = \frac{1}{C}i_1 - \frac{1}{C}i_2 \tag{10.13d}$$

$$\frac{di_2}{dt} = \frac{1}{L}u_1 - \frac{R}{L}i_2 - \frac{1}{L}u_2 \tag{10.13e}$$

$$\frac{du_2}{dt} = \frac{1}{C}i_2 - \frac{1}{C}i_3 \tag{10.13f}$$

$$\vdots$$

$$\frac{di_n}{dt} = \frac{1}{L}u_{n-1} - \frac{R}{L}i_n - \frac{1}{L}u_n \tag{10.13g}$$

$$\frac{du_n}{dt} = \frac{1}{C}i_n - \frac{1}{R_l \cdot C}(u_n - u_l) \tag{10.13h}$$

$$\frac{du_l}{dt} = \frac{1}{R_l \cdot C_l}(u_n - u_l) \tag{10.13i}$$

Let us assume that the load parameters are $R_l = 1$ kΩ and $C_l = 1$ nF.

Since the load resistor is much bigger than the line resistors, the newly introduced term in Eq.(10.13h) won't influence the dynamics of the transmission line significantly, and we can expect the sub–system (10.13a–h) to exhibit a similar behavior to the one of System (10.10).

However, the last state equation, Eq.(10.13i), introduces a fast pole. The position of this pole is approximately located at:

$$\lambda_l \approx -\frac{1}{R_l \cdot C_l} = -10^6 \ \text{sec}^{-1} \tag{10.14}$$

on the negative real axis of the complex λ–plane.

This means that we would have to reduce the step size by about a factor of 1000 with respect to the previous example, in order to obtain a numerically stable result.

Unfortunately, such a solution is completely unacceptable in the context of a real–time simulation.

A first alternative might be to replace the Forward Euler algorithm by the semi–implicit Euler method studied earlier in this chapter. However, this is a system of order 13, and, leaving superstitions aside, we may not have the luxury of inverting a 13×13 matrix at each step.

A second alternative might be to inline the Backward Euler algorithm [10.16] and apply the tearing method to the resulting set of difference equations. Let us rewrite the model using the inling approach.

$$i_L = \text{pre}(i_L) + \frac{h}{L}u_C \tag{10.15a}$$

$$u_C = \text{pre}(u_C) + \frac{h}{C}(u_C - k \cdot u_C^3 - i_L - i_1) \tag{10.15b}$$

$$i_1 = \text{pre}(i_1) + \frac{h}{L}u_C - \frac{Rh}{L}i_1 - \frac{h}{L}u_1 \tag{10.15c}$$

$$u_1 = \text{pre}(u_1) + \frac{h}{C}i_1 - \frac{h}{C}i_2 \tag{10.15d}$$

$$i_2 = \text{pre}(i_2) + \frac{h}{L}u_1 - \frac{Rh}{L}i_2 - \frac{h}{L}u_2 \tag{10.15e}$$

$$u_2 = \text{pre}(u_2) + \frac{h}{C}i_2 - \frac{h}{C}i_3 \tag{10.15f}$$

$$\vdots$$

$$i_n = \text{pre}(i_n) + \frac{h}{L}u_{n-1} - \frac{Rh}{L}i_n - \frac{h}{L}u_n \tag{10.15g}$$

$$u_n = \text{pre}(u_n) + \frac{h}{C}i_n - \frac{h}{R_l \cdot C}(u_n - u_l) \tag{10.15h}$$

$$u_l = \text{pre}(u_l) + \frac{h}{R_l \cdot C_l}(u_n - u_l) \tag{10.15i}$$

The causalized structure digraph is shown in Fig. 10.6.

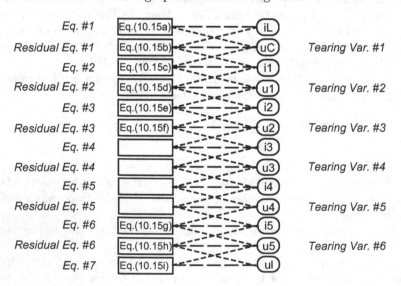

FIGURE 10.6. Causal structure diagram of electrical circuit.

Inlining did help indeed. We got away with six tearing variables. Instead of having to invert a 13×13 matrix in every step, we now must invert a

6×6 matrix. Since even the best linear sparse matrix solver grows at least quadratically with the size of the system in terms of its computational complexity, the savings were quite dramatic. The computations just got faster by about a factor of four.

Although inline integration had been developed for general simulation problems, it turns out that this method has become a quite powerful ally in dealing with real–time simulation as well [10.15].

But what, if the simulation is still too slow? What if the transmission line consists of 50 segments, instead of only 5 of them? Mixed–mode integration may be the answer to our needs.

10.7   Mixed–mode Integration

A more careful look at the system shows that there is no strong interaction between the subsystems of Eqs.(10.13a–h) and Eq.(10.13i). In fact, the fast dynamics can be explained by looking at the last equation alone.

Thus, it might be reasonable to use Backward Euler (or semi–implicit Euler) only in the last equation.

To this end, we inlined the equations once more, this time using the explicit Forward Euler algorithm everywhere except for the last equation, where we still used the implicit Backward Euler method.

The resulting inlined difference equation system no can be written as follows:

$$i_L = \text{pre}(i_L) + \frac{h}{L}\text{pre}(u_C) \tag{10.16a}$$

$$u_C = \text{pre}(u_C) + \frac{h}{C} = [\text{pre}(u_C) - k \cdot \text{pre}(u_C)^3 - \text{pre}(i_L)$$
$$\quad -\text{pre}(i_1)] \tag{10.16b}$$

$$i_1 = \text{pre}(i_1) + \frac{h}{L}\text{pre}(u_C) - \frac{Rh}{L}\text{pre}(i_1) - \frac{h}{L}\text{pre}(u_1) \tag{10.16c}$$

$$u_1 = \text{pre}(u_1) + \frac{h}{C}\text{pre}(i_1) - \frac{h}{C}\text{pre}(i_2) \tag{10.16d}$$

$$i_2 = \text{pre}(i_2) + \frac{h}{L}\text{pre}(u_1) - \frac{Rh}{L}\text{pre}(i_2) - \frac{h}{L}\text{pre}(u_2) \tag{10.16e}$$

$$u_2 = \text{pre}(u_2) + \frac{h}{C}\text{pre}(i_2) - \frac{h}{C}\text{pre}(i_3) \tag{10.16f}$$

$$\vdots$$

$$i_n = \text{pre}(i_n) + \frac{h}{L}\text{pre}(u_{n-1}) - \frac{Rh}{L}\text{pre}(i_n) - \frac{h}{L}\text{pre}(u_n) \tag{10.16g}$$

$$u_n = \text{pre}(u_n) + \frac{h}{C}\text{pre}(i_n) - \frac{h}{R_l \cdot C}[\text{pre}(u_n) - \text{pre}(u_l)] \tag{10.16h}$$

$$u_l \;=\; \text{pre}(u_l) + \frac{h}{R_l \cdot C_l}(u_n - u_l) \tag{10.16i}$$

All equations are now explicit, except for the very last equation, Eq.(10.16i), which is implicit in the variable u_l. Furthermore, Eq.(10.16i) can only be computed after $u_n(t)$ has been evaluated first from Eq.(10.16h). Thus, the size of the Jacobian is now 1×1.

We simulated the system using the same approach as before, i.e., we applied a step size of 10^{-4} seconds to the two oscillator equation, whereas we used a step size of 10^{-3} seconds on all of the other equations, including the implicit load equation. The simulation results are shown in Figure 10.7.

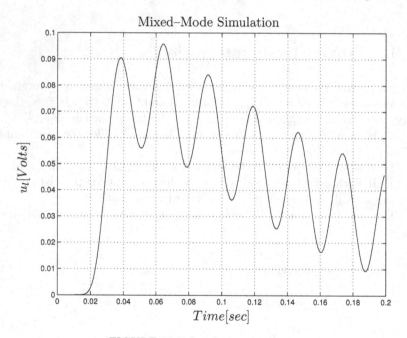

FIGURE 10.7. Load output voltage.

In more general terms, given a system like Eqs.(10.11a–b), the Backward–Forward Euler Mixed–Mode integration scheme is given by the formula:

$$\mathbf{x_s}(t_{k+1}) \;=\; \mathbf{x_s}(t_k) + h \cdot \mathbf{f_s}(\mathbf{x_f}(t_k), \mathbf{x_s}(t_k), t_k) \tag{10.17a}$$
$$\mathbf{x_f}(t_{k+1}) \;=\; \mathbf{x_f}(t_k) + h \cdot \mathbf{f_f}(\mathbf{x_f}(t_{k+1}), \mathbf{x_s}(t_{k+1}), t_{k+1}) \tag{10.17b}$$

Thus, the algorithm starts by computing explicitly the value of $\mathbf{x_s}(t_{k+1})$. It then uses this value to evaluate $\mathbf{x_f}(t_{k+1})$ either implicitly or in a semi–implicit fashion.

Mixed–mode integration as presented in this section was first introduced by Krebs [10.29] for an entirely different purpose, namely to resolve the problem of *conditional index changes* once and for all in a systematic and algorithmic fashion.

The technique was rediscovered independently by Schiela [10.39] for the purpose of speeding up real–time simulation. Schiela proposed the use of linearization and eigenvalue analysis to discern, which of the integrators should be inlined using Forward Euler, and which should be inlined using Backward Euler, i.e., for determining the slow and fast sub–systems.

The advantage of solving the implicit equation only for the components $\mathbf{x_f}(t_{k+1})$ can turn out to be very important in systems, such as the one presented here, where the length of vector $\mathbf{x_f}$ is considerably smaller than that of $\mathbf{x_s}$.

In our (rather academic) example, the reduction in the number of calculations is huge. In more realistic applications, the literature reports speed–up factors of 4 to 16 [10.39].

Mixed–mode versions of higher order Runge–Kutta methods and approaches also combine mixed–mode and multi–rate integration techniques have also been reported in the literature [10.42].

In fact, we used a mixture of multi–rate and mixed–mode integration in our example, as we used a ten times smaller step size of 10^{-4} seconds for the integration of the two oscillator equations.

Both multi–rate integration and mixed–mode integration assume that there indeed exist two distinct and discernable sub–systems. This may not always be the case. For example, the real–time simulation of a distributed parameter system described by a parabolic PDE, such as for the purpose of optimal control of a space heating system, does not share this property. The eigenvalues are simply spread out. Also, if a system is highly nonlinear, the concept of looking at eigenvalues by itself become dubious, as eigenvalues can only be defined for the linearized system. In a sufficiently nonlinear system, the eigenvalues of the linearized system move around as a function of time, which again may prevent us from subdividing the system into two distinct and time–invariant sub–systems, one fast, the other slow.

10.8 Discontinuous Systems

Real–time simulation of discontinuous models is highly problematic, as state events happen asynchronously. Event handling invariably causes overhead that needs to be accounted for. Thus, if until now, it may have been acceptable to have the computations associated with the simulation of a single step occupy somewhere around 80% of the allotted time, we can no longer do so if the model to be simulated is discontinuous. In the case of discontinuous models being simulated in real time, it is prudent to dimension the computer system such that regular steps occupy no more than about 20% of the allotted time. This will grant us the additional time needed to handle no more than one state event per step.

State–event handling in real–time simulation is simplified, when comparing it to the techniques introduced in the previous chapter, by two factors:

1. As we are using low–order integration techniques, we can also use low–order event localization algorithms.

2. Since we use much smaller step sizes, the precise localization of state events becomes less critical, and there shouldn't occur as often multiple state events within a single integration step.

Since we must control the total amount of computations performed within an integration step, iterative techniques for localizing state events are out. We must rely on interpolation alone.

Yet, as we are using low–order integration techniques, the former iterative algorithms can now be employed as interpolators. For example, if we integrate by inlining a first–order accurate algorithm, i.e., either Forward Euler or Backward Euler [10.16], we can use a single step of *Regula Falsi* to locate the event as accurately as we can hope to accomplish with such a crude integration algorithm. If we decide to inline the third–order accurate Radau–IIA algorithm [10.5, 10.7], a single step of cubic interpolation will localize the discontinuity as accurately as can be done using such an integration method.

Of course, it may be possible to reduce the residual on the zero–crossing function further by iteration, but this does not necessarily imply that we would thereby locate the event more accurately, as already the previous integration steps are contaminated by numerical errors.

Let us discuss, how event handling may proceed. We start out by performing a regular integration step, advancing the simulation from time t_n to time t_{n+1}. At the end of the step, we discover that a zero crossing has taken place. We interpolate to the next event time, t_{next}. Since we don't have dense output [10.13] available, as this would be too expensive to compute in real time, we shall have to repeat the last integration step to advance the entire state vector from time t_n to time t_{next}. We then perform the actions associated with the event, and compute a new consistent initial state. Starting from that new initial state, we perform another partial state advancing the state vector from time t_{next} to time t_{n+1}. The solution obtained in this way can then be pushed out through the D/A–converters and communicated back to whatever hardware needs it.

As no iteration takes place, the amount of work, i.e., the total number of floating–point operations needed, can be estimated accurately. Assuming that only one state event is allowed to occur within a single integration step, we can thus calculate, how much extra time we need to allot, in order to handle single state events within an integration step adequately.

Unfortunately, the extra amount of work for event handling is non–negligible. We perform three integration steps instead of only one, and we have to accommodate the additional computations needed to process the event actions themselves. Thus, the total effort grows by about a factor of four. This is the reason, why we wrote earlier that the allowed resource utilization for regular integration steps needs to drop from about 80% to about 20%.

10.9   Simulation Architecture

We haven't yet discussed, how the simulation engine is physically connected to the hardware. Although it would be possible to connect directly the output signals of the *sensor units* with the input of the *A/D–converters*, which form part of the simulation engine, and the outputs of the *D/A–converters*, also integrated with the simulation engine, with the input signals of the *actuator units*, this is hardly ever done in today's world.

Instead, commercial converters have their own computer chips built in, that perform the necessary computations and store the digital signals in *mailboxes*. Thus, an A/D–converter is really a converter together with a built–in *zero–order hold (ZOH)* unit. Once the analog signal has been converted, it is available for whichever process needs it, until it is overridden by the next *sample–and–hold (S/H)* cycle. A D/A–converter doesn't take its data from the simulation directly, but instead, takes it out of its own mailbox. Even the sensor and actuator units contain their own hardware–built sample–and–hold equipment.

Handshaking mechanisms are needed to prevent the simulator from replacing the data in the mailbox of the D/A–converter, while the converter tries to read out the data from its own mailbox. Similarly, handshaking mechanisms are needed to prevent the simulator from reading the data from the mailbox of the A/D–converter, while these data are in the process of being updated by the converter.

A possible physical configuration of a *hardware–in–the–loop (HIL)* simulation is shown in Fig. 10.8.

Protocols have been designed to ensure that these handshaking mechanisms always work correctly. To this end, the *High–Level Architecture (HLA)* standard was created in the U.S. [10.44], whereas Europe developed its own standard with CORBA [10.37].

Consequently, Fig. 10.1 needs to be modified. The time needed for the A/D–convertions and D/A–conversions are no longer part of the computational load associated with advancing the simulation by one time step, as these activities are performed in parallel by separate units. Instead, we must include the time needed for the read and write requests from and to the mailboxes across the architecture.

Since the total time needed for computing all activities associated with a single integration step must be known, both HLA and CORBA offer

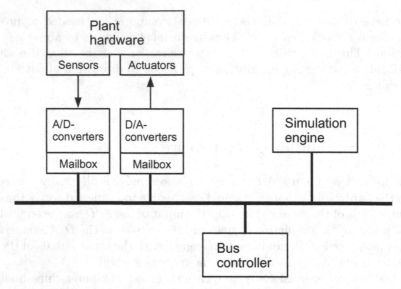

FIGURE 10.8. Physical configuration of HIL simulation.

mechanisms for specifying the *maximum allowed latency* in answering requests for information transfer across the architecture using the established communication channels and protocols.

10.10 Overruns

Overruns are defined as situations, where, in spite of our best efforts, the simulation engine is unable to perform all of the required computations in time to advance its state to the next clock time, before the real–time clock interrupt is received.

This may happen, because it cannot be guaranteed that no more than one state event will ever occur within a single integration step. As all events must be processed, it can happen that the simulation falls behind. Most real–time simulations specify the maximum percentage of overruns as e.g. 1% or 2%.

What happens, when the simulation falls behind? Thanks to the buffers implemented in the form of the mailboxes, the hardware will hardly notice it. It simply receives the same actuator values for a second time in a row.

For the simulation software, the situation may be worse, because it may need to know, what time it is. Thus, the following procedure is recommended in the case of an overrun. If the next real–time interrupt arrives, before the computations have been completed, the subsequent integration step is doubled in length to catch up with real time. In this way, we allow one integration step to be computed less accurately once in a while, in order to stay synchronous with the real–time clock.

10.11   Summary

In this chapter, we have attempted to paint, using a fairly wide brush, a picture of some of the requirements associated with real–time simulation of physical systems. It's a difficult problem to cope with, as the information available on this topic is widely scattered in the literature and hardly available in a concise and consistent fashion.

Just like in the case of the distributed parameter systems, we do not claim that we have been able to create here a body of knowledge that is exhaustive by any standard. We do not claim that you, the reader, will be able to successfully build a real–time simulator after having read this chapter.

Naturally, as this book concerns itself primarily with topics surrounding the numerical integration of ODE and DAE systems, we have focused our emphasis on issues related to the special demands of real–time simulation on the integration algorithms.

We only just mentioned the available literature on simulation speed–up by means of efficient function generation [10.21, 10.28], and we didn't talk at all about the use of special–purpose simulation hardware, a fashionable topic in the 1960's to 1980's. These systems have largely been overcome by events, as conventional digital hardware became faster and faster.

We barely scratched the surface of issues concerning the *simulation architecture*. There is a substantial body of knowledge available on this subject, although it concerns itself more with *discrete event simulation* in general, than with *physical system simulation*.

We didn't even mentioned the topic of *distributed real–time simulation* [10.8, 10.31], where the execution speed of the real–time simulation is increased by distributing the computations necessary to complete an integration step over multiple computers communicating with each other across the simulation architecture.

Notice that even Fig. 10.8 does not reflect the full real–time architecture needed to perform distributed simulation experiments. Both HLA and CORBA were developed to support *distributed processing*. Whereas CORBA was designed primarily for *instrumentation*, HLA has a strong emphasis on *distributed simulation*. In Fig. 10.8, the simulation is still "in charge" of the overall operations. All other units are essentially subservient to the simulation.

If we allow the simulation to be distributed over multiple processors working in parallel on a demanding simulation task, this approach won't work any longer. Figure 10.9 shows a once more enhanced architecture that supports distributed simulation.

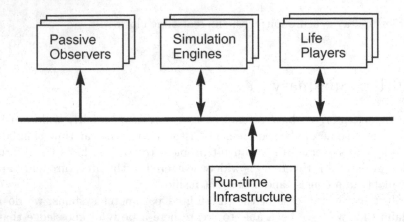

FIGURE 10.9. The HLA architecture.

Figure 10.9 depicts the overall HLA architecture [10.9, 10.10] for distributed simulation. Here, the former *bus controller* is replaced by the *Real-time Infrastructure (RTI)* [10.20], a distributed operating system that coordinates the activities of the various participants in the simulation. Each participant is responsible for finishing its assigned tasks within the allotted time slot and for returning the results in a timely fashion to the RTI.

¿From the perspective of the architecture, there is essentially no difference between *simulators* and *life players*, i.e., hardware–in–the–loop. *Passive observers* were added as an additional type of participants. Since passive observers never return any data to the architecture, it is not essential that they operate in a time–synchronous fashion. They can complete their tasks on an "as–fast–as–possible" basis.

10.12   References

[10.1] Jeff R. Cash. A Semi–implicit Runge–Kutta Formula for the Integration of Stiff Systems of Ordinary Differential Equations. *Chemical Engineering J.*, 20(3):219–224, 1980.

[10.2] François E. Cellier, Larry C. Schooley, Malur K. Sundareshan, and Bernard P. Zeigler. Computer–aided Design of Intelligent Controllers: Challenge of the Nineties. In *Recent Advances in Computer Aided Control Systems Engineering*, pages 53–80, Amsterdam, the Netherlands, 1992. Elsevier Science Publishers.

[10.3] François E. Cellier, Larry C. Schooley, Bernard P. Zeigler, Adele Doser, Glenn Farrenkopf, JinWoo Kim, YaDung Pan, and Brian Willams. Watchdog Monitor Prevents Martian Oxygen Production Plant from Shutting Itself Down During Storm. In *Proceedings IS-*

RAM'92, ASME Conference on Intelligent Systems for Robotics and Manufacturing, pages 697–704, Santa Fe, N.M., 1992.

[10.4] François E. Cellier. *Continuous System Modeling*. Springer Verlag, New York, 1991. 755p.

[10.5] François E. Cellier. Inlining Step–size Controlled Fully Implicit Runge–Kutta Algorithms for the Semi–analytical and Semi–numerical Solution of Stiff ODEs and DAEs. In *Proceedings 5th Conference on Computer Simulation*, pages 259–262, Mexico City, Mexico, 2000.

[10.6] Richard J. Charron and Min Hu. A–contractivity of Linearly Implicit Multistep Methods. *SIAM Journal on Numerical Analysis*, 32(1):285–295, 1995.

[10.7] Christoph Clauss, Hilding Elmqvist, Sven Erik Mattsson, Martin Otter, and Peter Schwarz. Mixed Domain Modeling in Modelica. In *Proceedings FDL'02, Forum on Specification and Design Languages*, Marseille, France, 2002.

[10.8] Rémi Cozot. From Multibody Systems Modeling to Distributed Real–Time Simulation. In *Proceedings Simulation'96 IEEE Conference*, pages 234–241, 1996.

[10.9] Judith S. Dahmann, Frederick Kuhl, and Richard Weatherly. Standards for Simulation: As Simple As Possible But Not Simpler – The High Level Architecture For Simulation. *Simulation*, 71(6):378–387, 1998.

[10.10] Judith S. Dahmann. The High Level Architecture and Beyond: Technology Challenges. In *Proceedings PADS'99, 13th Workshop on Parallel and Distributed Simulation*, pages 64–70, Atlanta, Georgia, 1999.

[10.11] Álvaro de Albornoz Bueno and François E. Cellier. Qualitative Simulation Applied to Reason Inductively About the Behavior of a Quantitatively Simulated Aircraft Model. In *Proceedings QUARDET'93, IMACS International Workshop on Qualitative Reasoning and Decision Technologies*, pages 711–721, Barcelona, Spain, 1993.

[10.12] Álvaro de Albornoz Bueno and François E. Cellier. Variable Selection and Sensor Fusion in Automatic Hierarchical Fault Monitoring of Large Scale Systems. In *Proceedings QUARDET'93, IMACS International Workshop on Qualitative Reasoning and Decision Technologies*, pages 722–734, Barcelona, Spain, 1993.

[10.13] John R. Dormand and Peter J. Prince. Runge–Kutta Triples. *J. of Computational and Applied Mathematics*, 12A(9):1007–1017, 1986.

[10.14] Eduard Eitelberg. *Modellreduktion linearer zeitinvarianter Systeme durch Minimieren des Gleichungsfehlers.* PhD thesis, University of Karlsruhe, Karlsruhe, Germany, 1979.

[10.15] Hilding Elmqvist, Sven Erik Mattsson, and Hans Olsson. New Methods for Hardware–in–the–loop Simulation of Stiff Models. In *Proceedings Modelica'2002 Conference*, pages 59–64, Oberpfaffenhofen, Germany, 2002.

[10.16] Hilding Elmqvist, Martin Otter, and François E. Cellier. Inline Integration: A New Mixed Symbolic/Numeric Approach for Solving Differential–Algebraic Equation Systems. In *Proceedings European Simulation Multiconference*, pages xxiii–xxxiv, Prague, Czech Republic, 1995.

[10.17] Hilding Elmqvist and Martin Otter. Methods for Tearing Systems of Equations in Object–oriented Modeling. In *Proceedings European Simulation Multiconference*, pages 326–332, Barcelona, Spain, 1994.

[10.18] Javier Garcia de Jalón and Eduardo Bayo. *Kinematic and Dynamic Simulation of Multibody Systems –The Real–Time Challenge–.* Wiley, 1994.

[10.19] Kjell Gustafsson and Gustaf Söderlind. Control Strategies for the Iterative Solution of Nonlinear Equations in ODE Solvers. *SIAM Journal on Scientific Computing*, 18(1):23–40, 1997.

[10.20] Frank Hodum and David Edwards. Time Management Services in the RTI–NG. In *Proceedings SIW'01, Fall Simulation Interoperability Workshop*, paper 01F–SIW–090, 2001.

[10.21] Robert M. Howe and Kuo-Chin Lin. The Use of Function Generation in the Real–time Simulation of Stiff Systems. In *AIAA Flight Simulation Technologies Conference and Exhibit*, pages 217–224, Dayton, Ohio, 1990.

[10.22] Robert M. Howe. The Use of Mixed Integration Algorithms in State Space. *Transactions of the Society for Computer Simulation*, 7(1):45–66, 1990.

[10.23] Robert M. Howe. On–line Calculation of Dynamic Errors in Real–time Simulation. In *Proceedings of SPIE*, volume 3369, pages 31–42, 1998.

[10.24] Robert M. Howe. Real–Time Multi–Rate Asynchronous Simulation with Single and Multiple Processors. In *Proceedings of SPIE*, volume 3369, pages 3–14, 1998.

[10.25] Thomas Kailath. *Linear Systems*. Prentice–Hall, Englewood Cliffs, N.J., 1980.

[10.26] Granino A. Korn and John V. Wait. *Digital Continuous–System Simulation*. Prentice–Hall, Englewood Cliffs, N.J., 1978.

[10.27] Granino A. Korn. *Interactive Dynamic–System Simulation*. McGraw–Hill, New York, 1989.

[10.28] Granino A. Korn. *Neural–Network Experiments on Personal Computers*. MIT Press, Cambridge, Mass., 1991.

[10.29] Matthias Krebs. Modeling of Conditional Index Changes. Master's thesis, Dept. of Electrical & Computer Engineering, University of Arizona, Tucson, Ariz., 1997.

[10.30] John Laffitte and Robert M. Howe. Interfacing Fast and Slow Subsystems in the Real–time Simulation of Dynamic Systems. *Transactions of SCS*, 14(3):115–126, 1997.

[10.31] Nicolas Léchevin, Camille Alain Rabbath, and Paul Baracos. Distributed Real–time Simulation of Power Systems Using Off–the–shelf Software. *IEEE Canadian Review*, pages 5–8, 2001. summer edition.

[10.32] Kuo-Chin Lin and Robert M. Howe. Simulation Using Staggered Integration Steps — Part I: Intermediate–Step Predictor Methods. *Transactions of the Society for Computer Simulation*, 10(3):153–164, 1993.

[10.33] Kuo-Chin Lin and Robert M. Howe. Simulation Using Staggered Integration Steps — Part II: Implementation on Dual–Speed Systems. *Transactions of the Society for Computer Simulation*, 10(4):285–297, 1993.

[10.34] Francisco Mugica and François E. Cellier. Automated Synthesis of a Fuzzy Controller for Cargo Ship Steering by Means of Qualitative Simulation. In *Proceedings ESM'94, European Simulation MultiConference*, Barcelona, Spain, 1994.

[10.35] Martin Otter, Hilding Elmqvist, and François E. Cellier. 'Relaxing' – A Symbolic Sparse Matrix Method Exploiting the Model Structure in Generating Efficient Simulation Code. In *Proceedings Symposium on Modeling, Analysis, and Simulation, CESA'96, IMACS Multi-Conference on Computational Engineering in Systems Applications*, volume 1, pages 1–12, Lille, France, 1996.

[10.36] Olgierd A. Palusinski. Simulation of Dynamic Systems Using Multirate Integration Techniques. *Transactions of SCS*, 2(4):257–273, 1986.

[10.37] José Ignacio Rodríguez, José Manuel Jiménez, Francisco Javier Fu-
nes, and Javier Garcia de Jalón. Dynamic Simulation of Multi-Body
Systems on Internet Using CORBA, Java and XML. *Multibody System
Dynamics*, 10(2):177–199, 2003.

[10.38] Anton Schiela and Folkmar Bornemann. Sparsing in Real Time
Simulation. *ZAMM, Zeitschrift für angewandte Mathematik und
Mechanik*, 83(10):637–647, 2003.

[10.39] Anton Schiela and Hans Olsson. Mixed-mode Integration for Real-
time Simulation. In *Modelica Workshop 2000 Proceedings*, pages 69–
75, Lund, Sweden, 2000.

[10.40] Anton Schiela. Sparsing in Real Time Simulation. Diploma
Project, Technische Universität München, 2002. 75p.

[10.41] Michael C. Schweisguth and François E. Cellier. A Bond Graph
Model of the Bipolar Junction Transistor. In *Proceedings SCS* 4th
International Conference on Bond Graph Modeling and Simulation,
pages 344–349, San Francisco, California, 1999.

[10.42] Siddhartha Shome. *Dual–rate Integration Using Partitioned
Runge–Kutta Methods for Mechanical Systems With Interacting Subsystems*. PhD thesis, The University of Iowa, 2000.

[10.43] Jon M. Smith. *Mathematical Modeling and Digital Simulation for
Engineers and Scientists*. John Wiley & Sons, New York, second edition, 1987.

[10.44] Simon J.E. Taylor, Jon Saville, and Rajeev Sudra. Developing
Interest Management Techniques in Distributed Interactive Simulation
Using Java. In *Proceedings WSC'99, Winter Simulation Conference*,
pages 518–523, 1999.

[10.45] Pentti J. Vesanterä and François E. Cellier. Building Intelligence
into an Autopilot – Using Qualitative Simulation to Support Global
Decision Making. *Simulation*, 52(3):111–121, 1989.

[10.46] Jörg Wensch, Karl Strehmel, and Rüdiger Weiner. A Class of
Linearly–Implicit Runge–Kutta Methods for Multibody Systems. *Applied Numerical Mathematics*, 22(1–3):381–398, 1996.

[10.47] Bernard P. Zeigler, Larry C. Schooley, François E. Cellier, and
FeiYue Wang. High–Autonomy Control of Space Resource Processing
Plants. *IEEE Control Systems*, 13(3):29–39, 1993.

10.13   Bibliography

[B10.1] Hilding Elmqvist, Sven Erik Mattsson, Hans Olsson, Johan Andreasson, Martin Otter, Christian Schweiger, and Brück, Dag . Realtime Simulation of Detailed Automotive Models. In *Proceedings 2003 Modelica Conference*, Linköping, Sweden, 2003.

[B10.2] José Manuel Jiménez Bascones. *Formulaciones cinemáticas y dinámicas para la simulación en tiempo real de sistemas de sólidos rígidos*. PhD thesis, Universidad de Navarra, San Sebastián, Spain, 1993.

[B10.3] Sean Murphy, Jonathan Labin, and Robert Lutz. Experiences Using the Six Services of the IEEE 1516.1 Specification: A 1516 Tutorial. In *Proceedings SIW'04, Spring Simulation Interoperability Workshop*, paper 04S–SIW–056, 2004.

[B10.4] Shinichi Soejima and Takashi Matsuba. Application of Mixed Mode Integration and New Implicit Inline Integration at Toyota. In *Proceedings 2002 Modelica Conference*, Oberpfaffenhofen, Germany, 2002.

10.14   Homework Problems

[H10.1] Semi–implicit Trapezoidal Rule

Derive a semi–implicit version of the trapezoidal rule

$$\mathbf{x}(t_{k+1}) = \mathbf{x}(t_k) + \frac{h}{2}[\mathbf{f}(\mathbf{x}(t_k), t_k) + \mathbf{f}(\mathbf{x}(t_{k+1}), t_{k+1})] \qquad \text{(H10.1a)}$$

Hint: Approximate $\mathbf{f}(\mathbf{x}(t_{k+1}), t_{k+1})$ using the ideas developed in Section 10.4.

Show that the stability domain of the method coincides with that of the fully–implicit trapezoidal rule (i.e., show that also the semi–implicit trapezoidal rule is F–stable).

[H10.2] Pendulum

Using the semi–implicit trapezoidal formula, simulate a pendulum motion that can be described by the state–space model:

$$\dot{x}_1 = x_2$$
$$\dot{x}_2 = -\sin(x_1) - b \cdot x_2$$

assuming that the friction parameter is $b = 0.02$.

Start from the initial condition $x_0 = (0.5 \ 0.5)^T$ using a step size of $h = 0.5$, and simulate until $t_f = 500$ time units.

Repeat the experiment using the fully–implicit trapezoidal rule.

Compare the results as well as the number of floating–point operations required for the two simulations.

[H10.3] Frictionless Pendulum

Repeat problem H10.2 with $b = 0$ (i.e., without friction).

Compare the results obtained with the two integration methods. Explain the differences.

[H10.4] Hydraulic Motor

We wish to simulate a position control system involving a hydraulic motor. Figure H10.4a shows the schematic of a hydraulic motor with two chambers and a set of flows.

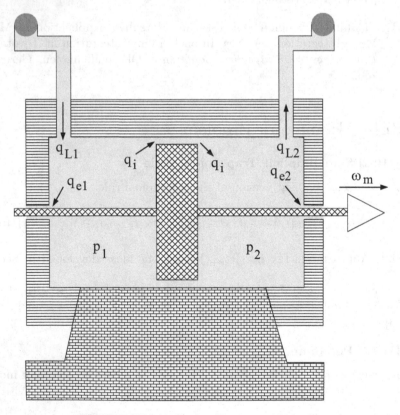

FIGURE H10.4a. Hydraulic motor schematic.

The physics behind the hydraulic motor model are explained in the companion book [10.4].

Due to the compressibility of the fluid, the change in the pressures of the two chambers is proportional to the flow balance in and out of these chambers:

$$\dot{p}_1 = c_1 \cdot (q_{L1} - q_i - q_{e1} - q_{ind}) \qquad \text{(H10.4a)}$$

$$\dot{p}_2 = c_1 \cdot (q_{ind} + q_i - q_{e2} - q_{L2}) \qquad \text{(H10.4b)}$$

where $c_1 = 5.857 \times 10^{13}$ kg m^{-4} sec^{-2} is the inverse of the compressibility constant.

There exist several laminar leakage flows in this model. The flow q_i is the internal leakage flow between the two chambers:

$$q_i = c_i \cdot p_L = c_i \cdot (p_1 - p_2) \qquad \text{(H10.4c)}$$

where p_L is the load pressure of the motor, and $c_i = 0.737 \times 10^{-13}$ kg^{-1} m^4 sec is the internal leakage coefficient. The flows q_{e1} and q_{e2} are external leakage flows:

$$q_{e1} = c_e \cdot (p_1 - p_0) \qquad \text{(H10.4d)}$$

$$q_{e2} = c_e \cdot (p_2 - p_0) \qquad \text{(H10.4e)}$$

where $p_0 = 1.0132 \times 10^5$ N m^{-2} is the ambient air pressure, and $c_e = 0.737 \times 10^{-12}$ kg^{-1} m^4 sec is the external leakage coefficient.

The *load pressure*, p_L, causes a mechanical torque, τ_m, on the motor block, which makes the motor spin, ω_m, and move forward. Thereby an *induced flow*, q_{ind}, is generated. In the process, some of the hydraulic power, $p_L \cdot q_{ind}$, is converted into mechanical power, $\tau_m \cdot \omega_m$. The equations of transformation can be written as:

$$\tau_m = \psi_m \cdot p_L \qquad \text{(H10.4f)}$$

$$q_{ind} = \psi_m \cdot \omega_m \qquad \text{(H10.4g)}$$

where $\psi_m = 0.575 \times 10^{-5}$ m^3.

On the mechanical side, the motor experiences inertia and friction. Newton's law can be formulated as follows:

$$J_m \cdot \dot{\omega}_m = \tau_m - \rho_m \cdot \omega_m \qquad \text{(H10.4h)}$$

where $J_m = 0.08$ kg m^2 is the inertia of the motor block, and $\rho_m = 1.5$ kg m^2 sec^{-1} is the friction constant of the motor.

The load flows, q_{L1} and q_{L2}, in and out of the hydraulic motor are controlled by the four–way servo valve shown in Fig. H10.4b.

The tongue position, x, is normalized such that, $x = 1$ corresponds to the orifices of the valve being entirely open. In the central position, $x = 0$, all four orifices are 5% open, i.e., the servo valve has an underlap of $x_0 = 0.05$.

The flows through the orifices are turbulent. Consequently, they observe a square–root law, as shown in Fig. H10.4c.

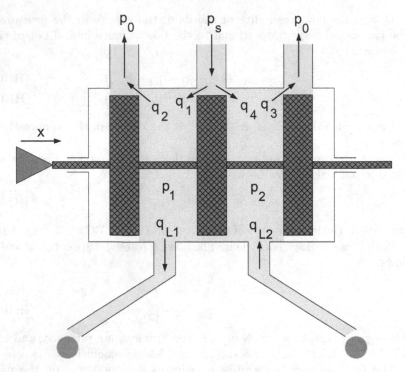

FIGURE H10.4b. Four–way servo valve schematic.

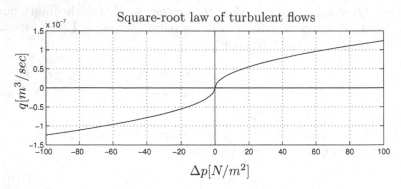

FIGURE H10.4c. Square–root law of turbulent flows of a liquid substance through a narrow orifice.

Thus, each of the four turbulent flows, $q_1 \ldots q_4$, satisfies an equation of the type:

$$q = k \cdot \Delta x \cdot \text{sign}(\Delta p) \cdot \sqrt{|\Delta p|} \qquad (H10.4i)$$

where $k = 0.248 \times 10^{-6}$ kg$^{-1/2}$ m$^{7/2}$, Δx is the relative opening of the orifice, i.e. $\Delta x = x_0 \pm x$ limited between zero and one, and Δp is the pressure drop across the orifice. $p_s = 0.137 \times 10^8$ N m^{-2} is the *line pressure* of the hydraulic motor.

¿From Fig. H10.4b, we conclude that:

$$q_{L1} = q_1 - q_2 \tag{H10.4j}$$

$$q_{L2} = q_3 - q_4 \tag{H10.4k}$$

The tongue of the valve is moved by the servo, an electro–mechanical device, depicted in Fig. H10.4d.

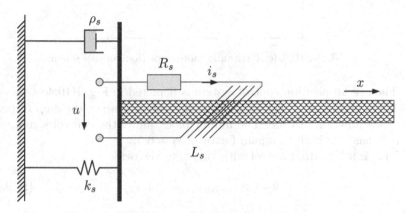

FIGURE H10.4d. Servo schematic.

On the electrical side, the applied voltage, u, causes a current, i_s to flow through a coil. The coil exhibits a resistance, R_s, and an inductance, L_s. Thus:

$$u = R_s \cdot i_s + L_s \cdot \frac{di_s}{dt} + u_{ind} \tag{H10.4l}$$

where $R_s = 1.25 \times 10^{-5}$ Ω, and $L_s = 10^{-9}$ H are the normalized resistance and inductance of the coil. The current i_s causes a force, F_s, in the tongue, which makes the tongue move. The velocity of the tongue, v, causes an induced voltage, u_{ind} on the electrical side. In the process, some of the electrical power, $u_{ind} \cdot i_s$, is converted to mechanical power, $F_s \cdot v$. The equations of transformation are:

$$F_s = \psi_s \cdot i_s \tag{H10.4m}$$

$$u_{ind} = \psi_s \cdot v \tag{H10.4n}$$

where $\psi_s = 0.005$ Volts m^{-1} sec $= 0.005$ N Amps^{-1}.

On the mechanical side, the motion of the tongue is opposed by a spring and a damper. Thus, Newton's law can be formulated as follows:

$$m_s \cdot \dot{v} = F_s - k_s \cdot x - \rho_s \cdot v \tag{H10.4o}$$

where $m_s = 0.01$ kg is the normalized mass, $k_s = 400$ N m^{-1} is the normalized spring constant, and $\rho_s = 2$ N m^{-1} sec is the normalized damper coefficient.

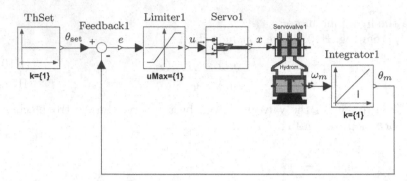

FIGURE H10.4e. Hydraulic motor position control scheme.

The overall position control system is depicted in Fig. H10.4e.

We apply a step of $\theta_{\text{set}} = 1$ rad, and want to observe the step response θ_m as a function of time. The limiter block limits the control signal, u to ± 1. It has also built in a gain factor of $k_l = 0.5$.

This is a 7th–order model with the state vector:

$$\mathbf{x} = (\theta_m, \omega_m, p_1, p_2, x, v, i_s)^T \qquad \text{(H10.4p)}$$

We simulate the model until $t_f = 0.5$ sec. You can set the initial values of all state variables equal to zero, except for the two pressures, p_1 and p_2, which you should set initially both equal to the arithmetic mean value between the line pressure, p_s, and the ambient air pressure, p_0.

As we wish to prepare this model for real–time simulation, we choose to simulate the model using a fixed–step FE algorithm.

Determine experimentally the largest step size, $h_{\max}$, for which the simulation remains numerically stable. Reduce the step size until the solution is sufficiently accurate. As a criterion for accuracy, we shall compare solutions $\theta_m(t)$ found once with the step size h and once with the step size $h/2$. When the two solutions no longer vary by more than 0.1%:

$$err = \max_{\forall t}(|\theta_m(t)_{[h]} - \theta_m(t)_{[h/2]}|) \leq 0.001 \qquad \text{(H10.4q)}$$

we consider the solution sufficiently accurate. Find h_{acc}, the step size that produces an accurate solution, and plot the output variable, θ_m as a function of time.

[H10.5] Algebraic Differentiation

For the hydraulic motor of Hw.[H10.4], compute symbolically the Jacobian of the model. Determine its eigenvalues for the initial state, and explain, on the basis of these eigenvalues, the largest step size, $h_{\max}$ found experimentally in Hw.[H10.4].

What can you conclude about the eigenvalue distribution?

[H10.6] Multi–rate Integration

We shall once more consider the hydraulic motor problem of Hw.[H10.4]. We noticed in Hw.[H10.5] that the electrical time constant of the servo valve is faster than the mechanical and hydraulic time constants by several orders of magnitude.

If the step size is indeed dictated by accuracy considerations, we may simply be out of luck. Yet, we may not really require an accuracy of 0.1%. Let us assume that an accuracy of 1% is acceptable. In that case, the step size is limited by the numerical stability rather than by accuracy considerations.

We now wish to implement a multi–rate integration scheme. We keep the step size of the electrical time constant at the maximum level determined earlier, $h_{\max}$, and we increase the step sizes of the other six integrators by making them multiples of $h_{\max}$.

We shall use the single–rate simulation as a reference solution. Determine experimentally, how many time less frequently you may sample the other six integrators, until the multi–rate solution starts differing from the reference solution by more than 1%.

[H10.7] Mixed–mode Integration

We wish to look at the hydraulic motor problem of Hw.[H10.4] once again. This time, we shall replace the FE algorithm of the electrical inductor by the semi–implicit version of the BE algorithm.

Using the same technique proposed earlier to compare the solution computed for step size h with that computed for step size $h/2$, determine the largest step size, h_{acc}, using a mixed–mode integration approach that will offer an accuracy of 1%.

[H10.8] Deep–sea Oil Drilling

We wish to study a deep–sea oil drilling operation. Figure H10.8a shows a deep–sea oil drilling platform with a pipe hanging from it.

The problem was taken from Eitelberg's Ph.D. dissertation [10.14]. The pipe has a length of $\ell = 5$ km. It has a diameter of $\phi = 0.5$ m. The pipe experiences forces from the sea. The inputs, $u(z)$, represent the forces per meter of exposed pipe at a given depth, z.

In accordance with [10.14], the displacement of the pipe, $x(t, z)$, can be modeled as follows:

$$
\begin{aligned}
\frac{\partial^2 x}{\partial t^2} =& 2 \cdot \frac{\mu_F \cdot v_F}{\mu} \cdot \frac{\partial^2 x}{\partial z \partial t} - \frac{\beta_1}{\mu} \cdot \frac{\partial x}{\partial t} - \frac{\alpha}{\mu} \cdot \frac{\partial^4 x}{\partial z^4} + \left(\gamma(z) - \frac{\mu_f \cdot v_F^2}{2\mu} \right) \cdot \frac{\partial^2 x}{\partial z^2} \\
& - \frac{\bar{\mu}_R \cdot g}{\mu} \cdot \frac{\partial x}{\partial z} + \frac{1}{\mu} \cdot u(z)
\end{aligned}
\tag{H10.8a}
$$

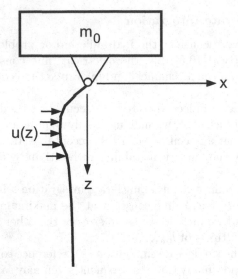

FIGURE H10.8a. Deep–sea oil drilling platform with pipe.

with the abbreviations:

$$\gamma(z) = \frac{g}{\mu} \cdot (m_L - \bar{\mu}_R \cdot (\ell - z)) \qquad \text{(H10.8b)}$$

$$\mu = \mu_R + \mu_F \qquad \text{(H10.8c)}$$

where $\mu_R = 173$ kg m^{-1} is the specific mass of the pipe, $\bar{\mu}_R = 150$ kg m^{-1} is a reduced specific mass of the pipe, $\mu_F = 180$ kg m^{-1} is the specific mass of the oil in the pipe, $\alpha = 142 \times 10^6$ kg m^3 sec^{-2} is the torsion stiffness, $\beta_1 = 20$ kg m^{-1} sec^{-1} is the damping coefficient, $v_F = 5$ m sec^{-1} is the velocity of the oil in the pipe, and $m_L = 10^4$ kg is the mass of the weight at the lower end of the pipe.

The boundary conditions can be specified as follows:

$$x\bigg|_{z=0} = 0 \qquad \text{(H10.8d)}$$

$$\frac{\partial^2 x}{\partial z^2}\bigg|_{z=0} = 0 \qquad \text{(H10.8e)}$$

$$\frac{\partial^2 x}{\partial z^2}\bigg|_{z=\ell} = 0 \qquad \text{(H10.8f)}$$

$$\frac{\partial^2 x}{\partial t^2}\bigg|_{z=\ell} = \frac{\alpha}{m_L} \cdot \frac{\partial^3 x}{\partial z^3} - g \cdot \frac{\partial x}{\partial z} + \frac{1}{m_L} \cdot u_L \qquad \text{(H10.8g)}$$

where u_L is the force tugging at the weight.

We shall convert this hyperbolic partial differential equation to a set of ordinary differential equations using the Method–of–Lines approach. To this end, we shall cut the pipe into 50 segments of $\Delta x = 100$ m length each. Thus, we end up with 100 first–order differential equations.

We shall use 4th–order accurate central differences for the first spatial derivatives, 5th–order accurate central differences for the second spatial derivatives, and 5th–order accurate central differences for the fourth spatial derivatives. Towards the boundary, we shall need to use 4th–order accurate biased formulae for the spatial derivatives.

For the second boundary condition at the lower end, we shall use 4th–order accurate biased formulae for the first and third spatial derivatives.

Why did we choose such high–order formulae? We really didn't have any choice. The fourth spatial derivative cannot be discretized, unless we use at least a formula that is 4th–order accurate. Since we use central differences, we get one additional order of accuracy for free. Since we have to use high–order for the discretization of the fourth spatial derivative, we might just as well do the same with the lower–order spatial derivatives, as we get these approximations for free.

Find the Jacobian of the resulting ODE system, and plot its eigenvalues in the complex λ–plane.

[H10.9] Inline Integration

We shall once more consider the oil–drilling operation of Hw.[H10.8]. Since we now know that the eigenvalues are spread up and down, a little to the left of the imaginary axis, we choose the F–stable trapezoidal rule for integration.

Inline the trapezoidal rule, and determine a suitable set of tearing variables. How many tearing variables do you need?

[H10.10] Inline Integration

We shall once more consider the oil–drilling operation of Hw.[H10.8]. this time around, rather than simulating the model using the trapezoidal rule of Hw.[H10.9], we shall inline the Newmark integration algorithm, introduced in Chapter 5 of this book:

$$\mathbf{v_{k+1}} = \mathbf{v_k} + h \cdot [(1 - \gamma) \cdot \mathbf{a_k} + \gamma \cdot \mathbf{a_{k+1}}] \tag{H10.10a}$$

$$\mathbf{x_{k+1}} = \mathbf{x_k} + h \cdot \mathbf{v_k} + \frac{h^2}{2} \cdot [(1 - 2\beta) \cdot \mathbf{a_k} + 2\beta \cdot \mathbf{a_{k+1}}] \tag{H10.10b}$$

where $\mathbf{x_k}$, $\mathbf{v_k}$, and $\mathbf{a_k}$ are approximations to the positions, velocities, and accelerations at time step k. We shall implement the method with the parameter values $\beta = 1/4$, and $\gamma = 1/2$.

How many tearing variables do you need now?

10.15 Projects

[P10.1] Helicopter Control

According to Kailath [10.25], the flight of a helicopter near hover conditions
can be described by the linear model:

$$\dot{\mathbf{x}} = \begin{pmatrix} -0.02 & -1.4 & 9.8 \\ -0.01 & -0.4 & 0 \\ 0 & 1 & 0 \end{pmatrix} \cdot \mathbf{x} + \begin{pmatrix} 0.9 \\ 6.3 \\ 0 \end{pmatrix} \cdot u$$

$$\mathbf{y} = \begin{pmatrix} 1 & 0 & 0 \end{pmatrix} \cdot \mathbf{x} \tag{P10.1a}$$

where x_1 is the horizontal velocity, x_2 is the pitch rate, and x_3 is the pitch
angle. The input, u, is the rotor tilt angle. The eigenvalues of the helicopter
model are located at $\lambda_1 = -0.6565$, and $\lambda_{2,3} = 0.1183 \pm 0.3678 \cdot j$. Thus,
the uncontrolled helicopter is unstable.

   We wish to design a stabilizing controller using output feedback and a
full–order Luenberger observer [10.25]. The control structure is shown in
Fig. P10.1a.

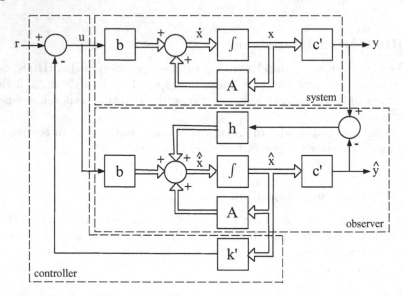

FIGURE P10.1a. Output feedback with full–order Luenberger observer.

   The controller works as follows. We can stabilize the helicopter easily by
a full state feedback. We can place the poles of the controlled helicopter at
$\lambda_{1,2} = -1 \pm j$ and $\lambda_3 = -2$ by multiplying the state vector $\mathbf{x}$ from the left
with the vector $\mathbf{k}' = (0.0627, 0.4706, 1)$. The controller gains can be easily
computed using any one of a number of *pole placement* algorithms [10.25].

   Unfortunately, we don't have the full state available, as only the output
variable, y, is being measured. Thus, we run a model of the helicopter in

parallel, producing the output, $\hat{y}$. We can obtain a stabilization of the observing model by subtracting the observed output, $\hat{y}$, from the true output, y, and then multiplying that error signal with a vector $\mathbf{h}$, which generates a set of signals that are then fed back to the summing point of the observer model. We can place the observer poles at $\lambda_{1,2} = -2 \pm 2 \cdot j$ and $\lambda_3 = -4$ by choosing $\mathbf{h} = (7.58, 2.2695, 2.4644)^T$.

Since we don't have a helicopter to play around with, we shall create a model of the helicopter that we simulate using a fixed–step RK4 algorithm. We shall implement that model on a computer reading the input signal, u, from an input port, and putting out the output, y, through an output port.

We implement the controller including the observer model on a second computer using an inlined version of the BE algorithm for the observer states. This model reads in the output, y, as input from an input port, and puts out the output, u, through an output port. The second input, r, is implemented in the form of an asynchronous time event, i.e., r remains constant until the user of the system decides to set a new value.

Both computers operate on the same real–time clock. The step size is set such that neither computer experiences overruns.

Real control engineers would go two steps further. They would reduce the order of the observer as much as possible. In the given system, we could get away with a second–order observer. They would then convert the resulting analog controller to an equivalent digital controller designed in the z–domain. We shall not do any of this, as this book is about simulation, and not about control.

We wish to implement the real–time simulation in MATLAB using the HLA architecture. To this end, several hurdles will have to be taken first. The real–time input and output ports are implemented using MATLAB's *Instrument Control Toolbox*, which can communicate with the outside world using a number of different protocols, including TCP/IP.

Yet, this won't solve all of our problems yet. Since MATLAB runs on general–purpose operating systems, such as MS/Windows or Linux, the execution speed of a MATLAB code cannot be guaranteed. None of these operating systems were designed for real–time applications.

However, MATLAB programs (M–files) can be translated to real–time executable C–code using MathWork's *Real-Time Workshop*. This code can then be ported over to a dedicated real–time system using MathWork's *xPC Target* software.

You will need a third computer to implement an HLA kernel that implements the basic RTI functionality and that can communicate with MATLAB's Instrument Control Toolbox.

10.16   Research

[R10.1] Real–time Simulation of Hyperbolic PDEs

We have looked in Hw.[H10.8] at a hyperbolic differential equation that we wished to simulate in real time. It didn't look good at all.

In Chapter 6 of this book, we have learnt that many researchers prefer explicit ODE solvers for dealing with hyperbolic PDEs. However, the FE algorithm won't cut the pie, because it will take incredibly small time steps to capture the eigenvalues close to the imaginary axis inside the stability region of the algorithm. Thus, we would need to use either an AB3 or an RK4 algorithm, which may still be the cheapest solution to the problem.

Multi–rate integration is out of the question, because all of the eigenvalues have similar real parts. For the same reason, also the mixed–mode integration won't work.

We tried inline integration of implicit F–stable algorithms instead, but weren't exactly successful with this approach either. The problem is that we need large numbers of tearing variables, i.e., the Hessian matrix in the Newton iteration is still unacceptably large.

This leads to an open research question: Can integration algorithms be found that are either F–stable of stiffly stable that would allow us to get away with a much smaller number of tearing variables?

Each explicit integrator breaks some loops, thereby reducing the size of the remaining Hessian matrix. Can we selectively turn some of the integrators into explicit integrators, while preserving the overall F–stable or stiffly–stable nature of the algorithm? Could we, for example, find an algorithm that is F–stable or stiffly–stable that integrates all of the velocities, $\mathbf{v_k}$, using an explicit algorithm, while all positions, $\mathbf{x_k}$, are being integrated using an implicit algorithm?

Eitelberg, in his dissertation, went a different route. First, he used non–equidistantly spaced intervals in his discretization scheme, making the intervals more narrow, where the pipe is bent the most, in order to reduce the number of segments needed for a sufficiently accurate representation of the pipe. He then used fifth–order accurate splines for the interpolation. In this way, Eitelberg ended up with a 40th–order model instead of a 100th–order model. Second, Eitelberg then studied systematic *model reduction* algorithms to find different models of lower orders that would still represent the most interesting solution, $x(z = \ell)$, accurately enough. In this way, he was able to reduce the order of the pipe model for control purposes down from forty to eight.

11

Discrete Event Simulation

Preview

This chapter explores a new way of approximating differential equations, replacing the time discretization by a quantization of the state variables. We shall see that this idea will lead us to discrete event systems in terms of the DEVS formalism instead of difference equations, as in the previous approximations.

Thus, before formulating the numerical methods derived from this approach, we shall introduce the basic definitions of DEVS. This methodology, as a general discrete event systems modeling and simulation formalism, will provide us the tools to describe and translate into computer programs the routines that implement a new family of methods for the numerical integration of continuous systems.

Further, the chapter explores the principles of quantization–based approximations of ordinary differential equations and their representation as DEVS simulation models.

Finally, we shall briefly introduce the QSS method in preparation for the next chapter, where we shall study this numerical method in more detail.

11.1 Introduction

In previous chapters, we studied many different methods for the simulation of continuous systems. In spite of their differences: explicit *vs.* implicit methods, fixed–step *vs.* variable–step, fixed–order *vs.* variable–order, all of these algorithms had something fundamental in common: given time t_{k+1}, a polynomial extrapolation is performed for the purpose of determining the values of all state variables at that time instant.

In this chapter, we shall pursue an entirely different idea. Rather than asking ourselves, what value a particular state variable assumes at any given point in time, we shall ask the question, at what time the state variable will deviate from its current value by more than ΔQ. Hence we wish to find the smallest time step, h, such that $x(t_k + h) = x(t_k) \pm \Delta Q$.

Evidently, such an integration algorithm will naturally be a variable–step method. During time periods, when the state variable changes its value slowly, the algorithm will compute using large step sizes, whereas it will use small step sizes, whenever the state variable exhibits a large either positive or negative gradient.

It should be remarked that, when **x** is a state vector, the resulting value of h will be different for each component of **x**. Then, we have two possibilities: we can either choose the smallest of these values as the next central step size, h, or we can use different values of h_i for different components of the state vector, x_i, leading to an *asynchronous simulation*, in which each state variable possesses its own simulation time.

The former of these two alternatives can be combined with any of the previously introduced integration algorithms. It simply represents a novel way of performing step–size control. It is an interesting concept, but shall not be pursued further in this chapter, as it doesn't really introduce any new challenges.

The latter idea looks much more revolutionary. At first glance, the resulting methods would consist in a sort of combination of multi–rate and variable–step algorithms.

Up to this point, all of the methods we saw can be described by *difference equations*. Such a representation makes no sense in a method, in which each component evolves following its own values of h_i.

A first consequence of this remark has to do with linearity. Given a linear time–invariant system:

$$\dot{\mathbf{x}} = \mathbf{A} \cdot \mathbf{x} \tag{11.1}$$

numerical integration using any of the previously introduced integration methods leads to a linear difference equation:

$$\mathbf{x_{k+1}} = \mathbf{F} \cdot \mathbf{x_k} \tag{11.2}$$

If we allow each component to follow its own step size, we not only lose the representation as a difference equation, but we also sacrifice linearity. Consequently, we can no longer hope to be able to draw a stability domain, as we have gotten accustomed to throughout this book.

¿From a system–theoretic point of view, we can say that all of the methods that we have looked at until now discretize time. In other words, the resulting simulation codes (i.e., the models executed by the simulation program) are always *discrete–time systems*. When we talk about discrete–time systems, we refer to systems that change their states synchronously in time.

Our proposed approach produces algorithms that are entirely different, as they operate in a completely asynchronous fashion. New problems arise that shall have to be dealt with. We shall need to discuss both *stability* and *accuracy* of these algorithms in a new light, as our previously used techniques break down in the context of these algorithms. Also, we shall need to discuss *synchronization mechanisms*, a problem that we had not encountered so far. As most state equations depend on more than one state variable, the values of which are now known at different time points, we shall need to analyze, how we can synchronize the state variables for the purpose of computing state equations under controlled accuracy conditions.

Yet, these new algorithms not only cause new difficulties. They also offer important simplifications and potential savings.

A first simplification relates to the handling of state events. As we mentioned in Chapter 9, the integration method must evaluate the discontinuous states at event times. Since those event times usually do not coincide with the discrete time instants prescribed by the integration method, we had to modify the method to hit the events with a given accuracy. This requirement implied adding iteration techniques that not only complicated the algorithms but also increased the number of computations per elapsed simulation time. Moreover in the context of real–time simulation, we may not be able to afford those iterations without losing the race against time. We shall learn that the newly proposed algorithms do not call for the iteration of state events, and therefore, may be better suited for the simulation of discontinuous systems, especially in the context of real–time simulation.

Another potential simplification results in the case of large systems of ODEs, as they arise when discretizing hyperbolic partial differential equations using the method–of–lines approach. Hyperbolic PDEs frequently lead to shock waves that travel through space with time. Consequently at any point in time, the gradients of these waves will be steep at some point in space, whereas they are flat in all other regions. Using a synchronous integration algorithm, the step size of all states will have to be adjusted to the steepest gradient, so that small step sizes will have to be used on all differential equations at all times. In contrast, the newly proposed algorithms will enable us to use large step sizes on most state variables most of the time.

11.2   Space Discretization: A Simple Example

Returning to our idea of designing integration methods, in which the steps are ruled by changes in states rather than in time, we shall introduce an example that shows some of the basic principles of these integration techniques.

Consider the first order system:

$$\dot{x}_a(t) = -x_a(t) + 10 \cdot \varepsilon(t - 1.76) \qquad (11.3a)$$

where $\varepsilon(t)$ represents the unit step function, i.e., $\varepsilon(t-1.76)$ describes a unit step taking place at $t = 1.76$.

We shall consider the initial condition

$$x_a(t_0 = 0) = 10 \qquad (11.3b)$$

If we try to simulate this model using a fixed–step method with a step size of $h = 0.1$, which would be appropriated in accordance with the rate, at which the single state variable changes its value, the time step would occur at an instant of time that does not coincide with the discrete time instants prescribed by the integration algorithm.

Let us now consider what happens with the following *continuous–time system*:

$$\dot{x}(t) = -\text{floor}[x(t)] + 10 \cdot \varepsilon(t - 1.76) \tag{11.4a}$$

or:

$$\dot{x}(t) = -q(t) + 10 \cdot \varepsilon(t - 1.76) \tag{11.4b}$$

where $q(t) \triangleq \text{floor}[x(t)]$ denotes the integer part of the positive real–valued variable $x(t)$.

Although the system defined by Eq.(11.4) is nonlinear and does not satisfy the analytical properties that we like (it is highly discontinuous), it can be easily solved.

When $0 < t < 1/9$, we have $q(t) = 9$ and $\dot{x}(t) = -9$. During this interval, $x(t)$ decreases linearly from 10.0 to 9.0. Then, during the interval $1/9 < t < 1/9 + 1/8$, we have $q(t) = 8$ and $\dot{x}(t) = -8$. During this time interval, $x(t)$ decreases linearly from 9.0 to 8.0.

Continuing with this analysis, we find that $x(t)$ reaches a value of 3.0 at time $t = 1.329$. If no time event were to occur, $x(t)$ would reach a value of 2.0 at time $t = 1.829$. However at time $t = 1.76$, when $x = 2.138$, the input changes, and from that moment on, we have $\dot{x}(t) = 8$. The variable $x(t)$ increases its value again linearly with time, until it reaches a value of 3.0 at time $t = 1.8678$.

The real–valued $x(t)$ variable continues to grow, until the system reaches $x(t) = q(t) = 10$, at which time the derivative $\dot{x}(t)$ becomes zero, and the system will not change its state any longer.

Figure 11.1 shows the trajectories of $x(t)$ and $q(t)$.

We completed this simulation using 17 steps and, ignoring the round–off problems, we obtained the *exact solution* of Eq.(11.4). All computations required to obtain this solution were trivial, because the state derivative *remains constant* in between event times, which enabled us to compute the real–valued variable $x(t)$ analytically.

The solution $x(t)$ and the solution of our original system of Eq.(11.3), $x_a(t)$, are compared in Fig.11.2.

The solutions of the original and the quantized system are definitely related to each other. By replacing the state variable $x(t)$ by $q(t) = \text{floor}[x(t)]$ on the right hand side of a first–order differential equation, we found an explicit method to simulate the quantized model.

The question naturally arises, whether we might not be able to generalize this approach to n^{th}–order systems by quantizing all state variables on the right hand side of all state equations. Unfortunately, we are not ready to answer this question yet. To this end, we shall first need to explore the

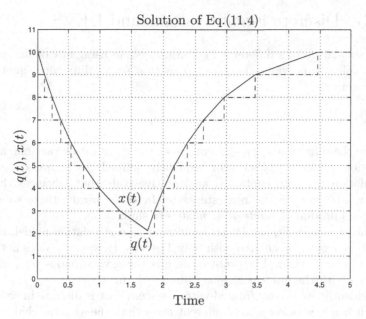

FIGURE 11.1. Variable trajectories of the system of Eq.(11.4).

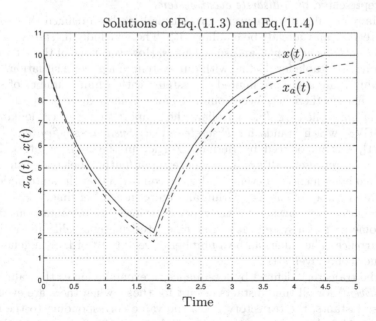

FIGURE 11.2. State trajectories of the systems of Eq.(11.3) and Eq.(11.4).

discrete nature of the system of Eq.(11.4) and introduce some tools for its representation and simulation.

11.3   Discrete Event Systems and DEVS

The simulation of a differential equation system using any of the methods we studied in previous chapters led us to a set of difference equations of the form:

$$\mathbf{x}(t_{k+1}) = \mathbf{f}(\mathbf{x}(t_k), t_k) \qquad (11.5)$$

where the difference $t_{k+1} - t_k$ can be either constant or variable, and the function $\mathbf{f}$ can be explicitly or implicitly defined. As a consequence, the simulation program contained an iterative code that advances the time in accordance with the next step size. In other words, those simulation methods produce *discrete–time models* of simulation.

The system of Eq.(11.4) can be viewed as a simulation model, because it can be exactly simulated with only 17 steps. However, it does not fit the format of Eq.(11.5). The problem here is the asynchronous way, in which it deals with the input change at time $t = 1.76$.

Evidently, we are confronted with a system that is discrete in some way, which however belongs to a different class than the systems characterized as discrete–time systems. As we shall see soon, our new approximation can be represented by a *discrete event system*.

Many popular discrete event formalisms have been defined that some of our readers may already be familiar with. These include the *state automata*, *Petri nets*, *event graphs*, and *state charts*. However, none of these representations is suitable for dealing with our system in a general situation. These graphical languages are limited to systems with a finite number of states. Fortunately, there has been found a more general discrete event system formalism, called DEVS, that offers the support that we were looking for.

DEVS, which stands for *Discrete EVent System specification* [11.14, 11.11], was introduced by Bernard Zeigler in the mid seventies. DEVS allows to represent all systems, whose input/output behavior can be described by sequences of events under the condition that the state undergoes a finite number of changes within any finite interval of time.

In our context, an *event* is the representation of an instantaneous change in some part of a system. It can be characterized by a value and a time of occurrence. The value can be a number, a vector, a word, or in general, an element of a given set.

The trajectory defined by a sequence of events assumes the value ϕ (or *No Event*) for all time instants except for those, when there are events. In those instants, the trajectory takes the value corresponding to the event. Figure 11.3 shows an event trajectory that takes the value x_2 at time t_1, then the value x_3 at time t_2, etc.

A DEVS model processes an input event trajectory and, according to that trajectory and its own initial conditions, provokes an output event trajectory. This input/output behavior is depicted in Fig.11.4.

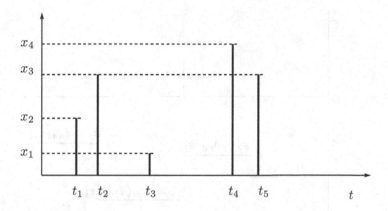

FIGURE 11.3. An event trajectory.

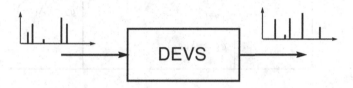

FIGURE 11.4. Input/output behavior of a DEVS model.

The behavior of a DEVS model is expressed in a way that is quite common in automata theory. This kind of representation consists in enumerating some sets and functions that define the system dynamics in accordance with certain rules. Since the rules are always the same in a given formalism, they are not mentioned in each model.

Following this idea, an *atomic* DEVS model is defined by the following structure:

$$M = (X, Y, S, \delta_{\text{int}}, \delta_{\text{ext}}, \lambda, ta)$$

where:

- X is the set of input event values, i.e., the set of all possible values that an input event can assume;

- Y is the set of output event values;

- S is the set of state values;

- $\delta_{\text{int}}, \delta_{\text{ext}}, \lambda$ and ta are functions that define the system dynamics.

Figure 11.5 illustrates the behavior of a DEVS model.

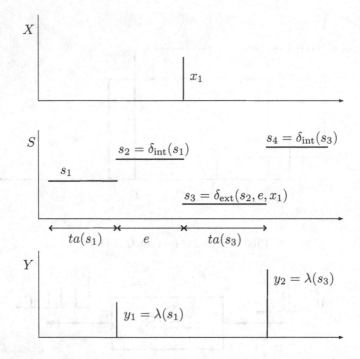

FIGURE 11.5. Trajectories in a DEVS model.

Each possible state s ($s \in S$) has an associated *time advance* calculated by the *time advance function $ta(s)$* ($ta(s) : S \to \mathbb{R}_0^+$). The time advance is a non–negative real number, determining how long the system remains in a given state in absence of input events.

Thus, if the state adopts the value s_1 at time t_1, after $ta(s_1)$ units of time (i.e., at time $t_1 + ta(s_1)$), the system performs an *internal transition*, taking it to a new state s_2. The new state is calculated as $s_2 = \delta_{\text{int}}(s_1)$. Function δ_{int} ($\delta_{\text{int}} : S \to S$) is called the *internal transition function*.

When the state changes its value from s_1 to s_2, an output event is produced with the value $y_1 = \lambda(s_1)$. Function λ ($\lambda : S \to Y$) is called the *output function*. In this way, the functions ta, δ_{int} and λ define the autonomous behavior of a DEVS model.

When an input event arrives, the state changes instantaneously. The new state value depends not only on the value of the input event, but also on the previous state value and the elapsed time since the last transition. If the system assumes the state value s_2 at time t_2, and subsequently, an input event arrives at time $t_2 + e < ta(s_2)$ with value x_1, the new state is calculated as $s_3 = \delta_{\text{ext}}(s_2, e, x_1)$. In this case, we say that the system performs an *external transition*. Function δ_{ext} ($\delta_{\text{ext}} : S \times \mathbb{R}_0^+ \times X \to S$) is called the *external transition function*. No output event is produced during an external transition.

Let us consider the following simple example: A system receives positive numbers in an asynchronous way. After it received a number x, it generates an output event with the number $x/2$ after $3 \cdot x$ time units. A DEVS model that correctly represents this behavior is the following:

$$M_1 = (X, Y, S, \delta_{\text{int}}, \delta_{\text{ext}}, \lambda, ta), \text{ where}$$
$$X = Y = S = \mathbb{R}^+$$
$$\delta_{\text{int}}(s) = \infty$$
$$\delta_{\text{ext}}(s, e, x) = x$$
$$\lambda(s) = s/2$$
$$ta(s) = 3 \cdot s$$

Observe that the state can assume a time advance equal to ∞. When this occurs, we say that the system is in a passive state, since it will no longer change its state, unless and until it receives an input event.

Let us analyze what happens with the model M_1 when it receives an input event trajectory. Consider for instance that input events occur at times $t = 1$, $t = 3$, and $t = 10$ with the values 2, 1, and 5, respectively. Suppose that initially we have $t = 0$, $s = \infty$ and $e = 0$. Then, the following behavior would be observed:

time $t = 0$:
    $s = \infty$
    $e = 0$
    $ta(s) = ta(\infty) = \infty$

time $t = 1^-$:
    $s = \infty$
    $e = 1$

time $t = 1$:
    $s = \delta_{\text{ext}}(s, e, x) = \delta_{\text{ext}}(\infty, 1, 2) = 2$

time $t = 1^+$:
    $s = 2$
    $e = 0$
    $ta(s) = ta(2) = 6$

time $t = 3^-$:
    $s = 2$
    $e = 2$

time $t = 3$:
    $s = \delta_{\text{ext}}(s, e, x) = \delta_{\text{ext}}(2, 2, 1) = 1$

time $t = 3^+$:
    $s = 1$
    $e = 0$

$$ta(s) = ta(1) = 3$$

time $t = 6$:
  output event with value $\lambda(s) = \lambda(1) = 0.5$
  $s = \delta_{\text{int}}(s) = \delta_{\text{int}}(1) = \infty$

time $t = 6^+$:
  $s = \infty$
  $e = 0$
  $ta(s) = ta(\infty) = \infty$

time $t = 10^-$:
  $s = \infty$
  $e = 4$

time $t = 10$:
  $s = \delta_{\text{ext}}(s, e, x) = \delta_{\text{ext}}(\infty, 4, 5) = 5$

time $t = 10^+$:
  $s = 5$
  $e = 0$
  $ta(s) = ta(5) = 15$

time $t = 25$:
  output event with value $\lambda(s) = \lambda(5) = 2.5$
  $s = \delta_{\text{int}}(s) = \delta_{\text{int}}(5) = \infty$

time $t = 25^+$:
  $s = \infty$
  $e = 0$
  $ta(s) = ta(\infty) = \infty$

According to the above model, when a new state arrives through an input event before the previous state has expired, the system assumes the new state value and forgets the previous one. In the above example, this happens at time $t = 3$. A different scenario might require that arriving input events are to be ignored, while the system is busy. This modified behavior can be generated using the following DEVS model:

$$M_2 = (X, Y, S, \delta_{\text{int}}, \delta_{\text{ext}}, \lambda, ta), \text{ where}$$
$$X = Y = \mathbb{R}^+$$
$$S = \mathbb{R}^+ \times \mathbb{R}_0^+$$
$$\delta_{\text{int}}(s) = \delta_{\text{int}}(z, \sigma) = (\infty, \infty)$$
$$\delta_{\text{ext}}(s, e, x) = \delta_{\text{ext}}(z, \sigma, e, x) = \tilde{s}$$
$$\lambda(s) = \lambda(z, \sigma) = z/2$$
$$ta(s) = ta(z, \sigma) = \sigma$$

where:

$$\tilde{s} = \begin{cases} (x, 3 \cdot x) & \text{if } z = \infty \\ (z, \sigma - e) & \text{otherwise} \end{cases}$$

In this new model, we included the variable σ in the state s. People working routinely with DEVS almost always introduce the variable σ, set equal to the time advance, as this generally facilitates the modeling task.

11.4   Coupled DEVS Models

As mentioned before, DEVS is a general formalism that can be used to describe highly complex systems. However, the representation of a complex system based only on transition and time advance functions is rather difficult. The reason is that in those functions we have to imagine and describe all possible situations in the system.

Complex systems can usually be thought of as the coupling of simpler systems. Through the coupling, the output events of some subsystems are converted to input events of other subsystems. The DEVS methodology guarantees that the coupling of atomic DEVS models defines new DEVS models, i.e., DEVS is closed under coupling. For this reason, complex systems can be represented by DEVS in a hierarchical fashion [11.11].

There are two different ways, in which DEVS models may be coupled. The first approach is the most general one. It uses *translation functions* between subsystems. The second approach is based on the use of input and output ports. We shall use the latter approach, since it is simpler and more adequate in the context of continuous system simulation.

The use of ports requires adding to the input and output events a new number, word, or symbol, representing the port, through which the event is arriving. It suffices to enumerate the connections describing the couplings between different systems. An *internal connection* involves an input and an output port belonging to subsystems. However in the context of hierarchical coupling, there also exist connections from the output ports of subsystems to the output ports of the system. These are called *external output connections*. There also exist connections from the input ports of the systems to input ports of subsystems. These are referred to as *external input connections*.

Figure 11.6 shows a coupled DEVS model N that is the result of coupling the models M_a and M_b. There, the output port 1 of M_a is connected to the input port 0 of M_b. This connection can be represented by the pair $[(M_a, 1), (M_b, 0)]$. Other connections are $[(M_b, 0), (M_a, 0)]$, $[(N, 0), (M_a, 0)]$, $[(M_b, 0), (N, 1)]$, etc. According to the closure property of DEVS, the model N can itself be used in exactly the same way, as an atomic DEVS model would be used, and it can be coupled with other atomic and/or coupled models.

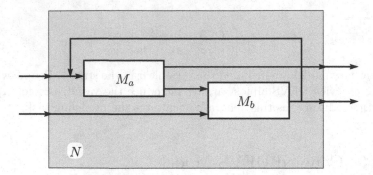

FIGURE 11.6. Coupled DEVS model.

Note that the input and output ports are numbered using integer numbers starting from 0. The DEVS methodology allows using any word to identify a port. However in the context of this book, we shall always use integer numbers starting from 0, because we shall work with a software tool that defines the ports in this fashion [11.7].

Consider for example a system that calculates a static function $f(u_0, u_1)$, where u_0 and u_1 are real–valued piecewise constant trajectories generated by other subsystems. We can represent piecewise constant trajectories by sequences of events, if we relate each event to a change in the trajectory value. Using this idea, we can build the following atomic DEVS model:

$$M_3 = (X, Y, S, \delta_{\text{int}}, \delta_{\text{ext}}, \lambda, ta), \text{ where}$$
$$X = Y = \mathbb{R} \times \mathbb{N}_0$$
$$S = \mathbb{R}^2 \times \mathbb{R}_0^+$$
$$\delta_{\text{int}}(s) = \delta_{\text{int}}(u_0, u_1, \sigma) = (u_0, u_1, \infty)$$
$$\delta_{\text{ext}}(s, e, x) = \delta_{\text{ext}}(u_0, u_1, \sigma, e, x_v, p) = \tilde{s}$$
$$\lambda(s) = \lambda(u_0, u_1, \sigma) = (f(u_0, u_1), 0)$$
$$ta(s) = ta(u_0, u_1, \sigma) = \sigma$$

where:

$$\tilde{s} = \begin{cases} (x_v, u_1, 0) & \text{if } p = 0 \\ (u_0, x_v, 0) & \text{otherwise} \end{cases}$$

Here, each input and output event includes an integer number, indicating the corresponding input or output port. In the input events (cf. the definition of $\tilde{s}$ in function δ_{ext}), the port p can be either 0 or 1. In the output events (cf. function λ), the output port is always 0.

Now, if we want to represent another system that calculates the function $f[f(u_0, u_1), u_2]$, we must couple two models of the M_3 class with a connection from the output port of the first subsystem to the input port

0 of the second subsystem. The system output must be taken from the second model. Thus, calling the subsystems A and B, respectively, and the overall system N, the connections can be expressed as: $[(A, 0), (B, 0)]$, $[(N, 0), (A, 0)]$, $[(N, 1), (A, 1)]$, $[(N, 2), (B, 1)]$, and $[(B, 0), (N, 0)]$.

The DEVS methodology uses a formal structure for representing coupled DEVS models with ports. The structure includes the subsystems, the connections, the system input and output sets, and a *tie–breaking function* to govern the occurrence of *simultaneous events*. The connections are divided into three sets: one set composed by the connections between subsystems (internal connections), another set that contains the connections from the system to the subsystems (external input connections), and a final set that lists the connections from the subsystems to the system (external output connections).

The use of the aforementioned tie–breaking function can be avoided with *Parallel–DEVS*, which is an extension of the DEVS formalism that allows dealing with simultaneous events.

We shall not develop these latter concepts, neither the coupled DEVS formal structure nor the parallel–DEVS formalism, any further, since our aim is not the introduction of the complete DEVS methodology here. We are only interested in using DEVS as a tool for continuous system simulation. For a more complete coverage of DEVS methodology, we refer the reader to Zeigler's book [11.11].

11.5   Simulation of DEVS Models

One of the most important features of DEVS is that very complex models can be simulated in an easy and efficient manner.

DEVS models can be simulated with a simple ad–hoc program written in any language. In fact, the simulation of a DEVS model is not much more complicated than that of a discrete–time model.

A basic algorithm that may be used for the simulation of a coupled DEVS model can be described by the following steps:

(a). Identify the atomic model that, according to its time advance and elapsed time, is the next to perform an internal transition. Call the system d^*, and let t_n be the time of the aforementioned transition.

(b). Advance the simulation clock t to $t = t_n$, and execute the internal transition function of model d^*.

(c). Propagate the output event produced by d^* to all atomic models connected to it through its output ports, while executing the corresponding external transition functions. Then, return to step (a) above.

One of the simplest ways for implementing these steps is by writing a program with a hierarchical structure equivalent to the hierarchical structure of the model to be simulated. This is the method developed in [11.11], where a routine called *DEVS–simulator* is associated with each atomic DEVS model, and a different routine called *DEVS–coordinator* is related to each coupled DEVS model. At the top of the hierarchy, there is a routine called *DEVS–root–coordinator* that manages the global simulation time. Figure 11.7 illustrates this simulation technique for a coupled DEVS model:

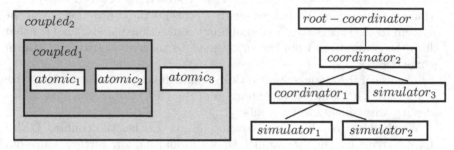

FIGURE 11.7. Hierarchical model and simulation scheme.

The simulators and coordinators of consecutive layers communicate with each other through messages. The coordinators send messages to their children, triggering the execution of their transition functions. When a simulator goes through a transition, it calculates its next state and, when the transition is internal, sends the output value to its parent coordinator. The simulator state coincides with its associated atomic DEVS model state.

When a coordinator executes a transition function, it sends messages to some of its children, triggering the execution of their own transition functions. When an output event produced by one of its children has to be propagated outside the coupled model, the coordinator sends a message to its own parent coordinator, carrying the output value.

Each simulator or coordinator has a local variable t_{n_i}, indicating the time instant, when its next internal transition is scheduled to occur. In a simulator, the value of that variable is calculated using the time advance function of the corresponding atomic model. In a coordinator, it is calculated as the minimum t_{n_i} of its children. Thus, the t_{n_i} of the coordinator at the root of the tree is the time instant, at which the next event of the entire system will occur. The root coordinator is responsible for advancing the global time t to that value.

At the beginning of the simulation, a message of initialization is sent from the root to the leaves of the tree structure.

The following pseudo–code corresponds to a simulator associated with a generic atomic model. There, the i–message, $*$–message, and x–message represent the initialization, internal transition, and input message, respectively. These messages are sent from a parent to its children. Similarly, the

y–message is an output message, sent from a child to its parent.

DEVS–simulator
   variables:
      t_l // time of last event
      t_n // time of next event
      s // state of the DEVS atomic model
      e // elapsed time in the actual state
      $y = (y.value, y.port)$ // current output of the DEVS atomic model
   when i–message (i, t) is received at time t
      $t_l = t - e$
      $t_n = t_l + ta(s)$
   when $*$–message $(*, t)$ is received at time t
      $y = \lambda(s)$
      send y–message (y, t) to parent coordinator
      $s = \delta_{\text{int}}(s)$
      $t_l = t$
      $t_n = t + ta(s)$
   when x–message (x, t) is received at time t
      $e = t - t_l$
      $s = \delta_{\text{ext}}(s, e, x)$
      $t_l = t$
      $t_n = t + ta(s)$
end DEVS–simulator

The routine corresponding to a coordinator can be written as follows:

DEVS–coordinator
   variables:
      t_l // time of last event
      t_n // time of next event
      $y = (y.value, y.port)$ // current output of the DEVS coordinator
      D // list of children
      IC // list of connections of the form $[(d_i, port_x), (d_j, port_y)]$
      EIC // list of connections of the form $[(N, port_x), (d_j, port_y)]$
      EOC // list of connections of the form $[(d_i, port_x), (N, port_y)]$
   when i–message (i, t) is received at time t
      send i–message (i, t) to all children
      $d^* = \arg[\min_{d \in D}(d.t_n)]$
      $t_l = t$
      $t_n = d^*.t_n$
   when $*$–message $(*, t)$ is received at time t
      send $*$–message $(*, t)$ to d^*
      $d^* = \arg[\min_{d \in D}(d.t_n)]$
      $t_l = t$
      $t_n = d^*.t_n$
   when x–message $((x.value, x.port), t)$ is received at time t
      $(v, p) = (x.value, x.port)$
      for each connection $[(N, p), (d, q)]$
         send x–message $((v, q), t)$ to child d
      $d^* = \arg[\min_{d \in D}(d.t_n)]$
      $t_l = t$
      $t_n = d^*.t_n$

```
when y–message ((y.value, y.port),t) is received from d*
    if a connection [(d*, y.port), (N, q)] exists
        send y–message ((y.value, q),t) to parent coordinator
    for each connection [(d*, y.port), (d, q)]
        send x–message ((y.value, q),t) to child d
end DEVS–coordinator
```

Finally, the root coordinator executes the following routine:

```
DEVS–root–coordinator
    variables:
        t // global simulation time
        d // child (coordinator or simulator)
    t = t_0
    send i–message (i,t) to d
    t = d.t_n
    loop
        send *–message (*,t) to d
        t = d.t_n
    until end of simulation
end DEVS–root–coordinator
```

There are many other possibilities for implementing a simulation engine for DEVS models. The main problem with the routines outlined is that, due to their hierarchical structure, we may observe a significant traffic of messages passing from higher to lower layers of the architecture. All of these messages and their corresponding computational time can be avoided if a flat simulation structure is being used. Hierarchical DEVS simulation architectures can be converted to flat DEVS simulation architectures fairly easily [11.4]. In fact, most of the software tools mentioned before implement the simulation based on a flat code.

Although the implementation of the pseudo code shown above is fairly straightforward, practical models are usually composed of many subsystems, and therefore, ad–hoc programming of all of these models may become very time–consuming.

In recent years, a number of different software tools for the simulation of DEVS models have been developed. Some of these tools offer software libraries, graphical user interfaces, and a variety of other facilities that are designed to support the user in the modeling task.

A number of software packages for DEVS simulation have been placed in the public domain, including DEVS–Java [11.13], DEVSim++ [11.5], DEVS–C++ [11.1], CD++ [11.10], and JDEVS [11.2].

In this book, we shall focus on PowerDEVS [11.7, 11.8, 11.9], a software that –in spite of being a general–purpose DEVS simulator– is a software environment that was specifically conceived for facilitating the simulation of continuous systems.

As we already mentioned, this textbook is not geared towards a general course on DEVS, and we do not expect the reader to become an expert on DEVS. Our aim is to provide enough information about DEVS, such that a PowerDEVS user will understand enough of the underlying principles to be able to make use of PowerDEVS as a tool for continuous system simulation.

PowerDEVS is a tool that was designed with many different kinds of users in mind, ranging from mere beginners, who don't know anything about either DEVS or C++ programming, to experts in both domains.

PowerDEVS offers a convenient *graphical user interface* that permits creating coupled DEVS models using the typical drag and drop tools. A number of DEVS atomic model definitions have been predefined and stored in a PowerDEVS model library.

Atomic models can be easily created and modified using an *atomic model editor*, where the user has to define the transition, output, and time advance functions using C++ syntax.

11.6   DEVS and Continuous System Simulation

In the first example of section 11.4, we saw that piecewise constant trajectories can be represented by sequences of events. This simple idea constitutes the basis for the use of DEVS in the simulation of continuous systems.

In that example, we also showed that a DEVS model can represent the behavior of a static function with piecewise constant input trajectories. The problem is that the continuous system trajectories are usually far from being piecewise constant. However, we can alter the system, such that it exhibits these kinds of trajectories. In fact, that is what we did to the system of Eq.(11.3), where we used the floor function to convert it to the system of Eq.(11.4).

We can split Eq.(11.4) in the following way:

$$\dot{x}(t) \;\; = \;\; d_x(t) \tag{11.6a}$$
$$q(t) \;\; = \;\; \text{floor}[x(t)] \tag{11.6b}$$

and:

$$d_x(t) = -q(t) + u(t) \tag{11.7}$$

where $u(t) = 10 \cdot \varepsilon(t - 1.76)$.

We can represent this system using the block diagram of Fig.11.8.

As we mentioned before, the subsystem of Eq.(11.7) –being a static function– can be represented by the DEVS model M_3 presented in Section 11.4.

The subsystem of Eq.(11.6) is a dynamic system having a piecewise constant input trajectory $d_x(t)$ and a piecewise constant output trajectory $q(t)$. We can represent it exactly using the following DEVS model:

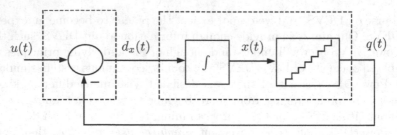

FIGURE 11.8. Block diagram representation of Eqs.(11.6–11.7).

$$M_4 = (X, Y, S, \delta_{\text{int}}, \delta_{\text{ext}}, \lambda, ta), \text{ where}$$
$$X = Y = \mathbb{R} \times \mathbb{N}$$
$$S = \mathbb{R}^2 \times \mathbb{Z} \times \mathbb{R}_0^+$$
$$\delta_{\text{int}}(s) = \delta_{\text{int}}(x, d_x, q, \sigma) = (x + \sigma \cdot d_x, d_x, q + \text{sign}(d_x), \frac{1}{|d_x|})$$
$$\delta_{\text{ext}}(s, e, x) = \delta_{\text{ext}}(x, d_x, q, \sigma, e, x_v, p) = (x + e \cdot d_x, x_v, q, \tilde{\sigma})$$
$$\lambda(s) = \lambda(x, d_x, q, \sigma) = (q + \text{sign}(d_x), 0)$$
$$ta(s) = ta(x, d_x, q, \sigma) = \sigma$$

where:

$$\tilde{\sigma} = \begin{cases} \dfrac{q + 1 - x}{x_v} & \text{if } x_v > 0 \\[2mm] \dfrac{q - x}{x_v} & \text{if } x_v < 0 \\[2mm] \infty & \text{otherwise} \end{cases}$$

Now, if we want to simulate the system of Eqs.(11.6–11.7) using PowerDEVS, we can translate the generic DEVS atomic models, M_3 and M_4, into corresponding PowerDEVS atomic models.

A PowerDEVS atomic model corresponding to M_3 looks, in the atomic model editor, as follows:

ATOMIC MODEL STATIC1
State Variables and Parameters:
    float $u[2]$,$sigma$; //states
    float y; //output
    float inf; //parameters

Init Function:
    $inf = 1\text{e}10$;
    $u[0] = 0$;
    $u[1] = 0$;
    $sigma = inf$;
    $y = 0$;

Time Advance Function:
    return $sigma$;

Internal Transition Function:
    $sigma = inf$;

External Transition Function:
    float xv;
    $xv = $ *(float*)$(x.value)$;
    $u[x.port] = xv$;
    $sigma = 0$;

Output Function:
    $y = u[0] - u[1]$;
    return Event($\&y$,0);

The conversion of the DEVS model M_3 to the corresponding PowerDEVS model STATIC1 is straightforward.

Similarly, the DEVS model M_4 can be represented in PowerDEVS as follows:

ATOMIC MODEL NHINTEGRATOR

State Variables and Parameters:
    float X, dX, q, $sigma$; //states
    //we use capital X because x is reserved
    float y; //output
    float inf; //parameters

Init Function:
    va_list $parameters$;
    va_start($parameters$, t);
    $X = $ va_arg($parameters$, double);
    $dX = 0$;
    $q = $floor$(X)$;
    $inf = 1e10$;
    $sigma = 0$;
    $y = 0$;

Time Advance Function:
    return $sigma$;

Internal Transition Function:
    $X = X + sigma * dX$;
    if $(dX > 0)$ {
        $sigma = 1/dX$;
        $q = q + 1$;
        }
    else {
        if $(dX < 0)$ {
            $sigma = -1/dX$;
            $q = q - 1$;

```
        }
    else {
        sigma = inf;
    };
};
```

External Transition Function:
```
float xv;
xv = *(float*)(x.value);
X = X + dX * e;
if (xv > 0) {
    sigma = (q + 1 - X)/xv;
    }
else {
    if (xv < 0) {
        sigma = (q - X)/xv;
        }
    else {
        sigma = inf;
    };
};
dX = xv;
```

Output Function:
```
if (dX == 0) {y = q;} else {y = q + dX/fabs(dX);};
return Event(&y,0);
```

Again, the translation was fairly direct. However, we added a few new items to the init function. The first two lines are automatically included by the atomic model editor, when a new model is being edited. They declare a variable *parameters*, where the graphical user interface puts the model parameters. In our case, we defined the initial condition in variable X as a parameter. The third line in the init function just takes the first parameter of the block and places it in X.

Then in the graphical user interface, we can just double click on the block and change the value of that parameter (i.e., we can change the initial condition without changing the atomic model definition).

The other change with respect to model M_4 is also related to the inclusion of initial conditions. At the beginning of the simulation, we force the model to provoke an event, so that the initial value of the corresponding quantized variable q becomes known to the rest of the system. We shall write more about this topic in the next chapter.

The coupled PowerDEVS model generated using the graphical model editor is shown in Fig.11.9.

In that model, beside from the atomic models STATIC1 and NHINTE-GRATOR, we included three more blocks: a STEP block that produces an event with value 10 at time $t = 1.76$, and two additional models that save and display the simulation results. These last two models are being

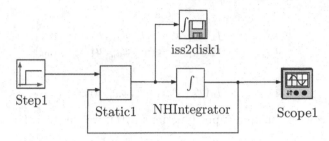

FIGURE 11.9. Coupled PowerDEVS model of Eqs.(11.6–11.7).

included with the standard PowerDEVS library.

Using the PowerDEVS model of Fig.11.9, we obtained the data plotted in Fig.11.1. The plot of that figure was generated by MATLAB, using the data saved in an appropriate format on a file by the PowerDEVS block iss2dsk.

The subsystem of Eq.(11.6) corresponds to the integrator together with the staircase block in the block diagram of Fig.11.8. It is equivalent to DEVS model M_4, represented by the NHINTEGRATOR model in PowerDEVS.

This is, what Zeigler called the *quantized integrator* [11.12, 11.11]. There, the function floor acts as a *quantization function*. A quantization function maps real–valued numbers onto a discrete set of real values.

A system that relates its input and output by any type of quantization function shall henceforth be called a *quantizer*. Thus, our staircase block is a particular case of a quantizer with uniform quantization.

A quantized integrator is an integrator concatenated with a quantizer that may employ either a uniform or a non–uniform quantization scheme.

Similarly, model M_3 models a static function with its input vector in $\mathbb{R}^2$. The corresponding STATIC1 PowerDEVS model implements a particular case of such a static function, namely the function: $f(u_0, u_1) = u_0 - u_1$. A DEVS model for generic static functions with their input vector in $\mathbb{R}^n$ can easily be built and programmed in PowerDEVS (cf. Hw.[H11.4] for the general linear case).

In the same way, we can obtain a DEVS model representation of general quantized integrators that can be employed, whenever their input trajectories are piecewise constant, and it is also evident that we can build a generic DEVS model of an arbitrary static function, as long as its input trajectories are piecewise constant.

Thus, if we have a general time–invariant system[1]:

---

[1]We shall use x_a to denote the state variables of the original system, so that $x_a(t)$ is the analytical solution.

$$\begin{aligned}
\dot{x}_{a_1} &= f_1(x_{a_1}, x_{a_2}, \cdots, x_{a_n}, u_1, \cdots, u_m) \\
\dot{x}_{a_2} &= f_2(x_{a_1}, x_{a_2}, \cdots, x_{a_n}, u_1, \cdots, u_m) \\
&\vdots \\
\dot{x}_{a_n} &= f_n(x_{a_1}, x_{a_2}, \cdots, x_{a_n}, u_1, \cdots, u_m)
\end{aligned} \tag{11.8}$$

with piecewise constant input functions $u_j(t)$, we can transform it into:

$$\begin{aligned}
\dot{x}_1 &= f_1(q_1, q_2, \cdots, q_n, u_1, \cdots, u_m) \\
\dot{x}_2 &= f_2(q_1, q_2, \cdots, q_n, u_1, \cdots, u_m) \\
&\vdots \\
\dot{x}_n &= f_n(q_1, q_2, \cdots, q_n, u_1, \cdots, u_m)
\end{aligned} \tag{11.9}$$

where $q_i(t)$ is related to $x_i(t)$ by some quantization function.

The variables q_i are called *quantized variables*. This system of equations can be represented by the block diagram of Fig.11.10, where **q** and **u** are the vectors formed by the quantized variables and input variables, respectively.

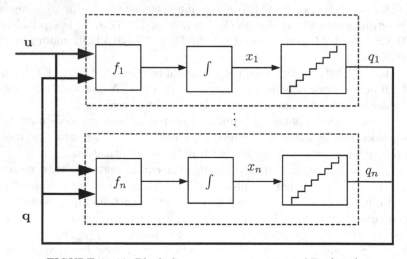

FIGURE 11.10. Block diagram representation of Eq.(11.9).

Each subsystem in Fig.11.10 can be represented by a DEVS model exactly, since all of the subsystems are composed either by a static function or by a quantized integrator. These DEVS models can then be coupled, and, due to the aforementioned closure under coupling property, the coupled system also forms a DEVS model.

Thus, when a system is modified by adding quantizers to the outputs of all integrators, the resulting system is equivalent to a coupled DEVS model that can be simulated, assuming that all of the input functions are piecewise constant as well.

This idea formed the first approximation to a discrete event–based method for continuous system simulation. With this method, we can simulate exactly –ignoring round–off errors– the system of Eq.(11.9), which seems to be a reasonable approximation to that of Eq.(11.8), while avoiding any kind of time discretization. The time discretization was replaced by the quantization of the state variables.

Unfortunately, there is a problem with the *legitimacy* of the resulting DEVS model. A DEVS model is said to be legitimate if it cannot perform an infinite number of transitions in a finite interval of time [11.11].

Although it can be easily verified that the subsystems in Fig.11.10 are legitimate, the legitimacy property is not closed under coupling.

In fact, this simulation method will lead to illegitimate DEVS models in most cases. The simulation of an illegitimate DEVS model gets stuck, when the number of state transitions per time unit grows to infinity.

The reason for the illegitimacy of the DEVS model is related to the solution of Eq.(11.9). There, the trajectories of $q_i(t)$ are not necessarily piecewise constant. Sometimes, they can exhibit an infinite number of state changes within a finite time interval, which produces an infinite number of events in the corresponding DEVS model. Due to this problem, we cannot claim that the use of a simple quantization in the state variables constitutes a general method for the simulation of continuous systems.

We can observe this problem in the system of Eqs.(11.6–11.7) by changing the input function to $u(t) = 10.5 \cdot \varepsilon(t - 1.76)$. The trajectories until time $t = 1.76$ are exactly the same as those shown in Fig.11.1. Once the step has been applied, the trajectory starts growing a bit faster than shown in Fig.11.1. When $x(t) = q(t) = 10$, the state derivative doesn't become zero, however. Instead, the trajectory continues to grow with a slope of $\dot{x}(t) = 0.5$. Then after 2 more time units, we obtain $x(t) = q(t) = 11$. At this point in time, the slope becomes negative. $x(t)$ now decreases with a slope of $\dot{x}(t) = -0.5$. Thus, $q(t)$ immediately returns to 10, the state derivative becomes again positive, and $x(t)$ starts growing again. We obtain a cyclic behavior with infinite frequency.

This anomalous and annoying behavior can also be observed in the resulting DEVS model. When the DEVS model corresponding to the integrator performs an internal transition, it produces an output event that represents the change in $q(t)$. This event is propagated through the internal feed-back loop (cf. Figs. 11.8–11.9), and produces a new external transition in the integrator that changes the time advance to zero. Consequently, the integrator undergoes another internal transition, and the cycle continues forever.

The reader may wonder why we introduced a method that only works in a very few cases. Yet, we had a very good reason for doing so. It turns out that, by adding only a small and very simple modification to the method, a general simulation method can indeed be designed that is based on the previously introduced simulation approach, yet avoids the illegitimacy problem that has plagued us so far. This new method is called *quantized state systems method* (*QSS method* for short) [11.6], and we shall dedicate the final part of this chapter to introducing this new simulation algorithm. The study of its theoretical and practical properties as well as some of its extensions shall be left to the next and final chapter of this book.

11.7   Quantized State Systems

If we try to analyze the infinitely fast oscillations in the system of Eqs.(11.6–11.7), we can see that they are caused by the changes in $q(t)$. An infinitesimally small variation in $x(t)$ can produce, due to the quantization, a significant oscillation with an infinitely fast frequency in $q(t)$.

A possible solution might consist in adding some delay after a change in $q(t)$ to avoid those infinitely fast oscillations. However, adding such delays is equivalent, in some way, to introducing time discretization. During the delays, we lose control over the simulation, and we have to deal with the problems associated with discrete–time algorithms once again.

A different solution is based on the use of hysteresis in the quantization. If we add hysteresis to the relationship between $x(t)$ and $q(t)$, the oscillations in $q(t)$ can only be produced by *large* oscillations in $x(t)$ that cannot occur instantaneously, as long as the state derivatives remain finite.

Therefore, before introducing the QSS method formally, we shall define the concept of a *hysteretic quantization function*.

Let $Q = \{Q_0, Q_1, ..., Q_r\}$ be a set of real numbers, where $Q_{k-1} < Q_k$ with $1 \leq k \leq r$. Let Ω be the set of piecewise continuous real–valued trajectories, and let $x \in \Omega$ be a continuous trajectory. Let b be a mapping $b : \Omega \to \Omega$, and let $q = b(x)$, where the trajectory q satisfies:

$$q(t) = \begin{cases} Q_m & \text{if } t = t_0 \\ Q_{k+1} & \text{if } x(t) = Q_{k+1} \quad \wedge \quad q(t^-) = Q_k \quad \wedge \quad k < r \\ Q_{k-1} & \text{if } x(t) = Q_k - \varepsilon \quad \wedge \quad q(t^-) = Q_k \quad \wedge \quad k > 0 \\ q(t^-) & \text{otherwise} \end{cases} \quad (11.10)$$

and:

$$m = \begin{cases} 0 & \text{if } x(t_0) < Q_0 \\ r & \text{if } x(t_0) \geq Q_r \\ j & \text{if } Q_j \leq x(t_0) < Q_{j+1} \end{cases}$$

Then, the map b is a hysteretic quantization function.

The discrete values Q_i are called *quantization levels*, and the distance $Q_{k+1} - Q_k$ is defined as the *quantum*, which is usually constant. The width of the hysteresis window is ε. The values Q_0 and Q_r are the lower and upper saturation values. Figure 11.11 shows a typical quantization function with uniform quantization intervals.

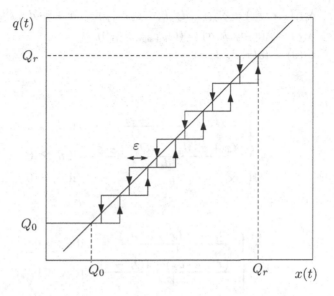

FIGURE 11.11. Quantization function with hysteresis.

Now, we are ready to define the QSS method:

> Given a system such as that of Eq.(11.8), the QSS method
> transforms the system to a system similar to that of Eq.(11.9),
> where the variables $x_i(t)$ and $q_i(t)$ are related by hysteretic
> quantization functions. The resulting system is called a *quan-
> tized state system (QSS)*.

In [11.6], it is shown that the quantized and state variable trajectories of
Eq.(11.9) are always piecewise constant and piecewise linear, respectively.
Hence a QSS can be simulated exactly by a legitimate DEVS model.

A legitimate DEVS model can be built as the coupling of subsystems
corresponding to static functions and *hysteretic quantized integrators*.

The hysteretic quantized integrators are quantized integrators, where
the simple memoryless quantization functions have been replaced by hysteretic quantization functions. This is equivalent to replacing Eq.(11.6b)
by Eq.(11.10) in the system of Eq.(11.6).

With this modification, the hysteretic quantized integrator constituted
by Eq.(11.6a) and Eq.(11.10) can be represented by the DEVS model:

$$M_5 = (X, Y, S, \delta_{\text{int}}, \delta_{\text{ext}}, \lambda, ta), \text{ where}$$
$$X = Y = \mathbb{R} \times \mathbb{N}$$
$$S = \mathbb{R}^2 \times \mathbb{Z} \times \mathbb{R}_0^+$$
$$\delta_{\text{int}}(s) = \delta_{\text{int}}(x, d_x, k, \sigma) = (x + \sigma \cdot d_x, d_x, k + \text{sign}(d_x), \sigma_1)$$

$$\delta_{\text{ext}}(s, e, x_u) = \delta_{\text{ext}}(x, d_x, k, \sigma, e, x_v, p) = (x + e \cdot d_x, x_v, k, \sigma_2)$$
$$\lambda(s) = \lambda(x, d_x, k, \sigma) = (Q_{k+\text{sign}(d_x)}, 0)$$
$$ta(s) = ta(x, d_x, k, \sigma) = \sigma$$

where:

$$\sigma_1 = \begin{cases} \dfrac{Q_{k+2} - (x + \sigma \cdot d_x)}{d_x} & \text{if } d_x > 0 \\[2mm] \dfrac{(x + \sigma \cdot d_x) - (Q_{k-1} - \varepsilon)}{|d_x|} & \text{if } d_x < 0 \\[2mm] \infty & \text{if } d_x = 0 \end{cases}$$

and:

$$\sigma_2 = \begin{cases} \dfrac{Q_{k+1} - (x + e \cdot d_x)}{x_v} & \text{if } x_v > 0 \\[2mm] \dfrac{(x + e \cdot d_x) - (Q_k - \varepsilon)}{|x_v|} & \text{if } x_v < 0 \\[2mm] \infty & \text{if } x_v = 0 \end{cases}$$

The QSS method then consists in choosing the quantization levels (Q_0, Q_1, ..., Q_r) and the hysteresis width ε to be used in each state variable. This choice automatically defines DEVS models of the M_5 class for each resulting hysteretic quantized integrator. Representing the static functions $f_1, \ldots, f_n$ with different DEVS models similar to M_3 and coupling them, the system of Eq.(11.9) can be exactly simulated (ignoring round–off problems). As mentioned above, the resulting coupled DEVS model is legitimate, and the simulation will consume a finite amount of time.

Figure 11.12 shows the block diagram representation of a generic QSS.

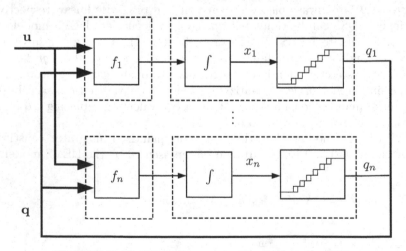

FIGURE 11.12. Block diagram representation of a QSS.

The hysteretic quantized integrator M_5 can be implemented in PowerDEVS as follows:

ATOMIC MODEL HINTEGRATOR
State Variables and Parameters:
```
float X, dX, q, sigma; //states
float y; //output
float epsilon, inf; //parameters
```

Init Function:
```
va_list parameters;
va_start(parameters, t);
dq = va_arg(parameters, double);
epsilon = va_arg(parameters, double);
X = va_arg(parameters, double);
dX = 0;
q =floor(X/dq) * dq;
inf = 1e10;
sigma = 0;
```

Time Advance Function:
```
return sigma;
```

Internal Transition Function:
```
X = X + sigma * dX;
if (dX > 0) {
    sigma = dq/dX;
    q = q + dq;
    }
else {
    if (dX < 0) {
        sigma = -dq/dX;
        q = q - dq;
        }
    else {
        sigma = inf;
    };
};
```

External Transition Function:
```
float xv;
xv = *(float*)(x.value);
X = X + dX * e;
if (xv > 0) {
    sigma = (q + dq - X)/xv;
    }
else {
    if (xv < 0) {
        sigma = (q - epsilon - X)/xv;
        }
    else {
        sigma = inf;
    };
};
```

$dX = xv;$

Output Function:
if $(dX == 0)$ $\{y = q;\}$ else $\{y = q + dq * dX/\text{fabs}(dX);\};$
return Event($\&y,0$);

Here we changed some things with respect to model M_5. In this model, we used a uniform quantum ΔQ, and we replaced variable k (the index of the quantization levels) by q (the quantized variable).

It has now become clear how the QSS method can be applied to systems such as those of Eq.(11.8). We only have to build a block diagram in PowerDEVS using atomic models such as HINTEGRATOR and STATIC1.

However, we must not forget that the result that we obtain is the solution of Eq.(11.9). Thus, the accuracy of the simulation will be connected to the similarity between this system and the original system of Eq.(11.8).

Taking into account that the only difference between both systems is the presence of the quantization functions, we expect that the error depends on the size of the quantization intervals. As we shall explain in the next chapter, this is indeed the case, and this dependence will provide us with a rule for choosing the quantization levels and the hysteresis width.

Let us illustrate the method by means of a simple example. Consider the second order system:

$$
\begin{aligned}
\dot{x}_{a_1}(t) &= x_{a_2}(t) \\
\dot{x}_{a_2}(t) &= 1 - x_{a_1}(t) - x_{a_2}(t)
\end{aligned}
\tag{11.11}
$$

with initial conditions:

$$
x_{a_1}(0) = 0, \quad x_{a_2}(0) = 0
\tag{11.12}
$$

We shall use a uniform quantum $Q_{k+1} - Q_k = \Delta Q = 0.05$ and a hysteresis width of $\varepsilon = 0.05$ for both state variables.

Thus, the resulting quantized state system:

$$
\begin{aligned}
\dot{x}_1(t) &= q_2(t) \\
\dot{x}_2(t) &= 1 - q_1(t) - q_2(t)
\end{aligned}
\tag{11.13}
$$

can be simulated using a coupled DEVS model, composed by two atomic models of the M_5 class, corresponding to the quantized integrators, and two atomic models similar to M_3 that calculate the static functions $f_1(q_1, q_2) = q_2$ and $f_2(q_1, q_2) = 1 - q_1 - q_2$. Figure 11.13 represents the coupled system.

Observe that, due to the fact that function f_1 does not depend on variable q_1, there is a connection that is not necessary. Moreover, taking into account that $f_1(q_1, q_2) = q_2$ the subsystem F_1 can be replaced by a direct connection from QI_2 to QI_1. These simplifications can reduce considerably the computational cost of the implementation.

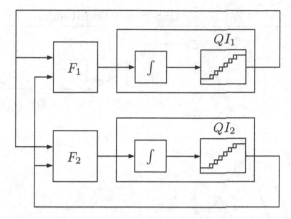

FIGURE 11.13. Block diagram representation of Eq.(11.13).

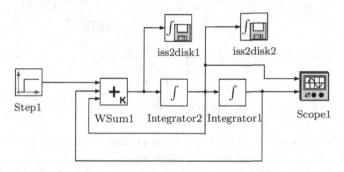

FIGURE 11.14. PowerDEVS model.

In fact, when drawing the PowerDEVS block diagram, we automatically make these simplifications (cf. Fig.11.14).

In the PowerDEVS model of Fig.11.14, there appears a new atomic block that calculates a weighted sum. The reader should be able to imagine, what this block does, and what the hidden DEVS model may look like (cf. Hw.[H11.4]).

The simulation results are shown in Fig.11.15. The first simulation was completed using 30 internal transitions at each quantized integrator, which gives a total of 60 steps. We can see in Fig.11.15 the piecewise linear trajectories of $x_1(t)$ and $x_2(t)$, as well as the piecewise constant trajectories of $q_1(t)$ and $q_2(t)$.

The presence of the hysteresis can be easily noticed where the slope of a state variable changes its sign. Near those points, we can observe different values of q for the same value of x.

The simplifications we mentioned in the connections can be applied to general systems, where few of the static functions do depend on all of the state variables. In this way, the QSS method can exploit the structural properties of the system to reduce the computational burden. When the

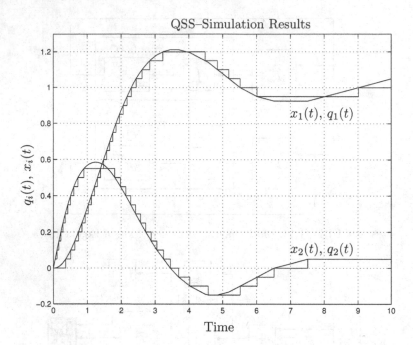

FIGURE 11.15. Trajectories of the system of Eq.(11.13).

system is *sparse*, QSS simulations are particularly efficient, since each step involves calculations at few integrators only.

Discrete–time algorithms can also exploit sparsity properties. However, these techniques require specific sparse matrix algorithms to do so. In the QSS method, the exploitation of sparsity is an intrinsic property.

11.8   Summary

In this chapter, we studied the basic principles of discrete event simulation under the DEVS formalism and their applications to continuous systems simulation.

We introduced the concept of state variable quantization and, based on this concept, we showed how to build a DEVS model that exactly represents the dynamics of general time–invariant continuous systems with quantization in their state variables. We saw that the use of simple memoryless quantization can produce illegitimacy in the DEVS model. We then demonstrated that the use of hysteretic quantization solves this problem. Making use of these techniques, we introduced the QSS method that allows the simulation of general time–invariant continuous system.

In the next chapter, we shall show and discuss the main theoretical properties and practical applications of the QSS method and its extensions. For now, it suffices to mention that the DSS method and its extension provide efficient simulations of discontinuous systems and sparse problems, and that these techniques are of particular interest in the context of real–time simulation.

We might mention further that quantization–based methods are not the only possible discrete event approaches to continuous system simulation. A different idea is based on the event representation of trajectories and the definition of GDEVS [11.3]. However, this *solution–based* approximation requires the knowledge of the continuous system response to some particular input trajectories, which is not available in most cases. For this reason, GDEVS does not constitute a general continuous simulation method.

For this reason, we shall not introduce GDEVS in this book. Yet, it should be acknowledged that some of the ideas behind GDEVS were used in the design of the second–order accurate QSS2 method that shall be introduced in the next chapter.

Finally, the reader might notice that this chapter does not offer a broad basis of references and bibliographic pointers. The reason for this is simply that discrete event simulation of continuous systems is a fairly recently developed topic. In fact, the first references that we know of are from the late nineties. This implies that these methods are not yet completely developed and optimized, which makes them a fertile field for research.

11.9   References

[11.1] Hyup Cho and Young Cho. *DEVS–C++ Reference Guide.* The University of Arizona, 1997.

[11.2] Jean Baptiste Filippi, Marielle Delhom, and Fabrice Bernardi. The JDEVS Environmental Modeling and Simulation Environment. In *Proceedings of IEMSS 2002*, volume 3, pages 283–288, Lugano, Switzerland, 2002.

[11.3] Norbert Giambiasi, Bruno Escude, and Sumit Ghosh. GDEVS: A Generalized Discrete Event Specification for Accurate Modeling of Dynamic Systems. *Transactions of SCS*, 17(3):120–134, 2000.

[11.4] Kihyung Kim, Wonseok Kang, and Hyungon Seo. Efficient Distributed Simulation of Hierarchical DEVS Models: Transforming Model Structure into a Non–Hierarchical One. In *Proceedings of Annual Simulation Symposium*, 2000.

[11.5] Tag Gon Kim. *DEVSim++ User's Manual. C++ Based Simulation with Hierarchical Modular DEVS Models.* Korean Advanced Institute of Science and Technology, 1994. Available at http://www.acims.arizona.edu/.

[11.6] Ernesto Kofman and Sergio Junco. Quantized State Systems: A DEVS Approach for Continuous System Simulation. *Transactions of SCS*, 18(3):123–132, 2001.

[11.7] Ernesto Kofman, Marcelo Lapadula, and Esteban Pagliero. PowerDEVS: A DEVS–based Environment for Hybrid System Modeling and Simulation. Technical Report LSD0306, LSD, UNR, 2003. Submitted to *Simulation*. Available at http://www.fceia.unr.edu.ar/lsd/powerdevs.

[11.8] Esteban Pagliero, Marcelo Lapadula, and Ernesto Kofman. PowerDEVS. Una Herramienta Integrada de Simulación por Eventos Discretos. In *Proceedings of RPIC03*, volume 1, pages 316–321, San Nicolas, Argentina, 2003.

[11.9] Esteban Pagliero and Marcelo Lapadula. Herramienta Integrada de Modelado y Simulación de Sistemas de Eventos Discretos. Diploma Work. FCEIA, UNR, Argentina, September 2002.

[11.10] Gabriel Wainer, Gastón Christen, and Alejandro Dobniewski. Defining DEVS Models with the CD++ Toolkit. In *Proceedings of ESS2001*, pages 633–637, Marseille, France, 2001.

[11.11] Bernard Zeigler, Tag Gon Kim, and Herbert Praehofer. *Theory of Modeling and Simulation. Second edition.* Academic Press, New York, 2000.

[11.12] Bernard Zeigler and Jong Sik Lee. Theory of Quantized Systems: Formal Basis for DEVS/HLA Distributed Simulation Environment. In *SPIE Proceedings*, pages 49–58, 1998.

[11.13] Bernard Zeigler and Hessam Sarjoughian. *Introduction to DEVS Modeling and Simulation with JAVA: A Simplified Approach to HLA-Compliant Distributed Simulations.* Arizona Center for Integrative Modeling and Simulation. Available at http://www.acims.arizona.edu/.

[11.14] Bernard Zeigler. *Theory of Modeling and Simulation.* John Wiley & Sons, New York, 1976.

11.10   Bibliography

[B11.1] Christos Cassandras. *Discrete Event Systems: Modeling and Performance Analysis.* Irwin and Aksen, 1993.

[B11.2] Ernesto Kofman. *Discrete Event Simulation and Control of Continuous Systems.* PhD thesis, Universidad Nacional de Rosario, Rosario, Argentina, 2003.

11.11   Homework Problems

[H11.1] Achilles and the Tortoise

Consider the second order system:

$$\begin{aligned}
\dot{x}_{a_1} &= -0.5 \cdot x_{a_1} + 1.5 \cdot x_{a_2} \\
\dot{x}_{a_2} &= -x_{a_1}
\end{aligned} \tag{H11.1a}$$

Apply the memoryless quantization function:

$$q_i = 2 \cdot \text{floor}(\frac{x_i - 1}{2}) + 1 \tag{H11.1b}$$

to both state variables, and study the solutions of the quantized system:

$$\begin{aligned}
\dot{x}_1 &= -0.5 \cdot q_1 + 1.5 \cdot q_2 \\
\dot{x}_2 &= -q_1
\end{aligned} \tag{H11.1c}$$

from the initial condition $x_{a_1} = 0$, $x_{a_2} = 2$.

(a). Show that the simulation time cannot advance more than 5 seconds.

(b). Draw the state–space trajectory $x_1(t)$ vs. $x_2(t)$.

[H11.2] DEVS Behavior

Using the DEVS model M_2, repeat the simulation *by hand* that was performed with model M_1 on page 527. Use the same input trajectory and
compare the evolution obtained with the evolution of M_1.

[H11.3] DEVS Demultiplexer

The use of ports in DEVS gives rise to a difficulty. After an internal transition took place, a model with ports produces an event that carries a value
at one specific output port. This is a limitation, because it is not difficult
to imagine a situation, in which the event value contains a vector, and each
component should be sent to a different sub–model.

This problem does not appear in the general definition of DEVS (coupling
without ports). However, even when using ports, the problem can be solved
with the addition of a DEVS model that demultiplexes events.

A DEVS demultiplexer receives input events carrying a vector value
through its input port, decomposes the vector into individual scalar values,
and sends those out immediately through different output ports.

Build a DEVS demultiplexer that receives events with values in $\mathbb{R}^k$. After
receiving an event, the model should send k events through its k output
ports, carrying the corresponding scalar component values.

[H11.4] Linear Static Function

Obtain a DEVS atomic model of a static function $f_i : \mathbb{R}^n \to \mathbb{R}$ defined as

$$f(u_0, u_1, \cdots, u_{n-1}) = \sum_{k=0}^{n-1} a_k \cdot u_k \qquad \text{(H11.4a)}$$

where $a_0, \cdots, a_{n-1}$ are known constants.

Then, program the model in PowerDEVS so that the constants and the number of inputs are parameters.

Hint: You will have to limit the number of inputs to a fixed number (10 for instance).

[H11.5] DEVS Delay Function

Consider a function that represents a fixed delay time T,

$$f(u(t)) = u(t - T) \qquad \text{(H11.5a)}$$

Consider the input $u(t)$ to be piecewise constant, and obtain a DEVS model of this function.

Create a PowerDEVS block of this function, where the delay time T is a parameter.

Hint: Assume that the number of state changes in $u(t)$ during a time period of T is limited to a fixed number (1000 for instance).

[H11.6] Achilles and the Tortoise Revisited

Modify the PowerDEVS atomic model NHINTEGRATOR of page 537, so that the quantizer satisfies Eq.(H11.1b). Then, use this new atomic model, and build the block diagram corresponding to Eq.(H11.1a). Verify by simulation the prediction made in Hw.[H11.1].

We suggest using a final simulation time such as 4.999 for example.

[H11.7] Varying Quantum and Hysteresis

Obtain the exact solution of the system of Eq.(11.11), and then repeat the QSS simulation using the following quantization and hysteresis:

(a). $\Delta Q_1 = \Delta Q_2 = \varepsilon = 0.01$

(b). $\Delta Q_1 = \Delta q_2 = \varepsilon = 0.05$

(c). $\Delta Q_1 = \Delta q_2 = \varepsilon = 0.1$

(d). $\Delta Q_1 = \Delta Q_2 = \varepsilon = 1$

Compare the results, and use them to hypothesize about the effects of the quantization and hysteresis on error and stability.

11.12 Projects

[P11.1] Grouping Models in the QSS Method

The division between quantized integrators and static functions in building the coupled DEVS model that implements the QSS method simplifies considerably the atomic models.

However, this division is not necessary. Indeed, we already mentioned that DEVS is closed under coupling, and therefore, it must be possible to define a unique atomic DEVS model that simulates the entire system. In this way, the number of events is reduced (we do not have to transmit events between components), and the computational efficiency is improved.

Of course, finding this atomic DEVS model may be quite difficult, and even if we find it, we might lose the possibility of implementing the simulation in a parallel fashion.

An intermediate solution for the QSS method, that probably represents the best compromise, consists in grouping each quantized integrator with the static function that calculates its derivative. In this way, the number of events is reduced to less than the one half.

Using this idea, propose an atomic DEVS model that represents simultaneously a static function and a quantized integrator. Program that model in PowerDEVS, and couple two of these models to simulate the system of Eq.(11.11).

Compare the total number of internal and external transitions performed by this coupled DEVS model with that obtained by simulating the model composed of separate quantized integrators and static functions. Compare also the execution time of the two simulations.

Repeat the experiment with other models, and try to determine, under which conditions the grouping of models yields noticeable advantages.

Conclude on the convenience of using grouped models, taking into account the trade–off between simplicity and execution time.

[P11.2] DEVS and Multi–Rate Integration

Build a PowerDEVS model of a *forward Euler integrator*, i.e., a model that receives input events with scalar values f_k and produces scalar output values:

$$x_{k+1} = x_k + h \cdot f_k \tag{P11.2a}$$

where the step–size h is a parameter.

Invoke that model multiple times together with the static function model to simulate some higher–order differential equation models using the FE method in PowerDEVS.

Then, using different values of h for different integrators, perform some *multi–rate integration* experiments.

We suggest that you reproduce the example of Section 10.5, given by Eq.(10.10).

Study the possibility of building integrators corresponding to higher-order algorithms (RK, AB, etc.).

Conclude about the advantages and disadvantages of using DEVS in the context of discrete–time integration algorithms.

12

Quantization–based Integration

Preview

This chapter focuses on the *Quantized State Systems (QSS)* method and its extensions. After a brief explanation concerning the connections between this discrete event method and perturbation theory, the main theoretical properties of the method, i.e., convergence, stability, and error control properties, are presented.

The reader is then introduced to some practical aspects of the method related to the choice of quantum and hysteresis, the incorporation of input signals, as well as output interpolation.

In spite of the theoretical and practical advantages that the QSS method offers, the method has a serious drawback, as it is only first–order accurate. For this reason, a second–order accurate quantization–based method is subsequently presented that conserves the main theoretical properties that characterize the QSS method.

Further, we shall focus on the use of both quantization–based methods in the simulation of DAEs and discontinuous systems, where we shall observe some interesting advantages that these methods have over the classical discrete–time methods.

Finally, and following the discussion of a real–time implementation of these methods, some drawbacks and open problems of the proposed methodology shall be discussed with particular emphasis given to the simulation of stiff system.

12.1 Introduction

In Chapter 2, we introduced two basic properties of numerical methods: the *approximation accuracy* and the *numerical stability*. If we want to rely on the simulation results generated by a method, we must know something about these properties in the context of the application at hand.

The conventional tools for the analysis of numerical stability are based on the discrete–time systems theory. The basic idea is to obtain the difference equations corresponding to a given method applied to a linear time–invariant autonomous system, and then to relate the eigenvalues of the $\mathbf{F}$–matrix of the transformed discrete–time system to those of the $\mathbf{A}$–matrix of the original continuous–time system.

This technique, that we have applied throughout this book to the analysis of discrete–time methods, cannot be extended to the QSS method, because the resulting simulation model is a discrete event system that does not possess an **F**–matrix.

Since linear stability theory is such a convenient tool, we might be inclined to attempt tackling this problem by looking for a discrete event systems theory that would support this kind of stability analysis. In fact, there exists a nice mathematical theory based on the use of *max–plus algebra* that permits expressing some discrete event systems through difference equations in the context of that algebra [12.1]. This theory also arrives at stability results based on the study of eigenvalues and is completely analogous to the discrete–time systems theory. However, it can only be applied to systems described by Petri nets, and unfortunately, the QSS method produces DEVS models that do not have Petri net equivalents.

A different approach to studying the QSS dynamics might be to compare directly the results obtained when simulating the original continuous–time model of Eq.(11.8) with those obtained when simulating its QSS approximation of Eq.(11.9).

Let us rewrite these two representations using vector notation. The original continuous system may be written as follows:

$$\dot{\mathbf{x}}_{\mathbf{a}}(t) = \mathbf{f}(\mathbf{x}_{\mathbf{a}}(t), \mathbf{u}(t)) \tag{12.1}$$

and the resulting quantized state system can be written as:

$$\dot{\mathbf{x}}(t) = \mathbf{f}(\mathbf{q}(t), \mathbf{u}(t)) \tag{12.2}$$

where $\mathbf{x}(t)$ and $\mathbf{q}(t)$ are componentwise related by hysteretic quantization functions.

Let us define $\boldsymbol{\Delta}\mathbf{x}(t) = \mathbf{q}(t) - \mathbf{x}(t)$. Then, Eq.(12.2) can be rewritten as:

$$\dot{\mathbf{x}}(t) = \mathbf{f}(\mathbf{x}(t) + \boldsymbol{\Delta}\mathbf{x}(t), \mathbf{u}(t)) \tag{12.3}$$

and now, the simulation model of Eq.(12.2) can be interpreted as a *perturbed representation* of the original system of Eq.(12.1).

Hysteretic quantization functions have a fundamental property. If two variables $q_i(t)$ and $x_i(t)$ are related by a hysteretic quantization function, such as that of Eq.(11.10), then:

$$Q_0 < x_i(t) < Q_r \Rightarrow |q_i(t) - x_i(t)| \leq \max(\Delta Q, \varepsilon) \tag{12.4}$$

where $\Delta Q = \max(Q_{j+1} - Q_j)$, $0 \leq j \leq r - 1$, is the largest quantum.

The property given by Eq.(12.4) implies that each component of the perturbation $\boldsymbol{\Delta}\mathbf{x}$ is bounded by the corresponding hysteresis width and quantum size. Thus, the accuracy and stability analysis can be based on the effects of a bounded perturbation.

In Chapter 2, we mentioned that there exists a theory that allows studying the numerical stability of nonlinear systems. We shunned away from pursuing that theory any further, because the mathematical apparatus required to do so is quite formidable.

Unfortunately, we now have no choice but to go down that route, because the QSS representation even of a linear system is in fact nonlinear. Luckily, we shall see that many of the problems associated with general contractivity theory disappear in the special case of a QSS, because we can reduce the nonlinearity of the quantization to the special case of a linear perturbation, and perturbation analysis can be applied even to nonlinear systems quite easily. Furthermore, when the QSS method is applied to a linear system, it will also be possible to establish a global error bound, and the problem of approximation accuracy can then be dealt with not only locally, but even globally.

Despite these advantages, a new problem appears in the QSS method. Let us illustrate this problem by means of the following example. Consider the first–order system:

$$\dot{x}_a(t) = -x_a(t) + 9.5 \qquad (12.5)$$

with the initial condition $x_a(0) = 0$.

The results of a simulation using the QSS method with a quantum of $\Delta Q = 1$ and a hysteresis width of $\varepsilon = 1$ are shown in Fig.12.1

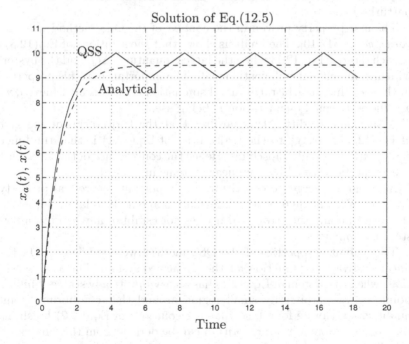

FIGURE 12.1. QSS simulation of Eq.(12.5).

Although the system of Eq.(12.5) is asymptotically stable, the QSS simulation ends in a limit cycle. The equilibrium point $\bar{x} = 9.5$ is no longer a stable equilibrium point in the resulting QSS, and we cannot claim that stability, in the sense of Lyapunov, is conserved by the method.

However, the QSS solution never deviates far from the exact solution, and it finishes with an oscillation around the equilibrium point. Taking into account that our goal was to just simulate the system and obtain some meaningful trajectories, this result is not bad.

The trajectory found by the QSS method is called *ultimately bounded* [12.8]. In general, the quantization based methods cannot ensure stability in accordance with the classical definition. Hence we shall talk, in the context of QSS simulations, about stability in terms of ultimate boundedness of the solutions obtained.

12.2   Convergence, Accuracy, and Stability in QSS

When a time–invariant system, such as that of Eq.(12.1), is simulated using the QSS method, we obtain an exact simulation, ignoring the roundoff problems, of the perturbed system of Eq.(12.3). Then, as it was mentioned above, the theoretical properties can be studied based on the effects of perturbation.

The first property proven in the context of the QSS method was that of *convergence* [12.10]. The analysis shows that the solutions of Eq.(12.3) approach those of Eq.(12.1) when the largest quantum, ΔQ, and the hysteresis width, ε, are chosen sufficiently small. The importance of this property lies in the fact that an arbitrarily small simulation error can be achieved, when a sufficiently small quantization is being used.

A sufficient condition that ensures that the trajectories of the system of Eq.(12.3) converge to the trajectories of Eq.(12.1) is that the function $\mathbf{f}(\mathbf{x}(t), \mathbf{u}(t))$ is locally Lipschitz[1]. Hence the convergence of the QSS method is a property satisfied by nonlinear systems in general.

Although convergence constitutes an important theoretical property, it does not offer any quantitative information about the relationship between the quantum and the error, and it does not establish any condition for the stability domain.

The stability properties of the QSS method were studied in [12.10] by finding a Lyapunov function for the perturbed system. The analysis shows that, when the system of Eq.(12.1) has an asymptotically stable equilibrium point, for any arbitrarily small region around that equilibrium point, a quantization can be found, so that the solutions of Eq.(12.2) finish inside that region. Moreover, an algorithm can be derived from this analysis that

---

[1]In a nutshell, this means that the function $\mathbf{f}$ must not escape to infinity within the range of interest.

allows calculating the appropriate quantum.

A sufficient condition for ensuring stability is that the function **f** be continuous and continuously differentiable. Hence the stability condition is a bit stronger than the convergence condition.

Thus, the QSS method offers tools, that can be applied to nonlinear systems, for choosing a quantum that ensures that the steady–state simulation error is smaller than a desired bound. Although this result represents an important advantage over the classical discrete–time methods, where stability is usually studied in the context of linear time–invariant (LTI) systems only, the algorithm is quite involved and requires the use of a Lyapunov function of Eq.(12.1) that cannot be easily derived in general cases. Thus, the importance of this stability analysis is, as before, more of a theoretical than a practical nature, and we shall refrain from delving into details about it.

Like in discrete–time methods, the most interesting qualities of QSS come from the analysis of its application to LTI systems. The main result of that analysis, performed in [12.13], states that the error in the QSS simulation of an asymptotically stable LTI system is always bounded. The error bound, which can be calculated from the quantum and some geometrical properties of the system, does not depend either on the initial condition or on the input trajectory and remains constant during the simulation.

Before explaining this fundamental property in more detail, we shall need to introduce some new notation, in order to be able to express the relationships between quanta and error bounds in terms of compact formulae.

We shall use the symbol $|\cdot|$ to denote the componentwise module of a vector or matrix. For instance, if **G** is a $j \times k$ matrix with complex components $g_{1,1}, \ldots, g_{j,k}$, then $|\mathbf{G}|$ is also a $j \times k$ matrix with the real positive components $|g_{1,1}|, \ldots, |g_{j,k}|$.

Similarly, we shall use the symbol "$\leq$" to perform a componentwise comparison between real–valued vectors of equal length. Thus, the expression $\mathbf{x} \leq \mathbf{y}$ states that $x_1 \leq y_1, \ldots, x_n \leq y_n$.

With these definitions, let $\mathbf{x_a}(t)$ be a solution of the LTI system:

$$\dot{\mathbf{x}}_\mathbf{a}(t) = \mathbf{A} \cdot \mathbf{x_a}(t) + \mathbf{B} \cdot \mathbf{u}(t) \tag{12.6}$$

Let $\mathbf{x}(t)$ be the solution, starting from the same initial condition, of its associated QSS:

$$\dot{\mathbf{x}}(t) = \mathbf{A} \cdot \mathbf{q}(t) + \mathbf{B} \cdot \mathbf{u}(t) \tag{12.7}$$

which can be written as:

$$\dot{\mathbf{x}}(t) = \mathbf{A} \cdot [\mathbf{x}(t) + \mathbf{\Delta x}(t)] + \mathbf{B} \cdot \mathbf{u}(t) \tag{12.8}$$

Let us define the error $\mathbf{e}(t) \triangleq \mathbf{x}(t) - \mathbf{x_a}(t)$. By subtracting Eq.(12.6) evaluated at $\mathbf{x_a}(t)$ from Eq.(12.8) evaluated at $\mathbf{x}(t)$, we find that $\mathbf{e}(t)$ satisfies the equation:

$$\dot{\mathbf{e}}(t) = \mathbf{A} \cdot [\mathbf{e}(t) + \mathbf{\Delta x}(t)] \qquad (12.9)$$

with $\mathbf{e}(t_0) = 0$ since both trajectories, $\mathbf{x_a}(t)$ and $\mathbf{x}(t)$, start out from identical initial conditions.

Let us start analyzing the simple scalar case:

$$\dot{e}(t) = a \cdot [e(t) + \Delta x(t)] \qquad (12.10)$$

For reasons that the reader will soon understand, we shall assume that a, e, and Δx belong to $\mathbb{C}$. We shall furthermore request $\mathbb{Re}\{a\}$ to be negative. We shall also ask that $|\Delta x| \leq w$, with w being some positive constant. $e(t)$ can be written in polar notation as:

$$e(t) = \rho(t) \cdot e^{j\theta(t)} \qquad (12.11)$$

where $\rho(t) = |e(t)|$.

Then, Eq.(12.10) becomes:

$$\dot{\rho}(t) \cdot e^{j\theta(t)} + \rho(t) \cdot e^{j\theta(t)} \cdot j \cdot \dot{\theta}(t) = a \cdot [\rho(t) \cdot e^{j\theta(t)} + \Delta x(t)] \qquad (12.12)$$

or:

$$\dot{\rho}(t) + \rho(t) \cdot j \cdot \dot{\theta}(t) = a \cdot [\rho(t) + \Delta x(t) \cdot e^{-j\theta(t)}] \qquad (12.13)$$

Let us take now only the real part of the last equation:

$$\begin{aligned} \dot{\rho}(t) &= \mathbb{Re}\{a\} \cdot \rho(t) + \mathbb{Re}\{a \cdot \Delta x(t) \cdot e^{-j\theta(t)}\} \\ &\leq \mathbb{Re}\{a\} \cdot \rho(t) + |a| \cdot |\Delta x(t)| \\ &\leq \mathbb{Re}\{a\} \cdot \left[\rho(t) - \left|\frac{a}{\mathbb{Re}\{a\}}\right| \cdot w\right] \end{aligned} \qquad (12.14)$$

Then, as $\mathbb{Re}\{a\}$ is negative and $\rho(0) = 0$, it will always be true that:

$$|e(t)| = \rho(t) \leq \left|\frac{a}{\mathbb{Re}\{a\}}\right| \cdot w \qquad (12.15)$$

since, when ρ reaches the upper limit, its derivative becomes negative (or zero).

Before proceeding to the most general situation, we shall apply this last result to the diagonal case. Let the $\mathbf{A}$–matrix in Eq.(12.9) be a diagonal matrix with complex diagonal elements $a_{i,i}$ with negative real parts. We shall also assume that:

$$|\mathbf{\Delta x}(t)| \leq \mathbf{w} \qquad (12.16)$$

Repeating the scalar analysis for each component of $\mathbf{e}(t)$, we find that:

$$|\mathbf{e}| \leq |\mathbb{Re}\{\mathbf{A}\}^{-1} \cdot \mathbf{A}| \cdot \mathbf{w} \qquad (12.17)$$

Now, let us return once more to Eq.(12.9), this time assuming that the A–matrix be Hurwitz and diagonalizable, i.e., that the original system is asymptotically stable and can be decoupled.

Let us assume that the quantum and hysteresis were adjusted such that:

$$|\mathbf{\Delta x}| = |\mathbf{q} - \mathbf{x}| \leq \mathbf{\Delta Q} \qquad (12.18)$$

where each element of vector $\mathbf{\Delta Q}$ contains the larger of the corresponding quantum and hysteresis width.

Let $\mathbf{\Lambda}$ be a diagonal eigenvalue matrix of $\mathbf{A}$, and let $\mathbf{V}$ be a corresponding right eigenvector matrix. The matrix $\mathbf{V}$ is sometimes also referred to as a right modal matrix. Then:

$$\mathbf{A} = \mathbf{V} \cdot \mathbf{\Lambda} \cdot \mathbf{V}^{-1} \qquad (12.19)$$

is the spectral decomposition of the A–matrix.

We introduce the variable transformation:

$$\mathbf{z}(t) = \mathbf{V}^{-1} \cdot \mathbf{e}(t) \qquad (12.20)$$

Using the new variable $\mathbf{z}$, Eq.(12.9) can be rewritten as follows:

$$\mathbf{V} \cdot \dot{\mathbf{z}}(t) = \mathbf{A} \cdot [\mathbf{V} \cdot \mathbf{z}(t) + \mathbf{\Delta x}(t)] \qquad (12.21)$$

and then:

$$\dot{\mathbf{z}}(t) = \mathbf{V}^{-1} \cdot \mathbf{A} \cdot [\mathbf{V} \cdot \mathbf{z}(t) + \mathbf{\Delta x}(t)] \qquad (12.22)$$

$$= \mathbf{\Lambda} \cdot [\mathbf{z}(t) + \mathbf{V}^{-1} \cdot \mathbf{\Delta x}(t)] \qquad (12.23)$$

¿From Eq.(12.18), it results that:

$$|\mathbf{V}^{-1} \cdot \mathbf{\Delta x}| \leq |\mathbf{V}^{-1}| \cdot \mathbf{\Delta Q} \qquad (12.24)$$

Taking into account that the $\mathbf{\Lambda}$–matrix is diagonal, it turns out that Eq.(12.23) is the diagonal case that we analyzed before, and consequently from Eq.(12.17), it results that:

$$|\mathbf{z}(t)| \leq |\mathbb{Re}\{\mathbf{\Lambda}\}^{-1} \cdot \mathbf{\Lambda}| \cdot |\mathbf{V}^{-1}| \cdot \mathbf{\Delta Q} \qquad (12.25)$$

and therefore:

$$|\mathbf{e}(t)| \leq |\mathbf{V}| \cdot |\mathbf{z}(t)| \leq |\mathbf{V}| \cdot |\mathbb{Re}\{\mathbf{\Lambda}\}^{-1} \cdot \mathbf{\Lambda}| \cdot |\mathbf{V}^{-1}| \cdot \mathbf{\Delta Q} \qquad (12.26)$$

Thus, we can conclude that:

$$|\mathbf{x}(t) - \mathbf{x_a}(t)| \leq |\mathbf{V}| \cdot |\mathbb{R}e\{\mathbf{\Lambda}\}^{-1} \cdot \mathbf{\Lambda}| \cdot |\mathbf{V^{-1}}| \cdot \mathbf{\Delta Q} \qquad (12.27)$$

Inequality Eq.(12.27) has strong theoretical and practical implications.

It can be easily seen that the error bound is proportional to the quantum and, for any quantum adopted, the error is always bounded.

It is also important to notice that the inequality of Eq.(12.27) is an analytical expression for the *global error bound*. Discrete–time methods lack similar formulae. The fact that Eq.(12.27) is independent of initial conditions and input trajectories promises additional important theoretical and practical advantages.

In some way, the QSS method offers an intrinsic error control without requiring the use of adaptation rules. Indeed:

> The QSS method is always stable without using implicit formulae at all.

The importance of this statement cannot be stressed enough. It revolutionizes the field of numerical ODE (and DAE) solution.

12.3   Choosing Quantum and Hysteresis Width

The QSS method requires the choice of an adequate quantum and hysteresis width. Although we mentioned that the error is, at least for LTI systems, always bounded, an appropriate quantization must be chosen in order to obtain decent simulation results.

The inequality of Eq.(12.27) can be used for quantization design. Given a desired error bound, it is not difficult to find an appropriate value of $\mathbf{\Delta Q}$ that satisfies that inequality. Let us illustrate this design in a simple example. Consider the system:

$$\begin{aligned} \dot{x}_{a_1} &= x_{a_2} \\ \dot{x}_{a_2} &= -x_{a_1} - x_{a_2} + u(t) \end{aligned} \qquad (12.28)$$

A set of matrices of eigenvalues and eigenvectors (calculated with MATLAB) are:

$$\mathbf{\Lambda} = \begin{pmatrix} -0.5 + 0.866j & 0 \\ 0 & -0.5 - 0.866j \end{pmatrix}$$

and:

$$\mathbf{V} = \begin{pmatrix} 0.6124 - 0.3536j & 0.6124 + 0.3536j \\ 0.7071j & -0.7071j \end{pmatrix}$$

Then:

$$\mathbf{T} \triangleq |\mathbf{V}| \cdot |\mathbb{R}e\{\mathbf{\Lambda}\}^{-1} \cdot \mathbf{\Lambda}| \cdot |\mathbf{V^{-1}}| = \begin{pmatrix} 2.3094 & 2.3094 \\ 2.3094 & 2.3094 \end{pmatrix} \qquad (12.29)$$

Let us consider that the goal is to simulate Eq.(12.28) for an arbitrary initial condition and input trajectory with an error less than or equal to 0.1 in each variable. Then, a quantum:

$$\Delta \mathbf{Q} = \begin{pmatrix} 0.05/2.3094 \\ 0.05/2.3094 \end{pmatrix} = \begin{pmatrix} 0.0217 \\ 0.0217 \end{pmatrix} \tag{12.30}$$

is sufficiently small to ensure that the error cannot exceed the given bound.

Although the inequality of Eq.(12.27) can be used to compute an upper bound for the error as a function of the quantum and the hysteresis width, the measure will often turn out to be quite conservative.

In fact, using quantum and hysteresis width equal to 0.05 in each variable of the system of Eq.(12.28) and applying $u(t) = 1$, we arrive at the simulation results shown in Fig.11.15. The predicted error bound is 0.23094 in each variable. However, the maximum error obtained in that simulation is considerably smaller than this bound.

Except in specific applications, where we would need to ensure a certain error bound, we do not want to calculate eigenvectors and eigenvalues before performing the simulation. A practical rule to avoid this is to use a quantum proportional to the estimated amplitude of each variable trajectory (assuming that we know in advance the order of magnitude of the values reached by each state variable).

The reader may have already noticed that, in all of the examples discussed so far, the hysteresis width was chosen to be equal to the quantum size. However, we did not provide any rationale for that choice.

The problem of hysteresis width selection is discussed in [12.11]. The conclusion is that it should be chosen equal to the quantum. The reason is that, in this way, the presence of hysteresis does not modify the error bound (cf. Eq.(12.4)), while the final oscillation frequency is being minimized.

The reduction of the oscillation frequency is due to the fact that the minimum time between successive changes in a quantized variable q_i is proportional to the inverse of the hysteresis (assuming that the quantum is greater or equal than the hysteresis width), as has been shown in [12.10].

Let us illustrate this idea with the simulation of the first–order system of Eq.(12.5), using a quantum equal to 1 and different hysteresis values.

Figure 12.2 shows the simulation results with a hysteresis width of $\varepsilon = 1$, $\varepsilon = 0.6$, and $\varepsilon = 0.1$, respectively. In all three cases, the maximum error is bounded by the same value. In theory, the bound is equal to 1, but in the simulations, we can observe that the maximum error is always 0.5.

However, the steady–state oscillation frequency increases as the hysteresis width becomes smaller. In fact, that frequency can be calculated as:

$$f = \frac{1}{2\varepsilon}$$

Then, it is clear that by choosing the hysteresis width equal to the quantum, the frequency is minimized without increasing the error. Reducing the

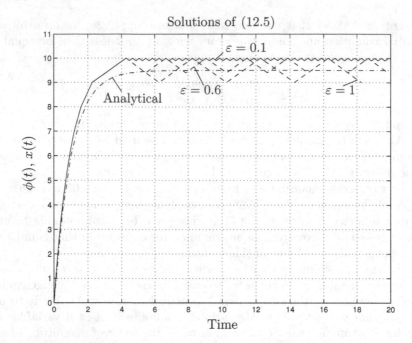

FIGURE 12.2. Simulation of Eq.(12.5) with different hysteresis values.

steady–state oscillation frequency reduces in the number of steps performed by the algorithm, and consequently reduces the computational cost.

12.4   Input Signals in the QSS Method

In the previous chapter, we mentioned that the QSS method allows the simulation of time–invariant systems with piecewise constant input signals. However, we did not say how these signals can be incorporated into the simulation model.

In the DEVS simulation model, each event represents a change in a piecewise constant trajectory. Consequently, input trajectories can be incorporated as sequences of events.

Looking at the block diagram of Fig.11.12, the input signals $\mathbf{u}(t)$ seem to come from the external world, and it is not clear, where the corresponding sequences of events should be generated.

In the context of a DEVS simulation, all events must emanate from an atomic DEVS model. Hence a new DEVS model class must be created that generates those sequences of events. The input function models must then be coupled with the rest of the system for the purpose of simulation.

Suppose that we have a piecewise constant input signal $u(t)$ that assumes the values v_1, v_2, $\ldots$, v_j, $\ldots$ at times t_1, t_2, $\ldots$, t_j, $\ldots$, respectively. A

DEVS model that produces events in accordance with this input signal can be specified as follows:

$$M_6 = (X, Y, S, \delta_{\text{int}}, \delta_{\text{ext}}, \lambda, ta), \text{ where}$$
$$X = \emptyset$$
$$Y = \mathbb{R} \times \mathbb{N}$$
$$S = \mathbb{N} \times \mathbb{R}_0^+$$
$$\delta_{\text{int}}(s) = \delta_{\text{int}}(j, \sigma) = (j + 1, t_{j+1} - t_j)$$
$$\lambda(s) = \lambda(j, \sigma) = (v_j, 0)$$
$$ta(s) = ta(j, \sigma) = \sigma$$

Notice that, in this model, the external transition function δ_{ext} is not defined, as it it will never be called, since the model is not designed to ever receive input events.

A particular case of model M_6 in PowerDEVS is the step function model that we invoked from within the models of Fig.11.9 and Fig.11.14.

ATOMIC MODEL STEP1
State Variables and Parameters:
    float *sigma*;
    int *j*; //states
    float *y*; //output
    float $T[3], v[3], inf$; //parameters

Init Function:
    va_list *parameters*;
    va_start(*parameters*, *t*);
    $inf = 1e10$;
    $T[0] = 0$;
    $T[1] = $ va_arg(*parameters*, double);
    $T[2] = inf$;
    $v[0] = $ va_arg(*parameters*, double);
    $v[1] = $ va_arg(*parameters*, double);
    $sigma = 0$;
    $j = 0$;

Time Advance Function:
    return *sigma*;

Internal Transition Function:
    $sigma = T[j + 1] - T[j]$;
    $j = j + 1$;

Output Function:
    $y = v[j]$;
    return Event(&*y*,0);

The parameters defined in the graphical block of this DEVS model are the step time, the initial value of the output trajectory, and the final value after the step.

An interesting advantage of the QSS method is that it deals with input trajectory changes in an asynchronous way. The event indicating a change in the signal is always processed at the correct instant of time, producing instantaneous changes in the slopes of the state variable trajectories that are directly affected.

This is an intrinsic characteristic of the method, and it is obtained without modifying the DEVS models corresponding to the quantized integrators and the static functions. In contrast, discrete–time methods require a special treatment in order to perform a step at the exact moment when an input change is supposed to occur. We shall revisit to this issue once more later in this chapter, demonstrating the advantages of the QSS method in the context of simulating discontinuous systems.

Up to this point, only piecewise constant input trajectories have been considered. In most applications, the input signals take on more general forms. However, these can be approximated by piecewise constant trajectories with the addition of quantization functions, and thus, they can be represented by DEVS models.

For example, the DEVS model M_7, shown below, generates an event trajectory that approximates a sine function with angular frequency ω and amplitude A using a quantum Δu.

$$M_7 = (X, Y, S, \delta_{int}, \delta_{ext}, \lambda, ta), \text{ where}$$
$$X = \emptyset$$
$$Y = \mathbb{R} \times \mathbb{N}$$
$$S = \mathbb{R} \times \mathbb{R}_0^+$$
$$\delta_{int}(s) = \delta_{int}(\tau, \sigma) = (\tilde{\tau}, \tilde{\sigma})$$
$$\lambda(s) = \lambda(\tau, \sigma) = (A \cdot \sin(\omega\tau), 1)$$
$$ta(s) = ta(\tau, \sigma) = \sigma$$

with:

$$\tilde{\sigma} = \begin{cases} \dfrac{\arcsin[\sin(\omega\tau) + \Delta u/A]}{\omega} - \tau & \text{if } (\sin(\omega\tau) + \Delta u/A \le 1 \wedge \cos(\omega\tau) > 0) \\ & \qquad \vee \sin(\omega\tau) - \Delta u < -1 \\[2ex] \dfrac{\pi \cdot \text{sign}(\tau) - \arcsin[\sin(\omega\tau) - \Delta u/A]}{\omega} - \tau & \text{otherwise} \end{cases}$$

and:

$$\tilde{\tau} = \begin{cases} \tau + \tilde{\sigma} & \text{if } \omega(\tau + \tilde{\sigma}) < \pi \\ \tau + \tilde{\sigma} - \dfrac{2\pi}{\omega} & \text{otherwise} \end{cases}$$

The DEVS model M_7 can be encoded in PowerDEVS as follows:

ATOMIC MODEL SINUS
State Variables and Parameters:
    float $tau, sigma$; //states
    float y; //output
    float A, w, phi, du, pi; //parameters

Init Function:
    va_list $parameters$;
    va_start($parameters, t$);
    $pi = 2 * asin(1)$;
    $A =$ va_arg($parameters$, double);
    $w =$ va_arg($parameters$, double)$*2 * pi$;
    $phi =$ va_arg($parameters$, double);
    $du =$ va_arg($parameters$, double);
    $sigma = 0$;
    $tau = phi/w$;

Time Advance Function:
    return $sigma$;

Internal Transition Function:
    if $(((\sin(w * tau) + du/A <= 1)$ && $(\cos(w * tau) > 0))$ || $(\sin(w * tau) - du/A < -1))$ {
        $sigma = \text{asin}(\sin(w * tau) + du/A)/w - tau$;
    }
    else {
        if $(tau > 0)$ {
            $sigma = (pi - \text{asin}(\sin(w * tau) - du/A))/w - tau$;
        }
        else {
            $sigma = (-pi - \text{asin}(\sin(w * tau) - du/A))/w - tau$;
        };
    };
$tau = tau + sigma$;
if $(tau * w >= pi)\{tau = tau - 2 * pi/w;\}$;

Output Function:
    $y = A * \sin(w * tau)$;;
    return Event(&y,0);

The trajectory generated by this model with parameters $A = 2.001$, $\omega = 0.5$, and $\Delta u = 0.2$ is shown in Fig.12.3.

A piecewise constant trajectory could also be obtained using a constant time step. However, the previously advocated approximation is better in QSS, since the quantization in the values ensures that the distance between the continuous signal an the piecewise constant signal is always less than the quantum. This fact can be easily noticed in Fig.12.3. In contrast, the maximum error would have depended on the relationship between the time step and the signal frequency, had a constant time step been used.

The input signal quantization introduces a new error to the simulation.

Quantized Sine

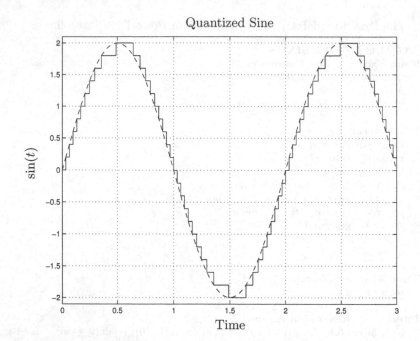

FIGURE 12.3. Piecewise constant sine trajectory.

In the particular case of LTI systems, the presence of the input quantization error transforms Eq.(12.27) into:

$$|\mathbf{x}(t) - \mathbf{x_a}(t)| \quad \leq \quad |\mathbf{V}| \cdot |\mathbb{Re}\{\mathbf{\Lambda}\}^{-1} \cdot \mathbf{\Lambda}| \cdot |\mathbf{V}^{-1}| \cdot \mathbf{\Delta Q}$$
$$+ |\mathbf{V}| \cdot |\mathbb{Re}\{\mathbf{\Lambda}\}^{-1} \cdot \mathbf{V}^{-1} \cdot \mathbf{B}| \cdot \mathbf{\Delta u} \qquad (12.31)$$

where $\mathbf{\Delta u}$ is the vector with the quanta adopted in the input signals. This formula can be derived following the approach developed in [12.18] (cf. Hw.[H12.2]).

12.5 Startup and Output Interpolation

The QSS method startup consists in assigning appropriate initial conditions to the atomic DEVS models that perform the simulation.

The quantized integrator state can be written as $s = (x, d_x, k, \sigma)$ (cf. model M_5 on page 543), where x is the state variable, d_x is its derivative, k is the index corresponding to the quantized variable q_k, and σ is the time advance.

It is clear that we must choose $x = x_i(t_0)$ and k such that $Q_k \leq x_i(t_0) \leq Q_{k+1}$. Finally, appropriate values of d_x and σ can be found from the corresponding state equation $f_i(\mathbf{q}, \mathbf{u})$.

Yet, there is a much simpler solution. If we choose $\sigma = 0$, all quantized integrators will perform internal transitions at the beginning of the simulation and send their initial values to the static functions.

Then, the models associated with the static functions, f_i, will calculate their output values, producing output events instantaneously. These output events will carry the values of the state derivatives. After that, the quantized integrators will receive the correct values of d_x, and they will calculate the corresponding σ in the external transition.

However, this behavior will only be observed if the quantized integrators receive the input events, arriving from the static functions, *after* they performed their own internal transitions. The problem is that, after the first quantized integrator performs its transition, not only the other quantized integrators but also some static models will have their σ set equal to 0, because they undergo an external transition due to the quantized integrator output event.

Thus, it is necessary to establish priorities between the components, in order to ensure that, when some models schedule their next transition for the same instant of time, the quantized integrators are those that perform it first. This can be easily accomplished using the tie–breaking function *Select* mentioned in Section 11.4.

These considerations also solve the problem of the initial conditions of the static model. They can be arbitrarily chosen, since the static models will receive input events arriving from the quantized integrators at the beginning of the simulation, and then their external transition functions will set the appropriate states.

Finally, the input signal generator models must start with their σ set equal to 0, and the rest of the states must be chosen such that the first output event corresponds to the initial value of the input signal.

In PowerDEVS, we treat these problems in the init function. The priorities between subsystems can be easily chosen from the *Edit* menu in the model editor.

Yet, there is another solution that avoids the need of priorities: We can check the external transition function of the quantized integrators for the condition $\sigma = 0$. If that condition is true, this means that an internal transition is going to occur. Thus, we can just leave $\sigma = 0$ in that case, and this solves the initialization problems without using priorities.

When it comes to output interpolation, we already know that the state trajectories in QSS assume particular forms: They are piecewise linear and continuous.

Hence if we know the values adopted by variable x_i at all event times, we can interpolate using straight–line segments, in order to obtain the exact solution of Eq.(12.2). In fact, to do that, we only require the values after external transitions, because the slope does not change during internal transitions.

Consequently, the problem of output interpolation has a straightforward solution in the QSS method.

It is important to remember that Eq.(12.27) and Eq.(12.31) are valid for all values of time t. Thus, we can ensure that the interpolated values remain inside their theoretical error bound, which is an interesting and unusual characteristic for a simulation method.

12.6   Second–order QSS

As we saw along this chapter, the QSS method exhibits strong theoretical and practical properties that make the method attractive for use in the simulation of continuous systems. Unfortunately, the method is only first–order accurate, and therefore, the simulation results obtained with this method cannot be very accurate.

The inequality of Eq.(12.27) states that the error bound grows linearly with the quantum. Thus, if we want to reduce the error bound by a certain amount, we have to reduce the quantum in the same proportion. The problem is that also the time interval between successive events, i.e., the time advance σ in the DEVS model, is proportional to the quantum (cf. model M_5 on page 543). Consequently, the error reduction results in a proportional increment in the number of computations.

To improve the situation, a second–order accurate QSS method was proposed in [12.13]. This new approximation, called *second–order quantized state systems method*, or QSS2 method in short, exhibits similar stability, convergence, and accuracy properties as the previously introduced QSS method.

The basic idea behind this second–order accurate method is the use of *first–order quantizers*, replacing the simple hysteretic quantizers that were used in the design of the QSS method.

A hysteretic quantizer with equal quantum and hysteresis width can be viewed as a system producing a piecewise constant output trajectory that only changes when the difference between output and input reaches a certain threshold level, i.e., the quantum.

Following this idea, a first–order quantizer has been defined as a system that produces a piecewise linear output trajectory having discontinuities only, when its difference with the input reaches the quantum. This behavior is illustrated in Fig.12.4.

Formally, we say that the trajectories $x_i(t)$ and $q_i(t)$ are related by a first–order quantization function, if they satisfy:

$$q_i(t) = \begin{cases} x_i(t) & \text{if } \quad t = t_0 \vee |q_i(t^-) - x_i(t^-)| = \Delta Q \\ q_i(t_j) + m_j \cdot (t - t_j) & \text{otherwise} \end{cases}$$

$$(12.32)$$

with the sequence $t_0, \ldots, t_j, \ldots$ defined as:

First Order Quantizer

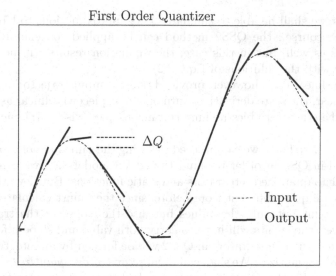

FIGURE 12.4. Input and output trajectories of a first–order quantizer.

$$t_{j+1} = \min(t), \quad \forall t > t_j \wedge |x_i(t_j) + m_j \cdot (t - t_j) - x_i(t)| = \Delta Q$$

and the slopes:

$$m_0 = 0; \quad m_j = \dot{x}_i(t_j^-), \quad j = 1, \ldots, k, \ldots$$

The QSS2 method then simulates a system like that of Eq.(12.2), where the components of $\mathbf{q}(t)$ and $\mathbf{x}(t)$ are related componentwise by first–order quantization functions. As a consequence, the quantized variable trajectories $q_i(t)$ are piecewise linear.

In QSS, we had to add hysteresis in order to ensure that the trajectories become piecewise constant avoiding illegitimacy. The reader may wonder why we did not need to add hysteresis here. The reason is that hysteresis is implicitly present in the definition of the first–order quantization function. Indeed, the absolute value in Eq.(12.32) expresses that hysteretic behavior.

Remember that, in QSS, not only the quantized variables, but also the state derivatives are piecewise constant. In this way, we were able to affirm that the state variables have piecewise linear trajectories, and we exploited this features in building the DEVS model.

Unfortunately, we cannot find this kind of particular trajectories in QSS2. Although the quantized variables are piecewise linear, this does not mean that the state derivatives are piecewise linear as well, even if all inputs possess piecewise linear trajectories. The reason is that a nonlinear function, f_i, applied to a set of piecewise linear trajectories does not necessarily result in a trajectory that is piecewise linear.

Thus, we shall be able to simulate QSS2 approximations to LTI systems only. Of course,, the QSS2 method can be applied to general nonlinear systems as well, but in this case, the simulation results will not coincide exactly with the solutions of Eq.(12.2).

In the linear case, however, provided that the input trajectories are piece-wise linear, the state derivatives turn out to be piecewise linear as well, and then, the state variables assume continuous piecewise parabolic trajecto-ries.

Using these facts, we can proceed following the same lines of thought that we used in QSS in order to build the DEVS model, i.e., we can split the model into quantized integrators and static functions. But now, the atomic models are quite different from before, since they must calculate and take into account not only the values but also the slopes of the trajectories. Moreover, the events will have to carry both value and slope information.

The quantized integrators in QSS2 will be formed by an integrator and a first–order quantizer. We shall call them *second–order quantized integrators*. The reason for this name is that they calculate the state trajectories using their first and second derivatives (i.e., state derivative values and their slopes).

In order to obtain a DEVS model of a second–order quantized integrator, we shall suppose that a state derivative is described, in a certain interval $[t_k, t_{k+1}]$, by:

$$\dot{x}(t) = d_x(t_k) + m_{d_x}(t_k) \cdot (t - t_k) \tag{12.33}$$

where $d_x(t_k)$ is the state derivative at time t_k, and $m_{d_x}(t_k)$ is the cor-responding linear slope. The slope of the state derivative , m_{d_x}, will, in general, be different from the slope of the quantized state variable, m_q.

Then, the state variable trajectory can be written as:

$$x(t) = x(t_k) + d_x(t_k) \cdot (t - t_k) + \frac{m_{d_x}(t_k)}{2} \cdot (t - t_k)^2 \tag{12.34}$$

If t_k is an instant, at which a change occurs in the quantized variable, i.e., t_k is the time instant of an internal transition, then $q(t_k)$ will have the same value and slope as $x(t_k)$:

$$q(t) = x(t_k) + d_x(t_k) \cdot (t - t_k) \tag{12.35}$$

and then, the time instant at which $x(t)$ and $q(t)$ differ from each other by ΔQ can be calculated as:

$$t = t_k + \sqrt{\frac{2 \cdot \Delta Q}{|m_{d_x}(t_k)|}} \tag{12.36}$$

If t_k is another instant in time, i.e., the time of an external transition, the time instant, at which $|q(t) - x(t)| = \Delta Q$ must be recalculated. Now, the quantized trajectory can be written as:

$$q(t) = q(t_k) + m_q(t_k) \cdot (t - t_k) \tag{12.37}$$

and then, we have to calculate the value of t, at which $|q(t) - x(t)| = \Delta Q$, by finding the roots of the corresponding quadratic polynomial.

A DEVS model that can represent the behavior of a second–order quantized integrator is presented below:

$M_8 = (X, S, Y, \delta_{\text{int}}, \delta_{\text{ext}}, \lambda, ta)$, where:

$X = \mathbb{R}^2 \times \mathbb{N}$

$S = \mathbb{R}^5 \times \mathbb{R}_0^+$

$Y = \mathbb{R}^2 \times \mathbb{N}$

$\delta_{\text{int}}(d_x, m_{d_x}, x, q, m_q, \sigma) = (d_x + m_{d_x} \cdot \sigma, m_{d_x}, \tilde{q}, \tilde{q}, d_x + m_{d_x} \cdot \sigma, \sigma_1)$

$\delta_{\text{ext}}(d_x, m_{d_x}, x, q, m_q, \sigma, e, x_v, m_{x_v}, p) = (x_v, m_{x_v}, \tilde{x}, q + m_q \cdot e, m_q, \sigma_2)$

$\lambda(d_x, m_{d_x}, x, q, m_q, \sigma) = (\tilde{q}, d_x + m_{d_x} \cdot \sigma, 0)$

$ta(d_x, m_{d_x}, x, q, m_q, \sigma) = \sigma$

where:

$$\tilde{q} = x + d_x \cdot \sigma + \frac{m_{d_x}}{2} \cdot \sigma^2; \quad \tilde{x} = x + d_x \cdot e + \frac{m_{d_x}}{2} \cdot e^2$$

$$\sigma_1 = \begin{cases} \sqrt{\dfrac{2 \cdot \Delta Q}{|m_{d_x}|}} & \text{if } m_{d_x} \neq 0 \\ \infty & \text{otherwise} \end{cases} \tag{12.38}$$

and σ_2 can be calculated as the smallest positive solution of:

$$\left| \tilde{x} + x_v \cdot \sigma_2 + \frac{m_{x_v}}{2} \cdot \sigma_2^2 - (q + m_q \cdot e + m_q \cdot \sigma_2) \right| = \Delta Q \tag{12.39}$$

The model M_8 represents a second–order quantized integrator with piecewise linear input trajectories exactly.

Equation (12.38) and Eq.(12.39) calculate the time advance, that is, the time instant at which the distance between the piecewise parabolic state trajectory $x(t)$ and the piecewise linear quantized trajectory $q(t)$ reaches the quantum ΔQ.

The corresponding PowerDEVS atomic model can be coded as follows:

ATOMIC MODEL QSS2INT
State Variables and Parameters:
  float $dx, mdx, X, q, mq, sigma$; //states
  float $y[2]$; //output
  float inf, dq; //parameters

Init Function:

```
va_list parameters;
va_start(parameters, t);
dq = va_arg(parameters, double);
X = va_arg(parameters, double);
inf = 1e10;
q = X;
dx = 0;
mdx = 0;
mq = 0;
sigma = 0;
```

Time Advance Function:
```
return sigma;
```

Internal Transition Function:
```
X = X + dx * sigma + mdx/2 * sigma * sigma;
q = X;
dx = dx + mdx * sigma;
mq = dx;
if (mdx == 0) {
    sigma = inf;
}
else
    sigma = sqrt(2 * dq/fabs(mdx));
};
```

External Transition Function:
```
float *xv;
float a, b, c, s;
xv = (float*)(x.value);
X = X + dx * e + mdx/2 * e * e;
dx = xv[0]; //input value
mdx = xv[1]; //input slope
if (sigma != 0) {
    q = q + mq * e;
    a = mdx/2;
    b = dx - mq;
    c = X - q + dq;
    sigma = inf;
    if (a == 0) {
        if (b != 0) {
            s = -c/b;
            if (s > 0) {sigma = s;};
            c = X - q - dq;
            s = -c/b;
            if ((s > 0) && (s < sigma)) {sigma = s;};
        };
    }
    else {
        s = (-b + sqrt(b * b - 4 * a * c))/2/a;
        if (s > 0) {sigma = s;};
        s = (-b - sqrt(b * b - 4 * a * c))/2/a;
        if ((s > 0) && (s < sigma)) {sigma = s;};
        c = X - q - dq;
```

```
        s = (−b + sqrt(b * b − 4 * a * c))/2/a;
        if ((s > 0) && (s < sigma)) {sigma = s;};
        s = (−b − sqrt(b * b − 4 * a * c))/2/a;
        if ((s > 0) && (s < sigma)) {sigma = s;};
    };
};
```

Output Function:
```
    y[0] = X + dx * sigma + mdx/2 * sigma * sigma;
    y[1] = u + mdx * sigma;
    return Event(&y[0], 0);
```

In a QSS2 simulation, we represent the integrators with models of the M_8 class, instead of using those of the M_5 class.

For representing static functions, we used models of the M_3 class in the QSS method. However, the M_3 model does not take into account the slopes. Thus, the representation of static functions in QSS2 requires using a different DEVS model as well.

Each component f_j of a static vector function $\mathbf{f}(\mathbf{q}, \mathbf{u})$ receives the piecewise linear trajectories of the quantized states and input variables.

Let us define $\mathbf{v} \triangleq [\mathbf{q}; \mathbf{u}]$. Each component of $\mathbf{v}$ has a piecewise linear trajectory:

$$v_j(t) = v_j(t_k) + m_{v_j}(t_k) \cdot (t - t_k)$$

Then, the output of the static function can be written as:

$$\dot{x}_i(t) = f_i(\mathbf{v}(t)) = f_i(v_1(t_k) + m_{v_1}(t_k) \cdot (t - t_k), \ldots, v_l(t_k) + m_{v_l}(t_k) \cdot (t - t_k))$$

where $l \triangleq n + m$ is the number of components of $\mathbf{v}(t)$.

Defining $\mathbf{m_v} \triangleq [m_{v_1}, \ldots, m_{v_l}]^T$, the last equation can be rewritten as:

$$\dot{x}_i(t) = f_i(\mathbf{v}(t)) = f_i(\mathbf{v}(t_k) + \mathbf{m_v}(t_k) \cdot (t - t_k))$$

which can be developed into a Taylor series as follows:

$$\dot{x}_i(t) = f_i(\mathbf{v}(t)) = f_i(\mathbf{v}(t_k)) + \left(\frac{\partial f_j}{\partial \mathbf{v}}(\mathbf{v}(t_k))\right)^T \cdot \mathbf{m_v}(t_k) \cdot (t - t_k) + \ldots \quad (12.40)$$

Then, a piecewise linear approximation of the output can be obtained by truncating the Taylor series after the first two terms of Eq.(12.40).

In the linear time–invariant case, we have $\mathbf{f}(\mathbf{v}(t)) = \mathbf{A} \cdot \mathbf{v}(t)$, and therefore: $f_i(\mathbf{v}(t)) = \mathbf{a_i}^T \cdot \mathbf{v}(t)$ where $\mathbf{a_i} \in \mathbb{R}^l$. Then:

$$\dot{x}_i(t) = \mathbf{a_i}^T \cdot \mathbf{v}(t_k) + \mathbf{a_i}^T \cdot \mathbf{m_v}(t_k) \cdot (t - t_k)$$

which means that the output value and its slope are obtained as linear combinations of the input values and their slopes, respectively.

The construction of the corresponding DEVS model is left to the reader (cf. Hw.[H12.4]).

The nonlinear case is a bit more complicated. The expression of Eq.(12.40) requires the knowledge of the partial derivatives of function f_i evaluated at the successive values of the quantized state and input variables. In a general case, we may not have a closed–form expression for these derivatives, in which case we shall need to approximate them numerically.

A DEVS model that follows this idea, representing a static nonlinear function $f_i(\mathbf{v}) = f(v_1, \ldots, v_l)$ and taking into account input and output values and their slopes can be coded as follows :

$$M_9 = (X, S, Y, \delta_{\text{int}}, \delta_{\text{ext}}, \lambda, ta), \text{ where:}$$
$$X = \mathbb{R}^2 \times \mathbb{N}$$
$$S = \mathbb{R}^{3l} \times \mathbb{R}_0^+$$
$$Y = \mathbb{R}^2 \times \mathbb{N}$$
$$\delta_{\text{int}}(\mathbf{v}, \mathbf{m_v}, \mathbf{c}, \sigma) = (\mathbf{v}, \mathbf{m_v}, \mathbf{c}, \infty)$$
$$\delta_{\text{ext}}(\mathbf{v}, \mathbf{m_v}, \mathbf{c}, \sigma, e, x_v, m_{x_v}, p) = (\tilde{\mathbf{v}}, \tilde{\mathbf{m}}_\mathbf{v}, \tilde{\mathbf{c}}, 0)$$
$$\lambda(\mathbf{v}, \mathbf{m_v}, \mathbf{c}, \sigma) = (f_i(\mathbf{v}), m_f, 0)$$
$$ta(\mathbf{v}, \mathbf{m_v}, \mathbf{c}, \sigma) = \sigma$$

where $\mathbf{v} = (v_1, \ldots, v_l)^T$ and $\tilde{\mathbf{v}} = (\tilde{v}_1, \ldots, \tilde{v}_l)^T$ are input values. Similarly, $\mathbf{m_v} = (m_{v_1}, \ldots, m_{v_l})^T$ and $\tilde{\mathbf{m}}_\mathbf{v} = (\tilde{m}_{v_1}, \ldots, \tilde{m}_{v_l})^T$ represent the corresponding input slopes.

The coefficients $\mathbf{c} = (c_1, \ldots, c_l)^T$ and $\tilde{\mathbf{c}} = (\tilde{c}_1, \ldots, \tilde{c}_l)^T$ estimate the partial derivatives $\frac{\partial f_i}{\partial v_j}$ that are used to calculate the output slope in accordance with:

$$m_f = \sum_{j=1}^{n} c_j \cdot m_{v_j}$$

When the system undergoes an external transition, the components of $\tilde{\mathbf{v}}$, $\tilde{\mathbf{m}}_\mathbf{v}$, and $\tilde{\mathbf{c}}$ are calculated using the equations:

$$\tilde{v}_j = \begin{cases} x_v & \text{if } p+1=j \\ v_j + m_{v_j} \cdot e & \text{otherwise} \end{cases}$$

$$\tilde{m}_{v_j} = \begin{cases} m_{x_v} & \text{if } p+1=j \\ m_{v_j} & \text{otherwise} \end{cases}$$

$$\tilde{c}_j = \begin{cases} \dfrac{f_i(\mathbf{v} + \mathbf{m_v} \cdot e) - f_i(\tilde{\mathbf{v}})}{v_j + m_{v_j} \cdot e - \tilde{v}_j} & \text{if } p+1=j \ \wedge \ v_j + m_{v_j} \cdot e - \tilde{v}_j \neq 0 \\ c_j & \text{otherwise} \end{cases}$$

$$(12.41)$$

where p denotes the port number, i.e., determines, which of the inputs is currently undergoing an external transition.

If function $f_i(\mathbf{v})$ is linear, this DEVS model represents the behavior of the system exactly, assuming that the components of $\mathbf{v}$ are piecewise linear. However, as we already mentioned, there exists a much simpler and more efficient solution in that case, since the coefficients c_j are constant and coincide with the entries $a_{i,j}$ of matrix $\mathbf{A}$ (cf. Hw.[H12.4]).

The PowerDEVS model for this general nonlinear static function can then be specified as follows:

ATOMIC MODEL STFUNCTION2

State Variables and Parameters:
```
float sigma, v[10], mv[10], c[10]; //states
float y[2]; //output
float inf;
int l;
```

Init Function:
```
va_list parameters;
va_start(parameters, t);
l = va_arg(parameters, double);
inf = 1e10;
sigma = inf;
for (int i = 0; i < l; i++) {
    v[i] = 0;
    mv[i] = 0;
};
```

Time Advance Function:
```
return sigma;
```

Internal Transition Function:
```
sigma = inf;
```

External Transition Function:
```
float *xv;
float fv, vaux;
xv = (float*)(x.value);
for (int i = 0; i < l; i++) {
    v[i] = v[i] + mv[i] * e;
};
fv = f_j(v);//put your function here
vaux = v[x.port];
v[x.port] = xv[0];
mv[x.port] = xv[1];
y[0] = f_j(v); //put your function here
if (vaux != v[x.port]) {
    c[x.port] = (fv - y[0])/(vaux - v[x.port]);
};
y[1] = 0;
for (int i = 0; i < l; i++) {
    y[1] = y[1] + mv[i] * c[i];
};
```

$sigma = 0;$

Output Function:
    return Event($\&y[0], 0$);

The only problem with the PowerDEVS model proposed above is that a new atomic model must be introduced for each distinct function f_j. However, this problem has already been solved in PowerDEVS by introducing a function that parses algebraic expressions. Thus, the corresponding nonlinear function block included in the *Continuous* library of PowerDEVS, offers a parameter consisting in a string that contains the algebraic expression describing the function. That expression, just like any other parameter, can be modified by double clicking on the block.

There are also some particular nonlinear functions, for which the partial derivatives can be calculated analytically and the coefficients c_j do not have to be computed using Eq.(12.41). Some examples of such functions are the sin() function, the multiplier, and the $(\cdot)^2$ block included in the *Continuous* library (cf. Hw.[H12.5]).

By coupling DEVS models of the M_8 and M_9 classes in the same way as we did in Fig.11.12, we can use the QSS2 method to simulate any time–invariant ODE system. Using PowerDEVS, we can simulate any time–invariant ODE system using the QSS2 method by building the block diagram from QSS2INT models, used in place of the quantized hysteretic integrators, and from STFUNCTION2 blocks, used instead of the static functions introduced earlier.

We already mentioned that the QSS2 method shares the main properties of QSS. The reasons behind this assertion can be easily explained. Two variables, $x_i(t)$ and $q_i(t)$, that are related by a first–order quantization function satisfy:

$$|q_i(t) - x_i(t)| \leq \Delta Q_i \quad \forall t \tag{12.42}$$

This inequality is just a particular case of Eq.(12.4) with a constant quantum equal to the hysteresis width. The QSS properties were derived using this inequality, which implies that the method only introduces a bounded perturbation in a system that can be represented in the form of Eq.(12.3). Taking into account that this representation is also valid for QSS2, we conclude that the QSS2 method satisfies the same convergence, stability, and accuracy properties as QSS.

However in nonlinear systems, the QSS2 definition does not coincide exactly with the DEVS simulation. Thus, our analysis only ensures in a strict sense that those properties hold true in the simulation of LTI systems. Although there are many good reasons that allow us to conjecture that the convergence and stability properties would hold true in the simulation of nonlinear systems as well, there has not yet been found a formal proof of this conjecture.

As far as the error bound is concerned, the inequalities of Eq.(12.27) and Eq.(12.31) hold true for the QSS2 method, since they were derived for LTI systems. Thus, if we use the same quantum in QSS and QSS2, we obtain the same error bound in both cases.

This last remark gives rise to a question: Where is the advantage of using the QSS2 method, if it offers the same error bound as the QSS method?

This question can be answered by Eq.(12.38) and Fig.12.4.

On the one hand in QSS, the time advance is proportional to the quantum and to the error. In QSS2 however, it is proportional to the square root of the quantum, as Eq.(12.38) shows. For this reason, we can reduce the quantum without obtaining a proportional increment in the number of calculations, when using the QSS2 method.

This fact can be clearly observed in Fig.12.4, where the use of a simple hysteretic quantizer instead of a first–order quantizer with the same quantum would have resulted in a much larger number of events.

On the other hand, each transition in the QSS2 method involves more computations than in QSS. Thus, if we are not interested in obtaining simulation results that are highly accurate, the QSS method may turn out to be more efficient.

All these facts are discussed in more detail in [12.13], where an experimental comparison between the execution times of both methods was presented, illustrating the characteristics of the two methods as outlined in the above paragraph.

Beside from the theoretical properties that were derived from perturbation analysis, we saw that the QSS method also exhibits practical advantages related to the incorporation of input signals and the exploitation of sparsity. Let us discuss then what happens with these practical issues in the QSS2 method.

The sparsity exploitation is straightforward. We conserve the same simulation structure as in QSS, and each transition only involves calculation at the integrators and static functions that are directly connected to the integrator that undergoes the transition.

When it comes to input signals, we have now further advantages. We not only ensure that changes are being processed as soon as they occur, but we are furthermore able to correctly represent piecewise linear instead of just piecewise constant input trajectories.

Let us illustrate these advantages in the following example, taken from [12.13].

The circuit of Fig.12.5 represents an RLC transmission line. A similar model had already been introduced in Chapter 10 of this book.

This model can be used to study the performance of integrated circuits transmitting data at a very fast rate. Although the wires are only a few centimeters long, the high frequency of the transmitted signal requires that the delays introduced by the wires must not be ignored, and transmission line theory must be applied.

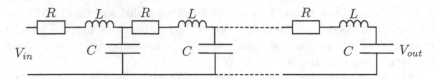

FIGURE 12.5. RLC transmission line.

Transmission lines are described as systems of partial differential equations. However, they can be approximated by lumped models, where the distributed effects of capacity, inductance, and resistance are approximated by a cascade of single capacitors, inductors, and resistors, as Fig.12.5 shows. In order to constitute a good approximation, the RLC model must be formed by several sections. As a consequence of this, the resulting model is a linear time–invariant system of ordinary differential equations with a sparse system matrix.

In [12.6], an example composed by five sections of an RLC circuit is introduced. The resistance, inductance, and capacitance values used in [12.6] can be considered realistic parameter values. The model obtained is a 10th–order linear time–invariant system with the following system matrix:

$$\mathbf{A} = \begin{pmatrix} -R/L & -1/L & 0 & 0 & 0 & 0 & 0 & 0 & 0 & 0 \\ 1/C & 0 & -1/C & 0 & 0 & 0 & 0 & 0 & 0 & 0 \\ 0 & 1/L & -R/L & -1/L & 0 & 0 & 0 & 0 & 0 & 0 \\ 0 & 0 & 1/C & 0 & -1/C & 0 & 0 & 0 & 0 & 0 \\ \vdots & \vdots & \vdots & \vdots & \vdots & \vdots & \vdots & \vdots & \vdots & \vdots \\ 0 & 0 & 0 & 0 & 0 & 0 & 0 & 0 & 1/C & 0 \end{pmatrix}$$

A typical input trajectory for these digital systems is a trapezoidal wave, representing the "0" and "1" levels, as well as the rising and falling edges. Since a trapezoidal wave is a piecewise linear trajectory, we can generate it exactly using the following DEVS model:

$$M_{10} = (X, S, Y, \delta_{\text{int}}, \delta_{\text{ext}}, \lambda, ta), \text{ where:}$$
$$X = \emptyset$$
$$S = \mathbb{N} \times \mathbb{R}_0^+$$
$$Y = \mathbb{R}^2 \times \mathbb{N}$$
$$\delta_{\text{int}}(k, \sigma) = (\tilde{k}, T_{\tilde{k}})$$
$$\lambda(k, \sigma) = (u_{\tilde{k}}, m_{u_{\tilde{k}}}, 1)$$
$$ta(k, \sigma) = \sigma$$

where $\tilde{k} = (k + 1 \mod 4)$ is the next cycle index, which has 4 phases: The low state (index 0), the rising edge (1), the high state (2), and the falling edge (3). The duration of each phase is given by the corresponding T_k value.

During the low state, the output is u_0, and during the high state, it is u_2. During these two phases, the slopes, m_{u_0} and m_{u_2}, are zero. During the rising edge, we have $u_1 = u_0$, and the slope is $m_{u_1} = (u_2 - u_0)/T_1$. Similarly, during the falling edge, we have $u_3 = u_2$ and $m_{u_3} = (u_0 - u_2)/T_3$.

The DEVS generator representing the input trajectory produces only four events in each cycle. This is an important advantage, since the presence of the input wave only adds a few extra calculations. Moreover, since the representation is exact, it does not introduce any error, i.e., we can estimate the error bound using the inequality of Eq.(12.27) instead of that of Eq.(12.31).

We performed the simulation using the parameter values $R = 80\ \Omega$, $C = 0.2\ pF$, and $L = 20\ nH$. These parameter values correspond to a transmission line of one centimeter length divided into five sections, where the line resistance, capacitance, and inductance values are $400\ \Omega/cm$, $1\ pF/cm$, and $100\ nH/cm$, respectively.

The trapezoidal input has rising and falling times of $T_1 = T_3 = 10\ psec$, whereas the durations of the low and high states are $T_0 = T_2 = 1\ nsec$. The low and high levels are $0\ V$ and $2.5\ V$, respectively.

The quantization adopted was $\Delta v = 4\ mV$ for the state variables representing voltages, and $\Delta i = 10\ \mu A$ for the state variables representing currents. This quantization, in accordance with Eq.(12.27), ensures that the maximum error is smaller than $250\ mV$ in the variable V_{out}.

The input and output trajectories are shown in Fig.12.6.

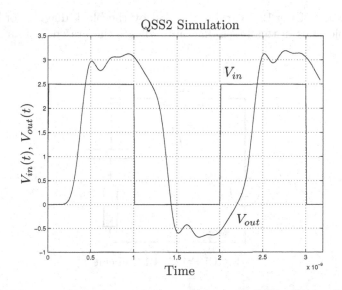

FIGURE 12.6. QSS2 simulation results in an RLC transmission line.

The simulation required a total of 2536 steps (between 198 and 319 internal transitions at each integrator) to obtain the first 3.2 *nsec* of the system trajectories.

The experiment was repeated using a 100 times smaller quantization, which ensures a maximum error on the output voltage, V_{out}, of 2.5 *mV*. This new simulation was performed consuming a total of 26.883 internal transitions. Here we can see the effects of the second–order approximation: when reducing the quantum by a factor of 100, the number of events grows only by a factor of 10, i.e., it grows inverse to the square root of the quantum.

We also compared the trajectories of both simulations, and the difference in V_{out} was always less than 14.5 *mV*. The conclusion is that the error in the first simulation was less than 17 *mV*, in spite of the theoretical bound of 250 *mV*. The error bound formula given by Eq.(12.27) often produces highly conservative results, especially when it is applied to high order systems.

Although the number of steps in the simulations is big, it is important to remember that each step only involves scalar calculations at three integrators, the integrator that undergoes the internal transition, and the two integrators that are directly connected to its output. This is due to the sparsity of the **A**–matrix.

12.7   Algebraic Loops in QSS Methods

The circuit of Fig.12.7 can be modeled by the block diagram of Fig.12.8. The bold lines in this block diagram indicate the presence of an algebraic loop.

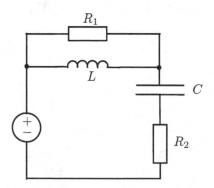

FIGURE 12.7. RLC circuit.

This algebraic loop expresses the algebraic restriction:

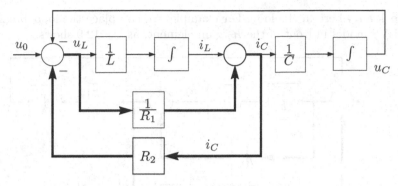

FIGURE 12.8. Block diagram representation of the RLC circuit.

$$i_C = i_L + \frac{1}{R_1}(u_0 - u_C - R_2 \cdot i_C) \qquad (12.43)$$

We can implement the QSS method by transforming the integrators into DEVS models, e.g. of the M_5 class (quantized hysteretic integrators), and the static functions into their DEVS equivalent representation (DEVS models such as M_3). Then, we can couple these DEVS models according to the coupling scheme of Fig.12.8. In fact, that is precisely what we did to convert the system of Eq.(12.1) into the DEVS representation of Eq.(12.2) (cf. Fig.11.12).

As you were taught in the companion book on *Continuous System Modeling* [12.2], block diagrams are not necessarily the most convenient tool for modeling physical systems, but they are in use widely, and some of the most popular continuous system simulation tools, in particular SIMULINK [12.3], are built on this modeling paradigm. Hence also the graphical user interface of PowerDEVS was built around block diagrams.

Although the block by block translation from a block diagram representation of a continuous system to its corresponding DEVS model may result in inefficient simulation code from the point of view of computational cost, it is a very simple procedure that does not require any kind of symbolic manipulations. Thus, if we do not want to perform the translation manually, and if we do not have another automatic tool, such as Dymola [12.4], available for generating a set of equations, like those of Eq.(12.1), from a higher–level graphical representation of the system to be simulated, the block by block translation may be the most convenient way for applying a QSS method to the simulation of a continuous–time system.

However, if we apply this procedure to the block diagram of Fig.12.8, we encounter a problem. Due to the algebraic loop, the resulting DEVS model will turn out to be illegitimate. When an event arrives at the loop, it propagates forever through the static functions around the algebraic loop.

As we do not want to drastically alter the block diagram or the atomic model definitions, we tackle the problem by adding a new loop–breaking block anywhere in the loop. For example, we can place the loop–breaking DEVS model in front of the R_2 gain element, as Fig.12.9 shows.

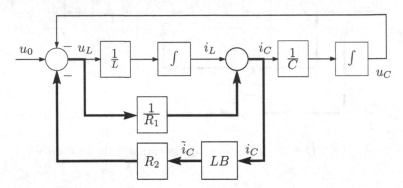

FIGURE 12.9. Addition of a loop-breaking model to the block diagram of Fig.12.8.

Now, Eq.(12.43) becomes:

$$i_C = i_L + \frac{1}{R_1}(u_0 - u_C - R_2 \cdot \tilde{i}_C) \tag{12.44}$$

where $\tilde{i}_C$ is the output value of the loop–breaking block.

Since this block is inside the loop, whenever it sends an event with a value $\tilde{i}_C$ out through its output port, it immediately (in terms of simulation time) receives an event with value i_C, calculated using Eq.(12.44), back through its input port.

If we want Eq.(12.44) to be equivalent to Eq.(12.43), we need to ensure that $\tilde{i}_C$ is equal to i_C. In other words, the value received by the loop–breaking model must be the same that it previously sent out.

Thus, the loop–breaking block could operate as follows: it sends $\tilde{i}_C$ out and receives i_C back. If the two signals differ from each other, it tries with a different $\tilde{i}_C$. Otherwise, the loop–breaking block becomes passive and doesn't send out any further events, until a new external event, caused by a transition of an integrator or an input function, arrives at the loop.

This technique should solve our problem. Notice that the proposed technique corresponds closely to the *tearing method* introduced in Chapter 7 of this book. i_C is a tearing variable. The user will need to introduce enough tearing variables (loop–breaking blocks) to break all algebraic loops in the system.

So far, we have not explained, how the value of $\tilde{i}_C$ is to be calculated. Yet, before providing an answer to this question, we need to reformulate our problem in a more general framework.

Let us call z the variable sent by the loop–breaking model. Then, when it sends an event with value z_1, it immediately receives a new event with value $h(z_1)$ calculated by the static functions.

Thus, the model should calculate a new value for z, let us call it z_2, that should satisfy:

$$h(z_2) - z_2 \triangleq g(z_2) \approx 0 \tag{12.45}$$

If $g(z_2)$ remains too large, the process must be repeated by sending a new value z_3.

Clearly, z_{i+1} must be calculated following some algorithm to find the solution of $g(z) = 0$. Taking into account that the loop–breaking block does not know the expression, and hence the derivative, of $g(z)$, a good alternative to Newton iteration is the use of the *secant method*.

Using this approach, z_{i+1} can be calculated as:

$$z_{i+1} = \frac{z_{i-1} \cdot g(z_i) - z_i \cdot g(z_{i-1})}{g(z_i) - g(z_{i-1})} \tag{12.46}$$

and since $g(z_i) = h(z_i) - z_i$, we obtain:

$$z_{i+1} = \frac{z_{i-1} \cdot h(z_i) - z_i \cdot h(z_{i-1})}{h(z_i) - h(z_{i-1}) + z_{i-1} - z_i} \tag{12.47}$$

Based on these ideas, the loop–breaking DEVS model can be represented as follows:

$$M_{11} = (X, Y, S, \delta_{\text{int}}, \delta_{\text{ext}}, \lambda, ta), \text{ where}$$
$$X = \mathbb{R} \times \mathbb{N}$$
$$Y = \mathbb{R} \times \mathbb{N}$$
$$S = \mathbb{R}^3 \times \mathbb{R}_0^+$$
$$\delta_{\text{ext}}(s, e, x) = \delta_{\text{ext}}(z_1, z_2, h_1, \sigma, e, x_v, p) = \tilde{s}$$
$$\delta_{\text{int}}(s) = \delta_{\text{int}}(z_1, z_2, h_1, \sigma) = (z_1, z_2, h_1, \infty)$$
$$\lambda(s) = \lambda(z_1, z_2, h_1, \sigma) = (z_2, 1)$$
$$ta(s) = ta(z_1, z_2, h_1, \sigma) = \sigma$$

where:

$$\tilde{s} = \begin{cases} (z_1, z_2, h_1, \infty) & \text{if} \quad |x_v - z_2| < tol \\ (z_2, \tilde{z}, x_v, 0) & \text{otherwise} \end{cases}$$

with:

$$\tilde{z} = \frac{z_1 \cdot x_v - z_2 \cdot h_1}{x_v - h_1 + z_1 - z_2} \tag{12.48}$$

The parameter *tol* represents the largest absolute error that we allow between z and h. Equation (12.48) is the result of applying the secant method to approximate $g(z) = 0$, where $g(z)$ is defined in accordance with Eq.(12.45). We can change the iteration algorithm by modifying Eq.(12.48).

A corresponding PowerDEVS model can be coded as follows:

ATOMIC MODEL LOOP-BREAK1
State Variables and Parameters:
  float $z1, z2, h1, sigma$; //states
  float y; //output
  float tol, inf;

Init Function:
  va_list $parameters$;
  va_start($parameters, t$);
  $tol =$ va_arg($parameters$, double);
  $inf = 1e10$;
  $sigma = inf$;
  $z1 = 0$;
  $z2 = 0$;
  $h1 = 0$;
  $y = 0$;
Time Advance Function:
  return $sigma$;

Internal Transition Function:
  $sigma = inf$;

External Transition Function:
  float xv;
  $xv =$*(float*)($x.value$);
  if (fabs($xv - z2$) $< tol$) {
    $sigma = inf$;
  }
  else{
    $z3 = z2$;
    if (($z1 ==0$) && ($z2 == 0$)){
      $z2 = xv$; //initial guess
    }
    else {
      $z2 = (z1 * xv - z2 * h1)/(xv - h1 + z1 - z2)$;
    };
    $z1 = z3$;
    $h1 = xv$;
    $sigma = 0$;
  };

Output Function:
  $y = z2$;
  return Event(&y, 0);

For the circuit example of Fig.12.7, we built a coupled DEVS model in accordance with Fig.12.9 and simulated it during 30 seconds using the QSS algorithm. We used the parameter values $R_1 = R_2 = L = C = 1$, and u_0 was chosen as a unit step. The quantum and hysteresis adopted were 0.01 in both state variables, and the error tolerance tol was chosen equal to 0.001.

The simulation, the results of which are shown in Fig.12.10, was completed after 118 and 72 internal transitions at each quantized integrator and a total of 377 iterations at the loop–breaking DEVS model.

QSS Simulation

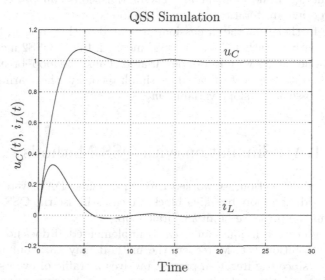

FIGURE 12.10. QSS simulation of the RLC circuit using a loop–breaking DEVS.

In this case, due to the linearity of the system, the secant method arrives at the exact solution of $g(z) = 0$ after only two iterations. This explains why the total number of iterations at the loop–breaking model was twice the total number of steps at both quantized integrators.

In a more general nonlinear case, more iterations per transition would probably be needed, and we would have to discuss the convergence criteria associated with the chosen iteration method.

An interesting observation is that the effect of the error in the calculation of i_C can be seen as an additional perturbation. If we can ensure that this error remains bounded, the perturbation is also bounded and can be seen as something equivalent to having a bigger quantum that only affects the error bound but does not modify the stability properties.

¿From the discussion in this section, the reader might reach the conclusion that the QSS methods deal with algebraic loops in the same, or at least a very similar, way as the discrete–time methods introduced in earlier chapters. However, such a conclusion would be totally wrong.

Almost all of the discrete–time methods presented earlier in this book are centralized integration schemes that require the iteration of *all* algebraic loops during *every* integration step, or even more accurately, during each function evaluation.

In contrast, the QSS methods operate in a completely asynchronous fashion. An algebraic loop will only be iterated upon when it gets triggered by a transition occurring either in a quantized integrator or in an input function. Due to the inherent sparsity property of large–scale physical systems, plenty of transitions may take place in a larger model that do not affect any of the algebraic loops at all. These transitions can proceed without ever triggering an iteration on any of the loops.

Notice further that the discussion, presented in this section, focused on the QSS method, rather than the QSS2 method. In the QSS2 method, each loop variable carries with it a slope variable. Thus, the loop–breaking block will need to iterate on two variables simultaneously: the tearing variable, z, and its associated slope variable, m_z.

12.8   DAE Simulation with QSS Methods

In the previous section, we worked with a particular DAE system. We saw that, by adding a loop–breaking block, we can still use the QSS simulation method in the presence of an algebraic loop.

Although this technique can be easily implemented, it does not constitute a general solution yet. Moreover, the method may turn out to be fairy inefficient, since the iteration process involves a traffic of events across all blocks that constitute the loop.

This section is aimed at introducing a more general case and a more efficient solution. As usual, we shall start by analyzing an example.

Figure 12.11 shows the transmission line model of Fig.12.5, modified by the addition of a load.

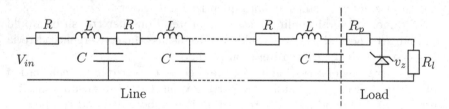

FIGURE 12.11. RLC transmission line with surge voltage protection.

The load is composed of a resistor R_l, possibly representing the gate of some electronic component, and a surge protection circuit formed by a Zener diode and a resistor R_p. The Zener diode satisfies the following nonlinear relationship between its voltage and its current:

$$i_z = \frac{I_0}{1 - (v_z/v_{br})^m} \tag{12.49}$$

where m, v_{br}, and I_0 are parameters, the values of which depend on the physical characteristics of the device.

If the transmission line is divided into five sections, as we did earlier, the following equations are obtained:

$$\frac{di_1}{dt} = \frac{1}{L} \cdot v_{in} - \frac{R}{L} \cdot i_1 - \frac{1}{L} \cdot u_1$$

$$\frac{du_1}{dt} = \frac{1}{C} \cdot i_1 - \frac{1}{C} \cdot i_2$$

$$\frac{di_2}{dt} = \frac{1}{L} \cdot u_1 - \frac{R}{L} \cdot i_2 - \frac{1}{L} \cdot u_2$$

$$\frac{du_2}{dt} = \frac{1}{C} \cdot i_2 - \frac{1}{C} \cdot i_3 \qquad (12.50)$$

$$\vdots$$

$$\frac{di_5}{dt} = \frac{1}{L} \cdot u_4 - \frac{R}{L} \cdot i_5 - \frac{1}{L} \cdot u_5$$

$$\frac{du_5}{dt} = \frac{1}{C} \cdot i_5 - \frac{1}{R_p C} \cdot (u_5 - v_z)$$

Here, the state variables, u_j and i_j, represent the voltage and current in the capacitors and inductors of the transmission line, respectively, and the output voltage, v_z, is an algebraic variable that satisfies the equation:

$$\frac{1}{R_p} \cdot u_5 - \left(\frac{1}{R_p} + \frac{1}{R_l} \right) \cdot v_z - \frac{I_0}{1 - (v_z/v_{br})^m} = 0 \qquad (12.51)$$

Thus, we are confronted with a DAE that cannot be converted into an ODE by symbolic manipulation, and the simulation using any discrete–time method will have to iterate on Eq.(12.51) during each step, in order to solve for the unknown variable v_z.

If we want to apply the QSS method to the system defined by Eq.(12.50) and Eq.(12.51), we can try to proceed as before, replacing the state variables i_j and u_j by their quantized versions.

In order to be able to use our standard notation, we define $x_{2j-1} \triangleq i_j$ and $x_{2j} \triangleq u_j$ for $j = 1, \ldots, 5$. As before, we shall refer to the quantized version of variable x_j as q_j.

Then, the use of QSS transforms Eq.(12.50) to:

$$\frac{dx_1}{dt} = \frac{1}{L} \cdot v_{in} - \frac{R}{L} \cdot q_1 - \frac{1}{L} \cdot q_2$$

$$\frac{dx_2}{dt} = \frac{1}{C} \cdot q_1 - \frac{1}{C} \cdot q_3$$

$$\frac{dx_3}{dt} = \frac{1}{L} \cdot q_2 - \frac{R}{L} \cdot q_3 - \frac{1}{L} \cdot q_4$$

$$\frac{dx_4}{dt} = \frac{1}{C} \cdot q_3 - \frac{1}{C} \cdot q_5 \qquad (12.52)$$

$$\vdots$$

$$\frac{dx_9}{dt} = \frac{1}{L} \cdot q_8 - \frac{R}{L} \cdot q_9 - \frac{1}{L} \cdot q_{10}$$

$$\frac{dx_{10}}{dt} = \frac{1}{C} \cdot q_9 - \frac{1}{R_p C} \cdot (q_{10} - v_z)$$

and the implicit part of the system, i.e., Eq.(12.51), turns into:

$$\frac{1}{R_p} \cdot q_{10} - \left(\frac{1}{R_p} + \frac{1}{R_l} \right) \cdot v_z - \frac{I_0}{1 - (v_z/v_{br})^m} = 0 \qquad (12.53)$$

Notice that Eq.(12.52) looks like a quantized state system, except for the presence of v_z. However, the variable v_z is algebraically coupled to q_{10}. Thus, each time that q_{10} undergoes a transition, we iterate on Eq.(12.53) to find the new value of v_z, and then use that value in Eq.(12.52), however and contrary to the previously proposed solution involving a loop–breaking block, this iteration occurs entirely within a single DEVS model, and therefore doesn't involve events being passed around between different blocks in a loop.

We can build a block diagram corresponding to Eq.(12.52) and Eq.(12.53) as follows:

- We start by representing system Eq.(12.52) with quantized integrators and static functions, treating v_z as if it were an external input.

- We then add a new atomic block that computes v_z as a function of q_{10}. Consequently, this block has q_{10} as an input and v_z as an output.

The latter block will be in charge of iteration to determine a new value of v_z, each time q_{10} changes. Ignoring round–off errors and assuming that the iteration block computes the value of v_z correctly, the resulting coupled DEVS model will exactly simulate the system defined by Eq.(12.52) and Eq.(12.53).

Notice that the iteration block is only activated by a change in q_{10}. In all other steps, i.e., when either one of the other nine quantized variables changes, or when the input changes, no iteration has to be performed, and the QSS models acts like in the case of an explicit ODE model. Clearly, QSS is still able to exploit the sparsity inherent in DAE models of large–scale physical systems.

Our original system, defined by Eq.(12.50) and Eq.(12.51), is a particular case of the implicit model introduced in Chapter 8:

$$\mathbf{f}(\mathbf{x}, \dot{\mathbf{x}}, \mathbf{u}, t) = 0 \qquad (12.54)$$

In that chapter, we studied methods for simulating this model without transforming it to explicit ODE form first.

Let us check whether we can apply similar ideas to the situation of a QSS simulation involving algebraic loops.

For simplicity, we shall only considering the time–invariant case, although the explicit inclusion of time is not problematic, especially in the context of performing a QSS2 simulation.

We can rewrite Eq.(12.54) as follows:

$$\tilde{\mathbf{f}}(\dot{\mathbf{x}}_{\mathbf{a}}, \mathbf{x}_{\mathbf{a}}, \mathbf{u}) = 0 \qquad (12.55)$$

As before, we call the state vector of the original system $\mathbf{x_a}$ to distinguish it from the quantized state vector.

Proceeding as we did before, we can modify Eq.(12.55) as follows:

$$\tilde{\mathbf{f}}(\dot{\mathbf{x}}, \mathbf{q}, \mathbf{u}) = 0 \qquad (12.56)$$

where $\mathbf{x}(t)$ and $\mathbf{q}(t)$ are related componentwise by hysteretic quantization functions.

Now, we can apply Newton iteration to Eq.(12.56) to solve for $\dot{\mathbf{x}}$. We shall assume that the perturbation index of Eq.(12.56) is 1. Otherwise, we shall apply the Pantelides algorithm first, in order to reduce the perturbation index of the DAE system to 1.

Once we have obtained numerical values for the state derivatives, they can be sent to the quantized integrators that perform the rest of the job, i.e., calculate the quantized variable trajectories. Each time a quantized variable changes, a new iteration process must be performed in order to recalculate $\dot{\mathbf{x}}$.

However in our previous example, we only needed to iterate, when q_{10} changed. For some reason, we lost the capability of exploiting sparsity in the compact notation of Eq.(12.56).

As we still want to be able to exploit sparsity, we need to rewrite Eq.(12.55) as follows:

$$\dot{\mathbf{x}}_\mathbf{a} = \mathbf{f}(\mathbf{x_a}, \mathbf{u}, \mathbf{z_a}) \qquad (12.57a)$$
$$0 = \mathbf{g}(\mathbf{x}_{\mathbf{a}_r}, \mathbf{u_r}, \mathbf{z_a}) \qquad (12.57b)$$

where $\mathbf{z_a}$ is a vector of tearing variables with dimension equal to or less than n. The vectors $\mathbf{x}_{\mathbf{a}_r}$ and $\mathbf{u_r}$ are reduced versions of $\mathbf{x_a}$ and $\mathbf{u}$, respectively.

A straightforward –but useless– way of transforming the system represented by Eq.(12.55) into the system of Eq.(12.57) is by defining $\mathbf{z_a} \triangleq \dot{\mathbf{x}}_\mathbf{a}$, which yields $\mathbf{x}_{\mathbf{a}_r} = \mathbf{x_a}$, $\mathbf{u_r} = \mathbf{u}$, and $\mathbf{g} = \mathbf{f} - \mathbf{z_a}$.

However in many cases, as in the case of the transmission line example, the dimensions of $\mathbf{x}_{\mathbf{a}_r}$ and $\mathbf{u_r}$ can be effectively reduced.

Equation (12.57b) expresses the fact that some state and input variables may not act directly on the algebraic loops.

Then, the use of the QSS methods transforms Eq.(12.57) into:

$$\dot{\mathbf{x}} = \mathbf{f}(\mathbf{q}, \mathbf{u}, \mathbf{z}) \qquad (12.58a)$$
$$0 = \mathbf{g}(\mathbf{q_r}, \mathbf{u_r}, \mathbf{z}) \qquad (12.58b)$$

and now, an iteration will only be performed, when components of either $\mathbf{q_r}$ or $\mathbf{u_r}$ change.

The use of QSS and QSS2 methods with DAE systems is quite similar. In order to simplify the derivation, we shall only consider the first–order method for now. We shall add remarks relating to the QSS2 method later.

When Eq.(12.58b) defines the value of **z**, we can see that the system of Eq.(12.58) defines something that behaves like a QSS. In fact, we can easily prove that the state and quantized variable trajectories correspond to a QSS (i.e., they are piecewise linear and piecewise constant, respectively). Moreover, the auxiliary variables **z** are also piecewise constant.

What still needs to be explained now is how an implicitly defined QSS can be translated into a DEVS model.

It is clear that Eq.(12.58a) can be represented by quantized integrators and static functions, as we did in Chapter 11. The only difference here is the presence of the auxiliary variables **z**, acting as inputs just like **u**. However, whereas the inputs **u** arrive from signal generators of the M_6 or M_7 class, the auxiliary variables must be calculated by solving the constraints of Eq.(12.58b).

Thus, a new DEVS model must be created. This DEVS model receives events with the values of either $\mathbf{q_r}$ or $\mathbf{u_r}$, and calculates a new value of **z** in return that it then sends out through its output port.

A DEVS model that solves a general implicit equation, such as:

$$\mathbf{g}(\mathbf{v}, \mathbf{z}) = \mathbf{g}(v_1, \ldots, v_m, z_1, \ldots, z_k) = 0 \qquad (12.59)$$

can be written as follows:

$$M_{12} = (X, Y, S, \delta_{\text{int}}, \delta_{\text{ext}}, \lambda, ta), \text{ where}$$
$$X = \mathbb{R} \times \mathbb{N}$$
$$Y = \mathbb{R}^k \times \mathbb{N}$$
$$S = \mathbb{R}^{m+k} \times \mathbb{R}_0^+$$
$$\delta_{\text{ext}}(s, e, x) = \delta_{\text{ext}}(\mathbf{v}, \mathbf{z}, \sigma, e, x_v, p) = (\tilde{\mathbf{v}}, \mathbf{h}(\tilde{\mathbf{v}}, \mathbf{z}), 0)$$
$$\delta_{\text{int}}(s) = \delta_{\text{int}}(\mathbf{v}, \mathbf{z}, \sigma) = (\mathbf{v}, \mathbf{z}, \infty)$$
$$\lambda(s) = \lambda(\mathbf{v}, \mathbf{z}, \sigma) = (\mathbf{z}, 1)$$
$$ta(s) = ta(\mathbf{v}, \mathbf{z}, \sigma) = \sigma$$

where:

$$\tilde{\mathbf{v}} = (\tilde{v}_1, \ldots, \tilde{v}_m)^T; \quad \tilde{v}_i = \begin{cases} x_v & \text{if} \quad p = i \\ v_i & \text{otherwise} \end{cases}$$

and the function $\mathbf{h}(\tilde{\mathbf{v}}, \mathbf{z})$ returns the result of applying a Newton iteration or some other type of iteration to find the solution of Eq.(12.59) using an initial value **z**.

When the size of **z** (i.e., k) is greater than 1, the output events of model M_{12} contain a vector. Thus, they cannot be sent to static functions such as M_3. However, we can use a DEVS *demultiplexer* (cf. Hw.[H11.3]), in order to tackle this problem.

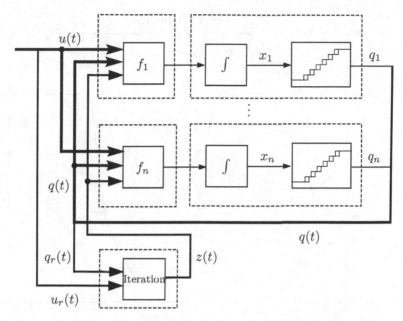

FIGURE 12.12. Coupling scheme for the QSS simulation of Eq.(12.57).

Figure 12.12 shows the new coupling scheme with the addition of a new DEVS model that calculates **z**.

When it comes to the QSS2 method, the same ideas can be applied, and analogous DEVS models to M_{12} can be built. The only difficulty is that now the trajectory slopes must be taken into account. This is not a problem in linear systems, but it becomes a bit more complicated in nonlinear systems, where estimations of the partial derivatives should be used. However, this problem has already been solved, and a DEVS model replacing M_{12} for the QSS2 method is provided in [12.14].

PowerDEVS also offers a nonlinear implicit model that solves a constraint of the form $g(\mathbf{v}, z) = 0$, calculating both the value and slope of z. The symbolic expression g is a string parameter that can be modified by double clicking on the implicit model icon.

We are now ready to return to the transmission line example. Using the ideas expressed above, we modified the simulation of page 581 by adding a PowerDEVS model like the one introduced in the previous paragraph that solves Eq.(12.51), taking into account the slope.

Letting $R_l = 100\ M\Omega$, $I_0 = 0.1\ \mu A$, $v_{br} = 2.5\ V$, and $m = 4$ without modifying the remaining parameters, we obtained the results shown in Fig.12.13.

The first $3.2ns$ of the simulation were completed after 2640 steps (between 200 and 316 steps at each integrator). The implicit model performed a total of 485 iterations using the secant method. The reason for this is that the quantized integrator that calculates u_5 only performed 200 inter-

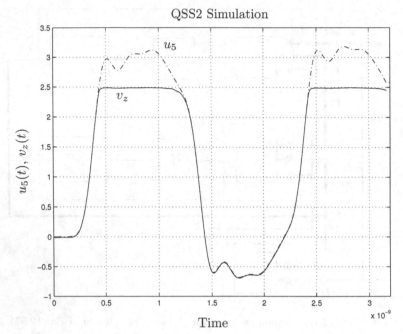

FIGURE 12.13. QSS2 simulation results in an RLC transmission line with surge protection.

nal transitions, and therefore, the implicit model received only 200 external events. The secant method needed between two and three iterations to find the solution of Eq.(12.51) with the required tolerance of $tol = 1 \times 10^{-8}$, which explains the fact that the total number of iterations was 485.

The advantages of the QSS2 method are evident in this example. In a discrete–time algorithm, the secant method would have been invoked at every integration step, whereas the QSS2 only called it after changes in u_5, i.e., about once every 13 steps. Thus, the presence of the implicit equation only adds a few calculations that do not affect significantly the total number of computations.

Returning once more to the issue of higher–index DAEs, we already mentioned that such DAEs can be simulated by reducing their perturbation index to index 1 first using the Pantelides algorithm, introduced in Chapter 7 of this book, and then applying the QSS approach, presented in this section, to the resulting index–1 DAE system. However, this may not be the only way to tackling higher–index DAE problems using a QSS method. An alternate solution was proposed in [12.9]. There, it is shown that applying the QSS method to a higher–index DAE results in a model that switches between two (or more) ODE systems. Thus, the simulation can be performed applying the same principles that rule the simulation of variable structure systems.

We shall not explore this idea any further here, since it was developed in the context of bond graph modeling only. Although we believe that this technique may also be applicable to general DAE systems, such a generalization would require more research.

12.9 Discontinuity Handling

In Chapter 9, we studied the simulation of discontinuous systems using discrete–time approaches. We remarked that numerical ODE solvers are based on Taylor–Series expansions, and therefore, their trajectories are approximated by polynomials or rational functions. Since neither of these functions exhibit discontinuities, the solvers will invariably be in trouble, when asked to integrate across discontinuous functions.

However, the same restriction does not hold in the case of the QSS and QSS2 methods. Here, the discontinuities in the input and quantized variable trajectories are in fact responsible for time advance. Simulation steps are only being calculated, when discontinuities are found in those trajectories.

Using discrete–time methods, the time instances of discontinuities had to be determined precisely, as we were unable to integrate across discontinuities accurately. Consequently, the primary difficulties in dealing with discontinuous functions using discrete–time methods are related to accurately detecting the time of occurrence of a discontinuity, and to designing suitable integration step–size control algorithms around them.

Two different kinds of discontinuities were distinguished in our analysis: *time events*, that could be scheduled ahead of time, as their time of occurrence was known in advance, and *state events*, that were specified indirectly by means of some threshold crossing function, which required the use of an iteration algorithm (a root solver algorithm) for locating them accurately in time.

As we shall discover soon, all of these problems disappear with the use of QSS and QSS2 methods.

The simulation of hybrid systems using QSS and QSS2 was studied in [12.15] and, as the reader may already have anticipated, the most important advantages of the discrete event approximations to continuous system simulation are to be found in these kinds of applications.

Let us begin by analyzing a simple example. The inverter circuit shown in Fig.12.14 is a device typically used to power electrical machines that are being operated off the grid.

The set of switches can assume two different positions. In the first position, switches 1 and 4 are closed, and consequently, the load receives a positive voltage. In the second position, switches 2 and 3 are closed, and the load sees a negative voltage accordingly.

The system can be represented by the following differential equation:

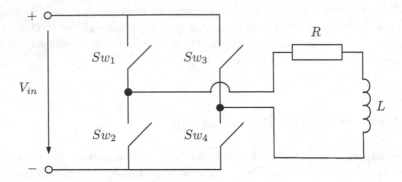

FIGURE 12.14. DC–AC full bridge inverter circuit.

$$\frac{di_L}{dt} = -\frac{R}{L} \cdot i_L + V_{in} \cdot s_w(t) \tag{12.60}$$

where s_w assumes a value of either $+1$ or -1, depending on the position that the four switches are operating in.

A typical way of controlling the switches in order to obtain an approximately sinusoidal current at he load is by using a *pulse width modulation (PWM)* strategy.

The PWM signal is obtained by comparing a triangular wave, the so–called carrier, with a modulating sinusoidal reference signal. The sign of the voltage to be applied, $+V_{in}$ or $-V_{in}$, and thereby the corresponding switch position, is determined by the sign of the difference between these two signals. Figure 12.15 illustrates this concept.

The difference between the carrier and modulation signal could be used as a zero–crossing function of a state–event description for the control strategy. However, since both the carrier signal and the modulating signal are simple static functions, the time of the next intersection between these two signals can easily be computed ahead of time, allowing an implementation of the PWM control strategy by means of time events.

The system can be thought of as the coupling of a continuous submodel, described by Eq.(12.60), and a discrete submodel that manages a sequence of time events for determining the correct value of s_w.

The discrete submodel can easily be represented as a DEVS model. It is a simple DEVS generator model of the M_6 class, introduced on page 565. The output alternates between $+1$ and -1, and the time elapse between commutations can be numerically computed ahead of time.

The continuous submodel can be approximated, using either the QSS or the QSS2 method, by transforming Eq.(12.60) to:

$$\frac{di_L}{dt} = -\frac{R}{L} \cdot q_{i_L} + V_{in} \cdot s_w(t) \tag{12.61}$$

where q_{i_L} is the quantized state associated with variable i_L.

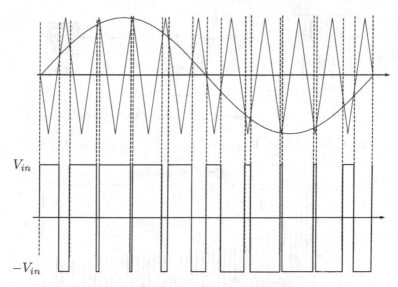

FIGURE 12.15. Pulse width modulation.

If we consider the last equation only, and if nobody tells us that $s_w(t)$ originates at a discrete submodel, we could think that Eq.(12.61) corresponds to the QSS or QSS2 approximation of a continuous model with an input trajectory $s_w(t)$.

Indeed, the QSS methods effectively treat $s_w(t)$ as an input without regard for where that signal originates. Since the changes in s_w are treated asynchronously, the QSS integration has no problems with accommodating this signal.

Thus, ignoring round–off errors, and assuming that the DEVS model generating $s_w(t)$ works properly, the system of Eq.(12.61) will be simulated exactly.

Moreover, as Eq.(12.60) is linear and $s_w(t)$ is piecewise constant, i.e., we can use the exact input trajectory, we can apply Eq.(12.27) to calculate the global error bound. Consequently, the error of i_L is bounded by the quantum used.

To corroborate these remarks, we simulated the system with the QSS2 method using PowerDEVS.

We first built a new block to provoke the correct sequence s_w. To this end, we assumed that the triangular carrier has a frequency of 1.6 kHz, and that the modulating sinusoidal signal has the same amplitude, but a frequency of 50 Hz.

Thus, the number of events per modulating cycle is $2 \cdot 1600/50 = 64$, which is sufficient for producing a fairly smooth sinusoidal current.

The PowerDEVS model is shown in Fig.12.16.

Employing the parameter values $R = 0.6\ \Omega$, $L = 100\ mH$, and $V_{in} = 300\ V$, the simulation starting from $i_L = 0$ and using a quantization of

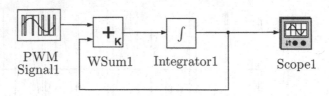

PWM
Signal1   WSum1   Integrator1   Scope1

FIGURE 12.16. PowerDEVS model of the inverter circuit.

$\Delta i_L = 0.01\ A$ produced the results shown in Figs.12.17–12.18.

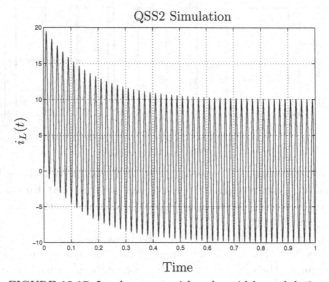

FIGURE 12.17. Load current with pulse width modulation.

The final time of the simulation chosen was 1 *sec*, and thus, the number of simulated cycles was 50. The discrete system underwent a total of 3200 changes in the switch positions, i.e., 3200 time events took place.

In spite of this large number of time events, the simulation was completed after only 3100 internal transitions at the second order quantized integrator. Thus, the total number of simulation steps was $3100 + 3200 = 6300$.

As mentioned before, the error is bounded by the quantum. The trajectories depicted in Figs.12.17–12.18 have an error that is no larger than 0.01 *A* at any instant of time, corresponding to roughly 0.1% of the oscillation amplitude.

The simulation of the same system with discrete–time methods, even using the most appropriate event handling techniques, requires many more integration steps, and there is little that we can say about the global error bound.

This example shows that time events are treated in a very natural way by QSS methods. The only new thing that we had to do is to express the discrete subsystem as a DEVS model and connect its output to the

QSS2 Simulation

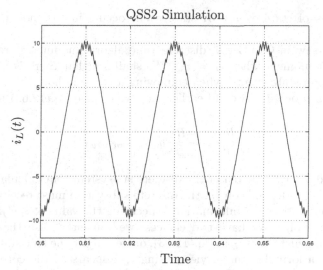

FIGURE 12.18. Steady–state behavior of load current.

continuous submodel.

Let us now discuss what happens in the presence of state events. To this end, we shall modify our previous example.

Due to failures (a short circuit in the load for instance) or during transients, the load current of the circuit in Fig.12.14 might take on values that are too large and thus could damage the components. To prevent this from happening, such circuits are usually protected.

A simple and cheap surge protection scheme consists in measuring the load current and, when it surpasses the allowed value, to close switches 2 and 4, so that the voltage applied to the load becomes 0. Then, when the current has returned to a level below the maximum allowed value, the circuit resumes its normal operation.

Applying this strategy to our example converts Eq.(12.60) to:

$$\frac{di_L}{dt} = -\frac{R}{L} \cdot i_L + V_{in} \cdot \tilde{s}_w(t) \tag{12.62}$$

where:

$$\tilde{s}_w(t) = \begin{cases} s_w(t) & \text{if } i_L(t) < i_M \\ 0 & \text{otherwise} \end{cases} \tag{12.63}$$

and i_M is the maximum allowed current.

In a real application, we should add hysteresis to the protection, and we should also prevent the condition $i_L < -i_M$. However, as this examples is introduced for illustrative purposes only, we shall limit our discussion to the simplified version.

Variable $\tilde{s}_w$ depends on the value of the state i_L. As the value of i_L cannot be computed analytically, the surge protection logic must be implemented by means of state events. A state event occurs in the model, whenever $i_L = i_M$.

As we saw in Chapter 9, discrete–time algorithms must iterate to find the exact moment when $i_L = i_M$. We shall see that there is no need for iterating on state events, when employing a QSS method.

We can proceed as we did earlier, i.e., we change Eq.(12.62) to:

$$\frac{di_L}{dt} = -\frac{R}{L} \cdot q_{i_L} + V_{in} \cdot \tilde{s}_w(t) \tag{12.64}$$

and build a DEVS model of the discrete subsystem that calculates $\tilde{s}_w(t)$.

Since $\tilde{s}_w(t)$ depends on i_L, the discrete subsystem must receive from the continuous subsystem information concerning the value of $i_L(t)$.

At this point, we have two choices: we can either use the quantized variable q_{i_L} instead of i_L in Eq.(12.63), or we can use the true state variable.

¿From a formal point of view, using i_L appears as the correct choice. Indeed, this is the idea followed in [12.15].

However from a practical point of view, it is much simpler to use q_{i_L}. Although we can obtain the successive values of the state variables (we already studied the problem of output interpolation), the quantized variables are directly seen at the output of the quantized integrators.

Moreover, the state and quantized variables never differ from each other by more than the quantum ΔQ. Thus, the replacement will not introduce a large error in general. If i_M is a hard limit, it would suffice to modify the state condition to $i_L = i_M - \Delta Q$ to ensure that the current i_L will never surpass the value of i_M.

Thus, for the moment, we shall use q_{i_L} instead of i_L for the discrete subsystem. We shall revisit this issue later on in the chapter.

The discrete subsystem can be formed by two atomic models. The first block generates $s_w(t)$ as before (the block PWM signal of Fig.12.16), and the second block sends events out with either the value s_w or 0 depending on the value of q_{i_L}. This second block, in order to decide the value to be sent out, must receive the previously calculated value of s_w and the successive values of q_{i_L} through its input ports.

If we are using the first–order accurate QSS method, q_{i_L} is piecewise constant. Provided that a quantization level equal to i_M exists, the detection of the condition $q_{i_L} = i_M$ is straightforward.

If such a level does not exists, we will not be able to detect the exact condition, since it will never occur. However, we can easily detect the crossings, because, when they occur, we have that $q_{i_L}(t_{k-}) < i_M$ and $q_{i_L}(t_k) > i_M$. Thus, the time of occurrence of the state event can be computed exactly.

In the case of the QSS2 method, the trajectory q_{i_L} will be piecewise linear. Thus, we can easily compute the precise instant in time, when q_{i_L} crosses i_M, by solving a linear equation. The time to the next crossing can be exactly calculated as:

$$\sigma = \begin{cases} (q_{i_L} - i_M)/m_q & \text{if } m_q \neq 0 \text{ and } (q_{i_L} - i_M)/m_q > 0 \\ \infty & \text{otherwise} \end{cases} \quad (12.65)$$

However, as q_{i_L} is discontinuous (cf. Fig.12.4), it can happen that q_{i_L} jumps over the event condition, and we detect a situation where $q_{i_L}(t_{k-}) < i_M$ and $q_{i_L}(t_k) > i_M$. In this case, the time of occurrence of the crossing can again be detected exactly.

PowerDEVS offers several blocks that detect and handle discontinuities in accordance with these concepts. For our example, we used a *Switch* block, that predicts the intersection of a piecewise parabolic trajectory with a given fixed threshold value. Then, when this trajectory, entering the block through the second input port, is greater than the threshold value, the output sends out the trajectory received through the first input port. Otherwise, it sends out the trajectory received through the third input port.

The PowerDEVS model of the circuit with the surge protection is shown in Fig.12.19. A *Delay* block was added, modeling the fact that the switches don't react instantaneously.

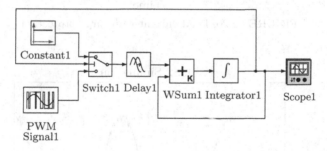

FIGURE 12.19. PowerDEVS model of inverter circuit with surge protection.

Had we not included the delay, the resulting model would have been illegitimate. The reason is that when the condition $i_L = i_M$ occurs, $\tilde{s}_w$ is set to 0, and consequently, the slope in i_L becomes negative. Thus, $\tilde{s}_w$ changes to 1 again, and a cyclic behavior is obtained without time advance.

As mentioned earlier in this chapter, the illegitimacy issue could also have been avoided by adding hysteretic behavior to the crossing condition. Such an approach might in fact be preferable to the delay solution, but the solution with the delay block suffices for illustrating, how PowerDEVS can deal with state events.

We simulated the model with the QSS2 method using the same parameters as before, while letting $i_M = 11$ A, and choosing a delay of $\Delta T = 1 \times 10^{-6}$ sec.

As in the previous example, the PWM block generated 3200 time events, but the switch now sent out 3288 events to the continuous subsystem. At least 88 state events must consequently have occurred during the simulation. The quantized integrator now performed 3188 steps.

Figures 12.20–12.21 display the simulation results.

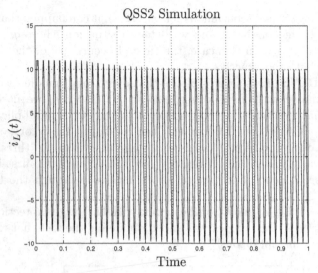

FIGURE 12.20. Load current with surge protection.

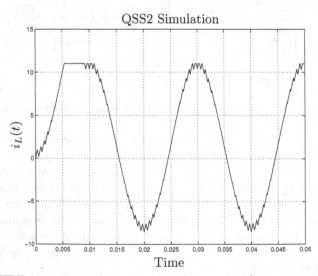

FIGURE 12.21. Initial behavior of load current with surge protection.

In this new situation, we cannot say anything about the global error bound. In the previous example, we knew that $s_w(t)$ was exactly generated. But now, $\tilde{s}_w(t)$ depends on i_L, which is not known exactly.

As mentioned above, we could have used i_L instead of q_{i_L} for the detection of the state events. However, the prediction is a bit more involved, since we must now solve a quadratic equation in Eq.(12.65). Moreover, in order to obtain the successive values of i_L, we need to look at the derivative of this signal and reintegrate it.

Thus, depending on the application, we can choose using either the continuous states or the quantized state variables for the detection of state events. Using the quantized state variables is more efficient computationally, but using the continuous states offers more accurate simulation results.

After these introductory examples, we are now ready to analyze a more general case.

We shall assume that the continuous subsystem can be represented by a set of DAEs, as specified below:

$$\dot{\mathbf{x}}_\mathbf{a}(t) = \mathbf{f}(\mathbf{x}_\mathbf{a}(t), \mathbf{u}(t), \mathbf{z}_\mathbf{a}(t), m_a(t)) \tag{12.66a}$$

$$0 = \mathbf{g}(\mathbf{x}_{\mathbf{a}_\mathbf{r}}(t), \mathbf{u}_\mathbf{r}(t), \mathbf{z}_\mathbf{a}(t), m_a(t)) \tag{12.66b}$$

where $m_a(t)$ is a piecewise constant trajectory emanating at the discrete subsystem that defines the different operational modes of the system (this was the role of s_w and $\tilde{s}_w$ in our first and second example, respectively). Thus for each value of $m_a(t)$, there is a different DAE representing the system dynamics.

As we did in the previous section, we shall assume that the implicit equation, Eq.(12.66b), has a solution for each value of $m_a(t)$, which implies that the system of Eq.(12.66) does not exhibit conditional index changes.

Independently of the way in which $m_a(t)$ is being calculated, the submodel corresponding to the continuous part can be built as before, considering $m_a(t)$ as an input.

Then, the QSS and QSS2 methods will transform Eq.(12.66) to:

$$\dot{\mathbf{x}}(t) = \mathbf{f}(\mathbf{q}(t), \mathbf{u}(t), \mathbf{z}(t), m(t)) \tag{12.67a}$$

$$0 = \mathbf{g}(\mathbf{q}_\mathbf{r}(t), \mathbf{u}_\mathbf{r}(t), \mathbf{z}(t), m(t)) \tag{12.67b}$$

with the same definitions that we used in Eq.(12.58). Consequently, the simulation scheme for the continuous subsystem will be identical to the one shown in Fig.12.12, but now, $m(t)$ must also be calculated and included with the input $\mathbf{u}(t)$.

The way, in which $m(t)$ is calculated, is determined by the discrete subsystem.

One of the most important features of DEVS is its capability for representing any kind of discrete system. Taking into account that the continuous subsystem has been approximated by a DEVS model, it is only natural to also represent the discrete subsystem by another DEVS model. Then, both DEVS models can be directly coupled building a single coupled DEVS model that approximates the entire system.

The DEVS model of the discrete subsystem will provoke events that carry the successive values of $m(t)$.

Taking into account the asynchronous fashion, in which the static functions and quantized integrators work, the events arriving from the discrete subsystem will be processed by the continuous subsystem as soon as they arrive, without a need of modifying anything in the QSS or QSS2 methods. Efficient event handling is an intrinsic characteristic of the QSS methods.

In the presence of state events, the discrete subsystem needs to detect the occurrence of such events by monitoring zero–crossing functions that are functions of inputs and state variables.

Here, the QSS and QSS2 methods have an even bigger advantage over discrete–time methods: input and state trajectories are known functions of time in a local context. They are either piecewise constant, or piecewise linear, or piecewise quadratic functions of time. For this reason, the time of occurrence of a state event that is about to take place can be calculated analytically, which makes it unnecessary to iterate on state events.

The only thing that has to be done is to provide those trajectories to the discrete subsystem, so that it can detect the occurrence of state events and calculate the trajectory $m(t)$, which it then passes on to the continuous subsystem. The continuous subsystem performs the state event handling in accordance with the changes in $m(t)$, as if these were mere time events.

The continuous state trajectories, $\mathbf{x}(t)$, are not directly available at the output of the quantized integrators. Only the quantized states, $\mathbf{q}(t)$, are generated by the quantized integrators. However, the state derivative functions are available. These can be integrated to obtain $\mathbf{x}(t)$. This is furthermore a very simple task that does not require much computational effort at all, since the state derivative trajectories are piecewise constant or piecewise linear (in QSS2), and obtaining their integrals only takes one or two simple calculations with the coefficients of the corresponding polynomials.

An alternative –and simpler– solution, which was the solution chosen in the introductory examples of this section, consists in using the quantized states instead of the continuous state variables in the discrete part. However, we may increase the error by doing so.

Using these ideas, the simulation model of a hybrid system, such as that of Eq.(12.66), using the QSS or QSS2 method will be a coupled DEVS with the structure shown in Fig.12.22.

Here the discrete part is a DEVS model that receives the events representing changes in the input trajectories and the quantized state variables. Alternatively to the quantized state variables, we could provide the state derivative signals to the discrete subsystem.

As state events can be detected before they occur, and since the discrete subsystem can set its time advance to that moment, ensuring the correct treatment of the corresponding event, there is no need for iterations or

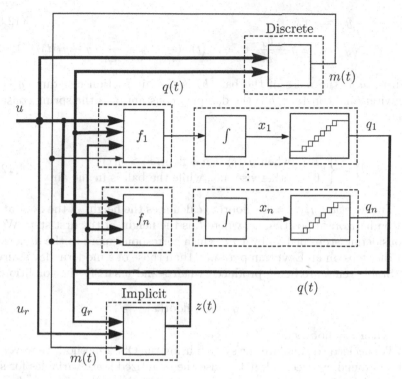

FIGURE 12.22. Coupling scheme for the QSS simulation of discontinuous systems.

back–stepping, in order to hit the event instants accurately. This is a very important advantage in the context of real–time simulation, as we shall discuss shortly.

In the following example, we shall see some further advantages of applying the QSS methods to the simulation of discontinuous systems.

A typical textbook example of a discontinuous system containing state events is the *bouncing ball* problem. We shall consider the case, where the ball moves in two directions, x and y, bouncing down a stairway. Thus, the bouncing condition depends on the two variables, x and y.

We postulate that the ball experiences air friction, which we shall assume linear for simplicity (in physics textbooks, air friction is usually assumed quadratic in the velocity), and we declare that, when the ball reaches the floor, it shall behave like a spring–damper system.

A corresponding differential equation model for the bouncing ball problem can be formulated as follows:

$$\dot{x} = v_x \tag{12.68a}$$

$$\dot{v}_x = -\frac{b_a}{m} \cdot v_x \tag{12.68b}$$

$$\dot{y} = v_y \qquad\qquad (12.68c)$$

$$\dot{v}_y = -g - \frac{b_a}{m} \cdot v_y - s_w(t) \cdot (\frac{b}{m} \cdot v_x + \frac{k}{m} \cdot (y - y_f(t))) \quad (12.68d)$$

where m is the mass of the ball, b_a is the air friction constant, g is the gravitational constant, b is the damping constant, k is the spring constant, and:

$$s_w(t) = \left\{ \begin{array}{ll} 1 & \text{when } y_f(t) \triangleq h - \text{floor}(x) > y(t) \\ 0 & \text{otherwise, i.e., while the ball is in the air} \end{array} \right. \qquad (12.69)$$

The function $y_f(t) = h - \text{floor}(x)$ calculates the height of the floor at any given horizontal position, x, where h is the height of the first step. We are considering steps of 1 m by 1 m, which correspond more to the dimensions of the steps on an Egyptian pyramid, than those of a modern–day stairway.

State events are being produced when x and y satisfy the condition:

$$y = h - \text{floor}(x)$$

or when $x = \text{floor}(x)$.

We decided to simulate the system using the QSS2 method. However this time around, we decided, not to use the quantized state variables for state event detection. Instead, we shall provide the derivatives of x and y, i.e., variables v_x and v_y to the discrete subsystem and re–integrate them.

The right hand side of Eq.(12.68), i.e., the continuous subsystem, depends only on the three variables v_x, v_y, and y. Consequently, the quantized state variable corresponding to x does not appear at all in the quantized system. Hence the quantized integrator that calculates it can be omitted.

Figure 12.23 shows the PowerDEVS model of the system. Here, the integrators to the left calculate v_y and y, whereas the integrator to the right computes v_x.

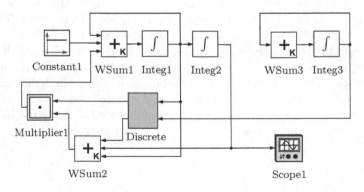

FIGURE 12.23. PowerDEVS bouncing ball model.

The discrete subsystem has been encapsulated. Figure 12.24 shows the complete discrete submodel.

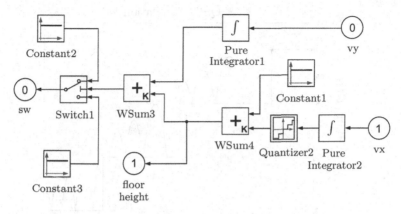

FIGURE 12.24. PowerDEVS bouncing ball model (discrete subsystem).

The discrete submodel contains two *Pure Integrator* blocks. These blocks compute exactly the integral of piecewise linear trajectories.

When one of these blocks receives an event with value $d_x(t_k)$ and linear slope $m_{d_x}(t_k)$, it immediately sends out an event with value $x(t_k) = x(t_{k-1}) + m_x(t_{k-1}) \cdot (t_k - t_{k-1}) + p_x(t_{k-1}) \cdot (t_k - t_{k-1})^2$, where the linear slope, m_x, is computed as $m_x(t_k) = d_x(t_k)$, and the quadratic slope, p_x, assumes a value of $p_x(t_k) = m_{d_x}(t_k)/2$. Then, it waits until a new input event arrives.

Thus, the trajectories calculated by these blocks are piecewise parabolic and continuous. In our example, they compute the state trajectories of $x(t)$ and $y(t)$.

The trajectory $x(t)$ is then sent to a *Quantizer* block that calculates floor(x) in order to evaluate the height of the floor. This is a block provided by the *Hybrid* library of PowerDEVS.

We already encountered the *Switch* block once before in this section. In the given model, its second input port receives the values of $y(t) - y_f(t)$, and predicts, when this signal crosses the threshold of 0 in either direction, in order to detect the next floor contact or floor separation event.

For the simulation, we used the parameter values $k = 100,000 \ N/m$, $b = 30 \ kg/s$, $ba = 0.1 \ kg/s$, and $m = 1 \ kg$. To simplify the logistics of the simulation (quite significantly, to tell the truth!), the ball is assumed to be infinitely small. Thus, the spring–damper system represents in fact the elasticity of the floor, rather than that of the bouncing ball.

A quantum of 0.01 m/s was chosen for the quantized integrators that calculate v_x and v_y, whereas a quantum of 0.0001 m was chosen for the quantized integrator that calculates y. The initial conditions were $x(0) = 0.575 \ m$, $v_x(0) = 0.5 \ m/s$, $y(0) = 10.5 \ m$ and $v_y = 0 \ m/s$. We simulated the system across 10 seconds of simulated time.

Figure 12.25 shows the simulation results for this system.

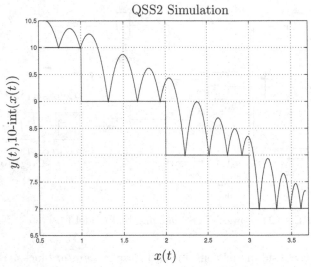

FIGURE 12.25. Ball bouncing down some stairs.

The integrator that calculates v_x required five steps, the integrator that calculates v_y required 518 steps, and the integrator that calculates y required 2413 steps. Hence the *Quantizer* block received only five external events. It detected four crossings by integer values, as the ball advanced four steps downward. Consequently, the routine that predicts crossings by solving a quadratic polynomial was invoked only nine times.

Similarly, the *Switch* block received 522 events through its second port, 518 of which originated at the quantized integrator of v_y, whereas the remaining four originated at the *Quantizer* block. It detected a total of 20 crossings, as each bounce results in two separate events, a floor contact event, and a floor separation event.

Notice that the second bounce was produced near the border of a step. Here, discrete–time methods will experience problems. Figure 12.26 details the result of simulating this system using a variable–step Runge–Kutta method (ode45 of MATLAB) with two different accuracy settings.

Using a sufficiently large tolerance value, the method skips the event. Since the time elapse between subsequent function evaluations is larger than the zone, in which the event condition is triggered, the algorithm does not recognize that it has passed through that zone.

An example of this problem was given in [12.5], where the authors proposed a solution based on decreasing the step size as the system approximates the discontinuity condition. In Chapter 9 of this book, we had proposed another solution involving the use of the derivative of the zero–crossing function as an additional (dummy) zero–crossing function.

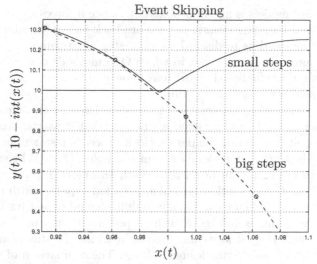

FIGURE 12.26. Event skipping in discrete–time algorithms.

In the QSS methods, this problem disappears. Provided that we use the continuous states instead of the quantized state variables for detecting the discontinuities, as we did in this example, it is impossible for QSS methods to skip *short–living state events* as discrete–time methods do.

12.10   Real–time Simulation

In Chapter 10, we studied, which algorithms were suitable in the context of real–time simulation. We concluded that fixed–step algorithms were necessary, and we only accepted explicit or semi–implicit methods. In the latter case, we tried furthermore to reduce the size of the matrix to be inverted using mixed–mode integration techniques. Thus at a first glance, QSS and QSS2 don't seem to fit the bill very well, since these are clearly variable–step algorithms.

We offered two reasons for discarding variable–step methods. The first reason was that we were not able to spend time discarding values and repeating steps, when we did not like the error obtained. The second reason was that the inputs and outputs of most real–time applications occur at fixed time intervals, which constitutes another good point in favor of fixed–step algorithms.

However, none of these reasons apply to quantization–based methods, since they never discard values, since they accept input changes at any point in time, and since their states are known at any instant of time also. QSS methods conveniently solve the *dense output* problem that we had encountered in Chapter 9 of this book.

We furthermore mentioned in Chapter 10 that real–time simulation of discontinuous models is highly problematic, since it forces us to spend time detecting and handling events, i.e., we are back facing the same problems that had convinced us to discard variable–step algorithms.

Taking into account the way, in which QSS methods deal with discontinuities, quantization–based algorithms appear to be a very good choice for handling discontinuities in real time.

Moreover, not all variables have to be calculated at the same rate. We can let each variable update itself at the rate that it wants. We only need to ensure that the overall simulation clock proceeds faster than real time.

Let us illustrate these ideas with an example. The PWM strategy introduced in Fig.12.15 can be used to synthesize more general signals than sinusoidal functions. Another typical application of pulse width modulation is the control of DC motors. Since it is difficult and expensive to generate a variable continuous voltage with high power output, the desired power signal is replaced by a switching signal, in which the duration of the *on* state is proportional to the desired voltage. The comparison of the desired voltage with a fast triangular waveform permits achieving such behavior.

Figure 12.27 shows the block diagram of a control system based on this technique. The controller compares the speed $\omega(t)$ with the input reference, and calculates the desired input voltage of the motor, u_{ref}. This value is then compared to a triangular waveform, obtaining the actual input voltage u_a that oscillates between two values.

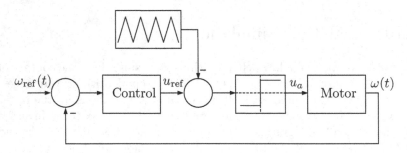

FIGURE 12.27. PWM controlled DC motor.

A real–time simulation of this kind of model may be needed in order to detect the presence of failures in the physical system that the model represents. As in the example of a watchdog monitor for a nuclear power station, mentioned in Chapter 10, the simulation output can be compared to the physical system output, and when the two signals differ significantly one from another, we can conclude that something went wrong with the system.

The simulation of this system requires accurate event detection. The moment, at which the triangular wave crosses the value given by u_{ref}, determines the actual voltage applied to the motor. A small error in the calculation of this time instant leads to a significant change in the output waveform.

An additional problem is that the triangular wave operates at a high frequency, and therefore, state events occur with a high rate. Thus, the efficient treatment of discontinuities becomes crucial.

The motor can be represented by a second order model of the form:

$$\frac{di_a}{dt} = \frac{1}{L_a} \cdot (u_a(t) - R_a \cdot i_a - k_m \cdot \omega)$$

$$\frac{d\omega}{dt} = \frac{1}{J} \cdot (k_m \cdot i_a - b \cdot \omega - \tau(t))$$

where $L_a = 3\ mH$ is the armature inductance, $R_a = 50\ m\Omega$ is the armature resistance, $J = 15\ kg \cdot m^2$ is the inertia, $b = 0.005\ kg \cdot m^2/s$ is the friction coefficient, and $k_m = 6.785\ V \cdot s$ is the electro–motorical force (EMF) constant of the motor. These parameter values correspond to those of a real system.

The inputs are the armature voltage, $u_a(t)$, and the torque load, $\tau(t)$, respectively. In the given example, $u_a(t)$ switches between $+500\ V$ and $-500\ V$ depending on the PWM control law. A torque step of $2500\ N \cdot m$ is applied after 3 seconds of simulation.

For the PWM law, we consider a triangular waveform of $1\ kHz$ frequency with an amplitude of $1.1\ V$. This triangular wave is compared to the error signal, saturated at a value of $1\ V$. The control is using a proportional law, i.e., u_{ref} is proportional to the error $\omega_{ref}(t) - \omega(t)$.

The angular velocity reference signal, $\omega_{ref}(t)$ is a ramp signal that increases from 0 to 60 rad/sec in 2 seconds.

As a first step, we performed an off–line simulation using the PowerDEVS model of Fig.12.28.

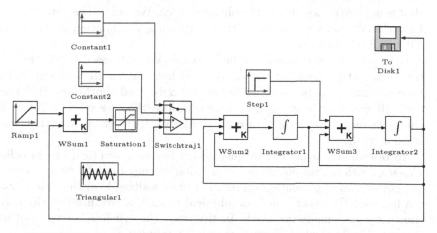

FIGURE 12.28. PowerDEVS model of the PWM controlled DC motor.

A simulation across 5 seconds of simulated time took about 0.63 seconds of real time on a 950 Mhz computer running under Windows XP. This allows to predict that we may be able to simulate the system in real time. We used QSS2 with a quantum of 0.01 for both state variables, which produced 1952 and 43,263 steps in the integrators corresponding to the angular velocity and current, respectively. An additional 10,000 events were caused by the triangular wave generator, and another 10,000 events were provoked by the switch. The saturation block also produced 398 events, which led to a total of 65,613 steps.

Before proceeding to the real time experiment, we need to discuss the real time requirements of the input and output signals. We shall assume that the angular velocity, $\omega(t)$, needs to be measured once each 10 $msec$, and that it suffices to provide the model with updated values of the input signal, $\omega_{ref}(t)$, at the same rate.

Thus, we added *sample and hold* blocks that provoke events every 10 $msec$ to both the input and output signals.

In principle, it would not have been necessary to provide clocked events for the inputs and outputs, as DEVS models require input values only, when a change occurs in the input pattern, and as they are able to compute dense outputs on their own, as all trajectories are known signals in a local context. For the given input signal, it would thus have sufficed to provide a single event at the time, when the ramp goes into saturation. However, it may be convenient to operate with clocked input and output signals in a real time environment, as this allows to synchronize model inputs with signals, whose trajectories are not known in advance, and as it allows to hook the output trajectories of the real–time simulation to equipment that does not know anything about DEVS models.

One way to simulate this system in real time is to synchronize the entire simulation with a physical clock, so that each event is performed at a time that is as close as possible to the physical clock. We can do this directly with PowerDEVS, since it has the option of running a simulation synchronized with a physical clock.

However, this is unnecessary in our case. We only need to ensure that the values at the exits of the sample and hold blocks are sent out at the correct time instants. In other words, we only need synchronization once each 10 $msec$. Of course, in order to achieve this, we need to finish all calculations corresponding to each period of 10 $msec$, before that time window ends.

To obtain this behavior, we just added two new identical blocks called *Clock wait* that send out an event, as soon as they receive one in terms of *simulated* time, but contain an internal busy waiting loop that waits with sending out the event, until the physical clock has advanced to the value shown by the simulation clock. In this way, the synchronization routines are only invoked, when and where they are needed.

Figure 12.29 shows the modified PowerDEVS model that can be used for real–time simulation.

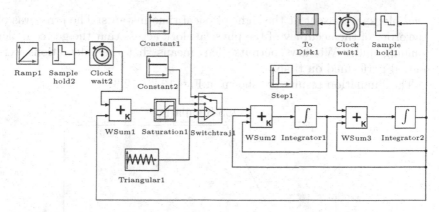

FIGURE 12.29. PowerDEVS model of the PWM controlled DC motor (RT).

The *Clock wait* block has been implemented using the following code in PowerDEVS:

ATOMIC MODEL CLOCK WAIT
State Variables and Parameters:
> float *sigma*;
> void *xv; //states
> void *y; //output
> float *itime*;

Init Function:
> $inf = 1e10$;
> $itime = 1.0 * clock()/CLOCKS\_PER\_SEC$;
> $sigma = inf$;

Time Advance Function:
> return *sigma*;

Internal Transition Function:
> $sigma = inf$;

External Transition Function:
> float *actime*;
> $xv = x.value$;
> $actime = 1.0 * clock()/CLOCKS\_PER\_SEC - itime$;
> while $(actime < t)$ {
> $actime = 1.0 * clock()/CLOCKS\_PER\_SEC - itime$;
> }
> $sigma = 0$;

Output Function:
> $y = xv$;
> return Event$(y, 0)$;

We simulated this new model, and saved the output data with both, the physical and the simulated time. All events at the outputs of the *Clock wait* blocks were sent at the right physical time instants. The error was of the order of the accuracy of the physical clock access that the gcc compiler running under Windows permits. This means that the calculations were indeed performed on time.

The simulation results are shown in Figs.12.30–12.31.

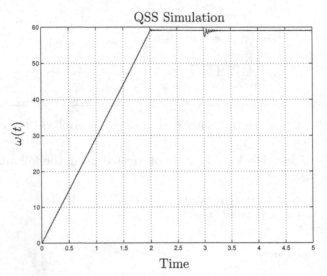

FIGURE 12.30. Simulation output of the DC motor.

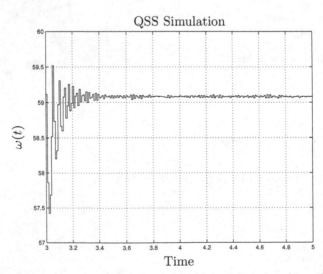

FIGURE 12.31. Simulation output of the DC motor (detail).

In this case, the most important advantage of using QSS2 is related to the efficient treatment of discontinuities. The off–line simulation of this system with any of MATLAB's integration methods requires more than 20 seconds of real time under identical operating conditions, which makes a real–time simulation impossible.

There is another advantage connected to the fact that the system variables are not updated simultaneously. For example in our simulation, $\omega(t)$ was only updated 1953 times. Unless we use a very sophisticated multi–rate algorithm, any discrete–time method would calculate $\omega(t)$ at least 20,000 times, as there are 20,000 events in the system. Thus, we would have to perform all calculations corresponding to each step in a period shorter than 0.25 *msec*.

In the QSS2 case, this period becomes about 10 times longer. It is true that we might have more calculations between two steps in ω. However, it is much easier to perform 1000 calculations in 10 *msec*, than performing 100 calculations in 1 *msec*.

Of course, the same idea of synchronizing only when and where it is necessary can also be used in the discrete–time case. However, this is not as easy in the case of the QSS2 method. Moreover, when we are forced to change the step size because of the presence of discontinuities, time synchronization becomes rather tricky, as we discussed in Chapter 10.

12.11   Open Problems in Quantization–based Methods

As we had mentioned in the previous chapter, the discrete event simulation of continuous systems is a newly developing area of research.

Despite the theoretical and practical advantages that we showed along this chapter, there are still many important problems that should be solved, before it can be claimed that quantization–based approaches represent a good choice for the simulation of general continuous systems.

The most important problem is probably related to the solution of *stiff systems*. Let us introduce the issue by means of a simple example. We shall consider the second–order ODE system given by:

$$\begin{aligned} \dot{x}_1 &= 100 \cdot x_2 \\ \dot{x}_2 &= -100 \cdot x_1 - 10,001 \cdot x_2 + u(t) \end{aligned} \tag{12.70}$$

The eigenvalues of this linear system are located at $\lambda_1 = -1$ and $\lambda_2 = -10,000$, which means that this system, in spite of its simplicity, is stiff.

We simulated the system across 10 seconds of simulated time using the QSS method with quantum sizes of 1×10^{-2} for x_1, and 1×10^{-4} for x_2. According to Eq.(12.27), this quantization ensures that the error in x_1 is bounded by 0.01, whereas the error in x_2 is bounded by 0.0003.

The initial conditions were both set equal to zero, and the input was chosen as a step function, $u(t) = 100 \cdot \varepsilon(t)$. The simulation was completed after 100 internal transitions in the quantized integrator that calculates x_1, and after 200 internal transitions in the quantized integrator that calculates x_2. The trajectory of the quantized state q_2 is shown in Figs.12.32–12.33.

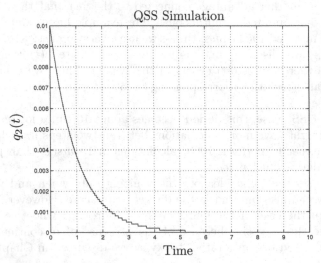

FIGURE 12.32. QSS simulation of the system of Eq.(12.70).

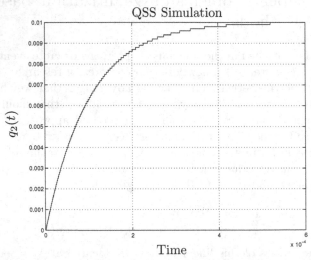

FIGURE 12.33. QSS simulation of the system of Eq.(12.70), startup period.

The stiffness of this system becomes evident, when we compare the time scale of Fig.12.32 with that of Fig.12.33. In this example, the performance of the QSS method is truly amazing. The method was able to adjust the step size in a very natural way, and performed the overall simulation in a surprisingly small number of integration steps. The number of transitions performed can be calculated here by dividing the amplitude of the trajectory by the quantum.

Looking at this result, we may think that the QSS technique is a wonderful method for solving stiff systems: it is an *explicit* method that allows simulating a stiff system much faster and more efficiently than the most complex implicit variable–step algorithms that we have met throughout this book.

This idea seems coherent with the fact that the QSS method is always "stable." A simulation method that good must surely turn the entire existing vault of theories concerning numerical ODE solutions upside down. Armies of applied mathematicians will have to rebuild from scratch the foundations of their trade.

Yet, we must not forget the meaning that we have given to the term "stable." The QSS method does not ensure asymptotic stability of a solution, but only its boundedness, and, unfortunately, this can become a problem when dealing with stiff systems.

To illustrate our point, let us check what happens if we introduce a small modification to the previously discussed example. We changed the input step function from $u(t) = 100 \cdot \varepsilon(t)$ to $u(t) = 99.5 \cdot \varepsilon(t)$, and repeated the simulation.

In the first five seconds of simulated time, the quantized integrator that calculates x_1 performed 100 internal transitions, i.e., the number stayed approximately the same as before, but the quantized integrator that calculates x_2 underwent a total of 25,057 transitions! The trajectory of q_2 is shown in Figs.12.34–12.35.

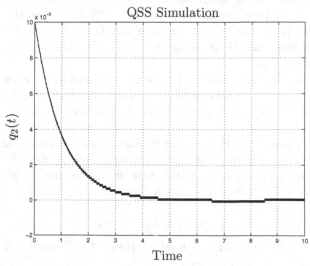

FIGURE 12.34. QSS simulation of the system of Eq.(12.70) with $u = 99.5$.

The results of this simulation are very similar to those obtained earlier.

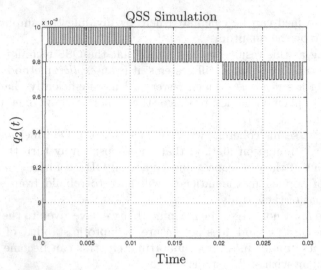

FIGURE 12.35. Details of the initial phase of Fig.12.34.

The error, in agreement with the theory, remains bounded, but the number of calculations is huge.

The reason is the appearance of ultra–fast oscillations in x_2 that ruined our triumph of having found an explicit, efficient, stable, and reliable method for dealing with stiff system. If those oscillations were not present, we could have calculated the number of transitions by dividing the trajectory amplitude by the quantum. However, this is not possible here.

Although these oscillations may have come as a surprise, they are in fact already familiar to us. Indeed, they are no different from the oscillations that we encountered in Fig.12.1 on page 557. There, the oscillations did not pose a problem, since their period was not much shorter than the settling time of the system.

Unfortunately, the oscillation frequency is related to the eigenfrequencies of the system, i.e., the location of its eigenvalues, which is bad news indeed.

How can we solve this problem? We do not have a final answer to this question yet.

An interesting trick that works in many cases is based on the use of concepts borrowed from the design of implicit ODE solvers. We have learnt very early on in this book that stiff systems require implicit algorithms for their solution. Whereas the FE algorithm:

$$\mathbf{x_{k+1}} = \mathbf{x_k} + h \cdot \mathbf{\dot{x}_k}$$

was unsuccessful in dealing with stiff systems, the implicit BE algorithm:

$$\mathbf{x_{k+1}} = \mathbf{x_k} + h \cdot \mathbf{\dot{x}_{k+1}}$$

could deal with them successfully. The difference between these two algorithms consists in the time instant, at which the state derivative is being considered in computing the next state.

We may be able to use the same idea in the design of an *Implicit Quantized State System (IQSS)* method. If only we knew, what the next state and the next input would be, we could compute the next state derivative:

$$\dot{x}_{i_{k+1}} = f_i(\mathbf{q_{k+1}}, \mathbf{u_{k+1}}) \tag{12.71}$$

and then we could determine the next quantized state, $q_{i_{k+1}}$, together with the time of occurrence of the next internal transition by making use of that future state derivative.

Explicit integration algorithms are loop breakers, whereas implicit algorithms are not. We saw this already in Chapter 7 in the discussion of the classical DAE solvers. Implicit algorithms invariably led to much larger algebraic loops.

The same observation is true for QSS methods as well. Until now, we were able to deal with the static functions independently of the hysteretic quantized integrators, because the integrators are breaking the algebraic loops. This is no longer the case, when we are looking at IQSS methods.

Yet, the problem may not be as bad as it looks. There are two points in our favor:

1. Although we don't know the value of the next quantized state, $q_{i_{k+1}}$, there are only two possible values that this state can assume: $q_{i_{k+1}} = q_{i_k} \pm \Delta Q$.

2. Looking at the integrator that is going to transition next, we only need to consider the future state of that integrator, as all other integrators still carry their previous value.

Given an nth–order system, we require an algorithm that can decide, which of the n integrators is going to transition next. Once we know this, we no longer need to iterate over n variables simultaneously. It then suffices to iterate over a single variable. Furthermore, "iteration" may be a rather fancy word for what needs to be done, since the future state can only assume one of two values.

The following algorithm could be used:

1. Given any state $q_i = Q$, where Q is the current value of the state q_i.

2. We replace Q by $Q + \Delta Q$ in the corresponding state equation: $\dot{x}_i = f_i(\mathbf{q}, \mathbf{u})$, assuming that the state variable x_i is about to increase.

3. We check, whether $\dot{x}_i$ is positive. If this is the case, we can now compute t_{k+1}, the next transition time. If $\dot{x}_i$ is negative, the assumption made was incorrect.

4. We now replace Q by $Q - \Delta Q$ in the corresponding state equation: $\dot{x}_i = f_i(\mathbf{q}, \mathbf{u})$, assuming that the state variable x_i is about to decrease.

5. We check, whether $\dot{x}_i$ is negative. If this is the case, we can now compute t_{k+1}, the next transition time. If $\dot{x}_i$ is positive, the assumption made was incorrect.

6. If one of the two assumptions turns out to be correct, whereas the other is incorrect, we know the next transition time for this state variable. If neither of the assumptions is correct, or if both are correct, we assume that the state isn't going to change on its own, and set the corresponding σ value to ∞.

7. We repeat the same algorithm for all integrators. The shortest t_{k+1} is the time of the next internal transition, and the corresponding state variable is the one that is going to transition next.

The algorithm is clearly more expensive than the explicit QSS algorithm, and it is more centralized. After each event, we need to recompute the time advance functions of all quantized integrators, and determine the one integrator that will transition next.

Notice that we should also have considered future values of the inputs in the calculation of the state derivative functions. However, the input values do not affect stability, and since our primary aim was to improve the stability behavior, i.e., get rid of the high–frequency oscillations, it may not be necessary to take future input values into account.

A much cheaper, and fully decentralized, version of this algorithm can also be proposed. In this new approach, we only look at the integrator that is undergoing an internal transition now, and compute its next value and time advance in the following way:

1. Let $q_i = Q$ be a quantized state that is currently going through an internal transition.

2. We replace Q by $Q + \Delta Q$ in the corresponding state equation: $\dot{x}_i = f_i(\mathbf{q}, \mathbf{u})$, assuming that the state variable x_i is about to increase.

3. We check, whether $\dot{x}_i$ is positive. If this is the case, we can now compute t_{k+1}, the next transition time. If $\dot{x}_i$ is negative, the assumption made was incorrect.

4. We now replace Q by $Q - \Delta Q$ in the corresponding state equation: $\dot{x}_i = f_i(\mathbf{q}, \mathbf{u})$, assuming that the state variable x_i is about to decrease.

5. We check, whether $\dot{x}_i$ is negative. If this is the case, we can now compute t_{k+1}, the next transition time. If $\dot{x}_i$ is positive, the assumption made was incorrect.

6. If one of the two assumptions turns out to be correct, whereas the other is incorrect, we know the next state value and the next internal transition time for this state variable. If neither of the assumptions is correct, or if both are correct, we assume that the state isn't going to change on its own, and set the corresponding σ value to ∞.

The cheap version of the IQSS algorithm is hardly more expensive than the explicit QSS method, yet it may improve the stability behavior of the method.

A remark here is that the quantized integrators have to be able to evaluate f_i, which means that we should no longer separate the quantized integrators from the static functions that calculate the derivatives.

Although the resulting atomic models are more complex, since they combine the features of a quantized integrator and a static function, the simulation becomes more efficient, since the number of events is reduced to less than one half, as we no longer have to transmit events from the static functions to the quantized integrators and from the quantized integrators back to the static functions that compute their own derivatives.

We implemented this idea for the above example, and repeated the simulation with $u(t) = 99.5 \cdot \varepsilon(t)$. The number of transitions was now approximately the same as when using the QSS method with $u(t) = 100 \cdot \varepsilon(t)$, i.e., the high–frequency oscillations have indeed disappeared.

However, we have not yet been able to prove that this approach works for all stiff systems, i.e., constitutes an efficient solution for stiff systems in general.

Furthermore, it may not be entirely trivial to generalize this technique to higher–order IQSS methods, as the number of combinations of states and slopes to be checked grows, and the search for consistent assumptions becomes quite a bit more involved.

Thus, in spite of the fact that we have been able to find an attractive and efficient quantization–based method for dealing with the above example, we cannot claim that the same approach will always work, and therefore, we consider that the problem is still open.

Another open problem in these methods has to do with the choice of the quantization. Although we mentioned algorithms and showed formulae for choosing the quantum in accordance with the desired error, this is not a completely satisfactory solution.

Except in applications, where we need to ensure a fixed error bound, which justifies performing a precise analysis prior to running the simulation, nobody wants to calculate a Lyapunov function or compute the eigenvalue and eigenvector matrices of a system to determine the quantum to be used.

An interesting idea would consist in developing quantum adaptation algorithms similar in concept to the step–size control algorithms of the variable–step discrete–time methods. Yet, there is no published research yet relating to this topic.

Another possibility might be to employ a logarithmic quantization scheme, so that the quantum becomes larger, when the variables assume bigger values. In that way, we might expect the algorithm to control the relative error, instead of the absolute error.

There are many other open problems that will probably be solved soon. We can easily imagine methods of orders greater than two, enjoying the same properties as QSS and QSS2, but reducing considerably the error bounds. A third–order accurate method (QSS3) has already been proposed and implemented in PowerDEVS [12.16, 12.17]. However, QSS3 is still not fully functional, as the currently implemented version does not work yet for general nonlinear blocks, and is not yet able to solve DAEs.

The use of high–order methods will also help with the choice of the quantum, since it will allow us to adopt a conservatively small quantum without a significant increase in the number of calculations. If we use a method of order five, for instance, the use of a quantum 1000 times smaller than the appropriate quantum would only increase the number of calculations by about a factor of four.

Another idea, mentioned in [12.13], is to apply QSS and QSS2 to the simulation of PDEs to exploit the natural sparsity of the resulting ODEs after applying the method of lines. In fact, we did something similar already in the transmission line examples.

The block diagrams of these ODEs have a very particular form, whereby a basic structure is repeated along the different spatial sections. Since we approximate each section by a DEVS model, we obtain in the process a sort of *cellular automaton*. Gabriel Wainer defined a particular formalism for describing cellular DEVS models, called *Cell–DEVS* [12.20], and he then combined it with QSS for the simulation of PDEs [12.21]. Similar approaches can also be found in [12.7] and [12.19].

Unfortunately, we saw that the method–of–lines approximation of PDEs of the parabolic type invariably leads to stiff ODEs. Thus, the problem of an efficient QSS simulation of stiff systems must be solved, in order to arrive at a general discrete event method for the simulation of distributed parameter systems.

Finally, another problem that has been treated recently is the application of QSS methods to marginally stable systems [12.12]. There, it was proven that, in the presence of purely imaginary eigenvalues, the error bound grows linearly with time and also depends linearly on the quantum.

Moreover, some simulation examples have shown that not only the error, but also the amplitude of the oscillations grows linearly with time, and hence the simulation becomes unstable.

However, as we saw, the use of QSS methods is equivalent to introducing a bounded perturbation. If that perturbation were not correlated with the state evolution, the presence of purely imaginary eigenvalues would not cause any problem. In fact, the response of a marginally stable system to a signal that does not contain spectral components at the resonance frequency is not unstable. Unfortunately, the presence of hysteresis introduces a large perturbation at the resonance frequency.

Thus, a modification of the quantized integrators, that attempts to eliminate these perturbation components, was proposed. Although the idea noticeably improves the results, a slowly increasing unstable term still remains.

Again, the usage of an IQSS algorithm might help solve this problem.

12.12 Summary

This chapter introduced the main theoretical and practical issues related to a new family of numerical methods.

Based on the idea of replacing time discretization by state quantization, two new ODE solvers, QSS and QSS2, were developed that exhibit theoretical properties, which differ noticeably from those of the classic discrete–time methods. The existence of a global error bound that can be explicitly calculated is probably the most interesting feature in this context.

The asynchronous nature of these methods and the knowledge of the state trajectories at any instant of time permits dealing with discontinuous systems in a very efficient fashion. In the presence of state and/or time events that occur with a frequency of the same order as the eigenfrequencies of the system, QSS and QSS2 can reduce significantly the simulation time with respect to conventional algorithms.

Further advantages can be observed in the simulation of DAE systems, and in the way of dealing with input signals.

In spite of all this, QSS and QSS2 exhibit a major drawback in the presence of stiff systems due to the frequent appearance of fast oscillations, but we should not be discouraged by these findings. First attempts at tackling the problem by introducing modified QSS algorithms that are implicit algorithm, and yet, don't require true iterations, led to very promising results.

To us, discrete event integration methods constitute one of the most exciting recent developments in the field of numerical ODE solutions. There are still lots of open problems that can constitute subjects of research for future MS theses and PhD dissertations, which should be good news for aspiring young applied mathematicians in search of a research topic.

QSS methods may look exotic and unfamiliar at a first glance, yet it is always the departure to new shores and unexplored lands that ultimately reaps the most benefit. It is the unknown and unexplored that keeps science alive.

12.13 References

[12.1] François Baccelli, Guy Cohen, Geert Jan Olsder, and Jean-Pierre Quadrat. *Synchronization and Linearity: An Algebra for Discrete Event Systems*. John Wiley & Sons, 1992. 485p.

[12.2] François E. Cellier. *Continuous System Modeling*. Springer–Verlag, New York, 1991. 755p.

[12.3] James B. Dabney and Thomas L. Harman. *Masterink SIMULINK 4*. Prentice Hall, Upper Saddle River, N.J., 2001.

[12.4] Hilding Elmqvist. *Dymola — Dynamic Modeling Language, User's Manual, Version 5.3*. DynaSim AB, Research Park Ideon, Lund, Sweden, 2004.

[12.5] Joel M. Esposito, R. Vijay Kumar, and George J. Pappas. Accurate Event Detection for Simulating Hybrid Systems. In *Proceedings of the 4th International Workshop on Hybrid Systems: Computation and Control*, volume 2034 of *Lecture Notes in Computer Science*, pages 204–217. Springer–Verlag, London, 2001.

[12.6] Yehea I. Ismail, Eby G. Friedman, and José Luis Neves. Figures of Merit to Characterize the Importance of On-chip Inductance. *IEEE Trans. on VLSI*, 7(4):442–449, 1999.

[12.7] Rajanikanth Jammalamadaka. Activity Characterization of Spatial Models: Application to Discrete Event Solution of Partial Differential Equations. Master's thesis, The University of Arizona, 2003.

[12.8] Hassan K. Khalil. *Nonlinear Systems*. Prentice–Hall, Upper Saddle River, N.J., 3rd edition, 2002. 750p.

[12.9] Ernesto Kofman and Sergio Junco. Quantized Bond Graphs: An Approach for Discrete Event Simulation of Physical Systems. In *Proceedings International Conference on Bond Graph Modeling and Simulation*, volume 33 of *Simulation Series*, pages 369–374. Society for Modeling and Simulation International, 2001.

[12.10] Ernesto Kofman and Sergio Junco. Quantized State Systems: A DEVS Approach for Continuous System Simulation. *Transactions of SCS*, 18(3):123–132, 2001.

[12.11] Ernesto Kofman, Jong Sik Lee, and Bernard Zeigler. DEVS Representation of Differential Equation Systems: Review of Recent Advances. In *Proceedings European Simulation Symposium*, pages 591–595, Marseille, France, 2001. Society for Modeling and Simulation International.

[12.12] Ernesto Kofman and Bernard Zeigler. DEVS Simulation of Marginally Stable Systems. In *Proceedings of IMACS 2005*, Paris, France, July 2005.

[12.13] Ernesto Kofman. A Second Order Approximation for DEVS Simulation of Continuous Systems. *Simulation*, 78(2):76–89, 2002.

[12.14] Ernesto Kofman. Quantization–based Simulation of Differential Algebraic Equation Systems. *Simulation*, 79(7):363–376, 2003.

[12.15] Ernesto Kofman. Discrete Event Simulation of Hybrid Systems. *SIAM Journal on Scientific Computing*, 25(5):1771–1797, 2004.

[12.16] Ernesto Kofman. A Third Order Discrete Event Simulation Method for Continuous System Simulation. Part I: Theory. In *Proceedings of RPIC'05*, Río Cuarto, Argentina, 2005.

[12.17] Ernesto Kofman. A Third Order Discrete Event Simulation Method for Continuous System Simulation. Part II: Applications. In *Proceedings of RPIC'05*, Río Cuarto, Argentina, 2005.

[12.18] Ernesto Kofman. Non–Conservative Ultimate Bound Estimation in LTI Perturbed Systems. *Automatica*, 41(10):1835–1838, 2005.

[12.19] James J. Nutaro, Bernard P. Zeigler, Rajanikanth Jammalamadaka, and Salil Akerkar. Discrete Event Solution of Gas Dynamics within the DEVS Framework. In *Proceedings of International Conference on Computational Science*, volume 2660 of *Lecture Notes in Computer Science*, pages 319–328, Melbourne, Australia, 2003. Springer–Verlag, Berlin.

[12.20] Gabriel Wainer and Norbert Giambiasi. Application of the Cell–DEVS Paradigm for Cell Spaces Modeling and Simulation. *Simulation*, 76(1):22–39, 2001.

[12.21] Gabriel Wainer. Performance Analysis of Continuous Cell–DEVS Models. In *Proceedings 18th European Simulation Multiconference*, Magdeburg, Germany, 2004.

12.14 Bibliography

[B12.1] Norbert Giambiasi, Bruno Escude, and Sumit Ghosh. GDEVS: A Generalized Discrete Event Specification for Accurate Modeling of Dynamic Systems. *Transactions of SCS*, 17(3):120–134, 2000.

[B12.2] Ernesto Kofman. *Discrete Event Simulation and Control of Continuous Systems*. PhD thesis, Universidad Nacional de Rosario, Rosario, Argentina, 2003.

[B12.3] Herbert Praehofer. *System Theoretic Foundations for Combined Discrete–Continuous System Simulation.* PhD thesis, Johannes Kepler University, Linz, Austria, 1991.

[B12.4] Bernard Zeigler and Jong Sik Lee. Theory of Quantized Systems: Formal Basis for DEVS/HLA Distributed Simulation Environment. In *SPIE AeroSense'98 Proceedings: Enabling Technology for Simulation Science (II)*, volume 3369, pages 49–58, Orlando, Florida, 1998.

12.15   Homework Problems

[H12.1] Error Bound in LTI Systems

Given the following LTI system:

$$\begin{aligned}
\dot{x}_{a_1} &= x_{a_2} \\
\dot{x}_{a_2} &= -2 \cdot x_{a_1} - 3 \cdot x_{a_2}
\end{aligned}$$

with initial conditions, $x_{a_1}(0) = x_{a_2}(0) = 1$.

1. Find the analytical solution, $x_{a_1}(t)$ and $x_{a_2}(t)$.

2. Obtain an approximate solution, $x_1(t)$ and $x_2(t)$, with the QSS technique, using a uniform quantization in both variables of $\Delta Q = \varepsilon = 0.05$.

3. Draw the error trajectories, $e_i(t) = x_i(t) - x_{a_i}(t)$, and compare them with the error bound given by Eq.(12.27).

[H12.2] Input Quantization Error

Prove the validity of Eq.(12.31).

[H12.3] Approximate Input Signals

Build DEVS models that generate events representing piecewise constant trajectories that approximate the following signals:

1. A ramp

2. A pulse

3. A square wave

4. A saw-tooth signal

5. A trapezoidal wave

Develop also corresponding PowerDEVS models, and try them out with some simple systems.

[H12.4] QSS2 Linear Static Function

Obtain a DEVS model for a linear static function:

$$f_i(\mathbf{z}) = \sum_{j=1}^{l} a_{i,j} \cdot z_j \tag{H12.4a}$$

that takes into account values and slopes.

Implement the model in PowerDEVS, and try it out, together with the second–order quantized integrator, to simulate the system of Eq.(11.11).

[H12.5] QSS2 Nonlinear Static Functions

Obtain DEVS models and the corresponding PowerDEVS models of the following static functions for QSS2:

1. $f_a(v) = v^n$, with n being an integer parameter.

2. $f_b(v) = e^v$.

3. $f_c(v) = \cos(2\pi \cdot f \cdot v + \phi)$, where f and ϕ are real–valued parameters.

4. $f_d(v_1, v_2) = v_1 \cdot v_2$

Exploit the fact that you can analytically calculate the partial derivatives of these functions.

[H12.6] Input Signals in QSS2

Repeat problem H12.3 by building signal generators for the QSS2 method, now considering piecewise linear trajectories.

[H12.7] Approximate Sinusoidal for QSS2

Propose a DEVS model that generates a piecewise linear approximation to a sinusoidal wave.

Can you obtain a solution similar to that of model M_7 provided on page 566? Explain the differences.

[H12.8] QSS2 Loop–breaking Block

Obtain a loop–breaking DEVS model for the QSS2 method analogous to M_{11}. Implement it in PowerDEVS, and simulate the circuit of Fig.12.7 using the QSS2 method.

[H12.9] Hybrid DAE Simulation

The circuit of Fig.H12.9a represents a modification of the example presented in Fig.12.14. Here, a resistor, R_p, and a nonlinear component were included to limit the voltage of the load resistor, R.

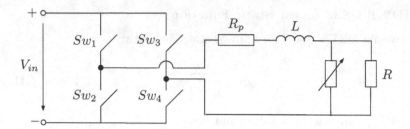

FIGURE H12.9a. DC–AC inverter with surge protection.

We shall assume that the nonlinear component is characterized by a varistor–like voltage–current relationship:

$$i(t) = k \cdot u(t)^{\alpha} \qquad \text{(H12.9a)}$$

Under this assumption, the equation describing the system dynamics becomes a nonlinear DAE:

$$\frac{di_L}{dt} = \frac{1}{L} \cdot (-R_p \cdot i_L - u + s_w \cdot V_{in}) \qquad \text{(H12.9b)}$$

where u must satisfy the nonlinear equation:

$$i_L - k \cdot u^{\alpha} - \frac{u}{R} = 0 \qquad \text{(H12.9c)}$$

Simulate this system in PowerDEVS using the QSS2 method. Use the same parameters as in the example of Fig.12.14, choosing in addition $R_p = 0.01\ \Omega$, $k = 5^{-7}\ mho$, and $\alpha = 7$.

Do you notice any further advantage of using quantization–based integration techniques in this example?

12.16   Projects

[P12.1] Pulse on a Transmission Line

Consider the transmission line model of Fig.12.5. It is composed of five sections of RLC circuits.

The goal of this project is to study the effects of varying the number of sections on the computational cost of the simulation.

To this end, a short pulse input will be considered. A pulse with a duration of $2 \times 10^{-10}\ sec$ may be appropriate.

Then, the idea is to start with a transmission line circuit consisting of a single section, and then to gradually increase the number of sections until at least 20.

In each case, measure the total number of transitions and the simulation time, and try to find a law that relates the computational cost to the number of sections.

Then, repeat the same experiment with longer pulses, and with periodic waves, such as the trapezoidal wave used a few times in the chapter.

If you find any difference in the relationship between the computational cost and the number of segments for any of the new inputs, try to explain the reason for this difference.

[P12.2] Logarithmic Quantization

Obtain a DEVS model of a logarithmically quantized integrator for the QSS and the QSS2 methods. To this end, use a quantum proportional to the state variable value. You will also need to define a minimum value for the quantum, as otherwise, you might obtain illegitimate behavior.

Once you have built the corresponding PowerDEVS models, take some of the examples presented in this chapter, and simulate them using the newly developed logarithmically quantized integrators. Study the advantages and disadvantages of this new approach.

Using logarithmic quantization, can you still guarantee stability as we did before?

Study the problems related to stability and global error bound from a theoretical point of view. Start with a first–order linear system, and then try to extend the analysis, if at all possible, to the general LTI case.

12.17   Research

[R12.1] Integration of PDEs

The use of the method of lines in PDEs produces a system of sparse ODEs or DAEs. As we saw, the QSS and QSS2 methods exploit sparsity in a very efficient fashion.

Study the advantages and disadvantages of using the quantization–based integration methods together with the method of lines for the simulation of PDEs. Hyperbolic PDEs may be of particular interest in this context. Do not forget to take into account the particular problems of shock waves and discontinuous PDEs.

[R12.2] QSS simulation of Stiff Systems

Investigate the possibility of introducing modifications to the QSS method, in order to improve their ability for dealing with stiff systems.

You can use the remarks of Section 12.11 as a starting point.

Index

Author Index